Edited by
Arnaud Brignon

**Coherent Laser Beam
Combining**

Related Titles

Sze, S.M., Ng, K.K.

Physics of Semiconductor Devices 3rd Edition

3rd Edition

2006

Print ISBN: 978-0-470-06832-8

Numai, T.

Laser Diodes and their Applications to Communications and Information Processing

2010

Print ISBN: 978-0-470-53668-1

Also available in digital format

Paschotta, R.

Encyclopedia of Laser Physics and Technology

2008

Print ISBN: 978-3-527-40828-3

Epstein, R., Sheik-Bahae, M. (eds.)

Optical Refrigeration

Science and Applications of Laser Cooling of Solids

2009

Print ISBN: 978-3-527-40876-4

Mandel, P.

Nonlinear Optics

An Analytical Approach

2010

Print ISBN: 978-3-527-40923-5

Rafailov, E.U., Cataluna, M.A., Avrutin, E.A.

Ultrafast Lasers Based on Quantum Dot Structures

Physics and Devices

2011

Print ISBN: 978-3-527-40928-0

Also available in digital format

Okhotnikov, O.G. (ed.)

Semiconductor Disk Lasers

Physics and Technology

2010

Print ISBN: 978-3-527-40933-4

Diels, J., Arissian, L.

Lasers

The Power and Precision of Light

2011

Print ISBN: 978-3-527-41039-2

Also available in digital format

Lüdge, K. (ed.)

Nonlinear Laser Dynamics

From Quantum Dots to Cryptography

2012

Print ISBN: 978-3-527-41100-9

Also available in digital format

Okhotnikov, O.G. (ed.)

Fiber Lasers

2012

Print ISBN: 978-3-527-41114-6

Also available in digital format

Edited by Arnaud Brignon

Coherent Laser Beam Combining

Verlag GmbH & Co. KGaA

Editor

Dr. Arnaud Brignon
Micro and Nano-Physics Lab.
Thales Research & Technology
91767 Palaiseau cedex
France

All books published by **Wiley-VCH** are carefully produced. Nevertheless, authors, editors, and publisher do not warrant the information contained in these books, including this book, to be free of errors. Readers are advised to keep in mind that statements, data, illustrations, procedural details or other items may inadvertently be inaccurate.

Library of Congress Card No.: applied for

British Library Cataloguing-in-Publication Data
A catalogue record for this book is available from the British Library.

Bibliographic information published by the Deutsche Nationalbibliothek
The Deutsche Nationalbibliothek lists this publication in the Deutsche Nationalbibliografie; detailed bibliographic data are available on the Internet at <http://dnb.d-nb.de>.

© 2013 Wiley-VCH Verlag GmbH & Co. KGaA, Boschstr. 12, 69469 Weinheim, Germany

All rights reserved (including those of translation into other languages). No part of this book may be reproduced in any form – by photoprinting, microfilm, or any other means – nor transmitted or translated into a machine language without written permission from the publishers. Registered names, trademarks, etc. used in this book, even when not specifically marked as such, are not to be considered unprotected by law.

Print ISBN: 978-3-527-41150-4
ePDF ISBN: 978-3-527-65280-8
ePub ISBN: 978-3-527-65279-2
mobi ISBN: 978-3-527-65278-5
oBook ISBN: 978-3-527-65277-8

Cover Design   Formgeber, Eppelheim
Typesetting   Thomson Digital, Noida, India
Printing and Binding   Markono Print Media Pte Ltd, Singapore

Printed in Singapore
Printed on acid-free paper

Contents

Preface *XV*
Acronyms *XVII*
List of Contributors *XXI*

Part One Coherent Combining with Active Phase Control *1*

1 Engineering of Coherently Combined, High-Power Laser Systems *3*
Gregory D. Goodno and Joshua E. Rothenberg

1.1	Introduction *3*	
1.2	Coherent Beam Combining System Requirements *5*	
1.3	Active Phase-Locking Controls *8*	
1.3.1	Optical Heterodyne Detection *11*	
1.3.2	Synchronous Multidither *13*	
1.3.3	Hill Climbing *14*	
1.4	Geometric Beam Combining *14*	
1.4.1	Tiled Aperture Combiners *15*	
1.4.2	Filled Aperture Combiners Using Diffractive Optical Elements *16*	
1.4.2.1	Overview of DOE Combiners *17*	
1.4.2.2	DOE Design and Fabrication *18*	
1.4.2.3	DOE Thermal and Spectral Sensitivity *20*	
1.5	High-Power Coherent Beam Combining Demonstrations *21*	
1.5.1	Coherent Beam Combining of Zigzag Slab Lasers *22*	
1.5.2	Coherent Beam Combining of Fiber Lasers *26*	
1.5.2.1	Phase Locking of Nonlinear Fiber Amplifiers *26*	
1.5.2.2	Path Length Matching with Broad Linewidths *30*	
1.5.2.3	Diffractive CBC of High-Power Fibers *31*	
1.5.2.4	CBC of Tm Fibers at 2 μm *37*	
1.6	Conclusion *39*	
	Acknowledgments *40*	
	References *40*	

2		**Coherent Beam Combining of Fiber Amplifiers via LOCSET** *45*
		Angel Flores, Benjamin Pulford, Craig Robin, Chunte A. Lu, and Thomas M. Shay
2.1		Introduction *45*
2.1.1		Beam Combination Architectures *46*
2.1.2		Active and Passive Coherent Beam Combining *47*
2.2		Locking of Optical Coherence by Single-Detector Electronic-Frequency Tagging *48*
2.2.1		LOCSET Theory *49*
2.2.2		Self-Referenced LOCSET *50*
2.2.2.1		Photocurrent Signal *50*
2.2.2.2		LOCSET Demodulation *53*
2.2.3		Self-Synchronous LOCSET *55*
2.3		LOCSET Phase Error and Channel Scalability *55*
2.3.1		LOCSET Beam Combining and Phase Error Analysis *55*
2.3.2		In-Phase and Quadrature-Phase Error Analysis *56*
2.3.3		Two-Channel Beam Combining *58*
2.3.4		16-Channel Beam Combining *60*
2.3.5		32-Channel Beam Combining *62*
2.4		LOCSET High-Power Beam Combining *63*
2.4.1		Kilowatt-Scale Coherent Beam Combining of Silica Fiber Lasers *64*
2.4.2		Kilowatt-Scale Coherent Beam Combining of Photonic Crystal Fiber Amplifiers *67*
2.5		Conclusion *71*
		References *71*
3		**Kilowatt Coherent Beam Combining of High-Power Fiber Amplifiers Using Single-Frequency Dithering Techniques** *75*
		Zejin Liu, Pu Zhou, Xiaolin Wang, Yanxing Ma, and Xiaojun Xu
3.1		Introduction *75*
3.1.1		Brief History of Coherent Beam Combining *75*
3.1.2		Coherent Beam Combining: State of the Art *76*
3.1.3		Key Technologies for Coherent Beam Combining *77*
3.2		Single-Frequency Dithering Technique *78*
3.2.1		Theory of Single-Frequency Dithering Technique *78*
3.2.2		Kilowatt Coherent Beam Combining of High-Power Fiber Amplifiers Using Single-Frequency Dithering Technique *85*
3.2.3		Coherent Polarization Beam Combining of Four High-Power Fiber Amplifiers Using Single-Frequency Dithering Technique *88*
3.2.4		Target-in-the-Loop Coherent Beam Combination of Fiber Lasers Based on Single-Frequency Dithering Technique *91*
3.3		Sine–Cosine Single-Frequency Dithering Technique *94*
3.3.1		Theory of Sine–Cosine Single-Frequency Dithering Technique *94*

3.3.2	Coherent Beam Combining of Nine Beams Using Sine–Cosine Single-Frequency Dithering Technique *97*	
3.4	Summary *99*	
	References *100*	

4	**Active Coherent Combination Using Hill Climbing-Based Algorithms for Fiber and Semiconductor Amplifiers** *103*	
	Shawn Redmond, Kevin Creedon, Tso Y. Fan, Antonio Sanchez-Rubio, Charles Yu, and Joseph Donnelly	
4.1	Introduction to Hill Climbing Control Algorithms for Active Phase Control *103*	
4.1.1	Conventional SPGD-Based Control Algorithm for Active Phase Control *104*	
4.1.2	Orthonormal Dither-Based Control Algorithm *106*	
4.1.3	Multiple Detector-Based Control Algorithm *114*	
4.2	Applications of Active Phase Control Using Hill Climbing Control Algorithms *117*	
4.2.1	Semiconductor Amplifier Active Coherent Combination *117*	
4.2.1.1	Introduction to SCOWA Semiconductor Waveguide and Phase Control *118*	
4.2.1.2	Tiled Array Beam Combination *120*	
4.2.1.3	Single-Beam Active Coherent Combination Using Diffractive Optical Elements *125*	
4.2.2	Fiber Amplifier Active Coherent Combination *128*	
4.2.2.1	Introduction to Fiber Amplifier Active Beam Combination Architectures *128*	
4.2.2.2	Tiled Array Beam Combination *129*	
4.2.2.3	Single-Beam Active Coherent Combination Using Diffractive Optical Elements *133*	
4.3	Summary *134*	
	Disclaimer *134*	
	References *135*	

5	**Collective Techniques for Coherent Beam Combining of Fiber Amplifiers** *137*	
	Arnaud Brignon, Jérome Bourderionnet, Cindy Bellanger, and Jérome Primot	
5.1	Introduction *137*	
5.2	The Tiled Arrangement *138*	
5.2.1	Calculation of the Far-Field Intensity Pattern *139*	
5.2.2	Influence of Design Parameters on the Combining Efficiency *141*	
5.2.2.1	Impact of the Near Field Arrangement *141*	
5.2.2.2	Impact of Collimation System Design and Errors *143*	

5.2.2.3	Impact of Phase Error	145
5.2.2.4	Impact of Power Dispersion	146
5.2.3	Beam Steering	146
5.3	Key Elements for Active Coherent Beam Combining of a Large Number of Fibers	147
5.3.1	Collimated Fiber Array	148
5.3.2	Collective Phase Measurement Technique	151
5.3.2.1	Principle of the Measurement	152
5.3.2.2	Implementation in the Experimental Setup	153
5.3.2.3	Phase Retrieval Techniques	153
5.3.3	Phase Modulators	155
5.4	Beam Combining of 64 Fibers with Active Phase Control	156
5.5	Beam Combining by Digital Holography	158
5.5.1	Principle	159
5.5.2	Experimental Demonstration	161
5.6	Conclusion	163
	Acknowledgments	164
	References	164

6 Coherent Beam Combining and Atmospheric Compensation with Adaptive Fiber Array Systems *167*

Mikhail Vorontsov, Thomas Weyrauch, Svetlana Lachinova, Thomas Ryan, Andrew Deck, Micah Gatz, Vladimir Paramonov, and Gary Carhart

6.1	Introduction	167
6.2	Fiber Array Engineering	168
6.3	Turbulence-Induced Phase Aberration Compensation with Fiber Array-Integrated Piston and Tip–Tilt Control	173
6.4	Target Plane Phase Locking of a Coherent Fiber Array on an Unresolved Target	175
6.4.1	Fiber Array Control System Engineering: Issues and Considerations	175
6.4.2	SPGD-Based Coherent Beam Combining: Round-Trip Propagation Time Issue	176
6.4.3	Coherent Beam Combining at an Unresolved Target over 7 km Distance	178
6.5	Target Plane Phase Locking for Resolved Targets	182
6.5.1	Speckle Metric Optimization-Based Phase Locking	183
6.5.2	Speckle Metrics	184
6.5.3	Experimental Evaluation of Speckle Metric-Based Phase Locking	186
6.6	Conclusion	188
	Acknowledgments	189
	References	189

7	**Refractive Index Changes in Rare Earth-Doped Optical Fibers and Their Applications in All-Fiber Coherent Beam Combining** *193*
	Andrei Fotiadi, Oleg Antipov, Maxim Kuznetsov, and Patrice Mégret
7.1	Introduction *193*
7.2	Theoretical Description of the RIC Effect in Yb-Doped Optical Fibers *194*
7.2.1	Introduction: Thermal and Electronic RIC Mechanisms *194*
7.2.2	Description of the Spectroscopic Properties of Yb-Doped Optical Fibers *195*
7.2.3	Description of the Electronic RIC Mechanism *195*
7.2.4	Description of the Thermal RIC Mechanism *200*
7.2.5	Comparison of Electronic and Thermal Contributions to the Pump-Induced Phase Shift *201*
7.2.6	Phase Shifts in the Case of Periodic Pulse Pumping and in the Presence of Amplified Signal *203*
7.2.7	Conclusion *205*
7.3	Experimental Studies of the RIC Effect in Yb-Doped Optical Fibers *205*
7.3.1	Previous Observations of the RIC Effect in Laser Fibers *205*
7.3.2	Methodology of Pump/Signal-Induced RIC Measurements *206*
7.3.3	Characterization of RIC in Different Fiber Samples *207*
7.3.4	Phase Shifts Induced by Signal Pulses *210*
7.3.5	Evaluation of the Polarizability Difference *212*
7.3.6	Comparison of the RIC Effects in Aluminum and Phosphate Silicate Fibers *213*
7.3.7	Conclusion *215*
7.4	All-Fiber Coherent Combining through RIC Effect in Rare Earth-Doped Fibers *215*
7.4.1	Coherent Combining of Fiber Lasers: Alternative Techniques *215*
7.4.2	Operation Algorithm and Simulated Results *217*
7.4.3	Environment Noise in Optical System to be Compensated *222*
7.4.4	Combining of Two Er-Doped Amplifiers through the RIC Control in Yb-Doped Fibers *223*
7.4.5	Extension Algorithm for Combining of N Amplifiers *224*
7.4.6	Conclusion *226*
7.5	Conclusions and Recent Progress *226*
	References *227*
8	**Coherent Beam Combining of Pulsed Fiber Amplifiers in the Long-Pulse Regime (Nano- to Microseconds)** *231*
	Laurent Lombard, Julien L. Gouët, Pierre Bourdon, and Guillaume Canat
8.1	Introduction *231*
8.2	Beam Combining Techniques *234*

8.2.1	Filled and Tiled Apertures 235
8.2.2	Locking Techniques 236
8.2.2.1	Direct Phase Locking Techniques 236
8.2.2.2	Indirect Phase Locking Techniques 237
8.2.3	Requirements of Various Techniques 239
8.2.3.1	Indirect Phase Locking Techniques 239
8.2.3.2	Direct Phase Locking Techniques 240
8.2.4	Case of Pulsed Laser 240
8.3	Amplification of Optical Pulse in Active Fiber 243
8.3.1	Approximations and Validity Domain of the Calculation 243
8.3.2	Pulse Propagation in the Resonant Medium 244
8.3.3	Practical Calculation of the Output Pulse Based on the CW Regime 245
8.3.4	Pulse Shape Distortion 246
8.3.5	Influence of the Amplified Spontaneous Emission 247
8.4	Power Limitations in Pulsed Fiber Amplifiers 248
8.4.1	Physical Principle of the Stimulated Brillouin Scattering 248
8.4.2	SBS Gain 249
8.4.3	SBS Threshold Input Power 250
8.4.4	SBS Reduction 251
8.4.5	Domain of SBS Predominance 251
8.4.6	Physical Principle of the Stimulated Raman Scattering 252
8.4.7	Maximum Peak Power Achievable 253
8.5	Phase Noise and Distortion in Fiber Amplifiers 253
8.5.1	Phase Noise Measurement 253
8.5.2	In-Pulse Phase Shift Measurement 258
8.5.3	In-Pulse Phase Shift Calculation 259
8.5.3.1	Kerr-Induced Phase Shift 260
8.5.3.2	Gain-Induced Phase Shift 262
8.6	Experimental Setup and Results of Coherent Beam Combining of Pulsed Amplifiers Using a Signal Leak between the Pulses 266
8.7	Alternative Techniques for Pulse Energy Scaling 269
8.8	Conclusion 271
	References 272

9 **Coherent Beam Combining in the Femtosecond Regime** 277
Marc Hanna, Dimitrios N. Papadopoulos, Louis Daniault, Frédéric Druon, Patrick Georges, and Yoann Zaouter

9.1	Introduction 277
9.2	General Aspects of Coherent Combining over Large Optical Bandwidths 278
9.2.1	Description and Propagation of Femtosecond Pulses 278
9.2.2	Coherent Combining over a Large Bandwidth 280
9.2.3	Influence of Spectral Phase Mismatch on the Combining Efficiency 281

9.2.4	Space–Time Effects *283*	
9.3	Coherent Combining with Identical Spectra: Power/Energy Scaling *284*	
9.3.1	Active Techniques *284*	
9.3.1.1	Experimental Implementations *284*	
9.3.1.2	Measurement of Spectral Phase Mismatch *287*	
9.3.2	Passive Coherent Combining Techniques: Path-Sharing Network *290*	
9.3.2.1	Principle *290*	
9.3.2.2	Experimental Demonstrations *292*	
9.4	Other Coherent Combining Concepts *295*	
9.4.1	Temporal Multiplexing: Divided Pulse Amplification *295*	
9.4.2	Passive Enhancement Cavities *296*	
9.4.3	Coherent Combining with Disjoint Spectra: Ultrafast Pulse Synthesis *298*	
9.5	Conclusion *299*	
	References *300*	

Part Two Passive and Self-Organized Phase Locking *303*

10	**Modal Theory of Coupled Resonators for External Cavity Beam Combining** *305*	
	Mercedeh Khajavikhan and James R. Leger	
10.1	Introduction *305*	
10.2	Coherent Beam Combining Requirements *306*	
10.3	General Mathematical Framework of Passive Laser Resonators *307*	
10.3.1	Coherent Beam Combining by a Simple Beam Splitter *308*	
10.3.2	Effect of Wavelength Diversity *311*	
10.4	Coupled Cavity Architectures Based on Beam Superposition *314*	
10.4.1	Generalized Michelson Resonators *314*	
10.4.2	Grating Resonators *318*	
10.5	Parallel Coupled Cavities Based on Space-Invariant Optical Architectures *321*	
10.5.1	Space-Invariant Parallel Coupled Resonators with Weakly Coupled Cavities *323*	
10.5.2	Spatially Filtered Resonators and the Effect of Path Length Phase Errors *325*	
10.5.3	Talbot Resonators *329*	
10.6	Parallel Coupled Resonators Based on Space-Variant Optical Architectures: the Self-Fourier Cavity *336*	
10.7	Conclusion *340*	
	Acknowledgments *341*	
	References *341*	

11 Self-Organized Fiber Beam Combining 345
Vincent Kermène, Agnès Desfarges-Berthelemot, and Alain Barthélémy

11.1 Introduction 345
11.2 Principles of Passively Combined Fiber Lasers 346
11.2.1 Different Configurations 346
11.2.2 Principles 347
11.3 Phase Coupling Characteristics 350
11.3.1 Power Stability 350
11.3.2 Cophasing Building Dynamics 351
11.3.3 Frequency Tunability 353
11.3.4 Effect of Laser Gain Mismatched on Combining Efficiency 354
11.3.5 Pointing Agility 355
11.3.6 Coherence Properties of Multiple Beams Phase Locked by Mutual Injection Process 356
11.4 Upscaling the Number of Coupled Lasers 360
11.4.1 Phasing Efficiency Evolution 360
11.4.2 Main Influencing Parameters 360
11.5 Passive Combining in Pulsed Regime 362
11.5.1 Q-Switched Regime 362
11.5.2 Mode-Locked Regime 365
11.6 Conclusion 367
References 368

12 Coherent Combining and Phase Locking of Fiber Lasers 371
Moti Fridman, Micha Nixon, Nir Davidson, and Asher A. Friesem

12.1 Introduction 371
12.2 Passive Phase Locking and Coherent Combining of Small Arrays 372
12.2.1 Efficient Coherent Combining of Two Fiber Lasers 372
12.2.2 Compact Coherent Combining of Four Fiber Lasers 375
12.2.3 Efficient Coherent Combining of Four Fiber Lasers Operating at 2 μm 376
12.3 Effects of Amplitude Dynamics, Noise, Longitudinal Modes, and Time-Delayed Coupling 377
12.3.1 Effects of Amplitude Dynamics 377
12.3.2 Effects of Quantum Noise 381
12.3.3 Effects of Many Longitudinal Modes 384
12.3.4 Effects of Time-Delayed Coupling 388
12.4 Upscaling the Number of Phase-Locked Fiber Lasers 391
12.4.1 Simultaneous Spectral and Coherent Combining 391
12.4.2 Phase Locking 25 Fiber Lasers 393
12.5 Conclusion 398
References 398

13	**Intracavity Combining of Quantum Cascade Lasers** *401*	

Guillaume Bloom, Christian Larat, Eric Lallier, Mathieu Carras, and Xavier Marcadet

13.1	Introduction *401*
13.2	External Cavity Passive Coherent Beam Combining *402*
13.2.1	Laser Scheme *403*
13.2.2	Modeling of the Coherent Beam Combining in External Cavity *404*
13.2.2.1	The Michelson Cavity *404*
13.2.2.2	General Case: The N-Arm Cavity *406*
13.2.3	Combining Efficiency in Real Experimental Conditions *408*
13.2.3.1	Influence of the Number of Arms N *408*
13.2.3.2	Influence of the Arm Length Difference ΔL *409*
13.3	Experimental Realization: Five-Arm Cavity with a Dammann Grating *410*
13.3.1	Dammann Gratings *410*
13.3.2	Quantum Cascade Lasers *413*
13.3.3	The Five-Arm External Cavity *414*
13.4	Subwavelength Gratings *418*
13.4.1	Principle *418*
13.4.2	Grating Design and Realization *419*
13.4.3	Antireflection Coating Design *422*
13.4.4	Calculated Performances *423*
13.5	Conclusion *423*
	References *424*

14	**Phase-Conjugate Self-Organized Coherent Beam Combination** *427*	

Peter C. Shardlow and Michael J. Damzen

14.1	Introduction *427*
14.2	Phase Conjugation *429*
14.2.1	Gain Holography *431*
14.2.2	Four-Wave Mixing within a Saturable Gain Media *433*
14.2.3	Self-Pumped Phase Conjugation *434*
14.2.3.1	Seeded Self-Pumped Phase-Conjugate Module *435*
14.2.3.2	Self-Starting Self-Adaptive Gain Grating Lasers *437*
14.3	PCSOCBC *438*
14.3.1	CW Experimental PCSOCBC *439*
14.3.2	Understanding Operation of PCSOCBC: Discussion *442*
14.3.3	Power Scaling Potential *444*
14.3.3.1	Scaling the Number of Modules *445*
14.3.3.2	Higher Power Modules *448*
14.3.3.3	Pulsed Operation *449*
14.4	Conclusions *450*
	References *451*

15		**Coherent Beam Combining Using Phase-Controlled Stimulated Brillouin Scattering Phase Conjugate Mirror** *455*
		Hong J. Kong, Sangwoo Park, Seongwoo Cha, Jin W. Yoon, Seong K. Lee, Ondrej Slezak, and Milan Kalal
15.1		Introduction *455*
15.2		Principles of SBS-PCM *456*
15.3		Reflectivity of an SBS-PCM *457*
15.4		Beam Combining Architectures *461*
15.5		Phase Controlling Theory *462*
15.6		Coherent Beam Combined Laser System with Phase-Stabilized SBS-PCMs *467*
15.6.1		Conventional Phase Fluctuation of SBS-PCM *467*
15.6.2		Phase Fluctuation without PZT Controlling *468*
15.6.3		Phase Fluctuation with PZT Controlling *471*
15.7		Conclusions *475*
		References *475*

Index *479*

Preface

Laser beam combining techniques allow increasing the power of lasers far beyond what it is possible to obtain from a single conventional laser. One step further, coherent beam combining (CBC) also helps to maintain the very unique properties of the laser emission with respect to its spectral and spatial properties. Such lasers are of major interest for many applications, including industrial, environmental, defense, and scientific applications. Recently, significant progress has been made in coherent beam combining lasers, with a total output power of 100 kW already achieved. Scaling analysis indicates that further increase of output power with excellent beam quality is feasible by using existing state-of-the-art lasers. Thus, the knowledge of coherent beam combining techniques will become crucial for the design of next-generation high-power lasers. The purpose of this book is to present the more recent concepts of coherent beam combining by world leader teams in the field.

It is now well established that the availability of high-power laser diodes has allowed the realization of efficient solid-state lasers. New configurations of diode pumping schemes permit to increase laser efficiency and extract more energy from the amplifying medium operating in the continuous-wave or pulsed regimes. However, the resulting thermal load induces a reduction of the source brightness and a degradation of the spatial beam profile. Recently, development of double-clad fiber lasers has allowed to solve these limitations and to maintain excellent beam quality with single-mode or large-mode area fibers. However, the fiber output power is inherently limited by its damage threshold or parasitic nonlinear effects, which spoil the laser emission. Coherent laser beam combining brings a solution to overcome all these limitations. The idea is to make a coherent summation of the power of individual laser sources or amplifiers, each modular element operating at a power or an energy under the threshold of all these parasitic effects.

The critical issue of CBC is to identify the most efficient architectures and techniques, depending on the requirements and the applications. The central theme of the book is thus to review the most advanced achievements in the field. This book intends to provide a guide for scientists and engineers working on, or with, lasers. Coherent beam combining techniques handle multidisciplinary fields such as laser physics, adaptive optics, electronics, optoelectronics, imaging, and nonlinear optics. In consequence, this book should interest a large community of people involved in these areas. Each technique has been presented in detail with an emphasis on practical implementations. Thus, both scientific and engineering aspects are addressed.

Part One of this book will review in detail the most promising CBC techniques with active phase control (Chapters 1–7). These techniques – strongly connected to adaptive optics – include heterodyne phase detection technique, stochastic parallel gradient descent algorithm phase locking, multidithering or single-frequency dithering techniques, imaging, and interferometric techniques. Active coherent phase combining may also offer unique functionalities such as nonmechanical beam steering and compensation of atmospheric turbulences. These aspects are discussed in Chapters 5 and 6. The use of coherent laser beam combining in the pulsed and the ultrafast regimes are also introduced in Chapters 8 and 9.

Part Two deals with passive phase locking of multiple gain media. Coherent combining is obtained through self-organization of the laser emission (Chapters 10–14) or by nonlinear interactions (Chapter 15). These more advanced concepts do not require electronic control and provide self-adaptive phase locking. Applications of these techniques to fiber lasers are highlighted in Chapters 11 and 12. Combining of quantum cascade lasers in the mid-infrared is detailed in Chapter 13.

This book represents a collective effort and I would like to express my warm thanks to all the authors for their very valuable contributions and cooperation during the preparation of this book.

Palaiseau, France *Arnaud Brignon*
May 2013

Acronyms

2D	two-dimensional
AC	alternating current
AO	adaptive optics
AOM	acousto-optic modulator
AR	antireflection
ASE	amplified spontaneous emission
BC	beam combiner
BCM	back-seeding concave mirrors
BEFWM	Brillouin-enhanced four-wave mixing
BGS	Brillouin gain spectrum
BQ	beam quality
BS	beam splitter
CBC	coherent beam combining
CCD	charge-coupled device
CCEPS	conduction-cooled, end-pumped slab
CMOS	complementary metal oxide semiconductor
COMD	catastrophic optical mirror damage
CPA	chirped-pulse amplifier
CPBC	coherent polarization beam combining
CW	continuous wave
DC	direct current
DF-SPGD	delayed-feedback stochastic parallel gradient descent
DFB	distributed feedback
DG	Dammann grating
DL	diffraction limited
DM	deformable mirror
DOE	diffractive optical element
DPA	divided pulse amplification
DSP	digital signal processor
DXRL	deep X-ray lithography
EDF	erbium-doped fiber
EDFA	erbium-doped fiber amplifier
EOM	electro-optic modulator

EYDF	erbium/ytterbium-doped fiber
FA	fiber amplifier
FBG	fiber Bragg grating
FC	fiber coupler
FF	far field
FPA	focal plane array
FPGA	field-programmable gate array
FR	Faraday rotator
FROG	frequency-resolved optical gating
FSM	fast steering mirror
FWHM	full width at half maximum
FWM	four wave mixing
GEV	generalized extreme value
GRIN	gradient index
HR	high-reflection
HWP	half-wave plate
ICP	inductively coupled plasma
IEC	individual external cavity
IR	infrared
JHPSSL	Joint High Power Solid State Laser
LC	liquid crystal
LIDAR	light detection and ranging
LMA	large mode area
LOCSET	locking of optical coherence by single-detector electronic-frequency tagging
LP	low-pass
MFD	mode field diameter
MIR	mid-infrared
MO	master oscillator
MOCVD	metalorganic chemical vapor deposition
MOPA	master oscillator power amplifier
MOPFA	master oscillator power fiber amplifier
NA	numerical aperture
NF	near field
NPC	nonphase conjugate
NPRO	nonplanar ring oscillator
NRTE	nonreciprocal transmission element
OC	output coupler
OHD	optical heterodyne detection
OPCPA	optical parametric chirped pulse amplification
OPD	optical path difference
OPO	optical parametric oscillator
OTDM	optical time division multiplexing
PBS	polarizing beam splitter
PC	personal computer

PCF	photonic crystal fiber
PCM	phase conjugate mirror
PCSOCBC	phase conjugate self-organized coherent beam combination
PD	photodiode
PI	proportional integrator
PIB	power-in-the-bucket
PLC	planar lightwave circuit
PLZT	Lanthanum-doped lead zirconate titanate
PM	polarization-maintaining
PMMA	polymethylmethacrylate
PoD	polarizability difference
POL	polarizer
PRF	pulse repetition frequency
PSD	power spectral density
PV	peak-to-valley
PZT	piezoelectric translator
QCL	quantum cascade laser
QCW	quasi-continuous wave
QW	quantum well
QWLSI	quadri-wave lateral shearing interferometer
QWP	quarter-wave plate
RCWA	rigorous coupled wave analysis
RF	radio frequency
RIC	refractive index change
RIN	relative intensity noise
RMS	root-mean-square
RT	room temperature
RWG	ridge waveguide
SBC	spectral beam combining
SBS	stimulated Brillouin scattering
SBS-PCM	stimulated Brillouin scattering phase conjugate mirror
SCOW	slab-coupled optical waveguide
SCOWA	slab-coupled optical waveguide amplifiers
SCOWL	slab-coupled optical waveguide laser
SCSFD	sine-cosine single-frequency dithering
SEM	scanning electron microscope
SESAM	semiconductor saturable absorber mirror
SF	single frequency
SFD	single-frequency dithering
SLM	spatial light modulator
SM	single mode
SMPL	speckle metric optimization-based phase locking
SOCBC	self-organized coherent beam combining
SPGD	stochastic parallel gradient descent
SPIDER	spectral phase interferometry for direct electric field reconstruction

SPM	self-phase modulation
SR	Strehl ratio
SRS	stimulated Raman scattering
SSL	solid-state laser
SWaP	size, weight, and power
TBP	time-bandwidth product
TDFA	Tm-doped fiber amplifier
TIL	target-in-the-loop
TIR	total internal reflection
UV	ultraviolet
VBG	volume Bragg grating
VBQ	vertical beam quality
VDL	variable delay line
WDM	wavelength-division multiplexer
WFS	wavefront sensor
YDF	ytterbium-doped fiber
YDFA	ytterbium-doped fiber amplifier

List of Contributors

Oleg Antipov
Institute of Applied Physics of the
Russian Academy of Science
Nizhny Novgorod 603950
Russia

Alain Barthélémy
Université de Limoges
CNRS
Xlim Institut de Recherche
87060 Limoges
France

Cindy Bellanger
Onera – The French Aerospace Lab
91761 Palaiseau cedex
France

Guillaume Bloom
Thales Research and Technology
1 avenue Augustin Fresnel
91767 Palaiseau cedex
France

Jérome Bourderionnet
Thales Research & Technology
1 avenue Augustin Fresnel
91767 Palaiseau cedex
France

Pierre Bourdon
Onera – The French Aerospace Lab
91761 Palaiseau cedex
France

Arnaud Brignon
Thales Research & Technology
1 avenue Augustin Fresnel
91767 Palaiseau cedex
France

Guillaume Canat
Onera – The French Aerospace Lab
91761 Palaiseau cedex
France

Gary Carhart
Computational and Information
Sciences Directorate
U.S. Army Research Laboratory
Intelligent Optics Laboratory
Adelphi, MD 20783
USA

Mathieu Carras
III-V Lab
Campus de Polytechnique
1 avenue Augustin Fresnel
91767 Palaiseau cedex
France

Seongwoo Cha
Korea Advanced Institute of Science
and Technology
Department of Physics
Daejeon 305-701
Republic of Korea

Kevin Creedon
Massachusetts Institute of Technology
Lincoln Laboratory
244 Wood Street
Lexington, MA 02420-9108
USA

Michael J. Damzen
Imperial College London
The Blackett Laboratory
London SW7 2AZ
UK

Louis Daniault
Université Paris-Sud
CNRS
Institut d'Optique
Laboratoire Charles Fabry
2 avenue Augustin Fresnel
91127 Palaiseau cedex
France

Nir Davidson
Weizmann Institute of Science
Department of Physics of Complex Systems
234 Herzl Street
Rehovot 76100
Israel

Andrew Deck
Optonicus
711 East Monument Avenue
Suite 101
Dayton, OH 45402
USA

Agnès Desfarges-Berthelemot
Université de Limoges
CNRS
Xlim Institut de Recherche
87060 Limoges
France

Joseph Donnelly
Massachusetts Institute of Technology
Lincoln Laboratory
244 Wood Street
Lexington, MA 02420-9108
USA

Frédéric Druon
Université Paris-Sud
CNRS
Institut d'Optique
Laboratoire Charles Fabry
2 avenue Augustin Fresnel
91127 Palaiseau cedex
France

Tso Y. Fan
Massachusetts Institute of Technology
Lincoln Laboratory
244 Wood Street
Lexington, MA 02420-9108
USA

Angel Flores
Air Force Research Laboratory
Directed Energy Directorate
Kirtland Air Force Base
Albuquerque, NM 87117
USA

Andrei Fotiadi
University of Mons
20, place du Parc
7000 Mons
Belgium

and

Ulyanovsk State University
Leo Tolstoy St.
432970 Ulyanovsk
Russia

and

Ioffe Physico-Technical Institute of the RAS,
194021 St. Petersburg
Russia

Moti Fridman
Weizmann Institute of Science
Department of Physics of Complex Systems
234 Herzl Street
Rehovot 76100
Israel

Asher A. Friesem
Weizmann Institute of Science
Department of Physics of Complex Systems
234 Herzl Street
Rehovot 76100
Israel

Micah Gatz
University of Dayton
School of Engineering
Intelligent Optics Laboratory
300 College Park
Dayton, OH 45469
USA

Patrick Georges
Université Paris-Sud
CNRS
Institut d'Optique
Laboratoire Charles Fabry
2 avenue Augustin Fresnel
91127 Palaiseau cedex
France

Gregory D. Goodno
Northrop Grumman Aerospace Systems
One Space Park Boulevard
Mail Stop ST71LK/R1184D
Redondo Beach, CA 90278
USA

Julien Le Gouët
Onera – The French Aerospace Lab
91761 Palaiseau cedex
France

Marc Hanna
Université Paris-Sud
CNRS
Institut d'Optique
Laboratoire Charles Fabry
2 avenue Augustin Fresnel
91127 Palaiseau cedex
France

Milan Kalal
Czech Technical University in Prague
Faculty of Nuclear Sciences and Physical Engineering
Brehova 7
115 19 Prague 1
Czech Republic

Vincent Kermène
Université de Limoges
CNRS
Xlim Institut de Recherche
87060 Limoges
France

Mercedeh Khajavikhan
University of Minnesota
Department of Electrical and Computer Engineering
Minneapolis, MN 55455
USA

Hong J. Kong
Korea Advanced Institute of Science and Technology
Department of Physics
Daejeon 305-701
Republic of Korea

Maxim Kuznetsov
Institute of Applied Physics of the Russian Academy of Science
Nizhny Novgorod 603950
Russia

Svetlana Lachinova
Optonicus
711 East Monument Avenue
Suite 101
Dayton, OH 45402
USA

Eric Lallier
Thales Research and Technology
91767 Palaiseau cedex
France

Christian Larat
Thales Research and Technology
91767 Palaiseau cedex
France

Seong K. Lee
Gwangju Institute of Science and Technology
Advanced Photonics Research Institute
Gwangju 500-712
Republic of Korea

James R. Leger
University of Minnesota
Department of Electrical and Computer Engineering
Minneapolis, MN 55455
USA

Zejin Liu
National University of Defense Technology
College of Opto-Electronic Science and Engineering
Deya road
Changsha 410073
China

Laurent Lombard
Onera – The French Aerospace Lab
91761 Palaiseau cedex
France

Chunte A. Lu
Air Force Research Laboratory
Directed Energy Directorate
Kirtland Air Force Base
Albuquerque, NM 87117
USA

Yanxing Ma
National University of Defense Technology
College of Opto-Electronic Science and Engineering
Deya road
Changsha 410073
China

Xavier Marcadet
III-V Lab
91767 Palaiseau cedex
France

Patrice Mégret
University of Mons
7000 Mons
Belgium

Micha Nixon
Weizmann Institute of Science
Department of Physics of Complex Systems
234 Herzl Street
Rehovot 76100
Israel

Dimitrios N. Papadopoulos
CNRS
Ecole Polytechnique
Laboratoire pour l'utilisation des lasers intenses
91128 Palaiseau cedex
France

Vladimir Paramonov
Optonicus
711 East Monument Avenue
Suite 101
Dayton, OH 45402
USA

Sangwoo Park
Korea Advanced Institute of Science and Technology
Department of Physics
Daejeon 305-701
Republic of Korea

Jérome Primot
Onera – The French Aerospace Lab
91761 Palaiseau cedex
France

Benjamin Pulford
Air Force Research Laboratory
Directed Energy Directorate
Kirtland Air Force Base
Albuquerque, NM 87117
USA

Shawn Redmond
Massachusetts Institute of Technology
Lincoln Laboratory
244 Wood Street
Lexington, MA 02420-9108
USA

Craig Robin
Air Force Research Laboratory
Directed Energy Directorate
Kirtland Air Force Base
Albuquerque, NM 87117
USA

Joshua E. Rothenberg
Northrop Grumman Aerospace Systems
One Space Park Boulevard
Mail Stop ST71LK/R1184D
Redondo Beach, CA 90278
USA

Thomas Ryan
Optonicus
711 East Monument Avenue
Suite 101
Dayton, OH 45402
USA

Antonio Sanchez-Rubio
Massachusetts Institute of Technology
Lincoln Laboratory
244 Wood Street
Lexington, MA 02420-9108
USA

Peter C. Shardlow
University of Southampton
Optoelectronics Research Centre
Southampton SO17 1BJ
UK

Thomas M. Shay
Air Force Research Laboratory
Directed Energy Directorate
Kirtland Air Force Base
Albuquerque, NM 87117
USA

Ondrej Slezak
Czech Academy of Sciences
Institute of Physics
HiLASE Project
Na Slovance 1999/2
182 21 Prague 8
Czech Republic

Mikhail Vorontsov
University of Dayton
School of Engineering
Intelligent Optics Laboratory
300 College Park
Dayton, OH 45469
USA

and

Optonicus
711 East Monument Avenue
Suite 101
Dayton, OH 45402
USA

Xiaolin Wang
National University of Defense Technology
College of Opto-Electronic Science and Engineering
Deya Road
Changsha 410073
China

Thomas Weyrauch
University of Dayton
School of Engineering
Intelligent Optics Laboratory
300 College Park
Dayton, OH 45469
USA

Xiaojun Xu
National University of Defense Technology
College of Opto-Electronic Science and Engineering
Deya Road
Changsha 410073
China

Jin W. Yoon
Institute for Basic Science
Daejeon 305-811
Republic of Korea

Charles Yu
Massachusetts Institute of Technology
Lincoln Laboratory
244 Wood Street
Lexington, MA 02420-9108
USA

Yoann Zaouter
Amplitude Systèmes
11 avenue de Canteranne
Cité de la Photonique
33600 Pessac
France

Pu Zhou
National University of Defense Technology
College of Opto-Electronic Science and Engineering
Deya Road
Changsha 410073
China

Part One
Coherent Combining with Active Phase Control

1
Engineering of Coherently Combined, High-Power Laser Systems
Gregory D. Goodno and Joshua E. Rothenberg

1.1
Introduction

In recent years, much effort has been expended toward scaling electric lasers to CW power levels on the order of 100 kW or greater [1]. The key challenge in such scaling is maintaining near-diffraction-limited (DL) beam quality (BQ) to enable tight focusing onto a distant target. Despite the maturation of scalable, diode-pumped laser amplifier technologies such as zigzag slabs [2] or fibers [3], thermal effects or optical nonlinearities currently limit near-DL output from single lasers to an order of magnitude lower power, around 10 kW.

Actively phase-locked coherent beam combination (CBC) of N laser amplifiers seeded by a common master oscillator (MO) represents an engineerable approach toward scaling laser brightness B (loosely defined here as $B \sim \text{power}/\text{BQ}^2$) beyond the limits of the underlying single-element laser technology. Ideally, the combined output behaves as if it were a single beam, and B is thereby increased by a factor of N over an unphased array or by a factor of N^2 over any individual laser [4].

A compelling architectural advantage of CBC systems in comparison to single-aperture lasers of comparable power is the graceful degradation in response to failure of any gain element. This feature can be elucidated from the scaling of $B \sim N^2$, so the relative rate of change in brightness as individual lasers fail is $1/B(dB/dN) = 2/N$. Hence, for large arrays, the drop in brightness is gradual. For example, failure of 1 out of $N = 100$ lasers would still allow a CBC system to continue operating at 98% of its original brightness.

Active CBC with servo-based phase locking can be straightforwardly engineered for very high channel counts and for very high-power laser gain elements. Recently, Northrop Grumman Aerospace Systems adopted an actively phase-locked approach to combine seven 15 kW Nd:YAG slab amplifier chains to demonstrate the world's first 100 kW electric laser with record-setting brightness [5]. As of this writing, work is underway to extend this technology to achieve similar power levels in a CBC array of fiber lasers with improved BQ and efficiency as well as reduced size and weight [6,7].

Coherent Laser Beam Combining, First Edition. Edited by Arnaud Brignon.
© 2013 Wiley-VCH Verlag GmbH & Co. KGaA. Published 2013 by Wiley-VCH Verlag GmbH & Co. KGaA.

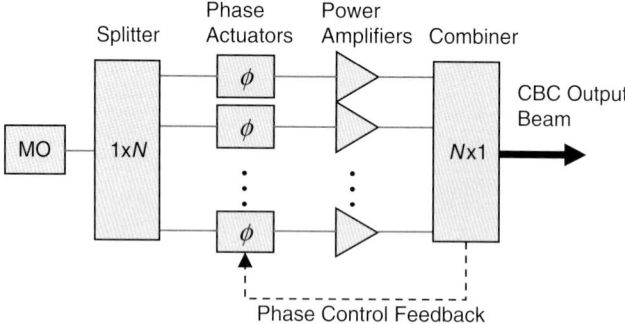

Figure 1.1 System-level block diagram for an actively phase-locked CBC master oscillator power amplifier (MOPA) array.

A canonical system-level architecture for a CBC laser array is shown in Figure 1.1. A single master oscillator is split to seed a number of N channels. Each channel contains a piston phase actuator capable of imposing at least one wave of phase and a coherence-preserving laser amplifier (or a chain of amplifiers) to boost the channel power to the limit of the laser technology. The high-power outputs from all N channels are geometrically combined, so they copropagate, either by using one or more beam splitters or by tiling side by side. The combined output beam is sampled optically to generate error signals for servo-based phase locking of all N channels up to a fraction of a wave.

From Figure 1.1, we can identify three key technologies that must be integrated to form an actively phase-locked, coherently combined, high-power laser system:

- laser amplifiers, preserving coherence properties of a common master oscillator while providing high gain and high output power;
- optical system, geometrically overlapping the amplified beams in the far field (FF) and for some implementations, in the near field (NF); and
- active control systems, cophasing the amplified output beams via closed-loop feedback.

In the remainder of this chapter, we review recent advances in these three technology areas. We begin by deriving engineering requirements on laser source uniformity, presented as trades against combining efficiency. This provides a framework to assess coherent combining technologies amenable to scaling to both high channel counts and high powers, including active piston control using optical heterodyne phase detection and geometric beam combining, using either tiled apertures or diffractive optical elements (DOEs). Finally, we review the engineering challenges, design, and CBC test results of two specific solid-state laser amplifier technologies, Nd:YAG zigzag slabs and Yb:SiO$_2$ fibers. These laser technologies are particularly well suited toward the demands of 100 kW level CBC owing to their scalability, high gain, high efficiency, and outstanding spatial and temporal coherence properties.

1.2
Coherent Beam Combining System Requirements

The primary requirement for high-efficiency CBC is that the combined beams must be mutually coherent in both space and time to allow complete constructive interference. This means the lasers must be spatially mode-matched and coaligned, power-balanced, copolarized, path length matched, and locked in phase with high precision. When these requirements are not perfectly met, combining efficiency suffers. For a large channel count CBC array, the coherence requirement can be expressed quantitatively and concisely in terms of statistical uniformity tolerances between the laser array elements [8].

We consider a large array of N input beams, combined in a filled aperture configuration using a beam splitter optic in reverse as a beam combiner (BC). This BC can represent, for example, a tapered fiber coupler, a DOE, or a cascade of free space or guided wave splitters. The BC has *a priori* unequal power splitting fractions D_n^2 over the desired channels $n = 1 - N$, where normalization $\sum_{n=1}^{\infty} D_n^2 = 1$ accounts for the possibility of coupling losses intrinsic to the BC into channels $n > N$ (Figure 1.2). The BC efficiency as a splitter is then $\eta_{\text{split}} = \sum D_n^2$, where the summation is over only the N channels of interest.

Operated as a $N \times 1$ combiner, the spatially resolved, time-averaged combining efficiency $\eta'(x)$ is the ratio of power in the desired output port to the total input power. It is straightforward to show [9] that

$$\eta'(x) = \left\langle \left| \sum D_n E_n(x,t) \right|^2 \right\rangle \Big/ \left\langle \sum |E_n(x,t)|^2 \right\rangle. \tag{1.1}$$

Here, the brackets denote time averaging and $E_n(x,t)$ are spatially and temporally nonuniform fields of the input beams. A simple illustration of Eq. (1.1) is shown in Figure 1.3 for the case of $N = 2$ beams combined on a 50/50 beam splitter with a small pointing misalignment. Due to the wavefront tilt between beams, the beams cannot interfere constructively over the entire aperture, leading to a spatially varying combining efficiency $\eta'(x)$.

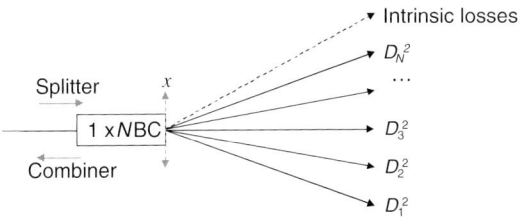

Figure 1.2 Power splitting ratios for a $1 \times N$ beam splitter/combiner.

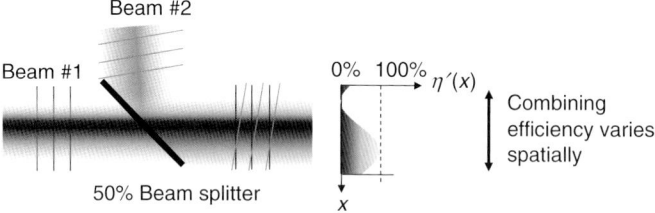

Figure 1.3 Illustration of spatially varying combining efficiency for two misaligned beams combined on a 50% beam splitter.

The overall combining efficiency η is the intensity-weighted average of $\eta'(x)$ over the combining aperture:

$$\eta = \frac{1}{P_{\text{in}}} \int \eta'(x) \left\langle \sum |E_n(x,t)|^2 \right\rangle dx = \frac{1}{P_{\text{in}}} \int \left\langle \left| \sum D_n E_n(x,t) \right|^2 \right\rangle dx, \qquad (1.2)$$

where $P_{\text{in}} = \int \langle \sum |E_n(x,t)|^2 \rangle dx$ is the total input power (up to a constant). We assume the beams are derived from a common single-mode (SM) CW master oscillator. The MO is assumed to be quasi-monochromatic with carrier frequency ω_0 and slowly time-varying phase modulation $\psi(t)$. Hence, the fields can be written in terms of spatially dependent amplitudes $A_n(x)$ and wavefronts $\phi_n(x)$: $E_n(x,t) = A_n(x)\cos(\chi_n)\exp[i\omega_0 t + i\psi(t) + i\phi_n(x)]$, where χ_n is the depolarization angle of the nth field from the array average.

Since our goal is to identify the effects of relatively small misalignments and aberrations of the input beams, we write each field perturbatively:

$$E_n(x,t) = [A(x) + \delta A_n(x)]\left(1 - \delta\chi_n^2/2\right)\exp\{i\omega_0(t + \delta\tau_n)$$

$$+ i\psi(t) + i\Delta\omega(t)(t + \delta\tau_n) + i[\varphi(x) + \delta\varphi_n(x)]\} \qquad (1.3)$$

Here, $\delta A_n(x)$ and $\delta\phi_n(x)$ are small deviations of amplitude in the nth field and wavefront distributions from their respective array average, and we have assumed small group delay mismatches $\delta\tau_n$ to allow substitution of the Taylor expansion $\psi(t + \delta\tau_n) \approx \psi(t) + \Delta\omega(t)(t + \delta\tau_n)$, where $\Delta\omega(t) \, d\psi(t)/dt$ is a small, time-dependent frequency perturbation. We also assume "quasi-uniform" BC splitting ratios, $D_n = (\eta_{\text{split}}/N)^{1/2} + \delta D_n$, where the amplitude split perturbations are $\delta D_n \ll N^{-1/2}$.

With these approximations, Eq. (1.2) can be evaluated by expanding the exponential, taking the modulus square, and neglecting perturbative terms higher than second order. The resulting expression for efficiency can be written compactly in terms of statistical parameter fluctuations across the array of N beams:

$$\eta = \eta_{\text{split}}\left[1 - (N/P_{\text{in}})\int \left(\sigma_{A(x)}^2 + A(x)^2\sigma_{\varphi(x)}^2 - 2\sqrt{N/\eta_{\text{BC}}}A(x)\sigma_{A(x),D}\right)dx\right.$$

$$\left. - \langle\Delta\omega(t)^2\rangle\sigma_\tau^2 - \sigma_\chi^2 - N\sigma_D^2/\eta_{\text{BC}}\right] \qquad (1.4)$$

Here, σ_u^2 represents the mean-square variance of the parameters $u = \{A(x), \phi(x), \tau, D, \chi\}$ across the array and $\sigma_{A(x),D}$ is the covariance of the input field amplitudes $A_n(x)$ with the corresponding splitting amplitudes D_n.

In the limit of perfectly coaligned, monochromatic plane waves with a uniform and lossless BC, Eq. (1.4) reduces to

$$\eta = 1 - (\sigma_A/A)^2 - \sigma_\phi^2. \qquad (1.5)$$

Equation (1.5) is the well-known Marechal approximation for the effects of amplitude imbalance and piston phase errors between beams [9,10–12]. One should note that since channel power $P \sim A^2$, the relative intensity noise (RIN) variance $(\sigma_P/P)^2 = 4(\sigma_A/A)^2$. Hence, the combining loss due to power imbalance between beams is $(\sigma_P/P)^2/4$ [11,12].

Several useful insights are immediately apparent from Eq. (1.4). First, in the limit of small uncorrelated misalignments, one can independently assess the impact of diverse physical effects such as wavefront errors, power imbalance, and group delay mismatch. The limits of validity of this approximation are illustrated in Figure 1.4 for the simple case of a lossless BC, with no correlation between the input channel powers and BC splitting fractions. The error in the efficiency calculation is <1% for normalized standard deviations in amplitude of <20% of the average across the array, corresponding to <40% variations in input power balance and BC power splitting fractions. Hence, while Eq. (1.4) is an approximation, it nevertheless yields a reliable lower bound on η that is quite accurate for most cases of practical interest (i.e., arrays of similar configuration lasers that are reasonably well aligned).

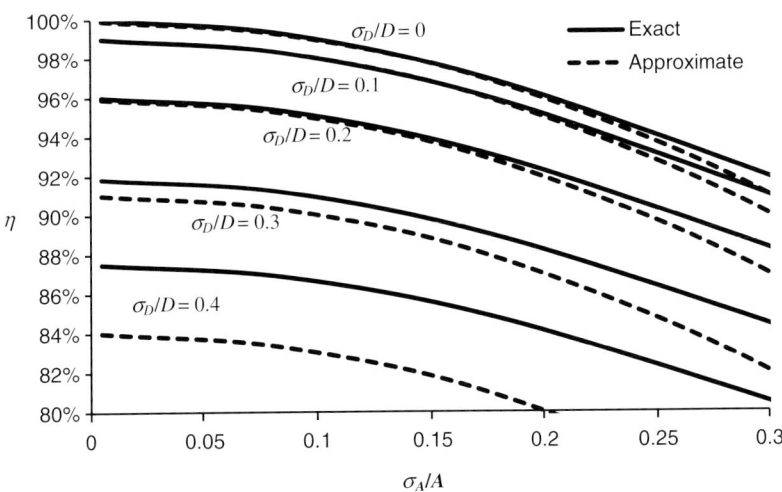

Figure 1.4 Comparison of the approximate [Eq. (1.4)] and exact [Monte Carlo model using Eq. (1.2)] combining efficiencies of large arrays ($N = 10^3$), with normalized standard deviations σ_A/A and σ_D/D of laser amplitude and BC splitting coefficients.

The ability to independently assess varied misalignments using Eq. (1.4) is quite useful toward guiding the design of a CBC laser array, enabling system designers to derive error budgets for the various subsystems and components needed to bring the beams into alignment. In particular, for Gaussian-shaped beams (such as those emitted from single-mode fiber laser amplifiers), Eq. (1.4) results in a very simple analytic expression for CBC loss in terms of Gaussian beam parameters given in Table 1.1. As will be discussed in more detail in the following sections, these expressions have been largely confirmed to be in agreement with experimental measurements of combining efficiency and alignment uniformity of fiber arrays. In general, beams must be spatially coaligned and mode matched to <10% of their Gaussian spot size or coherence length to maintain combining losses below 1%.

It is notable that the loss terms in Eq. (1.4) arise purely from noncommon path (uncorrelated) variations between beams. Adding a common path wavefront to each beam does not change the value of the mean-square variance $\sigma^2_{\phi(x)}$. Hence, common path wavefront aberrations have no impact on combining efficiency since there is no change in constructive interference between beams. Common wavefront errors can still degrade BQ, since they transmit through the BC onto the combined output.

Since the losses in Eq. (1.4) arise purely from uncorrelated aberrations between beams, the CBC process can be seen to serve as an effective "coherent filter." Whatever fraction of the total input power is successfully combined into the output beam will exhibit substantially reduced wavefront aberration, beam jitter, and phase or frequency noise. The physical intuition is that uncorrelated aberrations captured by the terms in Eq. (1.4) (e.g., pointing jitters, higher order wavefront errors, and nonuniform dispersion) are essentially removed upon coherent combining. This effect is clearly seen in experiments where the far-field pointing jitter and beam quality of the combined beam are in fact improved over the input beams, following a spatial filter to remove the uncombined light that appears outside the diffraction-limited central lobe [2,13]. This is essentially similar physics underlying Fabry–Pérot cavity-based optical mode cleaners [14], with the filter profile defined by the average input field $A(x)e^{i\phi(x)+i\psi(t)}$ rather than by the modes of a resonant cavity. Due to this coherent filtering effect, CBC appears a promising technology for applications requiring extremely high-quality, high-power beams, such as interferometers for gravity wave detection [15], or resonant cavity-enhanced high-harmonic generation [16].

1.3
Active Phase-Locking Controls

Much of the CBC research literature focuses exclusively on the means by which the lasers are locked in-phase modulo 2π. The advent of commercially available phase actuators based on fiber-coupled, waveguided electro-optic modulators (EOMs) with gigahertz bandwidths and multiwave strokes, coupled with modern RF electronic components and design tools, has resulted in a number of viable control methods, thereby leading a system designer to conduct meaningful trade studies to select the optimum approach for a given CBC system.

1.3 Active Phase-Locking Controls | 9

Table 1.1 Coalignment and uniformity tolerances for spatially and spectrally aligned Gaussian beams with 1% allowance for combining loss for each effect.

RMS parameter variation	Combining loss, $1-\eta$	Value for 1% loss	Schematic	Notes
Piston phase, σ_ϕ	σ_ϕ^2	$\sigma_\phi = 0.1$ rad		Also applies to noncommon path, intensity-weighted wavefront aberrations
Power fraction, σ_P/P	$\frac{1}{4}(\sigma_P/P)^2$	$\sigma_P/P = 20\%$		$P \sim A^2$, so $\delta P \sim 2\delta A$; hence $\sigma_P/P = 2\sigma_A/A$
Optical path mismatch, $c\sigma_\tau$	$\frac{\pi^2}{2\ln(2)}\Delta f_{FWHM}^2 \sigma_\tau^2$	$\Delta f_{FWHM} = 11$ GHz, $c\sigma_\tau = 1$ mm		Loss due to dephasing between beams
Polarization angle, σ_χ	σ_χ^2	$\sigma_\chi = 0.1$ rad		Equivalent to depolarized power loss
Fractional spot displacement, σ_x/w	$(\sigma_x/w)^2$	$\sigma_x/w = 10\%$		δx = beam position error, w = Gaussian beam radius ($1/e^2$ intensity)

(*continued*)

Table 1.1 (Continued)

RMS parameter variation	Combining loss, $1-\eta$	Value for 1% loss	Schematic	Notes
Fractional spot size, σ_w/w	$1/2(\sigma_w/w)^2$	$\sigma_w/w = 14\%$		δw = beam radius error, w = Gaussian beam radius ($1/e^2$ intensity)
Fractional pointing, σ_θ/θ	$(\sigma_\theta/\theta)^2$	$\sigma_\theta/\theta = 10\%$		$\delta\theta$ = beam pointing error, θ = Gaussian half-angle divergence ($1/e^2$ intensity)
Fractional divergence, σ_Θ/Θ	$1/2(\sigma_\Theta/\Theta)^2$	$\sigma_\Theta/\Theta = 14\%$		$\delta\Theta$ = divergence error, Θ = Gaussian half-angle divergence ($1/e^2$ intensity)

The BC is assumed to be lossless and uniform ($\eta_{\text{split}} = 1$ and $D_n = N^{-1/2}$).

1.3 Active Phase-Locking Controls

Table 1.2 Comparison of the three major classes of active phase control loops for CBC.

Method	OHD	Synchronous multidither (LOCSET)	Hill climbing
Channel count scaling, N	No limit; fully parallel system	$N = 32$ demonstrated [17]; more appears possible [18]	$N \sim 10$ with ~ 10 kHz control bandwidth
No. of detectors needed	N	1	1
Control bandwidth	>10 kHz	>10 kHz	Scales as N^{-1} [19]
Piston set point identification	Manual	Automatic	Automatic
Needs RF electronics?	Yes	Yes	No
Needs optical reference?	Yes	No	No
RMS phase errors	$\lambda/80$ [20]	$\lambda/70$ ($N=32$) [17]	$\lambda/40$ ($N=8$) [21]

Shaded cells indicate disadvantages, either with excess system complexity or limited scaling capability.

For a phase locking method to work successfully outside a laboratory environment, it must have an effective control bandwidth of multikilohertz to reject acoustically coupled phase noise and maintain RMS phase stability between channels within $\ll 1$ rad (Table 1.1). Hence, the primary requirement in selecting the phase control method is that it can achieve high-speed and high locking fidelity for the appropriate number of channels N. The most successful phasing methods fall into three broad classes summarized in Table 1.2. We will discuss these three methods in the following sections, concentrating on multichannel optical heterodyne detection (OHD) locking. More detailed descriptions of multidither and hill climbing approaches are presented in Chapters 2, 4, 6, and 8.

1.3.1
Optical Heterodyne Detection

In the OHD method, a reference beam derived from the MO is frequency shifted by $\Delta\omega$ upon passage through an acousto-optic modulator (AOM) before being interferometrically combined with a low-power sample of the beam array using a beam splitter (Figure 1.5). Within each individual beam footprint, a square-law photodetector senses the superimposed fields of the signal and reference beams and produces a time-dependent voltage

$$V(t) = |E_{sig}|^2 + |E_{ref}|^2 + 2|E_{sig}E_{ref}|\cos[\Delta\omega t + \varphi(t)]. \tag{1.6}$$

Here, E_{sig} and E_{ref} are the electric fields of the signal and reference beams and $\varphi(t)$ is the time-dependent optical phase jitter between signal and reference beams. OHD can alternatively be implemented in the spatial domain by imposing a tilt (spatial

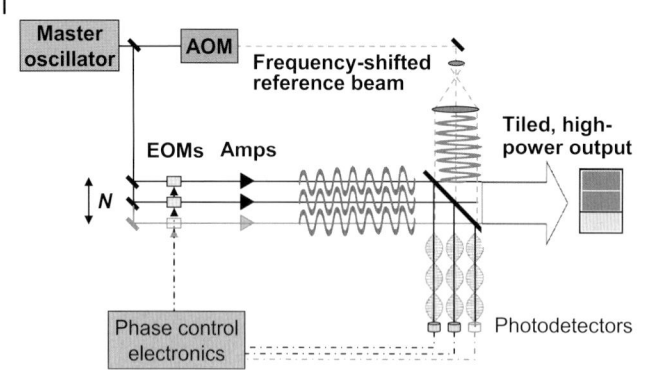

Figure 1.5 Actively phase-locked array of N lasers (N = 3 as shown) using the OHD method.

frequency shift) between the signal and reference beams, leading to a spatial interference pattern similar to Eq. (1.6) [22].

From Eq. (1.6), it is apparent that the detector voltage consists of a DC background modulated at the difference frequency owing to interference between the signal and reference beams. This sinusoidal beat waveform is squared up by passing it through a saturated amplifier and is then compared with a similar clock waveform derived from the RF drive voltage to the AOM. The time delay – or phase difference – between the edges of these two waveforms corresponds to the optical phase difference $\phi(t)$ between the signal and reference beams. Applying an exclusive OR function to the clock and heterodyne waveforms as shown in Figure 1.6 results in an output whose area is proportional to $\phi(t)$. This error signal is then fed back to an electro-optic modulator located in the low-power front end of the corresponding amplifier chain to control its phase within a multiple of 2π. In this manner, each beam is locked independently to the same phase as the reference beam and thus indirectly to one another.

The OHD phase locking method has been successfully implemented on numerous fiber- and slab-based laser arrays to demonstrate coherent beam combination [2,5,13,20]. As shown in Figure 1.7, control bandwidths in excess of 10 kHz are readily attainable with this approach, with RMS phase residuals $\sigma_\phi < 0.1$ rad in the presence of phase excursions exceeding 10^4 rad/s. From Eq. (1.5), the consequent CBC loss owing to imperfections in the phase control would be $\sigma_\phi^2 < 1\%$.

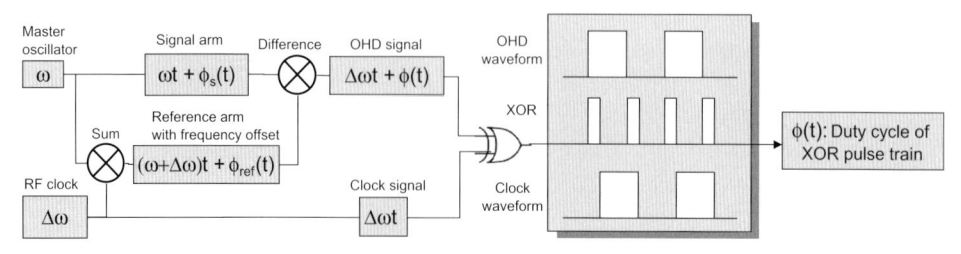

Figure 1.6 XOR logic to generate error signals for the OHD phasing control loop.

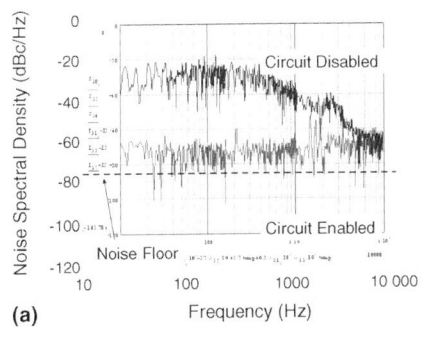

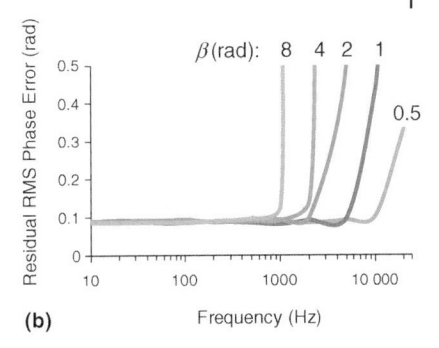

Figure 1.7 OHD phase control loop performance. (a) Noise reduction with OHD loop enabled with a 10 kW Nd:YAG slab amplifier chain [2]. (b) RMS phasing residuals as a function of applied sinusoidal phase noise $\beta\sin(2\pi ft)$ with varying frequency f and amplitude β.

The main benefit of the OHD phasing method is its scalability to large N. All beams are locked directly to a reference, not to each other. Hence, each beam's phase-locking behavior is independent of the presence or absence of other beams. In principle, there is no physical constraint on the number of beams that may be successfully combined.

Since the OHD approach requires one photodetector per beam, it tends to work best in the context of tiled array beam combiners where it is straightforward to overlay an expanded reference beam with a sample of the tiled composite output beam. The method has also been used with filled aperture beam combiners, but it requires reimaging of a common path beam sampled prior to the combiner element [13].

One drawback of the OHD approach is its lack of a direct method to ensure all N beams are locked to one another. Misalignments between the reference and signal beams, drifts in the RF electronics, or static transmissive phase shifts in the main output beam compared to the sampled beam can change the phase-locking set points between beams without changing the RF-detected phase $\phi(t)$. Such misalignments do not impact dynamic phase stabilization, but they do impose static phase errors between beams that must be removed via manual or automated set point adjustment.

1.3.2
Synchronous Multidither

A class of phase locking methods that similar to OHD also utilize heterodyne beats are synchronous multidither approaches and variants thereof [17,18,23–27]. The application of multidither methods for phasing arrays of high-power lasers was fully conceived by the 1970s in the context of atmospheric propagation [23], although the lack of high-speed modulators and sensing electronics severely limited the performance. More recent application of multidither control toward phasing high-power fiber amplifiers has been demonstrated in Ref. [18]. In a multidither phasing system, each laser channel is "tagged" with a small ($\ll 1$ rad) phase dither applied at a unique frequency, typically by superimposing with the control voltage on the EOM phase controller. A notable variant is the time sharing among channels of a single modulation frequency, with concurrent reduction in control bandwidth [26,27]. A single detector that samples the combined output beam will exhibit a superposition of beats owing to interference between the various beams. Application of standard RF demodulation techniques can extract unique error signals proportional to the phase error between each channel and the rest of the array. These error signals drive servo loops to cophase the beams.

The advantages of multidither are its utilization of a single detector for sensing the phase errors of an entire array and the avoidance of the phase set point ambiguity intrinsic to the OHD method. Minimization of the beat signals in the combined beams corresponds to the condition where all beams are in-phase. The main disadvantage of multidither is its relative electronic complexity and cost, which has at present limited demonstrations to 32 channels [17] despite clear potential for multihundred channel scalability based on signal-to-noise considerations [18].

1.3.3
Hill Climbing

Perhaps the simplest methods for cophasing beams are to maximize the combined power in the far-field central lobe using hill climbing algorithms [19,28], among which the most widely used is the stochastic parallel gradient descent (SPGD) method [29]. In these approaches, the phases of the entire array of beams are simultaneously changed by small, statistically uncorrelated amounts, and the corresponding change in far-field power (or combining efficiency) is sensed. The phase set points are then updated proportionally to the detected change in power, eventually arriving at the maximum power when all beams are in-phase.

Many variants of this class of methods are possible to optimize performance [30,31], but in general hill climbing methods suffer from limited scaling potential since for each added channel, an additional dimension in phase space must be dithered. Consequently, closed-loop bandwidths drop proportionally to $1/N$ [19,21]. Despite this limit, the avoidance of high-speed RF electronics makes this control method relatively low cost and simple to implement using programmable computers. Hence, it is an attractive path for systems with either low channel counts ($N \ll 100$) or low phase noise amplifiers that do not require high-speed phase control [28]. Hill climbing variants with nested loops appear promising to bypass the scaling limitations at some increased cost of complexity [31,32].

1.4
Geometric Beam Combining

In addition to being locked in-phase with high fidelity, the amplified beams must be geometrically overlapped so that they copropagate as a single beam. While the specific optical arrangements can take many forms, the geometric beam combiners in general can be broken into two classes: tiled aperture and filled aperture combiners [4].

1.4.1
Tiled Aperture Combiners

For tiled aperture beam combiners, the amplified beams are positioned side by side in the near field as close together as is feasible without excessive clipping losses. The beams are then pointed in the same direction so that their far fields overlap, synthesizing a composite beam. This approach has the advantage of simplicity and low loss. Moreover, for large arrays, this approach enables the prospect of purely electronic beam steering by controlling the relative phases of each beam, which can be advantageous by eliminating the need for a bulky gimbaled beam director telescope. Finally, since the beam footprints have minimal overlap with one another on optical surfaces, laser-induced damage on one beam footprint allows continued system operation (albeit at reduced efficiency) simply by turning off the beam in question. This lack of single-point failure on the high-power optics can be quite attractive for extremely high-power systems.

The principal drawback of side-by-side beam tiling is that near-field intensity nonuniformity owing to the tiling gaps between beams leads to far-field side lobes in the composite beam. In the limit of large N, the power fraction in the far-field central lobe can be expressed in terms of CBC loss owing to amplitude nonuniformity from Eq. (1.4):

$$1 - P_{\text{side lobes}} = 1 - \frac{N}{P_{\text{in}}} \int \sigma^2_{A(x)} dx. \tag{1.7}$$

While this term was derived for a filled aperture beam, for a tiled aperture $\int \sigma^2_{A(x)} dx$ it can be simply reinterpreted as the variance in field amplitude across the composite near field. For the simplified case of flattop near-field beams, Eq. (1.7) reduces to the fill factor of the composite near field. Hence, it is equivalent to the Strehl ratio S or the normalized far-field peak intensity [4]. As the tiled gaps between beams increase, more of the far-field power appears in side lobes away from the diffraction-limited spot size, reducing S and degrading BQ.

If the individual amplified beams have near-flattop intensity profiles, such as those generated from an array of slab lasers (cf., Section 1.5.1), this fill factor impact on BQ can be made arbitrarily small by minimizing the tiling gap between beams [2]. However, when the laser elements are single-mode fiber lasers, which generate near-Gaussian intensity profiles, the loss in BQ of a close-packed tiled array can be substantial. This is illustrated in Figure 1.8 in which only 63% of the power in a coherently combined 2 × 2 tiled fiber laser array is contained within the diffraction-limited far-field central lobe [20]. The fraction of power in the far-field side lobes can be reduced only by reducing the spacing between beams, which unavoidably reduces the overall combination efficiency owing to near-field clipping of the Gaussian beam wings. Gaussian beams can also be reshaped to a flattened profile with refractive or diffractive optics, but such an approach is limited in effectiveness and also leads to degradation of the beam wavefronts and thus to reduced coherent combination efficiency.

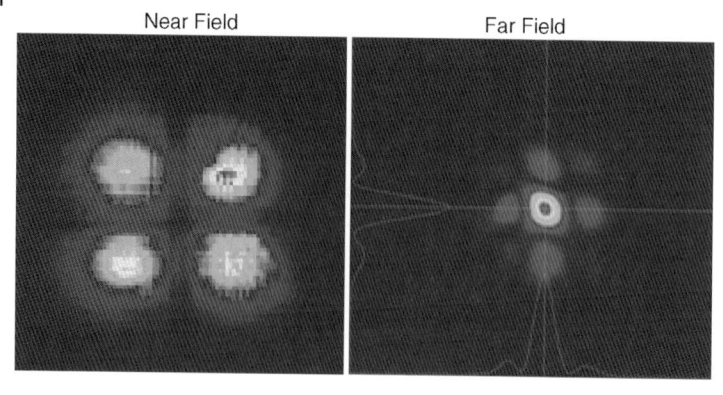

Figure 1.8 Near-field and far-field images of a tiled 2 × 2 CBC array of close-packed single-mode fiber lasers.

1.4.2
Filled Aperture Combiners Using Diffractive Optical Elements

Filled aperture beam combiners avoid this fill factor limitation by superimposing beams in the near field, using one or more beam splitter elements used in reverse as beam combiners. Examples of such elements are shown in Figure 1.9, and can include cascaded arrays of Fresnel beam splitters [33] or wave plate–polarizer pairs [34,35], tapered fiber couplers (either multichannel [36] or cascaded dual-channel [37]), Talbot-imaged waveguides [38,39], diffractive optical elements [9,13,40], and volume Bragg gratings (VBGs) [41]. As depicted in Figure 1.9, filled aperture combiners can be broadly classified as either free space or guided wave and as either dual-beam ports or multiple beam ports. While guided wave combiners fused to input delivery fibers are interesting for moderate power applications owing to their alignment insensitivity, it appears challenging to scale these components to extremely high ($\gg 10\,\text{kW}$) powers owing to the high intensities resulting from waveguide confinement. Dual-port combining elements require cascading to

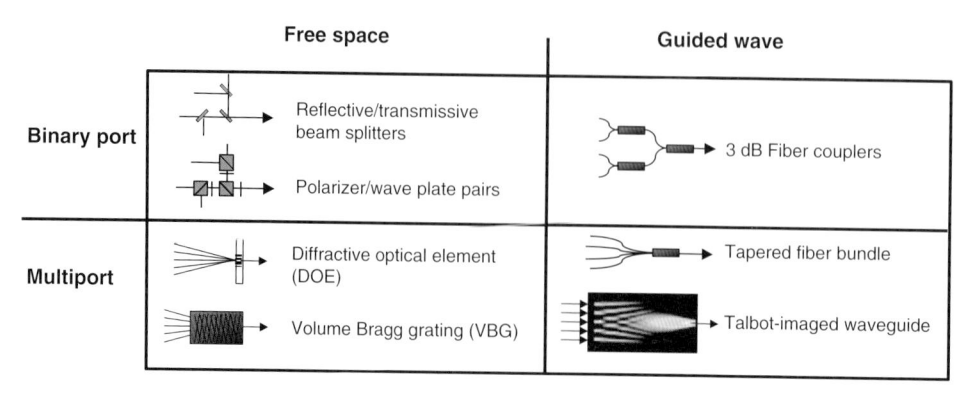

Figure 1.9 Optical configurations for filled aperture beam combining.

combine more than two beams; hence, these approaches tend to be lossy due to both direct transmission losses through each stage of combiner and the CBC losses from accumulated wavefront errors imposed by the upstream combiner elements.

From Figure 1.9, it is apparent that diffractive elements are uniquely well suited toward the demands of high-power, high channel count CBC. By virtue of their multiport and free space nature, both combining efficiency and power handling favorably compare to other filled aperture approaches, all of which involve multiple optical components and/or guided wave interactions. Of the two diffractive, free space–multiport approaches shown in Figure 1.9, power scaling of VBGs appears limited by thermal effects arising from trace absorption on the order of ∼100 ppm/cm at 1 μm laser wavelengths in the thick photothermorefractive glass medium [42]. In contrast, DOEs rely on a single low-loss optical surface interaction with an order of magnitude less absorption and the ability to be face-cooled from the rear for large-area heat sinking.

1.4.2.1 Overview of DOE Combiners

Used as a beam splitter, a DOE splits an input laser beam into multiple output beams at angles that represent the diffractive orders m of the structure shown in Figure 1.10. Proper design of the phase substructure within the primary DOE period Λ enables fine control over the amplitudes of the m orders, thereby allowing the power to be distributed nearly equally and with high efficiency among a desired number of orders [43]. Since light propagation is reciprocal, the same DOE splitter can also serve as a beam combiner. If mutually coherent beams are incident on the DOE at angles corresponding to the diffractive orders, the beams can constructively interfere to produce a single output beam, provided the phases of the input beams are locked to the correct modulo 2π values determined by the DOE design [9,44]. The combined beam exhibits substantially the same intensity profile as the individual beams, thus eliminating the far-field side lobe structure that is typically observed in tiled composite beam arrangements such as in Figure 1.8. An example of the output order from the CBC of a low-power, five-fiber array using a DOE is shown in Figure 1.11 [13]. The central DOE output order contained 91% of the input power, with the remaining power scattered into higher output orders.

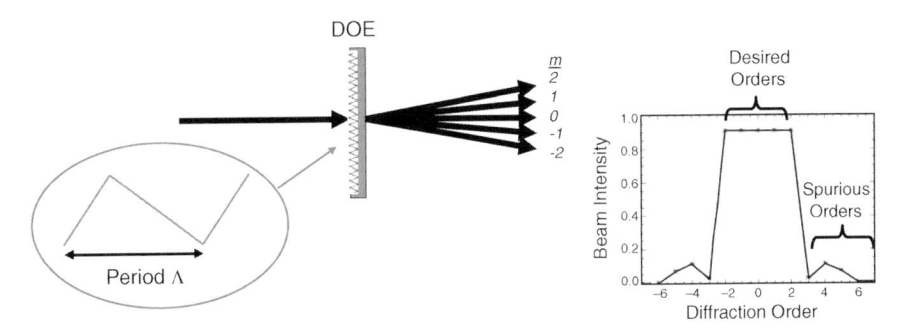

Figure 1.10 DOE periodic surface profiles create multiple grating orders as a splitter. The shape within each period defines the power distribution between orders.

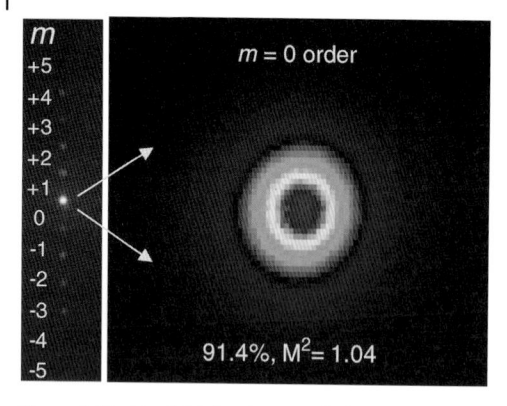

Figure 1.11 Far-field distribution of a five-channel phase-locked fiber array combined using a DOE.

1.4.2.2 DOE Design and Fabrication

Similar to traditional holographic diffraction gratings, DOEs used for CBC are fabricated by etching a continuous surface relief profile into a substrate, typically silicon or fused silica. A variety of DOE designs are possible to accommodate a wide range of channel counts and angular ranges of interest. For a surface relief structure with periodicity Λ and a near-normal incident beam, the diffraction angle θ_m of the mth order is given by the grating equation:

$$\sin(\theta_m) = m\lambda/\Lambda. \tag{1.8}$$

Hence, for a typical $\Lambda \approx 1$ mm DOE period and $\lambda = 1$ μm wavelength lasers, the angular separation between orders is $\lambda/\Lambda = 1$ mrad.

As illustrated in Figure 1.12, Λ of the DOE surface structures is typically about two orders of magnitude larger than the periods of high-dispersion diffraction gratings used for spectral beam combining (SBC) [45,46]. Consequently, the etched DOE surface is smooth with low aspect ratio. This yields a low-angle

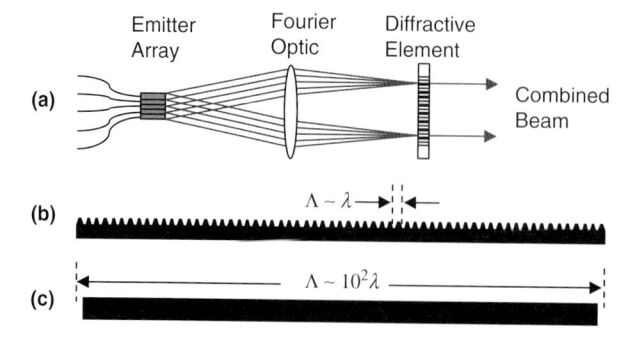

Figure 1.12 (a) Fourier optical configuration for SBC or CBC. (b) Typical surface profile for SBC diffraction gratings, with near-unity aspect ratio. (c) Typical surface profile for CBC DOE (25-beam DOE design shown to scale), with ~3° maximum surface slope.

surface profile that is easily coated with a high-quality optical coating, is cleaned with standard methods, and is resistant to thermal effects. Both antireflection (AR) and high-reflection (HR), ultralow absorption, multilayer dielectric coatings with reflectivities of 0.1% (AR) and 99.99% (HR) have been applied to fabricated DOEs without measurably affecting their diffraction properties, indicating that the surface profile is slightly changed by adding the thick dielectric coating stack. The photolithographic processes utilized in fabrication of the surface profile can be straightforwardly applied to substrates 10–15 cm in diameter to accommodate large beam footprints for power scaling with low damage risk.

The intrinsic beam splitter efficiency η_{split} is determined by the DOE design shape and manufacturing tolerances. The beam count N can be defined arbitrarily within the limits of surface shape manufacturing capability. To establish the capability to scale to large beam counts, numerous 1D and 2D DOEs have been designed and fabricated for a variety of channel counts, ranging from $N=3$ to $N=81$ beams with theoretical design values up to $\eta_{split}=99.4\%$ for $N=9$ (Table 1.3). Manufacturing tolerances are expected to slightly degrade the actual splitter efficiency by single-percent values as illustrated in Figure 1.13. By testing the fabricated DOEs' performance as beam splitters, we verified the capability of the as-fabricated DOEs to match the expected combination efficiency.

It is possible to optimize a DOE combiner design by relaxing the requirement that it split powers equally between the N channels of interest. With unequal power split fractions $|D_n|^2$, the RMS variation σ_D in amplitude transmission coefficients is nonzero. From Eq. (1.4), one can see that the efficiency of the DOE when used as a beam combiner, rather than as a splitter, will depend on the power balance among channels. The optimum case is when the input field amplitudes are perfectly correlated channel by channel with the DOE transmission coefficients ($A_n \propto D_n$). In this case, the loss term in Eq. (1.4) that is proportional to the covariance $\sigma_{A(x),D}$ exactly cancels the loss terms proportional to input power and splitter nonuniformities, $\sigma^2_{A(x)}$ and σ^2_D. In the absence of any beam

Table 1.3 Examples of DOEs designed for different numbers of beams, showing theoretical splitter efficiencies, along with the expected and as-fabricated efficiency as a combiner with perfectly aligned, equal power input beams.

DOE channel count N	Diffracted beam pattern $N_x \times N_y = N$	Designed splitter efficiency η_{split} (%)	Designed combiner efficiency $\eta_{split} - N\sigma_D^2$ (%)	As-fabricated combiner efficiency (%)
3	1 × 3	94.9	93.8	93
5	1 × 5	98.0	96.3	95.8
9	1 × 9	99.4	99.3	99.0
15	3 × 5	93.0	90.3	87.8
25	1 × 25	99.4	99.2	98.0
81	1 × 81	99.3	99.2	97
81	9 × 9	98.7	98.6	96

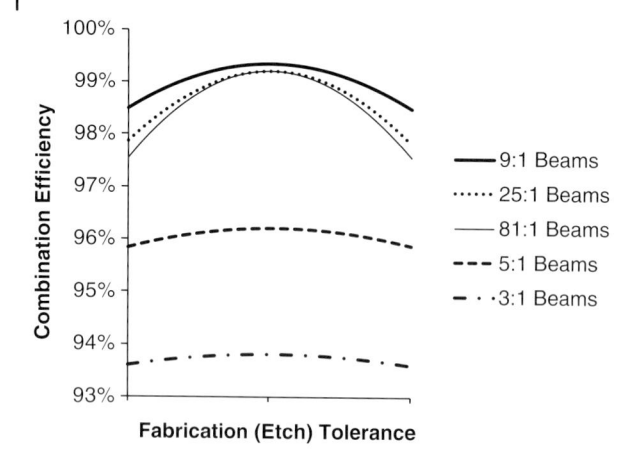

Figure 1.13 Dependence of DOE splitter efficiency on fabrication errors in the surface relief etch depth. The full range of the scale corresponds to typical manufacturing tolerances.

misalignments, the DOE combiner efficiency is then equal to its efficiency as a splitter η_{split}.

A more typical case is when the DOE is used to combine nominally equal power beams, such as generated from an array of identical lasers. In this case, $\sigma_{A(x)} = \sigma_{A(x),D} = 0$, and the BC efficiency is reduced from its limiting value as a splitter to

$$\eta = \eta_{\text{split}} - N\sigma_D^2 = \frac{1}{N}\left(\sum_n |D_n|\right)^2. \tag{1.9}$$

DOE design efficiencies as splitters and as combiners with equal power input beams for a variety of channel counts are shown in Table 1.3. It is notable that even with 20% RMS power split imbalances between channels, Eq. (1.9) predicts that combining efficiency is reduced by only 1% (cf., Figure 1.4). This is a testament to the well-known insensitivity of CBC to power imbalance among channels [8,11].

1.4.2.3 DOE Thermal and Spectral Sensitivity

Since angular dispersion is related to the groove density, the large DOE period results in greatly reduced sensitivity of DOE combiners to thermal distortions or to the input laser linewidths. For the typical case of small diffraction angles, the variation of diffraction angle θ_m with temperature of the DOE is approximately given by

$$\frac{d\theta_m}{dT} = -\alpha \tan(\theta_m) \approx -m\alpha\frac{\lambda}{\Lambda}, \tag{1.10}$$

where α is the thermal expansion coefficient of the DOE substrate. For a typical maximum diffraction angle of $\theta_{N/2} = \pm 25$ mrad (e.g., $N = 50$ beams separated by $\lambda/\Lambda = 1$ mrad in 1D, or $50^2 = 2500$ beams in 2D), one finds the maximum thermally induced angular shift is 1 μrad for an 80 °C temperature change of an SiO_2 substrate with $\alpha = 0.5$ ppm/°C. This shift is <10% of the natural divergence angle Θ of a diffraction-limited 10 cm diameter beam. As shown in Table 1.1, the resulting drop in combining efficiency would be substantially less than 1%.

To assess laser power handling, the surface temperature of an HR-coated, 81-beam DOE combiner has been measured under illumination with a 3.6 kW 1064 nm laser. The illuminated area was varied, and intensity of >20 kW/cm^2 was tolerated without damage. This suggests the DOE beam combination approach is scalable to megawatt-level with a combined beam diameter of <10 cm. A surface absorption coefficient of 17 ± 5 ppm was derived from the observed 3 °C steady-state temperature increase of the uncooled, illuminated DOE. A simple 1D thermal analysis of a back-cooled, 5 mm thick, silica DOE shows that with 20 ppm absorption and 10 kW/cm^2 irradiance, the DOE surface temperature rise is only 7 °C. This temperature change is more than an order of magnitude below the level at which thermal aberrations are expected to noticeably impact the combination efficiency of 10 cm diameter beams, indicating thermal issues should not limit scaling of this method to megawatt-level powers.

The spectral dispersion of a DOE is also very small and is approximately given by

$$\frac{d\theta_m}{df} = -\frac{\lambda}{c}\tan(\theta_m) \approx -m\frac{\lambda^2}{c\Lambda}, \tag{1.11}$$

which for a maximum diffraction angle of $\theta_{N/2} = \pm 25$ mrad yields $d\theta_{N/2}/df = 0.09$ μrad/GHz. Calculations based on Eqs. (1.4) and (1.11) show that for 10 cm beams, dispersion of laser bandwidths of 20 GHz will induce less than 1% impact on combination efficiency. This enables ready use of DOE combiners with >1 kW fiber lasers, which require multigigahertz frequency broadening to mitigate power limits from stimulated Brillouin scattering (SBS) [3,47,48].

1.5
High-Power Coherent Beam Combining Demonstrations

This section reviews the design and performance features of coherence-preserving high-power laser amplifiers suitable for integration with the phase control and geometric beam combining technologies described in the previous sections. General features required for the amplifiers are as follows:

- *High power.* For CW laser systems, the cost and complexity involved in CBC phase control and geometric beam combining typically make engineering sense only for scaling well beyond the power limits of the underlying amplifier technologies. As we will discuss subsequently, the present-day combinable CW fiber and free space laser amplifier limits are on the order of ~1 and ~10 kW, respectively.
- *High gain.* Low-noise master oscillators are typically limited to subwatt output powers. The available seed power for each amplifier channel is even lower, typically ~1–10 mW, owing to the attenuation by modulators and distribution splitting to each channel. Hence, amplifier gains often in excess of 50 dB are required to reach kilowatt-level output powers. Since such high levels of

amplification are impractical in a single amplifier stage, this leads to multistage master oscillator power amplifier (MOPA) architectures with gain stages separated by Faraday isolators.

- *High spatial coherence.* As can be seen in Eq. (1.4), CBC efficiency is degraded at >1% level when intensity-weighted, residual wavefront aberrations exceed 0.1 rad ($\lambda/60$ waves). Hence, each amplifier's output must possess near-perfect beam quality and very low pointing jitter to yield a low-aberration wavefront.
- *High temporal coherence.* Time-dependent phase changes can be imposed by thermal, acoustic, gain, or nonlinear dynamics in each amplifier. Low-frequency phase changes that fall within the servo bandwidth of the phase control system (typically <1–10 kHz) can be corrected with high fidelity and do not impact CBC. High-frequency phase changes must be constrained to low levels (again, about <0.1 rad RMS) to avoid impacting CBC efficiency.
- *Polarization.* The amplifier outputs must be copolarized to interfere constructively. Birefringence imposed by the amplifiers must either be kept to small levels or actively compensated.

In Sections 1.5.1 and 1.5.2, we review two laser amplifier technologies – zigzag slabs and fibers – that have been demonstrated to meet these requirements for brightness scaling via CBC. We describe integration of these amplifiers in high-power CBC demonstrations.

1.5.1
Coherent Beam Combining of Zigzag Slab Lasers

In 2009, Northrop Grumman Aerospace Systems demonstrated 105 kW continuous power output with record brightness from a solid-state laser system by coherently combining an array of seven Nd:YAG zigzag slab amplifier chains under the Joint High Power Solid State Laser (JHPSSL) program [5]. The high-stimulated emission cross section of Nd:YAG allowed efficient laser extraction and high amplifier gain with relatively low optical intensities. The consequent low nonlinearity led to outstanding temporal coherence of the amplified beams. The key challenge in this demonstration was maintaining high spatial coherence between amplifiers owing to thermally driven optical wavefront distortions. Low wavefront distortion was achieved through both the slab amplifier design and the use of adaptive optics.

The zigzag slab amplifier geometry shown in Figure 1.14 is, in principle, immune to wavefront distortions from thermal optic path differences (OPDs) [49]. In the zigzag slab concept, a tall, thin, uniformly pumped slab is cooled from both large-area surfaces [50]. Multiple total internal reflections (TIRs) from the cooled surfaces confine to and guide the extracting laser beam down the slab length. Since the extracting beam propagates at an angle relative to the primary thermal gradient, thermal lensing is eliminated as each part of the beam samples the entire slab thickness. The zigzag slab is also far less susceptible to thermal depolarization than other bulk gain architectures [51].

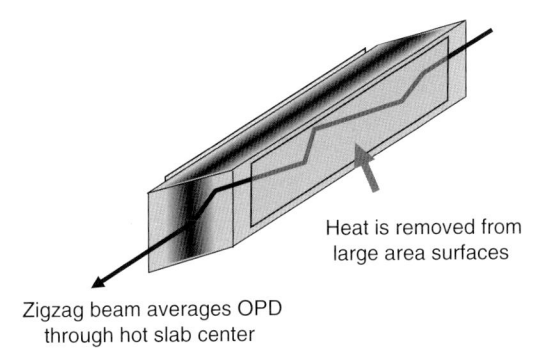

Figure 1.14 Zigzag slab amplifier concept.

In practice, nonuniformities in the slab pumping distribution and surface heat transfer coefficients, as well as edge and end effects, can impose substantial thermal OPD. Figure 1.15 shows a multikilowatt conduction-cooled, end-pumped zigzag slab (CCEPS) gain module that minimizes these effects [52,53]. By injecting diode pump light through the slab tips, the slab itself acts as a homogenizing waveguide as the pump light propagates via TIR down the length of the slab. This provides uniform pump excitation and minimizes any nonuniformities in volumetric heat generation. Careful engineering of a low thermal impedance conductive interface to copper microchannel coolers allows efficient and uniform heat removal. A 2–3 μm thick SiO_2 coating on the cooled faces of the slab contains evanescent fields to ensure near-lossless zigzag propagation of both the pump light and the high-power beam down the slab. The slab has undoped, diffusion-bonded, 45° cut end caps that protrude beyond the coolers to receive focused diode pump light. The high-power

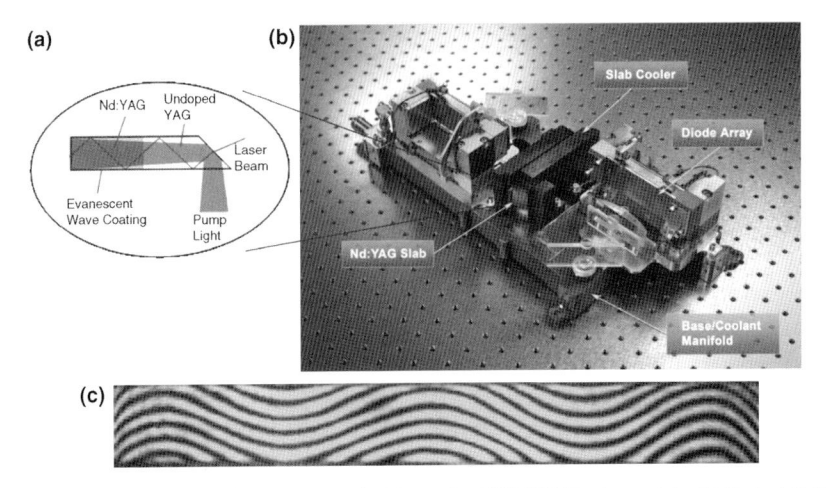

Figure 1.15 (a) CCEPS laser concept. (b) Photo of a 4 kW CCEPS gain module. (c) Typical 4 kW slab OPD measured using a Mach–Zehnder interferometer operating at 658 nm. The zigzag axis is vertical and the nonzigzag axis is horizontal.

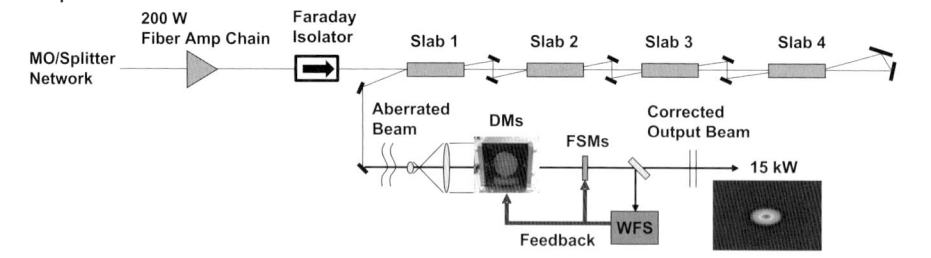

Figure 1.16 Schematic of 15 kW zigzag slab amplifier chain with adaptive optic wavefront correction. The far-field intensity distribution is shown at full power.

beam is injected into the slab at angles that are 20–30° from the normal to the input face to ensure that TIR occurs at the YAG–SiO$_2$ interface.

To achieve ∼15 kW power levels, four CCEPS gain modules seeded with an ∼200 W beam from a fiber amplifier chain were arranged in a double-pass serial configuration, as shown in Figure 1.16 [54,55]. The beam was image relayed from slab to slab to minimize geometric coupling losses. Double-passing each slab via angular multiplexing enabled good saturation and 30% optical extraction efficiency. Angular multiplexing of the slabs was straightforward by choosing different integral numbers of zigzag reflections on each pass [56]. After all eight amplification passes, the beam was amplified to 15 kW.

Despite the advantages of the CCEPS architecture in minimizing thermal aberrations, two or more waves of OPD were imposed on the extracting laser beam for each amplification pass (Figure 1.15c). This OPD arose from small residual pumping/cooling nonuniformities as well as from bulging of the TIR surfaces. Owing to these wavefront aberrations, the 15 kW output beam was unsuitable for coherent combining, necessitating the use of adaptive optics to recover a nearly flat wavefront. As shown in Figure 1.16, the aberrated, high-power beam was expanded to fill the active area of a continuous-facesheet deformable mirror (DM). The tilt was off-loaded to fast steering mirrors (FSMs) to conserve DM stroke and provide high-speed jitter stabilization. High-reflectivity dielectric coatings on the DM and SM enabled use of these elements in the 15 kW beam path. A sample of the output beam was directed to a Shack–Hartmann wavefront sensor (WFS), which generated error signals to drive both adaptive elements in a closed-loop configuration. The corrected output beam had a nearly flat wavefront, with beam quality <1.3 times the diffraction limit (×DL) [55].

Since the slab amplifiers naturally produced rectangular, near-flattop beams (Figure 1.17a), they were ideally suited for tiled aperture coherent combining with a high fill factor and relatively low far-field side lobes. The beams from $N = 7$ wavefront-corrected MOPA chains were tiled in a close-packed array configuration using scraper mirrors and were phase locked using the OHD phasing technique [20], as shown in Figure 1.5, to form a <3 times DL, 105 kW composite output beam. The far-field beam profiles displayed in Figure 1.17b and c illustrate the features of coherent beam combination [5,54]. Disabling the phase controller

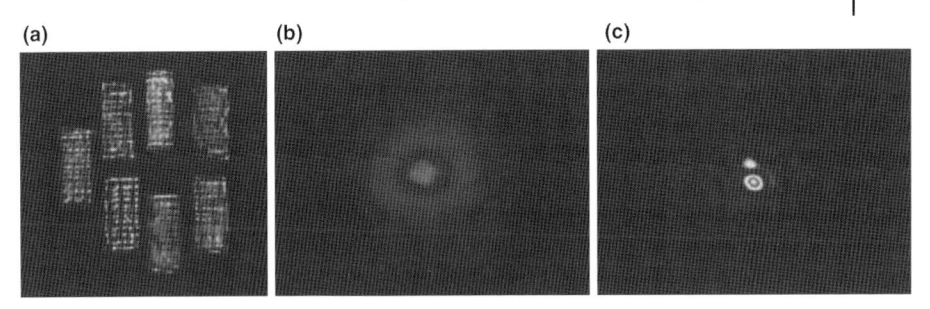

Figure 1.17 100 kW coherently combined beam from seven zigzag slab MOPA chains. (a) Tiled near field. (b) Far field with phasing control disabled. (c) Far field with phasing control enabled.

resulted in only a linear increase of the far-field peak intensity, as N was increased by turning on laser chains. Enabling the phase controller would theoretically increase the far-field intensity by another factor of N. Instead, the observed far-field brightness increased by a factor of 3.8, less than the ideal, owing to the imperfect spatial coherence from residual wavefront aberrations and jitter between amplifier chains.

The entire JHPSSL laser head, comprising the seven-slab MOPA chains and the beam combiner, was packaged in a single water-cooled enclosure with volume $\sim 10\,\mathrm{m}^3$ (Figure 1.18 a). The device was continuously operable for periods in excess of 300 s with no thermal degradation (Figure 1.18b). The parallel CBC architecture was essential toward achieving continuous run-times by virtue of distributing waste heat loads among multiple, spatially separated amplifiers. Table 1.4 summarizes the measured performance of the combined system [5]. In principle, brightness can be scaled indefinitely by adding more chains owing to the fully parallel beam tiling and OHD phase control architectures. To our knowledge, this remains the brightest, continuously operable laser system demonstrated to date.

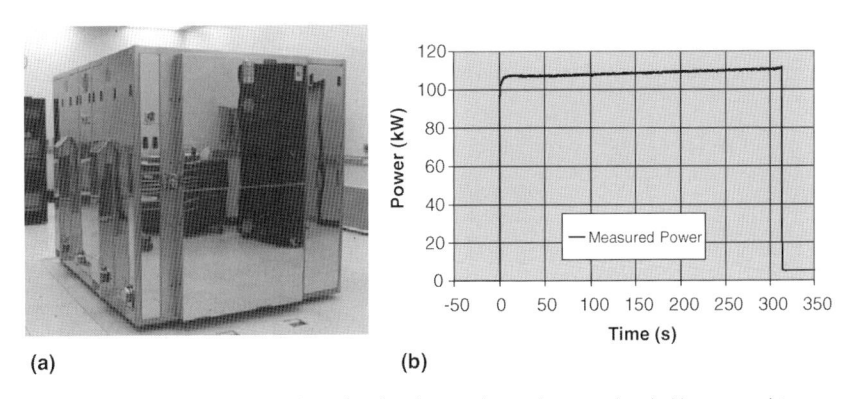

Figure 1.18 (a) 100 kW JHPSSL laser head with seven laser chains and a tiled beam combiner. (b) Output power over the course of a continuous 5 min shot.

Table 1.4 Measured performance of the 100 kW CBC JHPSSL slab laser system.

Parameter	Measured value
Power	105.5 kW
Beam quality	$2.9 \times$ DL
Run-time	313 s
Turn-on time	0.6 s
Electrical efficiency	19.3%

Electrical efficiency is defined as the ratio of CBC output power to diode electrical pump power.

1.5.2
Coherent Beam Combining of Fiber Lasers

While the slab-based JHPSSL system represents a significant advance in solid-state laser power scaling, a CBC array of Yb-doped fiber amplifiers (YDFAs) offers potential for improved efficiency, combined beam quality, and ease of packaging. YDFAs can exhibit optical conversion efficiencies near the quantum limit of $\sim$90% [47], compared to <50% typically achieved with bulk gain media. The kilowatt-level multistage fiber amplifiers can generate near-single-mode Gaussian-shaped beams with M^2 beam propagation parameters close to unity, greatly reducing CBC losses due to spatial wavefront aberrations and beam jitter. The mechanical flexibility of fiber gain media allows a high level of packaging and compaction. Ideally, only the free space optics in a CBC fiber system would reside in the beam combiner, where power levels preclude guided wave propagation.

CBC of fibers leads to distinct engineering challenges compared to CBC of slabs. Power levels attainable from modern coherence-preserving fiber amplifiers are an order of magnitude smaller than those from slab lasers, leading to $\sim$10$\times$ higher channel counts for comparable power output. Active polarization control may be required, since for typical fiber mode field diameters (MFDs) of tens of micrometers, the waveguide asymmetry of polarization-maintaining (PM) fibers makes single-mode output more difficult to obtain than from non-PM fibers. Finally, the combination of small core sizes, long amplifier lengths, and kilowatt-level power in the fiber can lead to significant nonlinear optical distortion of the seed and consequent loss of temporal coherence. In the remainder of this section, we describe recent experimental and analytic results probing the limits of fiber CBC and review recent demonstrations of high-power fiber CBC.

1.5.2.1 Phase Locking of Nonlinear Fiber Amplifiers

The primary concern for CBC with high-power YDFAs is preserving the temporal, rather than spatial, coherence properties of the MO to allow fully constructive interference of the amplified outputs. The high fiber nonlinearity at kilowatt-level powers makes it critical to avoid fast RIN (i.e., amplitude modulation) on the seed laser. In the presence of RIN, the Kerr nonlinearity (parameterized by the

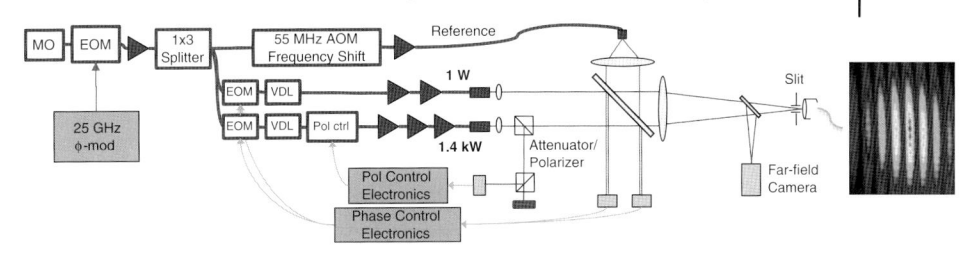

Figure 1.19 Schematic of high-nonlinearity fiber phase locking experiment.

nonlinear refractive index for silica fiber, $n_2 \approx 3 \times 10^{-20}\,\text{m}^2/\text{W}$ [57] will induce self-phase modulation (SPM) that may not be correctable with servo loops, degrading the temporal coherence and CBC efficiency [58]. For this reason, actively phase-locked CBC of fiber lasers is often implemented using a single-frequency (SF) MO with low RIN. This avoids the Kerr nonlinearity, but has been limited to ~150 W per fiber due to SBS [20]. An alternative method that avoids both Kerr nonlinearities and SBS is to broaden the MO linewidth using phase modulation. This has become the standard approach used for high-power CBC fiber demonstrations [21,59,60]. Typical modulation bandwidths range from one to several tens of gigahertz.

The effects of nonlinear temporal decoherence on CBC have been probed by integrating active phase control with a commercial high-power YDFA chain (Figure 1.19) [59]. A SF fiber MO (NP Photonics) operating at a wavelength $\lambda = 1064$ nm was phase modulated using an EOM to broaden the linewidth to 25 GHz FWHM for SBS suppression. Following the EOM, the output was amplified to 100 mW and split into three channels, one of which was frequency shifted by a 55 MHz AOM to serve as an OHD reference for phase locking. Each of the other two channels contained an EOM for piston phase actuation, a manually adjusted variable delay line (VDL) for path matching, and gain-staged YDFAs. The low-power channel contained two PM amplifiers to provide 1 W output power. The high-power channel contained a fiber polarization controller (General Photonics, POS-104) followed by a three-stage, non-PM YDFA chain (IPG Photonics) to boost power to 1.43 kW [48]. The final power amplifier stage was tandem-pumped by high-brightness 1018 nm fiber lasers [3,48]. The outputs from both the high- and low-power fiber amplifier channels were collimated and tiled side by side. The high-power beam was attenuated for amplitude equalization with low-power beam and polarization filtered to provide a feedback signal for the polarization controller. The frequency-shifted reference was combined interferometrically with the 2×1 tiled beam. Separate photodetectors in each channel sensed the phase of the 55 MHz OHD beat notes to provide error signals for phase locking of each beam to the reference with RMS phase fidelity of $\lambda/80$, with resulting beam-to-beam phasing errors of $\Delta\phi_0 = 2^{1/2}(\lambda/80) = 0.11$ rad.

The fiber nonlinear phase shift, or B-integral, was measured directly by partially amplitude-modulating the MO seed to generate a 100 ns "dark pulse" with 50%

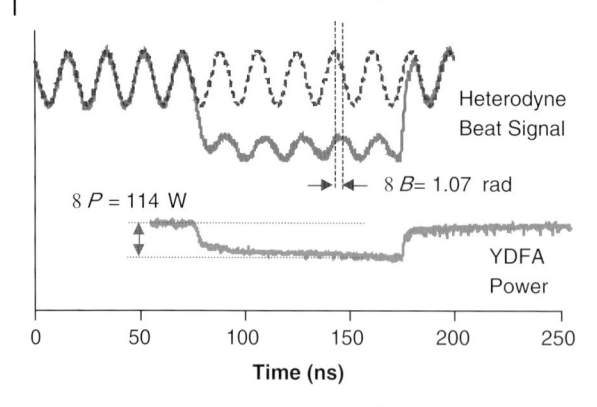

Figure 1.20 Heterodyne measurement of SPM in the high-power fiber amplifier.

lower power (Figure 1.20). This pulse was so fast that it transmitted through the entire amplifier chain without being affected by laser dynamics, and hence there was a substantial, transient drop in the output power ΔP that induced a nonlinear phase shift ΔB due to SPM:

$$\Delta B = \left(\frac{2\pi n_2}{\lambda}\frac{L_{\text{eff}}}{A_{\text{eff}}}\right)\Delta P. \tag{1.12}$$

Here, L_{eff} and A_{eff} are the effective power-weighted fiber length and mode field area, respectively. The term in parentheses in Eq. (1.12) is equivalent to $\Delta B/\Delta P$ and can be determined from the data in Figure 1.20 to be 9.4 ± 1.7 rad/kW from the 1.07 rad phase shift of the 55 MHz heterodyne beat note during the 114 W dark pulse. Hence, the B-integral at full power was $B = (\Delta B/\Delta P) \times 1.43$ kW $= 13.4 \pm 2.4$ rad.

Phase-locking effectiveness was quantified by focusing a low-power sample of the tiled beams onto a far-field camera to generate a stationary fringe pattern (Figure 1.19). A narrow slit whose width is ~5% of a fringe period provided a metric for mutual coherence between the two beams through the visibility [61]:

$$V = (I_{\max} - I_{\min})/(I_{\max} + I_{\min}). \tag{1.13}$$

Here, $I_{\max}$ and $I_{\min}$, respectively, are the intensities transmitted through the slit at a peak and a null of the far-field interference pattern, measured sequentially by applying a π-phase shift to the phase controller for one channel. With proper amplitude equalization between the two phase-locked channels, V is equivalent to the mutual coherence between the two beams and is representative of the coherent combining efficiency for coaligned beams.

Figure 1.21a shows V of the phase-locked beam as a function of fiber output power. Tiled aperture and filled aperture measurements yielded similar results within 1%. The low-power data agree with the expected limit based on the accuracy of active phase control [8], $1 - \Delta\phi_0^2 = 1 - (0.11 \text{ rad})^2 = 0.988$. As power was increased to 1.43 kW, V dropped to 0.90. This drop can be attributed to decoherence from SPM. Any RIN faster than the ~10 kHz closed-loop OHD phase control

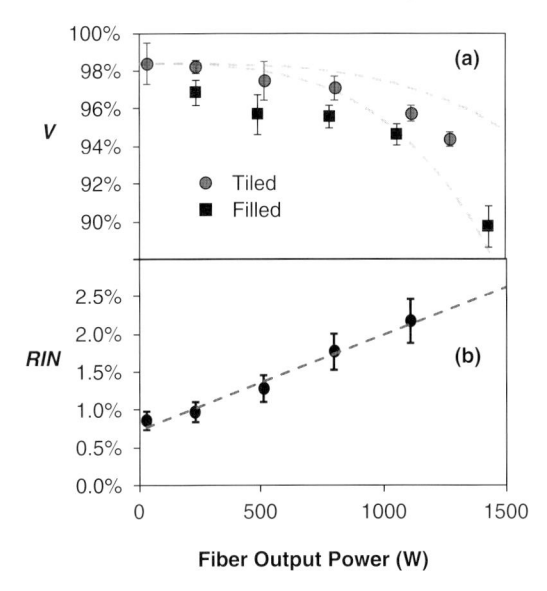

Figure 1.21 (a) Visibility measurements (symbols) and calculated impact of SPM from Eq. (1.14) with propagated uncertainty limits from B and RIN measurements represented by the dashed curves. (b) Measured RIN (symbols) with a linear fit (dashed curve) to provide values for the SPM calculation in (a).

bandwidth (Figure 1.7a) will result in RMS phase noise $B \cdot \text{RIN}$ that will be uncorrected and will contribute to decoherence and a drop in V:

$$V = \left(1 - \Delta\varphi_0^2\right)[1 - (B \cdot \text{RIN})^2/2]. \tag{1.14}$$

The factor of 2 in Eq. (1.14) arises because only one of the two channels suffers from SPM, so the phase noise variance $(B \cdot \text{RIN})^2$ of the single high-power channel is twice the ensemble phase noise variance of the two-beam array. RIN was measured to be a few percent with 10 kHz–2 GHz detection bandwidth (Figure 1.21b). Based on the measured RIN, the predicted values from Eq. (1.14) agree with the observed decoherence within the propagated uncertainty (Figure 1.21a, dashed curves).

To improve CBC efficiency at these power levels, RIN and/or B must be reduced. Standard noise reduction methods applied to the MO and pump components should enable reduction of RIN to the single percent level. The fiber nonlinearity can likely be reduced substantially by changing the amplifier fiber from tandem pumped to direct diode pumped to increase A_{eff} and decrease L_{eff}.

It is notable that high-efficiency CBC has also been demonstrated recently in pulsed fiber amplifier chains with $B = 38$ rad due to the high peak intensity [62]. This nonlinearity is approximately 3× greater than that of the 1.4 kW CW amplifier and approaches the threshold for stimulated Raman scattering [57]. Two amplified trains of 1 ns pulses at 25 kHz pulse repetition frequency were combined with 79% efficiency into a single beam with 0.42 mJ pulse energy. Precision matching of the pulse temporal intensity profiles and amplifier B-integrals was critical in achieving this result. It demonstrates that decoherence from active phase control and fiber nonlinearities are manageable in a CBC system despite the operation deep within the nonlinear regime.

1.5.2.2 Path Length Matching with Broad Linewidths

Owing to the relatively broad linewidths, $\Delta f = 25$ GHz, required for SBS suppression of the 1.4 kW fiber amplifier, optical paths in each amplifier channel must be equalized to a small fraction of the coherence length $L_{coh} = c/\Delta f$ to prevent significant combining loss due to dephasing [8]. A key question for practical operation of a large array of kilowatt-level fibers is whether the change in fiber path due to thermal expansion and index changes upon turn-on will result in significant dephasing loss. For the 1.4 kW amplifier, this path length change was determined to be approximately 1.5 mm [59]. From Table 1.1, one can calculate that for 25 GHz linewidths, paths must be matched to ±0.5 mm in order to keep CBC losses below 1%. This is approximately one-third of the turn-on transient, suggesting high-efficiency CBC of an array of such amplifiers would be feasible with modest attention to thermal uniformity between amplifiers.

For even broader linewidths, as may be required for SBS suppression at even higher powers or for CBC of ultrashort pulses [25,63], active controls may be warranted to automate path matching in a servo configuration. Figure 1.22 shows a particularly simple concept for implementing active path matching in a coherent array [64]. Since the accumulated phase errors between the path-mismatched beams are frequency dependent, spectrally filtering the combined output beam serves to optically transduce a group delay error into a frequency-dependent piston phase error. The transduced piston errors can then be nulled by duplicating the closed-loop phase detection electronics, thus bringing the beams into coalignment. It is worth noting that this Fourier domain filtering concept applies to the spatial domain as well, where it can be used for active coherent beam alignments [65].

By choosing the width and separation of the spectral filters in front of the phase and delay sensors, the locking range and precision can accommodate any laser linewidth, even over multiple coherence lengths. Figure 1.23 summarizes test results of simultaneous phase locking and group delay locking on a three-fiber CBC array with a 10.5 nm (2.8 THz) linewidth MO. Independently tunable ~1 nm FWHM spectral filters in front of both phase and delay sensors (Figure 1.23a)

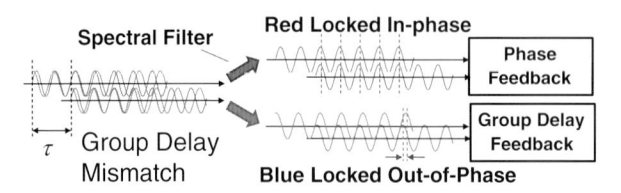

Figure 1.22 Fourier domain filtering concept for coherent detection and control of group delay errors between CBC beams.

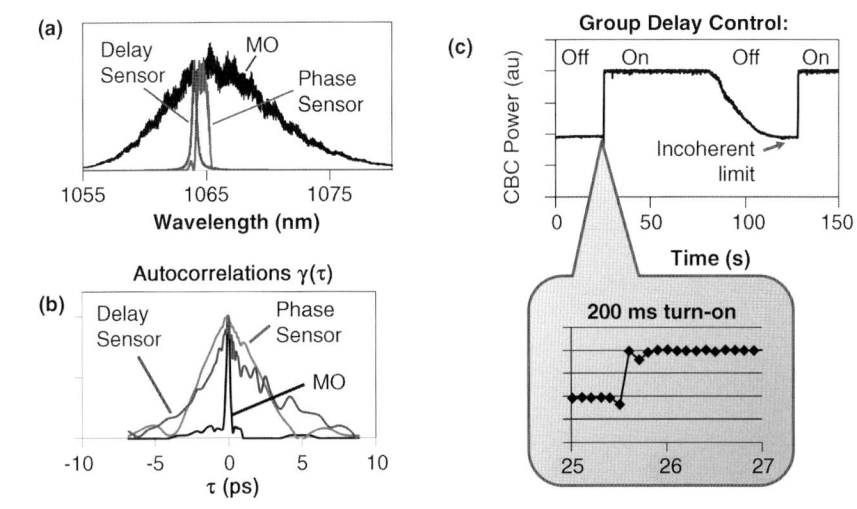

Figure 1.23 Demonstration of simultaneous active phase locking and group delay locking of a three-fiber coherent array. (a) Input and filtered spectra. (b) Linear field autocorrelation envelopes showing increased coherence time of the filtered light. (c) Closed-loop optimization of CBC efficiency when paths are actively controlled.

increased L_{coh} of the detected light from 100 μm to 1 mm (Figure 1.23b). Starting with fiber mismatches so large that the beams were completely dephased, the fibers were automatically coaligned to an absolute accuracy of ±6 μm, within 200 ms of engaging the closed loop (Figure 1.23c). Turning off the closed loop at 80 s in Figure 1.23c resulted in CBC power dropping to the incoherent limit due to fiber thermal drifts, which were corrected upon loop reengagement at 130 s. As long as the mismatches do not exceed the coherence length of the filtered light (which can be arbitrarily increased via narrower spectral filters), the closed loop will be within its control range even when the CBC output beam appears completely decoherent.

1.5.2.3 Diffractive CBC of High-Power Fibers

While the 1.4 kW phasing demonstration already described serves to clarify the nonlinear limitations on CBC of individual fiber amplifiers, a demonstration of actual beam combining requires multiple high-power fiber channels. In this section, we describe two recent demonstrations that highlight the scaling potential of fiber CBC using filled aperture DOE combiners at both high powers and high channel counts. These demonstrations also provide concrete examples of the utility of the perturbative model of Table 1.1 in identifying and quantifying sources of CBC loss.

1.5.2.3.1 CBC of a 1D Fiber Array into a 1.9 kW Beam

The highest power demonstration to date of filled aperture beam combining was performed in collaboration with MIT Lincoln Laboratory [66]. A coherent laser array comprised of five 500 W PM fiber amplifier chains was seeded by a common 10 GHz linewidth phase-modulated MO [21]. Each laser chain contained an EOM for piston phase control and a VDL for path

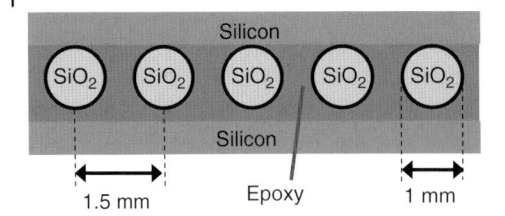

Figure 1.24 End-on face schematic of the Si V-groove exit array.

matching to <1 mm. Each chain's delivery fiber was spliced to a 1 mm diameter end cap and epoxied in a 1.5 mm pitch silicon V-groove array (Figure 1.24). Five adjacent beams from the array were directed onto a five-beam DOE using the Fourier optics geometry shown in Figure 1.12a, with the addition of a monolithic microlens array near the end cap facets that partially collimated each beam to allow an adjustable beam size on the DOE. The collimated Gaussian beam diameter on the DOE was approximately 3.3 mm. The DOE was fabricated on an SiO_2 substrate and was HR coated, with an angular spacing of 8.87 mrad between beams.

The DOE combined most of the incident power into the $m = 0$ diffractive order, with the uncombined power diffracted into higher $|m| > 0$ orders. The DOE was tilted slightly so that the combined beam was reflected outside the plane of diffraction for geometric separation from the input beams. Following the DOE, a loose aperture terminated any residual power left in the $|m| > 0$ orders. A portion of the combined beam was picked off by a beam sampler and sent to diagnostics and a detector to lock the beams in-phase using a hill climbing algorithm [21]. This automatically locked each beam at the DOE input to the conjugate of the phase imposed upon diffraction.

Figure 1.25a shows the combined power in the central $m = 0$ output order as the input power was increased. At low powers, the overall combining efficiency was 90%. This was less than the as-fabricated 96% DOE combining efficiency due to both transverse and rotational mounting errors of the fibers in the V-groove array, as well as a small fraction of output power contained in fiber of higher order modes [8,21]. The combining efficiency dropped near-quadratically with input power to 79% at 2.5 kW.

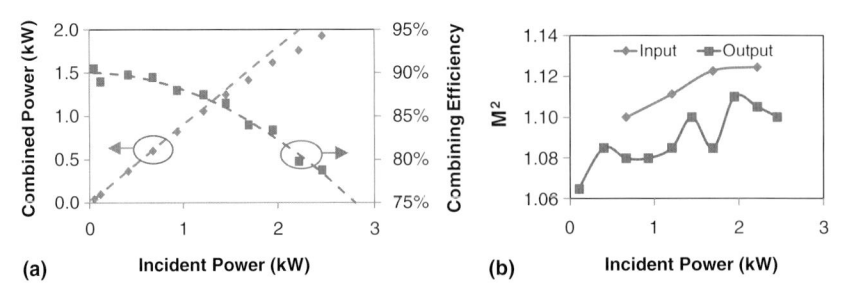

Figure 1.25 Power-dependence of 2.5 kW fiber DOE combining. (a) Combined power and efficiency. The dashed curves are fits to 90% slope efficiency and to Eq. (1.16). (b) Measured M^2 beam parameter.

The $m=0$ output beam was essentially diffraction limited, with $M^2 = 1.1$ (Figure 1.25b). A slight BQ improvement was seen in comparison to the input beams, which is a consequence of the coherent filtering effect discussed in Section 1.2. Any noncommon aberrations on the incident beams result in decreased combining efficiency as in Eq. (1.4), corresponding to an increase in power in the $|m| > 0$ output orders. This effectively filters out both amplitude and wavefront aberrations, $\delta A_n(x)$ and $\delta \phi_n(x)$, resulting in a near-diffraction-limited $m=0$ output beam. Owing to the space–time symmetry, the same filtering process also serves to clean up the temporal coherence (i.e., reduce time-dependent phase noise) of the output beam when coherently combining nonsingle-frequency lasers.

To diagnose the drop in combining efficiency at high powers, thermal images were recorded of the free space optical components under steady-state illumination at 2.5 kW. DOE heating was $\Delta T < 3\,°C$ with peak intensity $>30\,kW/cm^2$, consistent with the separately measured 17 ppm surface absorption described in Section 1.2. Using Eq. (1.10), the resulting thermal expansion of the DOE would lead to a maximum pointing shift of $(d\theta_m/dT)\Delta T = \pm 30$ nrad for the $m=\pm 2$ diffractive orders. This is ~ 4 orders of magnitude smaller than the diffraction-limited beam divergence of the 3.3 mm diameter beam. Hence, DOE power handling would not be expected to measurably impact CBC efficiency.

Thermal expansion of the fiber V-groove array would be expected to degrade the combining efficiency, since the fiber tip spacing (pitch, x) will no longer match the DOE angles of incidence after the Fourier optics. A maximum surface temperature rise of $\Delta T = 45\,°C$ was observed near the fiber tips with input power $P = 2.5$ kW. This suggests a heating coefficient $dT/dP \sim 0.018\,°C/W$ at the fiber–end cap interface. As can be seen from Figure 1.24, the undersized Si V-grooves only weakly constrain the fiber positions. Hence, power-dependent changes in pitch $dx/dP = \alpha t(dT/dP)$ are dominated by the $t = 500\,\mu m$ thick interstitial epoxy used to pot the end caps, which has a coefficient of thermal expansion $\alpha = 110\,ppm/°C$. For small pitch errors $\delta x = (dx/dP)P$, the resulting CBC loss from Table 1.1 is $L = (\sigma_x/w)^2$, where $w = 10$ μm is the mode field radius at the delivery fiber tip and σ_x^2 is the mean-square variance of the uniform distribution of tip position errors $\{0, 1, \ldots, N-1\}\cdot\delta x$ across the array [67]:

$$\sigma_x = \sqrt{\frac{N^2 - 1}{12}}\delta x. \quad (1.15)$$

This leads to an expected quadratic drop in CBC efficiency $\eta(P)$ with input power:

$$\eta(P) = \eta_0(1 - L) = \eta_0\left[1 - \frac{N^2 - 1}{12}\left(\frac{\alpha t\,dT}{w\,dP}\right)^2 P^2\right], \quad (1.16)$$

where η_0 is the combining efficiency at low power. Equation (1.16) is plotted numerically in Figure 1.25a and agrees with the observed quadratic power dependence and final $\sim 11\%$ drop in CBC efficiency at full power, corresponding to a pitch increase of $\delta x = 2.4\,\mu m$.

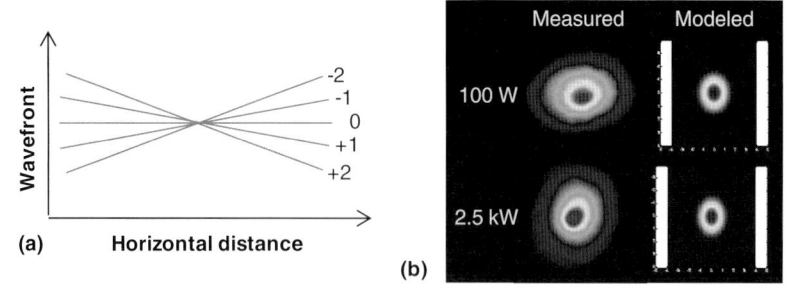

Figure 1.26 (a) Illustration of wavefront tilts across the DOE clear aperture due to thermal expansion of the exit array. (b) Measured and calculated intensity profiles of the CBC beam at the DOE.

Further evidence that thermal growth of the V-groove array is responsible for the drop in combining efficiency can be inferred from the near-field profiles of the combined beam at the DOE (Figure 1.26). As the exit array pitch grows with power, the Fourier lens converts fiber tip positioning errors into wavefront tilts on the DOE. As illustrated in Figure 1.26a, beams can fully interfere constructively only at the center of the DOE near field, since wavefront tilts impose phase errors that grow with distance from the center of the beam. Consequently, the near-field beam footprint of the combined output beam shrinks with power along the grating axis (horizontally in Figure 1.26b). Both measured and modeled beam profiles were consistent with the calculated growth of the exit array.

From these results, it is clear that DOE-based coherent combining is robust at the $\sim$2 kW level, with excellent combining efficiency, near-perfect output beam quality, and no indication of reaching DOE power handling limits. The observed drop in combining efficiency with power agreed well with calculations of the thermal expansion of the fiber exit array, which should be amenable to more robustly engineered designs.

1.5.2.3.2 CBC of a 2D Fiber Array into a 0.6 kW Beam

The use of multiple DOE orders opens the prospect of utilizing DOE structures that diffract light in two dimensions rather than in one dimension. This effectively squares the number of input orders and provides a means for scaling the number of input beams by one–two orders of magnitude over a linear array generator. Alternatively, for a given number of input beams, a 2D DOE can reduce the angular range of diffractive orders compared to a 1D element, providing added robustness against thermal distortions and enabling coherent combining of broad input linewidths by reducing distortions from angular dispersion. A further practical benefit of 2D DOEs is that more compact and lower aberration optical systems may be used to transform light from a large array of fiber tips onto the DOE.

By superimposing two orthogonal surface relief phase profiles for $1 \times M$ and $1 \times N$ 1D DOE beam splitters, linear patterns can be generated simultaneously in two axes to produce a 2D $M \times N$ grid pattern. For the resulting 2D DOE, the power

1.5 High-Power Coherent Beam Combining Demonstrations | 35

splitting efficiency of a single beam into the $(m \times n)$th order is $D_{m,n}^2 = D_m^2 D_n^2$, where $m = 1 \cdots M$, $n = 1 \cdots N$, and D_m^2 and D_n^2 are the power splitting efficiencies of the underlying 1D DOE designs. It is straightforward to show using Eq. (1.9) that the 2D DOE combining efficiency η_{MN} is simply the product of the underlying efficiencies along each axis:

$$\eta_{MN} = \eta_M \eta_N = \frac{1}{MN} \left(\sum_{m=1}^{M} D_m \right)^2 \left(\sum_{n=1}^{N} D_n \right)^2. \tag{1.17}$$

A 15-beam 2D DOE was designed with a surface relief phase profile based on 1D DOE designs with $M = 3$ and $N = 5$. After fabrication on an SiO_2 substrate, a low-absorption, multilayer dielectric HR coating at 1064 nm was applied. The fabricated DOE was first tested as a splitter to determine its intrinsic efficiency, producing a 3×5 rectangular grid with an angular separation of 8.87 mrad between adjacent beams. The diffraction pattern measured with the DOE used as a splitter is shown in Figure 1.27a, with the central 15 beams overexposed to highlight diffraction of ~10% of the incident power into higher orders. The predicted combining efficiencies for ideal 1×3 and 1×5 DOEs are $\eta_M = 93.8\%$ and $\eta_N = 96.3\%$, respectively, leading to a design efficiency of $\eta_{MN} = 90.3\%$. Based on the measured values for $D_{m,n}^2$ of all 15 beams, the as-fabricated combining efficiency was calculated to be $\eta_{MN} = 87.8\%$ using Eq. (1.17). This is 2.5% less than the design value owing to variations within the manufacturing tolerance range, illustrated in Figure 1.13, that produce a surface shape on the DOE that does not exactly match the design.

The 2D DOE was tested at the Air Force Research Laboratory's (AFRL) Advanced High Power Fiber Laser Testbed in Albuquerque, NM, to demonstrate coherent beam combination [68]. The AFRL Testbed consists of 16~100 W single-frequency Yb-doped fiber amplifiers seeded by a common master oscillator [17] and phase locked using the LOCSET approach [18]. The output of each amplifier was collimated, transmitted through a free space Faraday isolator, and directed onto the DOE with individual steering mirrors in each beam. Due to the layout of the facility, the amplifiers were not equidistant from the DOE, resulting in ~20% RMS mismatches

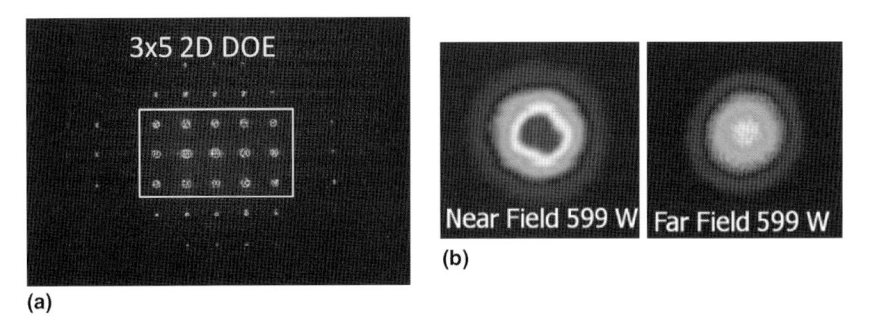

Figure 1.27 (a) Diffraction pattern produced using the 3×5 2D DOE as a beam splitter. (b) Output beam profiles using the 2D DOE as a combiner.

Table 1.5 3 × 5 2D DOE combining results and analysis.

In (W)	Out (W)	Measured η_{MN} (%)	WFS prediction of η_{MN} (%)
52	38.7	74.5	75.6
684	485	71.0	72.3
885	599	67.7	73.7

in beam sizes at the DOE. Fluctuations in the individual amplifiers limited power measurement accuracy to ±1%. Between each measurement, it was necessary to realign and recollimate each of the 15 beams due to thermal beam steering and focusing in the Faraday rotators. Thermal imaging of the DOE surface temperature showed no measurable increase even at full power.

Near-field and far-field images of the combined beam are shown in Figure 1.27b. M^2 of the combined beam was measured to be 1.1 at all power levels. Due to coherent filtering, this is substantially improved over the thermally aberrated input beams, whose M^2 values ranged from 1.1 to 1.4 with an average of 1.26.

Table 1.5 shows CBC results at three different input power levels up to 885 W in all beams. The measured combination efficiencies of ∼68–75% were 15–23% lower than expected for ideal beams. To better quantify the sources of loss, after each combination measurement, a Shack–Hartmann WFS captured the wavefront $\phi_{n,m}(x,y)$ and amplitude profiles $|A_{n,m}(x,y)|$ of each individual beam diffracted into the DOE output port. This field data allowed direct calculation of the expected combination efficiency using Eq. (1.2). The predicted values from this calculation are also listed in Table 1.5 and match the measured efficiencies for 52 W and 684 W input powers up to 2%. At the 885 W level, there is an ∼6% discrepancy between the measured and predicted efficiencies. Significantly, more thermal steering of the beams was observed at this power level, which, because the WFS data and efficiency measurements were not taken simultaneously, could lead to erroneous predictions due to dynamically varying misalignments.

Further analysis of the WFS data at 684 W, coupled with measurements of the LOCSET servo accuracy [69] and the input polarization extinction ratio, supports the conclusions drawn from the perturbative analysis (Table 1.6). After numerically

Table 1.6 Contributions to η_{MN} at 684 W input power.

Effect	Efficiency (%)	Calculation basis
Intrinsic DOE efficiency	87.8	DOE split ratios
Beam size and BQ	90.8	WFS data
NF and FF beam overlap	90.7	WFS data
Polarization	98	Measured
Piston phasing	>99.5	Estimate
Total calculated η_{MN}	70.5	
Measured η_{MN}	71.0	

removing near-field and far-field centroid overlap errors, the predicted efficiency increased from 72.3 to 79.6%, indicating that both beam centroid overlap and the remaining uncorrected errors (beam size variation and BQ) result in a 9% reduction in efficiency for each effect at this power level. Reducing the aberrations imposed by free space optics through the use of an integrated Fourier telescope combiner, as shown in Figure 1.12a, would be expected to bring combining efficiency closer to the intrinsic DOE limit of ∼88%. It should be noted that for larger numbers of beams, 2D DOEs are expected to be much more efficient than the 3×5 device demonstrated here. For example, an 81-beam 2D DOE with $M \times N = 9 \times 9$ is ideally expected to be 98.6% efficient, only 0.6% less than the 99.2% efficiency of a linear 1×81 DOE (Table 1.3).

1.5.2.4 CBC of Tm Fibers at 2 μm

In the interest of eye safety, it is desirable to develop laser sources at "retina-safe" wavelengths longer than 1.4 μm that are absorbed prior to being focused on the retina. Recent developments have shown Tm-doped fiber amplifiers (TDFAs) emitting in the 2 μm region to be a promising avenue for high-power scaling. Notably, an "all-fiber" TDFA has been demonstrated at 1 kW output power [70], highlighting the existence of suitable 790 nm pump diodes and high-power fiber-coupled components at the 2 μm wavelength.

For further power scaling via CBC, a 600 W purely single-frequency TDFA has also been demonstrated [71]. The longer lasing wavelength of the TDFA compared to YDFAs increases its SBS threshold through a combination of effects [72], enabling high-coherence SF output without requiring frequency broadening. The TDFA phase noise characteristics were quantified using the self-heterodyne configuration shown in Figure 1.28. Samples of the 600 W amplifier input and output beams were superimposed on a fast photodiode and the resultant electrical signal was passed through an adjustable low-pass filter. The maxima and minima I_{max} and I_{min} of the filtered signals were recorded by manually perturbing the reference fiber to slowly change its optical path length by a few waves, and the fringe visibility V was calculated using Eq. (1.13). Low V for a given cutoff frequency indicates the presence of integrated RMS phase noise σ_ϕ at higher frequencies that is a substantial fraction of a wave, according to the Marechal criterion $V = 1 - \sigma_\phi^2/2$. The factor of 2 is a worst-case assumption that all phase noise originates in the TDFA rather than in the reference arm. As can be seen from Figure 1.28b, $V > 95\%$ above 1 kHz, indicating $\sigma_\phi < 0.3$ rad above this frequency. It can also be seen from Figure 1.28b that phase noise depends only weakly on power. Shutdown of the coolant circulation pumps results in a dramatic decrease in the low-frequency phase noise, indicating the noise is dominated by vibrations coupled to the fiber.

These data indicate that piston phase locking systems with greater than kilohertz-level control bandwidths ought to enable further scaling via coherent combination of multiple fibers. Phase locking of the high-power TDFA output using the OHD technique was demonstrated by installing an AOM frequency shifter and a piezo-electric fiber stretcher in the reference leg of Figure 1.28a. The resulting phase noise

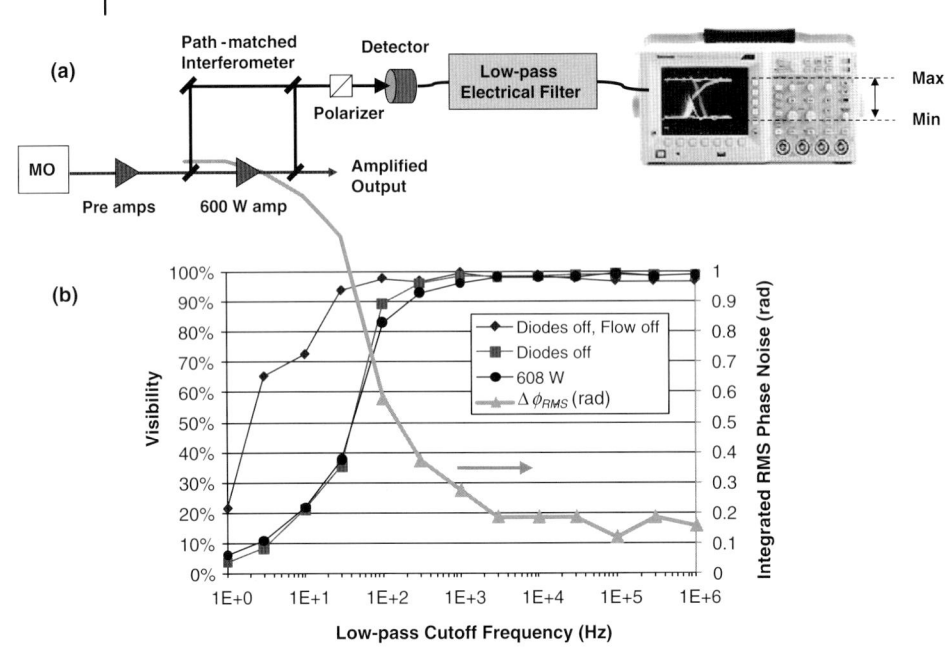

Figure 1.28 (a) Experimental configuration for homodyne fringe visibility measurements. (b) Measured visibility and inferred integrated RMS phase noise.

spectrum is shown in Figure 1.29 a for both open-loop and closed-loop operations at 430 W output (power was limited in these experiments by earlier failure of a diode pump module). Peak noise rejection was ~30 dB at low frequencies, and noise was reduced to ~10 kHz. The RMS phase noise residuals $\sigma_\phi \approx 0.18$ rad were nearly independent of amplifier power (Figure 1.29b). This performance leads to a predicted coherent combining efficiency of $1 - \sigma_\phi^2 = 97\%$ for a phase-locked array of such amplifiers. It should be noted that either the PM fibers or the active polarization control will be required to ensure constructive interference; both

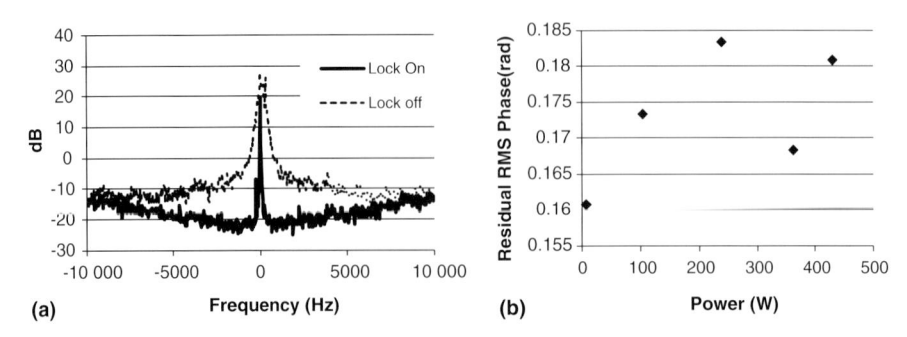

Figure 1.29 (a) Phase noise spectra at 430 W output power. (b) Residual RMS phasing errors with the servo loop engaged.

Table 1.7 Comparison of CBC tolerances between Tm and Yb fibers owing to the 2× longer wavelength.

Parameter	Tm versus Yb	Scaling
Self-phase modulation (nonlinear phase shift, B)	>8× Better (unrealized potential)	2–3× Lower B due to shorter gain lengths
		4× Lower B due to larger A_{eff}
		2× Lower B due to wavelength
Acoustic-induced phase noise	2× Better	Half the phase shift as per change in fiber length
Spatial beam alignments	2× Better	2× Larger diffraction limit doubles the misalignment tolerance

approaches appear readily extensible to Tm fibers. Recent work at milliwatt-level powers has shown coherent combining of two Tm fiber laser channels using both active and passive phase stabilization approaches [73].

From an engineering perspective, it is interesting to assess tolerances for fiber CBC with wavelength λ as a parameter. This allows a comparison of the relative difficulty of CBC with YDFAs and TDFAs, whose wavelengths are separated approximately by a factor of 2. Without question, doubling λ directly relaxes the tolerances required to coherently overlap the beams as summarized in Table 1.7. The benefits arise from improvements in both spatial and temporal coherences – which scale directly with λ – as well as from a lowering of the fiber nonlinearity B due to the λ-scaling of the fiber transverse waveguide dimension, which increases effective mode field area A_{eff} [cf., Eq. (1.12)]. While the full benefit of fiber A_{eff} scaling is yet to be realized owing to the relative immaturity of the Tm glass material and fiber drawing technology, progress in microstructured Tm-doped photonic crystal fibers [74] provides a technical path forward even in the absence of further material development.

1.6
Conclusion

Advances in the technologies of active phase control, geometric beam combination using diffractive optics, and high-coherence, high-power laser amplifiers have enabled unprecedented high-brightness laser demonstrations through coherent beam combining. CBC has allowed laser developers to replace physical constraints on laser power scaling with more traditional engineering constraints of complexity and cost. To date, tiled CBC of zigzag slab lasers has been demonstrated in excess of 100 kW, which represents not only the highest-power CBC demonstration but also the highest-brightness continuous laser source ever built. Filled aperture CBC of fiber lasers has also been demonstrated near 2 kW. While current CBC fiber powers are much lower than those from slabs, it appears likely that fiber-based systems will provide substantial benefits in size, weight, efficiency, and packaging despite their lower single-aperture power. Near-term demonstrations of Yb-doped fiber CBC at the 10–100 kW level are in process and appear readily achievable. Recent developments of high-coherence Tm-doped fiber amplifiers at the kilowatt level provide a technology roadmap toward retina-safe CBC laser systems, a key consideration for deployment and propagation through atmosphere.

Acknowledgments

This chapter is mainly the result of the efforts made by many talented and dedicated Northrop Grumman employees over much of the past decade. We would particularly like to express our gratitude to our colleagues Lewis Book, Eric Cheung, James Ho, Hagop Injeyan, Hiroshi Komine, Stuart McNaught, C. C. Shih, Peter Thielen, Mark Weber, Ben Weiss, and Michael Wickham for their many contributions to this work. We would also like to acknowledge our collaborators on the high-power fiber combining experiments, including Angel Flores, Benjamin Pulford, and Anthony Sanchez of the Air Force Research Laboratory and Steve Augst, T.Y. Fan, Shawn Redmond, Dan Ripin, and Charles Yu of MIT Lincoln Laboratory. Finally, we thank the organizations that have supported portions of this work, including the High Energy Laser Joint Technology Office, the US Army Space and Missile Defense Command/Army Forces Strategic Command, the US Air Force, the Defense Advanced Research Projects Agency, and the US Navy.

References

1 Hecht, J. (July 2009) Ray guns get real. *IEEE Spectrum*, **46**, 28–33

2 Goodno, G.D., Asman, C.P., Anderegg, J., Brosnan, S., Cheung, E.C., Hammons, D., Injeyan, H., Komine, H., Long, W., McClellan, M., McNaught, S.J., Redmond, S., Simpson, R., Sollee, J., Weber, M., Weiss, S.B., and Wickham, M. (2007). Brightness-scaling potential of actively phase-locked solid state laser arrays. *IEEE J. Select. Top. Quantum Electron.*, **13**, 460–472.

3 O'Connor, M. and Shiner, B. (2011) High power fiber lasers for industry and defense, in *High Power Laser Handbook* (eds H. Injeyan and G.D. Goodno), McGraw-Hill Professional, New York, pp. 517–532.

4 Fan, T.Y. (2005) Laser beam combining for high-power, high-radiance sources. *IEEE J. Select. Top. Quantum Electron.*, **11**, 567–577.

5 McNaught, S., Asman, C., Injeyan, H., Jankevics, A., Johnson, A., Jones, G., Komine, H., Machan, J., Marmo, J., McClellan, M., Simpson, R., Sollee, J., Valley, M., Weber, M., and Weiss, S. (2009) 100-kW coherently combined Nd: YAG MOPA laser array. Frontiers in Optics, San Jose, CA October 11, 2009, Paper FThD2.

6 http://www.as.northropgrumman.com/products/reli/assets/RELI_datasheet.pdf, (accessed May 14, 2012).

7 Wacks, M. (2011) Northrop Grumman coherently combined high-power fiber laser for the RELI program. 2nd Annual Advanced High-Power Laser Review, Santa Fe, NM.

8 Goodno, G.D., Shih, C.C., and Rothenberg, J.E. (2010) Perturbative analysis of coherent combining efficiency with

mismatched lasers. *Opt. Express*, **18**, 25403–25414.

9 Leger, J.R., Swanson, G.J., and Veldkamp, W.B. (1987) Coherent laser addition using binary phase gratings. *Appl. Opt.*, **26**, 4391–4399.

10 Nabors, C.D. (1994) Effect of phase errors on coherent emitter arrays. *Appl. Opt.*, **33**, 2284–2289.

11 Fan, T.Y. (2009) The effect of amplitude (power) variations on beam combining efficiency for phased arrays. *IEEE J. Select. Top. Quantum Electron.*, **15**, 291–293.

12 Liang, W., Satyan, N., Aflatouni, F., Yariv, A., Kewitsch, A., Rakuljic, G., and Hashemi, H. (2007) Coherent beam combining with multilevel optical phase-locked loops. *J. Opt. Soc. Am. B*, **24**, 2930–2939.

13 Cheung, E.C., Ho, J.G., Goodno, G.D., Rice, R.R., Rothenberg, J., Thielen, P., Weber, M., and Wickham, M. (2008) Diffractive optics-based beam combination of a phase-locked fiber laser array. *Opt. Lett.*, **33**, 354–356.

14 Willke, B., Uehara, N., Gustafson, E.K., Byer, R.L., King, P.J., Seel, S.U., and Savage, R.L. Jr. (1998) Spatial and temporal filtering of a 10-W Nd:YAG laser with a Fabry–Pérot ring-cavity premode cleaner. *Opt. Lett.*, **23**, 1704–1706.

15 Tünnermann, H., Pöld, J.H., Neumann, J., Kracht, D., Willke, B., and Weßels, P. (2011) Beam quality and noise properties of coherently combined ytterbium doped single frequency fiber amplifiers. *Opt. Express*, **19**, 19600–19606.

16 Pupeza, I., Eidam, T., Rauschenberger, J., Bernhardt, B., Ozawa, A., Fill, E., Apolonski, A., Udem, T., Limpert, J., Alahmed, Z.A., Azzeer, A.M., Tünnermann, A., Hänsch, T.W., and Krausz, F. (2010) Power scaling of a high-repetition-rate enhancement cavity. *Opt. Lett.*, **35**, 2052–2054.

17 Wagner, T.J. (2012) Fiber laser beam combining and power scaling progress: Air Force Research Laboratory Laser Division. *Proc. SPIE*, **8237**, 823718.

18 Shay, T.M. (2006) Theory of electronically phased coherent beam combination without a reference beam. *Opt. Express*, **14**, 12188–12195.

19 Zhou, P., Liu, Z., Wang, X., Ma, Y., Ma, H., Xu, X., and Guo, S. (2009) Coherent beam combining of fiber amplifiers using stochastic parallel gradient descent algorithm and its application. *IEEE J. Select. Top. Quantum Electron.*, **15**, 248–256.

20 Anderegg, J., Brosnan, S., Cheung, E., Epp, P., Hammons, D., Komine, H., Weber, M., and Wickham, M. (2006) Coherently coupled high-power fiber arrays, in *Fiber Lasers III: Technology, Systems, and Applications, Proceedings of SPIE* (eds A.J. Brown, J. Nilsson, D.J. Harter, and A. Tünnermann), Society of Photo Optical, p. 61020U–1.

21 Yu, C.X., August, S.J., Redmond, S.M., Goldizen, K.C., Murphy, D.V., Sanchez, A., and Fan, T.Y. (2011) Coherent combining of a 4kW, eight-element fiber amplifier array. *Opt. Lett.*, **36**, 2686–2688.

22 Yu, C.X., Kansky, J.E., Shaw, S.E.J., Murphy, D.V., and Higgs, C. (2006) Coherent beam combining of large number of PM fibres in 2-D fibre array. *Electron. Lett.*, **42**, 1024–1025.

23 O'Meara, T.R. (1977) The multidither principle in adaptive optics. *J. Opt. Soc. Am.*, **67**, 306–314.

24 Bourdon, P., Jolivet, V., Bennai, B., Lombard, L., Goular, D., Canat, G., and Vasseur, O. (2009) Theoretical analysis and quantitative measurements of fiber amplifier coherent combining on a remote surface through turbulence. *Proc. SPIE*, **7195**, 719527.

25 Daniault, L., Hanna, M., Lombard, L., Zaouter, Y., Mottay, E., Goular, D., Bourdon, P., Druon, F., and Georges, P. (2011) Coherent beam combining of two femtosecond fiber chirped-pulse amplifiers. *Opt. Lett.*, **36**, 621–623.

26 Ma, Y., Zhou, P., Wang, X., Ma, H., Xu, X., Si, L., Liu, Z., and Zhao, Y. (2010) Coherent beam combination with single frequency dithering technique. *Opt. Lett.*, **35**, 1308–1310.

27 Ma, Y., Zhou, P., Wang, X., Ma, H., Xu, X., Si, L., Liu, Z., and Zhao, Y. (2011) Active phase locking of fiber amplifiers using sine–cosine single-frequency dithering technique. *Appl. Opt.*, **50**, 3330–3336.

28 Levy, J. and Roh, K. (1995) Coherent array of 900 semiconductor laser amplifiers. *Proc. SPIE*, **2382**, 58–69.
29 Vorontsov, M.A. and Sivokon, V.P. (1998) Stochastic parallel-gradient-descent technique for high-resolution wave-front phase-distortion correction. *J. Opt. Soc. Am. A*, **15**, 2745–2758.
30 Weyrauch, T., Vorontsov, M.A., Carhart, G.W., Beresnev, L.A., Rostov, A.P., Polnau, E.E., and Liu, J.J. (2011) Experimental demonstration of coherent beam combining over a 7km propagation path. *Opt. Lett.*, **36**, 4455–4457.
31 Redmond, S.M., Creedon, K.J., Kansky, J.E., Augst, S.J., Missaggia, L.J., Connors, M.K., Huang, R.K., Chann, B., Fan, T.Y., Turner, G.W., and Sanchez-Rubio, A. (2011) Active coherent beam combining of diode lasers. *Opt. Lett.*, **36**, 999–1001.
32 Redmond, S.M. (2011) Active coherent combination of >200 semiconductor amplifiers using a SPGD algorithm. CLEO:2011 – Laser Applications to Photonic Applications, Paper CTuV1.
33 Andrews, J.R. (1989) Interferometric power amplifiers. *Opt. Lett.*, **14**, 33–35.
34 Dong, H., Li, X., Wei, C., He, H., Zhao, Y., Shao, J., and Fan, Z. (2009) Coaxial combination of coherent laser beams. *Chin. Opt. Lett.*, **7**, 1012–1014.
35 Uberna, R., Bratcher, A., and Tiemann, B.G. (2010) Coherent polarization beam combination. *IEEE J. Quantum Electron.*, **46**, 1191–1196.
36 Nelson, B.E., Shakir, S.A., Culver, W.R., Starcher, Y.S., Hedrick, J.W., and Bates, G.M. (2010) System and method for combining multiple fiber amplifiers or multiple fiber lasers. U.S. Patent Appl. 2010/0195195.
37 Bruesselbach, H., Jones, D.C., Mangir, M.S., Minden, M., and Rogers, J.L. (2005) Self-organized coherence in fiber laser arrays. *Opt. Lett.*, **30**, 1339–1341.
38 Christensen, S.E. and Koski, O. (2007) 2-Dimensional waveguide coherent beam combiner. Advanced Solid-State Photonics, Paper WC1.
39 Uberna, R., Bratcher, A., Alley, T.G., Sanchez, A.D., Flores, A.S., and Pulford, B. (2010) Coherent combination of high power fiber amplifiers in a two-dimensional re-imaging waveguide. *Opt. Express*, **18**, 13547–13553.
40 Leger, J., Swanson, G.J., and Veldkamp, W.B. (1986) Coherent beam addition of GaAlAs lasers by binary phase gratings. *Appl. Phys. Lett.*, **48**, 888.
41 Jain, A., Andrusyak, O., Venus, G., Smirnov, V., and Glebov, L. (2010) Passive coherent locking of fiber lasers using volume Bragg gratings. *Proc. SPIE*, **7580**, 75801S-1–75801S-9.
42 Lumeau, J., Glebova, L., and Glebov, L.B. (2011) Near-IR absorption in high-purity photothermorefractive glass and holographic optical elements: measurement and application for high-energy lasers. *Appl. Opt.*, **50**, 5905–5911.
43 Dammann, H. and Gortler, K. (1971) High-efficiency in-line multiple imaging by means of multiple phase holograms. *Opt. Commun.*, **3**, 312–315.
44 Hergenhan, G., Lücke, B., and Brauch, U. (2003) Coherent coupling of vertical-cavity surface-emitting laser arrays and efficient beam combining by diffractive optical elements: concept and experimental verification. *Appl. Opt.*, **42**, 1667–1680.
45 Madasamy, P., Jander, D., Brooks, C., Loftus, T., Thomas, A., Jones, P., and Honea, E. (2009) Dual-grating spectral beam combination of high-power fiber lasers. *IEEE J. Select. Top. Quantum Electron.*, **15**, 337–343.
46 Augst, S.J., Lawrence, R.C., Fan, T.Y., Murphy, D.V., and Sanchez, A. (2008) Characterization of diffraction gratings for use in wavelength beam combining at high average power. Frontiers in Optics 2008, Rochester, NY, October 19, 2008, Paper FWG2.
47 Khitrov, V., Farley, K., Leveille, R., Galipeau, J., Majid, I., Christensen, S., Samson, B., and Tankala, K. (2010) kW level narrow linewidth Yb fiber amplifiers for beam combining. *Proc. SPIE*, **7686**, 76860.
48 Shkurikhin, O., Gapontsev, V., and Platonov, N. (2009) Narrow-linewidth kilowatt-class cw diffraction-limited fiber lasers and amplifiers. 22nd Annual Solid State and Diode Laser Technology Review, Newton, MA.

49 Injeyan, H. and Goodno, G.D. (2011) Zigzag slab lasers, in *High Power Laser Handbook* (eds H. Injeyan and G.D. Goodno), McGraw-Hill Professional, New York, pp. 187–205.

50 Martin, W.S. and Chernoch, J.P. (1972) Multiple internal reflection face-pumped laser. U.S. Patent No. 3,633,126.

51 Ying, C., Bin, C., Patel, M.K.R., and Bass, M. (2004) Calculation of thermal-gradient-induced stress birefringence in slab lasers-I. *IEEE J. Quantum Electron.*, **40**, 909–916.

52 Goodno, G.D., Palese, S., Harkenrider, J., and Injeyan, H. (2001) Yb:YAG power oscillator with high brightness and linear polarization. *Opt. Lett.*, **26**, 1672–1674.

53 Injeyan, H. and Hoefer, C.S. (2000) End pumped zig-zag slab laser gain medium. U.S. Patent No. 6,094,297.

54 Goodno, G.D., Komine, H., McNaught, S.J., Weiss, S.B., Redmond, S., Long, W., Simpson, R., Cheung, E.C., Howland, D., Epp, P., Weber, M., McClellan, M., Sollee, J., and Injeyan, H. (2006) Coherent combination of high-power, zigzag slab lasers. *Opt. Lett.*, **31**, 1247–1249.

55 Redmond, S., McNaught, S., Zamel, J., Iwaki, L., Bammert, S., Simpson, R., Weiss, S.B., Szot, J., Flegal, B., Lee, T., Komine, H., and Injeyan, H. (2007) 15 kW Near-diffraction-limited single-frequency Nd:YAG laser. Conference on Lasers and Electro-Optics/Quantum Electronics and Laser Science, Paper CTuHH5.

56 Kane, T.J., Kozlovsky, W.J., and Byer, R.L. (1986) 62-dB-Gain multiple-pass slab geometry Nd:YAG amplifier. *Opt. Lett.*, **11**, 216–218.

57 Agarwal, G.P. (2007) *Nonlinear Fiber Optics*, 4th edn, Academic Press, New York.

58 Goodno, G.D. and Rothenberg, J.E. (2008) Advances and limitations in fiber beam combination. OSA Annual Meeting, Paper FTuW1.

59 Goodno, G.D., McNaught, S.J., Rothenberg, J.E., McComb, T., Thielen, P.A., Wickham, M.G., and Weber, M.E. (2010) Active phase and polarization locking of a 1.4-kW fiber amplifier. *Opt. Lett.*, **35**, 1542–1544.

60 Jones, D.C. and Scott, A.M. (2007) Characterisation and stabilising dynamic phase fluctuations in large mode area fibres. *Proc. SPIE*, **6453**, 64530Q.1–64530Q.10.

61 Goodman, J.W. (2000) *Statistical Optics*, Wiley-Interscience, pp. 163–165.

62 Palese, S., Cheung, E., Goodno, G., Shih, C., DiTeodoro, F., McComb, T., and Weber, M. (2012) Coherent combining of pulsed fiber amplifiers in the nonlinear chirp regime with intra-pulse phase control. *Opt. Express*, **20**, 7422–7435.

63 Klenke, A., Seise, E., Demmler, S., Rothhardt, J., Breitkopf, S., Limpert, J., and Tünnermann, A. (2011) Coherently-combined two channel femtosecond fiber CPA system producing 3mJ pulse energy. *Opt. Express*, **19**, 24280–24285.

64 Weiss, S.B., Weber, M.E., and Goodno, G.D. (2012) Group delay locking of coherently combined broadband lasers. *Opt. Lett.*, **37**, 455–457.

65 Goodno, G.D. and Weiss, S.B. (2012) Automated co-alignment of coherent fiber laser arrays via active phase-locking. *Opt. Express*, **20**, 14945–14953.

66 Redmond, S.M., Fan, T.Y., Ripin, D.J., Yu, C.X., Augst, S.J., Thielen, P.A., Rothenberg, J.E., and Goodno, G.D. (2012) Diffractive coherent combining of a 2.5-kW fiber laser array into a 1.9kW Gaussian beam. *Opt. Lett.*, **37**, 2832–2834.

67 Weisstein, E.W. (2012) Uniform Distribution. From MathWorld, A Wolfram Web Resource, http://mathworld.wolfram.com/UniformDistribution.html.

68 Thielen, P.A., Ho, J.G., Burchman, D.A., Goodno, G.D., Rothenberg, J.E., Wickham, M.G., Flores, A., Lu, C.A., Pulford, B., Hult, D., Rowland, K.B., and Robin, C. (2012) Two-dimensional diffractive coherent beam combining. Advanced Solid State Photonics, Paper AM3A.2.

69 Pulford, B.N. (2011) LOCSET phase locking: operation, diagnostics, and applications, Ph.D. dissertation, University of New Mexico.

70 Ehrenreich, T., Leveille, R., Majid, I., Tankala, K., Rines, G., and Moulton, P. (2010) 1 kW all-glass Tm:fiber laser. Photonics West, Session 16, Jan 28, 2010.

71 Goodno, G.D., Book, L.D., and Rothenberg, J.E. (2009) Low-phase-noise,

single-frequency, single-mode 608 W thulium fiber amplifier. *Opt. Lett.*, **34**, 1204–1206.

72 Goodno, G.D., Book, L.D., Rothenberg, J.E., Weber, M.E., and Weiss, S.B. (2011) Narrow linewidth power scaling and phase stabilization of 2-μm thulium fiber lasers. *Opt. Eng.*, **50**, 111608.

73 Zhou, P., Wang, X., Ma, Y., Ma, H., Han, K., Xu, X., and Liu, Z. (2010) Active and passive coherent beam combining of thulium-doped fiber lasers. *Proc. SPIE*, **7843**, 784307.

74 Modsching, N., Kadwani, P., Sims, R.A., Leick, L., Broeng, J., Shah, L., and Richardson, M. (2011) Lasing in thulium-doped polarizing photonic crystal fiber. *Opt. Lett.*, **36**, 3873–3875.

2
Coherent Beam Combining of Fiber Amplifiers via LOCSET

Angel Flores, Benjamin Pulford, Craig Robin, Chunte A. Lu, and Thomas M. Shay

2.1
Introduction

Fiber laser systems with a broad range of industrial [1], medical [2], and military [3] applications have evolved rapidly over the past decade. Generally, fiber lasers offer several advantages over conventional solid-state and chemical lasers, including compactness, near-diffraction-limited beam quality, superior thermal–optical properties, and high optical–optical conversion efficiencies. Despite their advantages and brisk development, fiber lasers are still behind both chemical and bulk solid-state lasers in terms of total output power. Currently, the intensity and hence power available from single-mode optical fibers are limited by optical surface damage, thermal loads, and nonlinear optical effects.

Due to small core sizes and long amplifier lengths, high-power, single-mode fiber lasers are limited by the onset of (power-dependent) detrimental effects such as stimulated Brillouin scattering (SBS) [4,5] and modal instabilities [6,7]. SBS is a third-order phase-matched nonlinear interaction that couples acoustic phonons to photons of the optical field and the associated backscattered Stokes light. Consequently, optical power is transferred from the laser field to the Stokes light, degrading amplification of the signal light and possibly damaging the fiber amplifier (FA) through pulsation. In addition to nonlinear effects, a recent phenomenon limiting the power scaling of large mode area (LMA) fiber amplifiers has been modal instabilities or the modal "hopping" of the fundamental mode (LP_{01}) into the next higher order mode (LP_{11}). A fundamental problem of LMA fibers is that they are inherently multimode. As a result, there have been reports of a sudden and dramatic loss in beam quality above a certain modal instability threshold [7]. Due to such constraints, single-mode fiber lasers do not meet requirements for future long-range directed energy (DE) applications. As a result, to scale the overall power and brightness, beam combining techniques where multiple lasers are efficiently combined into a single output beam while maintaining high beam quality (and brightness) are being actively researched.

Coherent Laser Beam Combining, First Edition. Edited by Arnaud Brignon.
© 2013 Wiley-VCH Verlag GmbH & Co. KGaA. Published 2013 by Wiley-VCH Verlag GmbH & Co. KGaA.

Table 2.1 Brief summary of major beam combining architectures and techniques.

Beam combining techniques	Advantages	Disadvantages	Combined power
Coherent beam combining (tiled aperture)	Atmospheric compensation possible	Active phase control required	1.4 kW with 16 elements, single frequency (AFRL) [13]
	Fine electronic beam steering	Low array fill factor limits PIB (50–70%) dependent on the number of array subapertures, subaperture packing, and optical fill of individual array subapertures	4 kW with eight elements, 10 GHz (MIT-LL) [14]
	N^2 irradiance scaling		
Coherent beam combining (filled aperture, DOE)	Excellent beam quality	Active phase control required	2 kW with five elements, 10 GHz linewidth (MIT-LL) [15]
	Power concentrated in single lobe	Combining element power handling	
Passive beam combination	No active phase control	Limited scalability	0.7 kW with four elements (Lockheed-Aculight) [16]
Spectral beam combining	No active phase control required	Scalability influenced by laser gain bandwidth	8 kW with four elements (Friedrich-Schiller University) [11]
	Excellent beam quality	Beam quality sensitive to laser linewidth	
	Power concentrated in single lobe		
Incoherent beam combination	Higher power from single amplifier (no SBS or linewidth limitations)	Steering optics required for each beam	3 kW with four elements (NRL) [10]
	Simplicity in design	Limited propagation range Requires a large platform	

2.1.1
Beam Combination Architectures

The major beam combining techniques can be broadly categorized into incoherent [8] and coherent [9] beam combining approaches. A brief summary of the major beam combining approaches is presented in Table 2.1. In incoherent beam combining, an array of lasers is superimposed in the far field without control of the relative spectra or phases of the different elements. Such beam combining has been demonstrated at a length of 1.2 km with powers up to 3 kW [10]. Similarly, in spectral beam combining (SBC), a separate class of incoherent beam combining, incoherent beams of different wavelengths are spatially overlapped (in the near field) to create a single beam of multiple colors. SBC has the advantage of not requiring active phase control or mutual temporal coherence of the individual beams. Although combined powers of 8 kW ($M^2 \sim 4$) have been reported [11], SBC channel scalability may be limited by the finite gain bandwidth and beam quality (as a function of linewidth) sensitivity of the combining gratings [12].

2.1.2
Active and Passive Coherent Beam Combining

In contrast, coherent beam combining (CBC) schemes require proper phase, frequency, and polarization relationships for efficient combination. CBC can be divided into techniques that use active or passive techniques to force coherence between all array elements. Active CBC uses electronic feedback to equalize and control the optical phase of the individual laser array elements, while passive CBC relies on self-phase locking via passive coupling mechanisms (i.e., fiber ring [17] and self-Fourier cavity [18]) to coherently combine multiple lasers. Despite bypassing complex phase controls required in active CBC, passive scaling to higher channel counts appears limited with maximal channel counts N_{max} of 10–12 elements predicted [19].

In comparison, active CBC with channel scalability of up to 64 elements has been reported [20] and channel counts up to 100 appear feasible [21]. Active CBC can be further divided into tiled aperture and filled aperture combining architectures. In tiled array formats, individual laser array elements are combined (interfered) in the far-field regime. Tiled array systems have several advantages such as fine electronic beam steering [22], potential N^2 irradiance scaling [23], distributed thermal load on final optics, and atmospheric turbulence compensation [24]. More importantly, tiled array systems are being investigated for extension to fiber-phased array platforms and remote target phase locking. Nevertheless, tiled array systems are limited by nonuniform fill factors that contribute to far-field side lobes that limit the optical power in the central lobe. In general, optimal close-packed hexagonal arrays of Gaussian filled circular subapertures predict power-in-the-central bucket values of $\sim$75% [23]. With regard to tiled apertures, combined single-frequency and narrow-linewidth fiber arrays of 1.45 [13] and 4 kW [14], respectively, have been reported.

Contrary to tiled arrays, filled aperture techniques are based on near-field beam combining of laser array elements. Similar to utilizing a beam splitter in reverse, a beam combining optic is required to overlap the lasers in the near field. Filled aperture beam combination has the advantage of maintaining the near-diffraction-limited beam quality of fiber lasers, with all power deposited into a single coherent beam. However, filled array systems require a combining element capable of managing the entire combined power. To date, combined powers of up to 2 kW [15] have been reported in filled array systems.

Regardless of the CBC architecture, active phase control of the array elements is required. As such several techniques have been developed for electronic phase control. In this chapter, we describe in detail the CBC of fiber amplifiers via locking of optical coherence by single-detector electronic-frequency tagging (LOCSET) [13,21,25,26,27]. LOCSET is a novel approach to electronic phase locking that eliminates the need for a reference beam and requires a single-detector for full-phase correction and beam combining. Here, we will discuss the theory of LOCSET operation and detail low-power LOCSET beam combining demonstrations of up to 32 elements with $\lambda/71$ average residual phase error. Moreover, high-power kilowatt-scale CBC of conventional silica fiber and photonic crystal fiber s (PCF) amplifier via LOCSET phase locking is reported. Successful tiled array beam combination of 16 single-frequency 100 W lasers into a kilowatt class (1.4 kW) laser beam is presented. In addition, kilowatt-scale filled aperture CBC of novel SBS-suppressive PCF amplifiers is detailed.

2.2
Locking of Optical Coherence by Single-Detector Electronic-Frequency Tagging

In order to coherently combine multiple lasers, accurate control of the optical phase is required. Some of the more prominent methods of active phase control include heterodyne [28,29], stochastic parallel gradient descent (SPGD) [30,31], and LOCSET [13,21,25,26,27] phase locking techniques. In heterodyne phase locking, each optical channel phase locks to a common frequency-shifted reference beam. While the scheme is noted for its simplicity and excellent phase error performance ($\lambda/80$ residual phase fluctuation for two-element beam combining reported) [32], the method is hindered by the detector array arrangement (N detectors needed to combine N beams) and its common reference beam requirement. Here, the elimination of the reference element would terminate the coherent beam combination.

In contrast, SPGD requires a single detector for CBC. During operation, a random optical phase perturbation is applied in parallel to each phase-controlled beam. Then an intensity-based metric algorithm is implemented to optimally combine (interfere) the beams in the near or far field. Notably, SPGD can be used for additional higher order wave front controls and 48-channel phase locking with $\lambda/30$ residual phase error [33] has been demonstrated. Nonetheless, SPGD may be limited by the inverse relationship between SPGD control loop bandwidth (BW_{SPGD}) and number of combined beams N ($BW_{SPGD} \propto 1/N$) [27].

Similar to SPGD, LOCSET utilizes a single photodetector for active phase locking. However, LOCSET is not a stochastic, intensity-based process. Based on coherent radio frequency (RF) demodulation, LOCSET electronics is capable of independently determining an error signal proportional to the optical phase difference of each beam measured with respect to every other beam in the array. Although LOCSET can

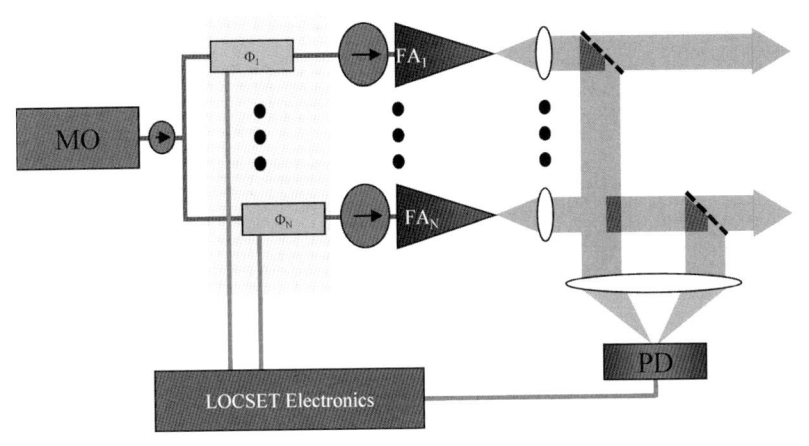

Figure 2.1 General LOCSET active CBC schematic. MO, master oscillator; FA, fiber amplifier; PD, photodiode.

only apply piston phase corrections, its excellent phase error performance at high channel counts (and high bandwidth) makes it attractive for active CBC.

The LOCSET system, as shown in Figure 2.1, typically employs a master oscillator power amplifier (MOPA) configuration where a narrow-linewidth laser is split N ways and seeds an array of fiber amplifier. Prior to amplification, each of the N beams passes through a phase modulator allowing the LOCSET control electronics the ability to apply piston phase corrections to each beam in the system. Each of the N beams is then amplified, collimated, and launched from the exit aperture of the system. Here the sampled light from a partial reflector is overlapped (interfered) onto a single photodetector that feeds into the LOCSET control electronics. To achieve optimal beam combination, each of the N beams is "tagged" with a small-amplitude phase dither at a unique radio frequency. These phase dithers are then measured with the photodetector as an intensity interference beat note that contains the phase information needed for CBC. Subsequently, in the next sections, we will study the general LOCSET theory and operation.

2.2.1
LOCSET Theory

There are two operational configurations of LOCSET: self-referenced and self-synchronous phase locking [25,27]. In self-referenced phase locking, N-1 beams are tagged with a unique RF phase dither that is used to demodulate the phase difference of a single beam with respect to all other beams in the system. Here the remaining unmodulated beam is used as a reference for each of the RF phase-modulated elements to minimize the phase difference between itself and the reference. It is important to note that a reference beam is not required in LOCSET. As we will show, if the amplitude of the unmodulated reference beam is set to zero, the expression for the phase error signal remains valid. This technique, known as self-synchronous LOCSET, determines the phase difference between itself and all other channels and applies the appropriate phase corrections. Because each channel is working toward minimizing the phase difference between itself and all other beams, the phase difference between the beams will converge to zero and establish optimal beam combination.

2.2.2
Self-Referenced LOCSET

In self-referenced LOCSET, there are N-1 phase-modulated beams, $E_i(t)$ and a single unmodulated reference beam $E_u(t)$, expressed as

$$E_u(t) = E_{u_0} \cos(\omega_L t + \phi_u(t)), \tag{2.1}$$

$$E_i(t) = E_{i_0} \cos(\omega_L t + \phi_i(t) + \beta_i \sin(\omega_i t)). \tag{2.2}$$

Here, E_{u_0} and E_{i_0} are the field amplitudes of the unmodulated and phase-modulated beams, respectively. The angular laser frequency is ω_L and $\phi_u(t)$ and $\phi_i(t)$ are the time-varying phase states of the unmodulated and modulated beams, respectively. We note that since $\phi_u(t)$ and $\phi_i(t)$ have much slower variations than both the optical laser and RF modulation frequencies, they will be treated as constants. For the phase-modulated beams, the third added term ($\beta_i \sin(\omega_i t)$) represents an applied sinusoidal phase modulation with amplitude β_i and RF modulation frequency ω_i. In practice, to minimize residual phase errors, the RF modulation amplitude is kept on the order of 1/10th of a radian or approximately 1/60th of the optical wavelength ($\lambda/60$).

2.2.2.1 Photocurrent Signal
A basic LOCSET signal processing diagram is presented in Figure 2.2, where the overlapped beams on the photodetector produce a combined electric field $E_T(t)$:

$$E_T(t) = E_u(t) + \sum_{i=1}^{N-1} E_i(t), \tag{2.3}$$

with the individual fields represented by Eqs. (2.1) and (2.2), respectively. Accordingly, the electric field produces a photocurrent at the detector ($i_{PD}(t)$):

$$i_{PD}(t) = R_{PD} \cdot A \cdot \frac{1}{2} \cdot \left(\frac{\varepsilon_0}{\mu_0}\right)^{1/2} \cdot E_T^2(t), \tag{2.4}$$

where R_{PD} is the responsivity of the photodetector and A is the active area of the photodetector. Next, substituting Eq. (2.3) into Eq. (2.4) with arbitrary summation indices j and k yields

$$i_{PD}(t) = R_{PD} \cdot A \cdot \frac{1}{2} \cdot \left(\frac{\varepsilon_0}{\mu_0}\right)^{1/2} \\ \cdot \left(E_u^2(t) + 2E_u(t) \cdot \sum_{j=1}^{N-1} E_j(t) + \left(\sum_{j=1}^{N-1} E_j(t)\right)\left(\sum_{j=1}^{N-1} E_k(t)\right) \right). \tag{2.5}$$

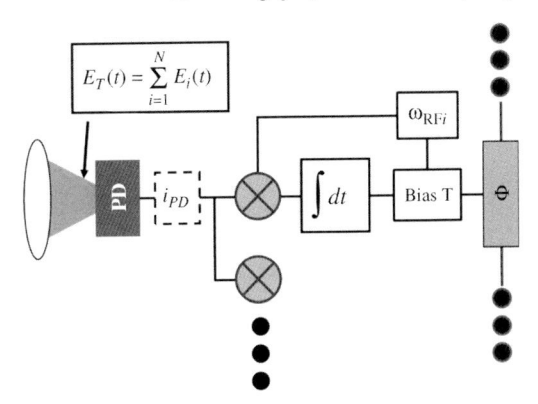

2.2 Locking of Optical Coherence by Single-Detector Electronic-Frequency Tagging

Figure 2.2 Basic LOCSET signal processing diagram. The phase-modulated beams are combined and incident on the photodetector. The photocurrent is then mixed with a unique RF tag and a phase error signal is generated via integration. The error signal and RF phase dither are then applied to a phase modulator keeping the ith beam in phase with all other beams. An identical control loop is applied to all N or $N-1$ channels.

Noticeably, the photocurrent in Eq. (2.5) can be divided into three components: a photocurrent due to unmodulated reference beam ($i_u(t)$), a photocurrent due to an unmodulated reference beam interfering with the phase-modulated beams ($i_{u_j}(t)$), and a photocurrent due to interference of each phase-modulated beam with all other modulated beams ($i_{jk}(t)$) or

$$i_{PD}(t) = i_u(t) + i_{u_j}(t) + i_{jk}(t). \tag{2.6}$$

A complete expression for the generated photocurrent can be derived by substituting Eqs. (2.1) and (2.2) into Eq. (2.5). This expression can then be partitioned into three photocurrent components. For example, the photocurrent due to the unmodulated reference beam is

$$i_u(t) = \frac{R_{PD} \cdot P_u}{2}(1 + \cos(2\omega_L t + 2\phi_u)) \approx \frac{R_{PD} \cdot P_u}{2}, \tag{2.7}$$

where P_u is the optical power of the unmodulated beam and the terms oscillating at the laser frequency are neglected (cannot be resolved by photodetector). Thus, this photocurrent contributes a DC bias to the total current. Similarly, the photocurrent due to unmodulated beam interfering with the modulated beams, after substitution, yields

$$i_{u_j}(t) = R_{PD} \cdot P_u^{1/2} \cdot \sum_{j=1}^{N-1} P_j^{1/2}$$
$$\cdot \left(\cos(\phi_u - \phi_j)\cos(\beta_j \sin(\omega_j t)) + \sin(\phi_u - \phi_j)\sin(\beta_j \sin(\omega_j t)) \right) \tag{2.8}$$

and can be simplified through Fourier series expansion as follows:

$$i_{u_j}(t) = R_{PD} \cdot P_u^{1/2} \cdot \sum_{j=1}^{N-1} P_j^{1/2} \cdot \left(\cos(\phi_u - \phi_j) \left(J_0(\beta_j) + 2 \sum_{n=1}^{\infty} J_{2n}(\beta_j) \cdot \cos(2n \cdot \omega_j t) \right) \right.$$

$$\left. + \sin(\phi_u - \phi_j) \left(2 \sum_{n=1}^{\infty} J_{2n-1}(\beta_j) \cdot \sin((2n-1) \cdot \omega_j t) \right) \right), \quad (2.9)$$

where J_n is a Bessel function of the first kind of order n. We note that the second term in the sum of Eq. (2.9) is proportional to the sine of the phase difference between the unmodulated and modulated reference beams ($\sin(\phi_u - \phi_j)$). Notably, this term is a characteristic error signal where minimizing the sinusoidal phase difference equalizes the individual phases for optimal phase locking.

The final photocurrent term quantifies the interference of each phase-modulated beam with the set of other phase-modulated elements. Here, after utilizing several trigonometric identities, Fourier series expansions, and neglecting laser frequency oscillation, the photocurrent can be expressed as [27]

$$i_{jk}(t) = \frac{R_{PD}}{2} \cdot \sum_{k=1}^{N-1} P_k^{1/2} \cdot \sum_{j=1}^{N-1} P_j^{1/2}$$

$$\cdot \left[\begin{pmatrix} \cos(\phi_k - \phi_j) \left(J_0(\beta_k) + 2 \sum_{n_k=1}^{\infty} J_{2n_k}(\beta_k) \cdot \cos(2n_k \cdot \omega_k t) \right) \\ \cdot \left(J_0(\beta_j) + 2 \sum_{n_j=1}^{\infty} J_{2n_j}(\beta_j) \cdot \cos(2n_j \cdot \omega_j t) \right) \end{pmatrix} \right.$$

$$- \begin{pmatrix} \sin(\phi_k - \phi_j) \left(2 \sum_{n_k=1}^{\infty} J_{2n_k-1}(\beta_k) \cdot \sin((2n_k-1) \cdot \omega_k t) \right) \\ \cdot \left(J_0(\beta_j) + 2 \sum_{n_j=1}^{\infty} J_{2n_j}(\beta_j) \cdot \cos(2n_j \cdot \omega_j t) \right) \end{pmatrix}$$

$$+ \begin{pmatrix} \sin(\phi_k - \phi_j) \left(J_0(\beta_k) + 2 \sum_{n_k=1}^{\infty} J_{2n_k}(\beta_k) \cdot \cos(2n_k \cdot \omega_k t) \right) \\ \cdot \left(2 \sum_{n_j=1}^{\infty} J_{2n_j-1}(\beta_j) \cdot \sin((2n_j-1) \cdot \omega_j t) \right) \end{pmatrix}$$

$$\left. + \begin{pmatrix} \cos(\phi_k - \phi_j) \left(2 \sum_{n_k=1}^{\infty} J_{2n_k-1}(\beta_k) \cdot \cos((2n_k-1) \cdot \omega_k t) \right) \\ \cdot \left(2 \sum_{n_j=1}^{\infty} J_{2n_j-1}(\beta_j) \cdot \cos((2n_j-1) \cdot \omega_j t) \right) \end{pmatrix} \right]$$

$$(2.10)$$

Once again, the characteristic phase error signals proportional to the sinusoidal phase difference between the kth and jth modulated beams are generated $[\sin(\phi_k - \phi_j)]$ allowing ideal phase optimization.

2.2.2.2 LOCSET Demodulation

Once the combined interference signal reaches the photodetector, demodulation of the phase error signal for each of the modulated beams occurs, as shown in Figure 2.2. Schematically, this is done via individual and independent control loops acting on each of the phase-modulated channels. Although each channel performs identical operations, each channel is distinguished by its unique RF modulation frequency ω_i. To this end, the coherent demodulation process involves multiplying (mixing) the sampled photocurrent with an RF demodulation signal $[\sin(\omega_c t)]$ and integrating over time τ. The subsequent RF demodulation can be expressed as

$$S_x = \frac{1}{\tau}\int_0^\tau i_{PD}(t)\cdot \sin(\omega_c t)\mathrm{d}t. \tag{2.11}$$

Here, ω_c represents the demodulation frequency of the xth LOCSET channel and S_x is the phase error correction signal of the xth modulated beam. Particularly, the demodulation frequency for each channel is chosen to equal that specific channel's RF phase dither frequency ($\omega_c = \omega_i = \omega_x$). In addition, the integration time is chosen such that the LOCSET control loop can isolate the phase error signal for all modulated beams (j and k), while remaining short enough to effectively cancel the phase disturbances of the system:

$$\tau \gg \frac{2\pi}{|\omega_j - \omega_k|}. \tag{2.12}$$

Similar to the photocurrent signal, the error signal can be divided into three components due to the interaction of the unmodulated beam, the unmodulated beam and all other phase-modulated elements, and interference of each phase-modulated beam with all other modulated beams or

$$S_x = S_u + S_{x_u} + S_{xj}. \tag{2.13}$$

Hence, to resolve the self-reference LOCSET phase error signal, Eqs. (2.7), (2.9), and (2.10) can be substituted into Eqs. (2.11) and (2.13). The first term S_u, due to the presence of the unmodulated beam, is zero since there are no interference terms in Eq. (2.7):

$$S_u = \frac{R_{PD}\cdot P_u}{2\tau}\int_0^\tau \sin(\omega_x\cdot t)\mathrm{d}t = \frac{R_{PD}\cdot P_u}{2}\left(\frac{1}{\omega_x\cdot\tau} - \frac{\cos\omega_x}{\omega_x\cdot\tau}\right) \approx 0. \tag{2.14}$$

Because there are no time-varying contributions solely due to the presence of the unmodulated beam, this is expected. Next, by substituting Eq. (2.9) into Eq. (2.11), the second term due to interference of the unmodulated beam with the xth phase-modulated beam can be expressed as

$$S_{x_u} = R_{PD} \cdot P_u^{1/2} \cdot P_x^{1/2} \cdot J_1(\beta_x) \cdot \sin(\phi_u - \phi_x), \tag{2.15}$$

where the remaining sinusoidal phase difference terms can be neglected due to the aforementioned long integration time τ (these integrals converge to zero) [27]. In addition, Bessel functions beyond the second order were neglected as they evaluate near zero for the small modulation depths (β) imposed in LOCSET [27].

Subsequently, the last signal term is derived from the interference of the xth modulated beam with all other phase-modulated beams ($j \neq x$), S_{xj}. Equation (2.10) is inserted into Eq. (2.11) with Bessel functions beyond the second order being neglected. Unfortunately, this yields an unwieldy solution due to the matrix of optical beam interactions involved. A complete analysis of this error signal derivation can be found in the literature where the following expression was obtained [27]:

$$S_{xj} = R_{PD} \cdot P_x^{1/2} \cdot J_1(\beta_x) \cdot \frac{1}{2} \sum_{\substack{j=1 \\ j \neq x}}^{N-1} P_j^{1/2} J_0(\beta_j) \sin(\phi_j - \phi_x). \tag{2.16}$$

Here, we note that S_{xj} adds to the overall robustness of the LOCSET system. Differing from schemes where the phase error signal is governed by the independent interaction of each phase-controlled beam with a common reference beam, LOCSET adds a measurement signal proportional to the sum of the phase difference of each beam with respect to all other beams in the system. Consequently, if the reference beam is lost, the LOCSET system continues to phase lock the remaining beams with graceful degradation. After deriving the individual phase error terms, a complete representation of the self-referenced LOCSET phase error signal can be determined according to Eqs. (2.13)–(2.15) [27]:

$$S_{SR_x} = R_{PD} \cdot P_x^{1/2} \cdot J_1(\beta_x) \left(P_u^{1/2} \sin(\phi_u - \phi_x) + \frac{1}{2} \sum_{\substack{j=1 \\ j \neq x}}^{N-1} P_j^{1/2} J_0(\beta_j) \sin(\phi_j - \phi_x) \right). \tag{2.17}$$

Fittingly, the phase error signal changes with slow variations in the optical phases of the combined beams. Due to the sinusoidal terms as long as the phase difference between each combined beam is zero, S_{SR_x} is also zero. If the xth beam drifts out of phase with the rest of the system, the phase error signal will be nonzero and carry a sign ($\pm$) indicating the phase drift direction. An error correction signal is then applied to the xth beam (through external phase modulators) to minimize S_{SR_x} and return the system to optimal phase locking. Likewise, independent LOCSET control loops for each additional phase-modulated element ensures constant phase locking for all laser array elements.

2.2.3
Self-Synchronous LOCSET

It is important to note that an unmodulated reference beam is not necessary for LOCSET operation. Referring back to self-referenced LOCSET and Eq. (2.17), only the first term within the parenthesis is influenced by the unmodulated reference beam $(P_u^{1/2})$. Thus, by setting the unmodulated reference beam to zero ($P_u = 0$), we obtain the following phase error expression for self-synchronous LOCSET:

$$S_{SS_x} = R_{PD} \cdot P_x^{1/2} \cdot J_1(\beta_x) \frac{1}{2} \sum_{\substack{j=1 \\ j \neq x}}^{N-1} P_j^{1/2} J_0(\beta_j) \sin(\phi_j - \phi_x). \quad (2.18)$$

Appreciably, LOCSET is capable of operating without a reference beam. Because it measures the relative phase error of a single beam with respect to every other beam in the system, the phase information of a given beam is known with respect to all others. Thus, a reference beam is no longer required. Similarly, the independent LOCSET control loops ensure persistent phase locking through minimization of the individual error signals S_{SS_x}.

2.3
LOCSET Phase Error and Channel Scalability

Scalability to high channel counts and excellent phase error performance are critical parameters for any active CBC system. Toward this end, LOCSET has exhibited outstanding phase error performance with little degradation in residual phase error as we scale to higher channel counts. Such performance is vital for CBC, where a phase error-dependent efficiency degradation η_ϕ is inherent [34]:

$$\eta_\phi \approx 1 - \Delta\bar{\phi}_{rms}^2. \quad (2.19)$$

For example, a residual phase error $\Delta\phi_{rms}$ of $\lambda/15$ results in a prohibitive drop in efficiency (18%), while a $\lambda/60$ phase error contributes a negligible loss in efficiency (<1%). Hence, to characterize LOCSET's beam combining performance, detailed low-power multibeam CBC and phase error analysis of 2, 16, and 32 lasers was performed [27].

2.3.1
LOCSET Beam Combining and Phase Error Analysis

The general LOCSET experimental arrangement for CBC with diagnostic phase error analysis [in-phase (I) and quadrature (Q) processing] [9,27,35] is depicted in Figure 2.3. The setup is based on a MOPA arrangement with light from the master oscillator split into $N+1$ channels with N (or $N-1$) phase-modulated laser beams. The remaining beam (top of Figure 2.3) is coupled to an acousto-optic modulator (AOM)

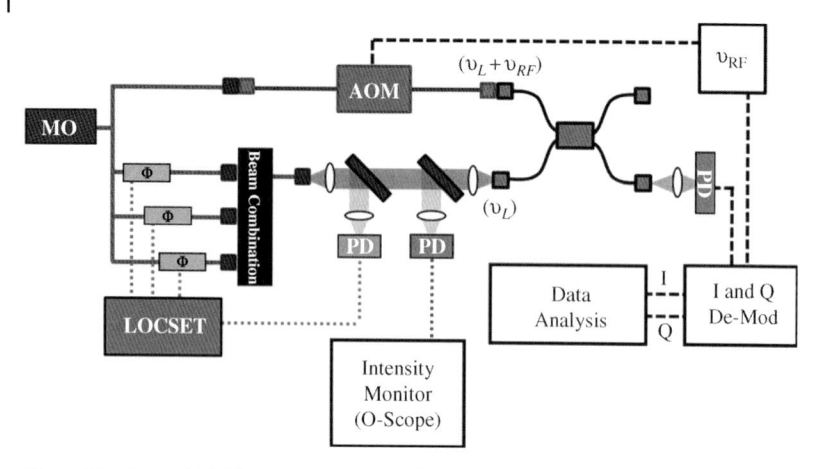

Figure 2.3 General LOCSET CBC experimental arrangement with in-phase (*I*) and quadrature (*Q*) demodulation for phase error analysis.

where the light undergoes a fixed frequency shift ($\nu_{RF} = 80$ MHz). Here, the frequency-shifted beam is used as a phase-stable reference for the phase error measurements. The combined output beam is then sampled with a 10% wedge, where the sampled light incident on a photodetector provides feedback to the LOCSET control electronics. After passing through another beam sampler (for intensity monitoring), the remaining light is coupled to a 2×2 fiber splitter/combiner where it is interfered with the frequency-shifted reference beam.

The combined output, a beat note caused by the different frequencies and phases of the beams, is incident on another photodetector that feeds into the *I* and *Q* data processing electronics. *I* and *Q* signals are then generated via coherent RF demodulation and measured as a function of time. Moreover, before initiating beam combining experiments, the system background phase behavior due to external disturbances was quantified. By choosing a data sampling window less than or equal to 1 ms, the behavior of the frequency-shifted reference beam was evaluated to be stable to within $\leq \lambda/450$, when measured relative to any beam in the CBC system. Thus, establishing our phase error measurement resolution at $\sim \lambda/450$ (~ 0.014 rad) that is significantly greater than the phase error tolerances needed to analyze LOCSET's phase error performance.

2.3.2
In-Phase and Quadrature-Phase Error Analysis

The *I* and *Q* phase error analysis is based on the optical phase difference $\Delta \phi$ between a phase-stable reference laser and the combined laser beam. Referring to Figure 2.3, the AC photocurrent generated by the interfering lasers can be expressed as

$$i_{AC}(t) = \chi \cdot \cos(\Delta \omega \cdot t + \Delta \phi), \tag{2.20}$$

where $\chi = R_{PD} \cdot A \cdot 2 \cdot Int_0$ is a constant term related to the optical intensity (Int_0), detector responsivity, and active area. Likewise, $\Delta\omega$ is proportional to the difference in frequency between the combined output beam and the frequency-shifted reference beam. Therefore, after trigonometric expansion, the I and Q phase components can be described as

$$i_{AC}(t) = I(\Delta\phi)\cos(\Delta\omega \cdot t) - Q(\Delta\phi)\sin(\Delta\omega \cdot t), \tag{2.21}$$

where

$$I(\Delta\phi) = \chi\cos(\Delta\phi), \tag{2.22}$$

$$Q(\Delta\phi) = \chi\sin(\Delta\phi). \tag{2.23}$$

Subsequently, the goal is to extract $I(\Delta\phi)$ and $Q(\Delta\phi)$ from Eq. (2.21) via coherent demodulation. We note that $I(\Delta\phi)$ and $Q(\Delta\phi)$ are the Fourier cosine and sine coefficients of $i_{AC}(t)$ at frequency $\Delta\omega$, respectively. These terms can be isolated by mixing $i_{AC}(t)$ with a sine or cosine demodulation signal at frequency $\Delta\omega$ and integrating over time T. The Fourier cosine component a_c can be expressed as

$$a_c = \frac{1}{T}\int_0^T i_{AC}(t)\cos(\Delta\omega \cdot t)dt, \tag{2.24}$$

where after substituting Eq. (2.21) into Eq. (2.24) and accounting for long integration times yields

$$a_c = \frac{I(\Delta\phi)}{T}\int_0^T \cos^2(\Delta\omega \cdot t)d \approx \frac{I(\Delta\phi)}{2}. \tag{2.25}$$

Similarly, the Fourier sine component a_s can be extracted and represented as

$$a_s = -\frac{Q(\Delta\phi)}{2}. \tag{2.26}$$

Equations (2.25) and (2.26) provide us with measurable quantities that are proportional to the optical phase difference $\Delta\phi$ between the reference field E_R and the field of interest E_i. Relating the ratio of a_s and a_c with Eqs. (2.23) and (2.24), we get an expression proportional to the tangent of the optical phase difference:

$$\Delta\phi = \tan^{-1}\left(-\frac{a_s}{a_c}\right). \tag{2.27}$$

Therefore, by measuring $I(\Delta\phi)$ and $Q(\Delta\phi)$, or more accurately a_s and a_c, we can extract the optical phase difference as a slowly varying function of time. We note that due to asymptotic nature of the arctangent term, additional I and Q phase data unwrapping and processing are required [27]. Nevertheless, in-phase and

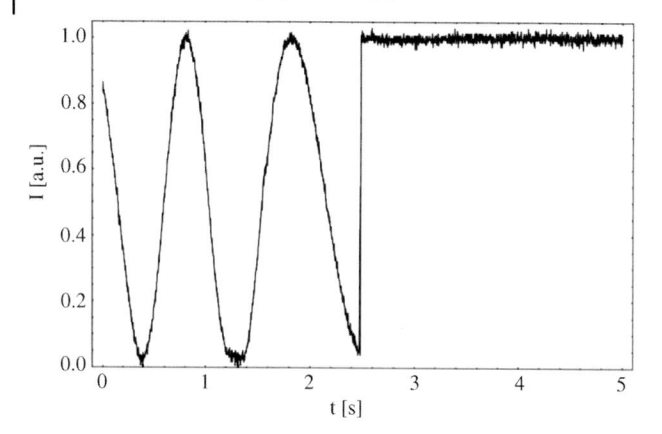

Figure 2.4 Time-varying intensity of the two-channel LOCSET CBC system. For 2.5 s of data acquisition, LOCSET electronics remain off [27].

quadrature-phase analysis provides an adequate tool for LOCSET beam combining performance evaluation.

2.3.3
Two-Channel Beam Combining

To highlight LOCSET's channel scalability and superior phase error performance, multichannel LOCSET beam combining of 2, 16, and 32 channels was performed. The initial two-channel experiment follows the general arrangement in Figure 2.3, with two low-power beams (~2 mW) in the self-referenced LOCSET configuration. The beams were combined with a 2 × 2 fiber splitter (filled aperture scheme) and monitored for both intensity and phase behavior performance.

The subsequent intensity behavior of our two-channel CBC system is shown in Figure 2.4. For the initial 2.5 s of data acquisition, the LOCSET electronics are off allowing the intensity to drift due to environmental disturbances. Then the LOCSET electronics were turned on to establish coherent combination of the two beams. While satisfactory beam combining is observed from Figure 2.4, it is difficult to extrapolate the phase behavior from the intensity data. For small phase errors ($<\lambda/25$), the phase behavior is lost in the measurement noise and the general interference expression used to extrapolate phase error may no longer be valid [$\text{Int}(\Delta\phi)/\text{Int}_0 = 2(1 + \cos(\Delta\phi))$]. As a result, in-phase and quadrature-phase analysis is performed.

We note that in addition to the phase noise of the combined beam, LOCSET control electronics contribute additional phase error that cannot be measured by the I and Q system. This is due to the RF phase dithers, with modulation depths β_i, that are applied to each beam. In the current case, a 100 MHz RF phase dither with modulation depth of 0.094 rad ($\lambda/67$) is applied to the phase modulator. Therefore, to obtain the root-mean-square (RMS) phase error of the entire CBC system, we must determine the phase error of the RF sinusoidal dither of each beam ($\Delta\phi_{\text{RF}}$) and combine it with extracted I and Q phase error ($\Delta\phi$). The RMS phase error contributed by the sinusoidal phase dither can be quantified as

$$\Delta\phi_{RF} = \frac{\sqrt{2}}{2}\beta_{P-P}, \tag{2.28}$$

where β_{P-P} is the peak-to-peak amplitude of the time-varying sinusoidal signal. Hence, for a phase modulation depth of 0.094 rad ($\lambda/67$), an RMS RF phase error of 0.067 rad is imparted. Furthermore, since the sinusoidal RF phase modulation is well defined and the phase errors of the coherently combined beam vary arbitrarily (due to environmental disturbances), the two signals are effectively uncorrelated. Because the two signals are uncorrelated, we can calculate the total RMS phase error signal as [27]

$$\Delta\phi_{RMS} = \sqrt{\Delta\phi^2 + \Delta\phi_{RF}^2}. \tag{2.29}$$

The ensuing average RMS phase error for two-channel beam combination was measured as $\Delta\phi_{RMS} \approx \lambda/66$ (0.095 rad), as shown in Figure 2.5 for multiple data sets. Notably, such phase error performance results in less than 1% drop in coherent combination efficiency. Though the data shown in Figure 2.5 was limited to a 5 min data collection period, such beam combination performance and stability were observed for hours at a time repeatedly.

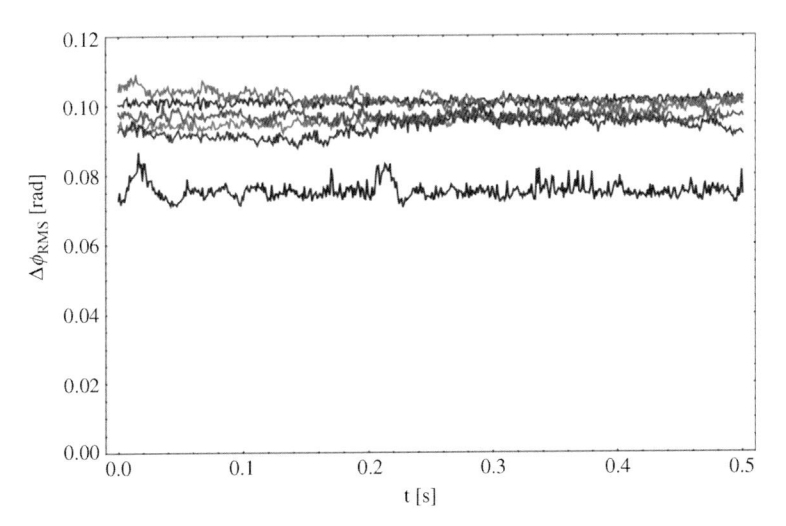

Figure 2.5 RMS phase error as a function of time for the two-channel LOCSET CBC system. Multiple data sets, taken during a single run of the two-channel system, were included to demonstrate consistency in beam combination performance. Average observed RMS phase error: $\sim\lambda/66$ [27].

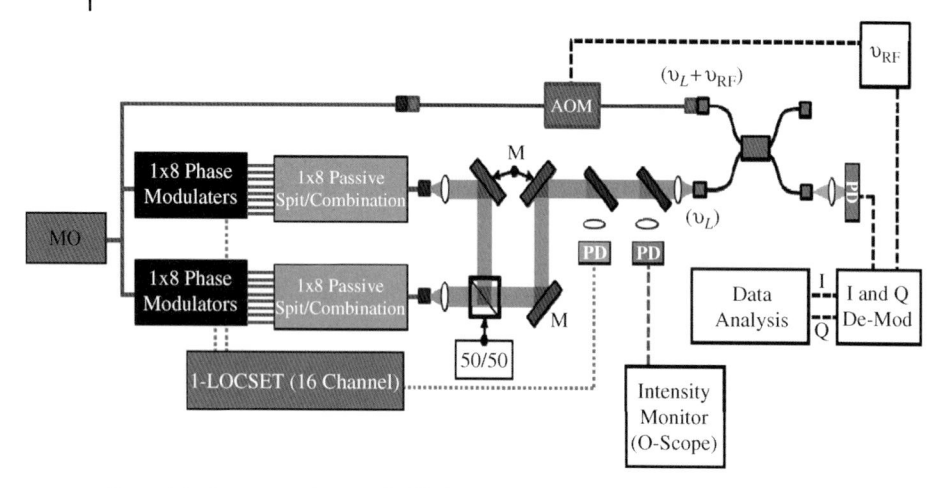

Figure 2.6 Experimental setup of a filled aperture 16-channel LOCSET CBC system. M, mirror. 50/50: 50% reflective/transmissive beam splitter.

2.3.4
16-Channel Beam Combining

The second LOCSET beam combination experiment combined 16 low-power beams in the self-referenced LOCSET configuration. The experimental setup, shown in Figure 2.6, begins with a single master oscillator (MO) split into three individual fiber channels. Two of the fiber channels are further cascaded into 16 individual fiber channels for coherent combination. These fiber channels are coupled into two 1×8 LiNbO$_3$ phase modulators, each converting a single input beam into eight phase controllable optical channels for a total of 16 beams. The outputs of each 1×8 LiNbO$_3$ module are then recombined via a passive 1×8 LiNbO3 fiber splitter/combiner, combining eight beams into one. The ensuing two beams, each consisting of eight individual phase-modulated beams, propagate in free space and combine at the interface of a 50/50 beam splitter in a filled aperture format. The remaining beam, as before, is frequency shifted via an AOM and used as a reference beam. Then, time-varying intensity measurements of the final combined beam were recorded with a fast photodiode, as shown in Figure 2.7.

Figure 2.7 illustrates the intensity disparity of the 16-channel LOCSET system when the system is locked and unlocked. When the LOCSET detector is blocked, the intensity of the combined output beam fluctuates due to the oscillating phase behavior of the 15 modulated beams. However, when the LOCSET detector is active, the intensity is stabilized at peak intensity (optimal beam combination). Subsequently, in-phase and quadrature-phase error analysis was performed to quantify the RMS phase behavior. The ensuing data shown in Figure 2.8 represent the RMS phase error as a function of time for the 16-channel LOCSET system. Here, we

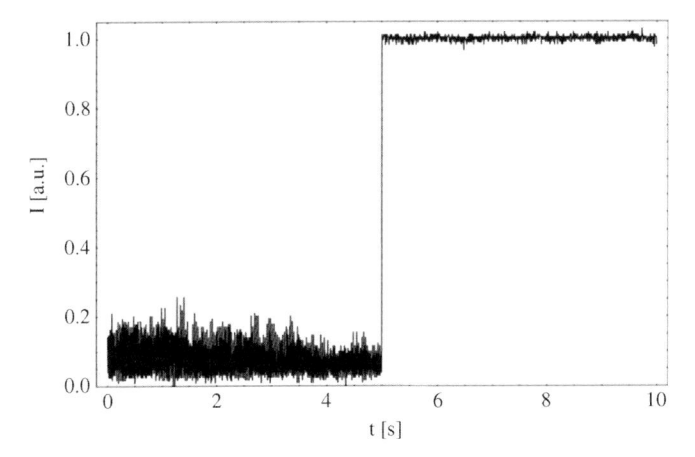

Figure 2.7 Time-varying intensity of the 16-channel LOCSET CBC system. LOCSET electronics remain on for the entire data acquisition time. LOCSET photodetector blocked for the first 5 s of the data acquisition. After 5 s, the LOCSET detector was unblocked allowing the system to phase lock [27].

observe rapid phase fluctuations due to the 2π phase reset voltages. In order to protect the phase modulators from voltage overloading, the phase is unwrapped continuously at $2N\pi$ intervals ($N=1$ for the current experiments). Therefore, 15 phase modulators randomly reset or unwrap according to the operating environment and thermal/vibrational disturbances. Nevertheless, the average RMS phase error performance of the 16-channel LOCSET CBC experiment was excellent – $\lambda/62$ (0.1 rad).

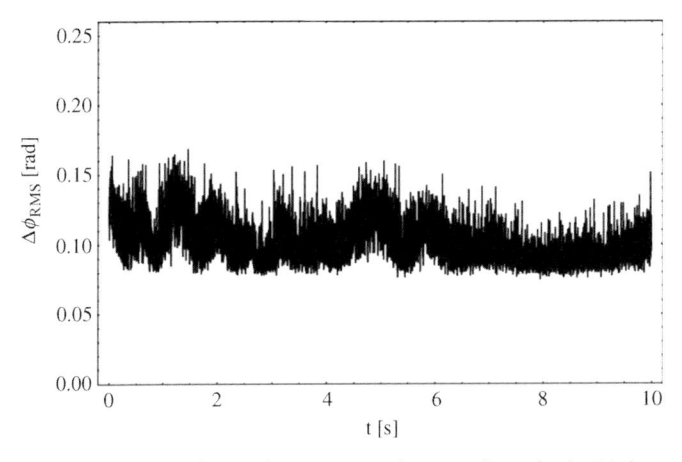

Figure 2.8 Measured RMS phase error as a function of time for the 16-channel LOCSET CBC experiment. RMS values calculated over a time period of 1 ms. Average observed RMS phase error for 16-channel system: $\sim\lambda/62$ (0.1 rad) [27].

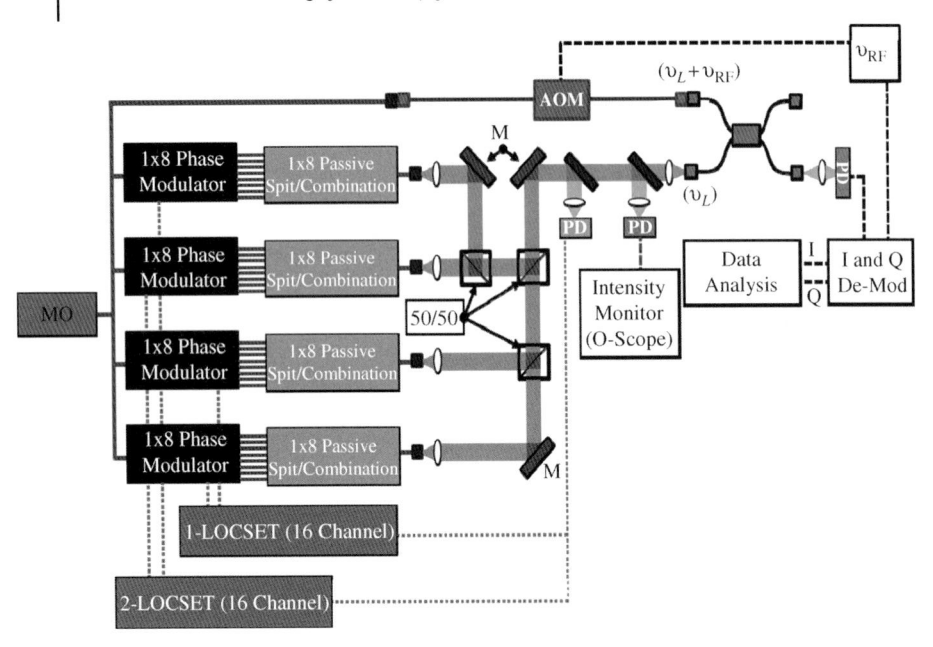

Figure 2.9 Experimental setup of a 32-channel LOCSET CBC system [27].

2.3.5
32-Channel Beam Combining

The third low-power beam combining experiment, shown in Figure 2.9, was a 32-channel self-referenced LOCSET demonstration. The experimental arrangement remains similar to the 16-channel setup, except that the MO is split into five ways with four of the channels further cascaded into 32 channels using 1×8 LiNbO3 phase modulators. The fifth channel is used as a frequency-shifted reference beam. The outputs of the LiNbO3 modules are then recombined via passive LiNbO3 fiber splitters. The ensuing four combined beams, each consisting of eight phase-modulated beams, propagate in free space and combine via a binary splitter tree (filled aperture configuration). The binary tree consists of three 50/50 beam splitters and combines the four free space beams into a single coherent beam. The final beam is then processed via the same optical setup providing the LOCSET error signal and beam combination performance measures.

The time-varying intensity of the 32-channel LOCSET beam combining system is shown in Figure 2.10. Once again successful coherent combination is observed with stable peak intensity. The I and Q phase error data shown in Figure 2.11 represent the RMS phase error of the 32-channel combined beam as a function of time. Yet again, the presence of random 2π phase resets in each of the 31 phase-controlled beams contributes to the fluctuating phase errors. Despite these fluctuations, the average RMS phase error of the 32-channel coherently combined beam was approximately $\lambda/71$ (∼0.09 rad). A promising result, but more importantly no phase

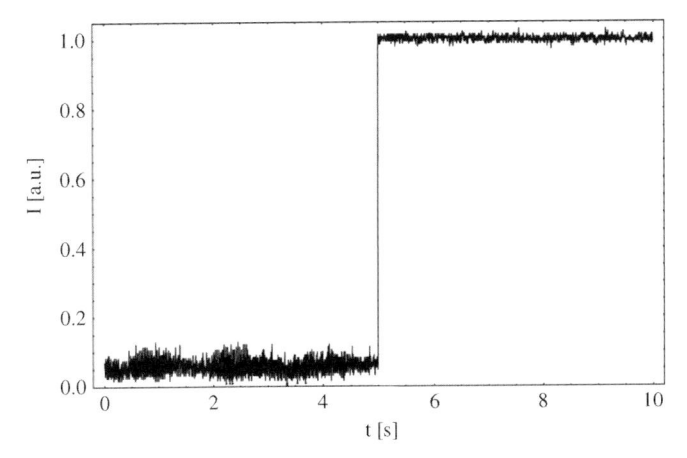

Figure 2.10 Time-varying intensity of the 32-channel LOCSET CBC system [27].

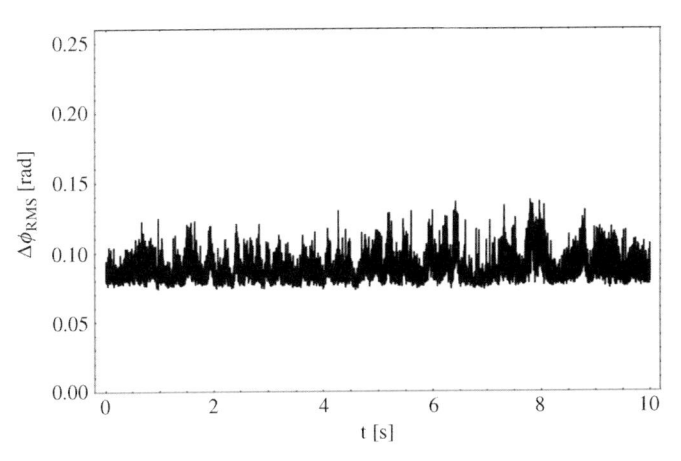

Figure 2.11 RMS phase error as a function of time for the 32-channel LOCSET system. RMS values calculated over a time period of 1 ms with an average RMS phase error of $\lambda/71$ (~0.09 rad) [27].

error degradation, was detected when scaling from 2 to 32 channels, as expected from previous simulations [21]. Thus, LOCSET appears readily scalable to more than 100 elements as the RMS phase error appears to be independent of the number of array elements in the system.

2.4
LOCSET High-Power Beam Combining

The LOCSET technique is a well-established and coherent combination of 32 lasers at low power has been demonstrated. More importantly, the ability of LOCSET to potentially scale to hundreds of elements using a single photodetector is promising.

Using electronic feedback to control the phases of the individual fiber amplifiers, we achieved an average RMS phase stability of $\lambda/71$. In addition, LOCSET phase locking has been utilized by other researchers for beam combining applications, such as polarization beam combining [36], two-dimensional waveguide beam combining [37], remote target phase locking [38], and DOE beam combining (1D and 2D) [39,40]. Recently, to further demonstrate combining at higher powers, coherent combination of fiber amplifiers via LOCSET was extended into the kilowatt regime.

2.4.1
Kilowatt-Scale Coherent Beam Combining of Silica Fiber Lasers

Coherent combination of 16 100 W single-frequency fiber amplifiers was investigated via LOCSET. Based on a standard MOPA configuration, a single-frequency nonplanar ring oscillator (NPRO) was used to seed 16 polarization-maintaining (PM) fiber amplifiers. Each amplifier chain consisted of three fiber amplifier stages and produced 100 W of near-diffraction-limited ($M^2 \sim 1.1$–1.2) output power. A complete block diagram of the monolithic fiber amplifier chain for one element is shown in Figure 2.12a. In addition, a schematic of the Nufern copumped 100 W main amplifier is presented in Figure 2.12b. Here, the power amplifier is seeded by the intermediate amplifier with 8–10 W of 1064 nm light. Next, six 50 W (976 nm) fiber-coupled diode pump lasers (LIMO) were fusion spliced onto a $6 \times 1 \times 1$ pump combiner. The ensuing output of the combiner is spliced onto a 5 m long double-clad PM ytterbium (Yb)-doped silica gain fiber, where the gain fiber is cladding pumped with the 976 nm light. While the preamplifier and intermediate gain stages are based on 6 (core)/125 μm (clad) and 10/125 μm diameter gain fibers, respectively, the power amplifier stage uses a 25/400 μm Nufern LMA fiber for SBS suppression.

To further mitigate SBS, a two-stage thermal gradient was applied to the power amplifier, as shown in Figure 2.12b. SBS suppression can be achieved through utilization of the steep temperature gradient at the fiber output end [41] or application of an external thermal gradient, which applies a Brillouin frequency shift. Here the gain fiber is divided equally into two separate spools, a cold spool (17 °C) and a hot spool (80 °C), and held at constant temperature via thermoelectric coolers. A plot of the main amplifier backward power reflectivity versus signal output power with (dotted) and without (dashed) an applied thermal gradient is presented in Figure 2.12c. As expected, the thermal gradient provides a nearly two times enhancement in SBS threshold and allow us to reach 100 W of single-frequency power. Powers in excess of 100 W can be attained through shortening the fiber length, although this is limited by increased unabsorbed pump powers and lower optical-to-optical efficiencies. Finally, to maintain the diffraction-limited beam quality, the fiber was coiled to suppress higher order mode content [42]. A cladding mode stripper was also implemented to remove any unabsorbed (stray) pump cladding light. It is important to note that although 300 W of pump power is available, only 100 W of single-frequency power is achieved due to SBS limitations. Nevertheless, we can mitigate SBS by broadening the pump spectral linewidth. Hence, we have recently utilized sinusoidal phase modulation to broaden the laser linewidth [43] and have attained pump-limited powers of >200 W at narrow linewidths (200 MHz).

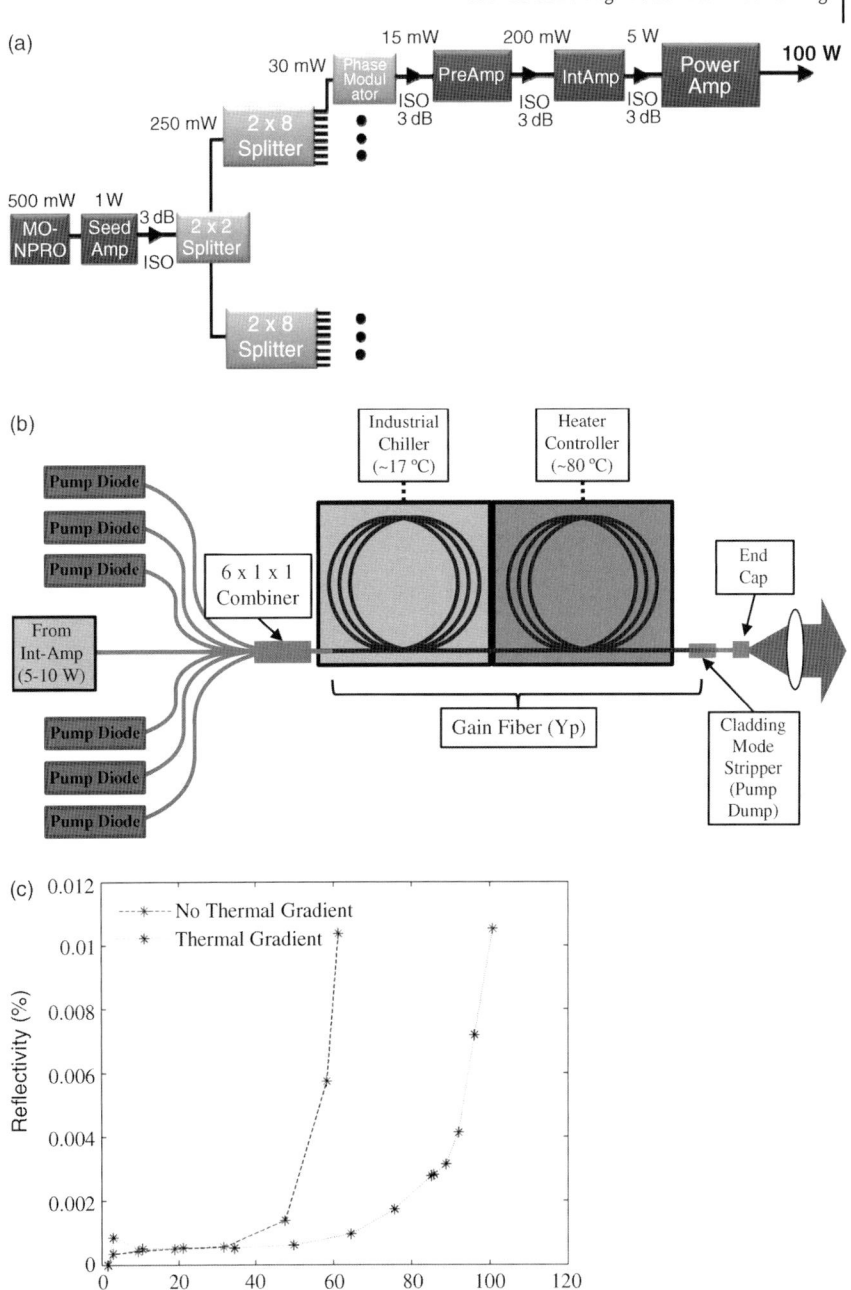

Figure 2.12 (a) Block diagram of the monolithic fiber amplifier chain. (b) Schematic of a copumped 100 W fiber amplifier built by NuFern; SBS suppression accomplished via the introduction of a thermal gradient in the gain fiber with hot and cold fiber spools. (c) Experimental reflectivity (%) versus signal power for main amplifier with (dotted) and without (dashed) applied thermal gradient. Plot shows power enhancement via thermal gradient.

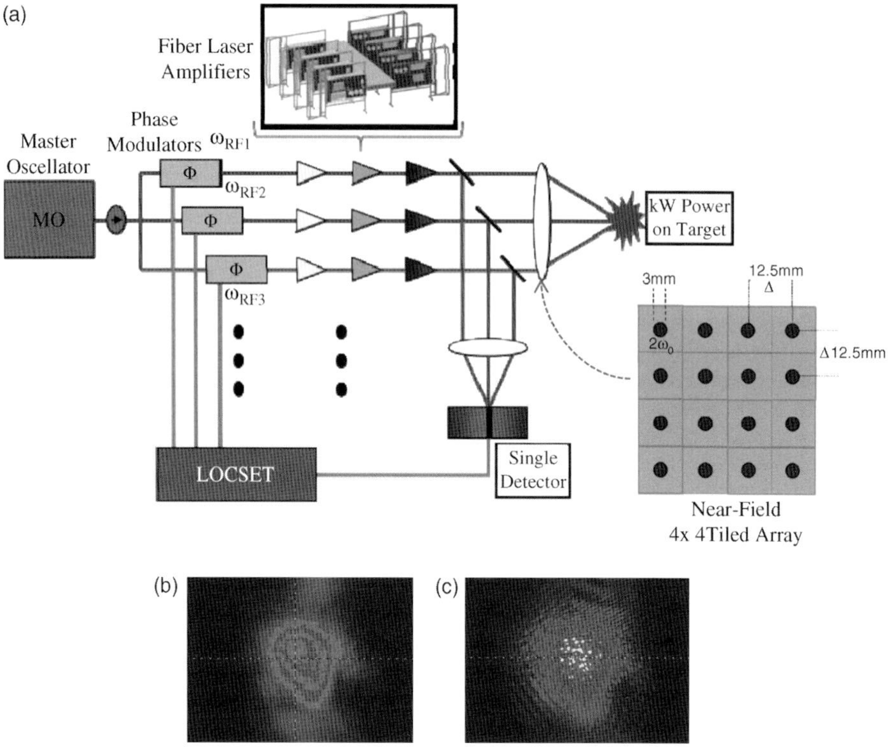

Figure 2.13 (a) Experimental arrangement for LOCSET CBC of 16 100 W lasers in a tiled 4 × 4 laser array ($2\omega_0/\Delta = 0.24$). We note that the fill factor is limited by the size of the turning prisms, causing substantial power in the side lobes. Resulting (b) unlocked and (c) phase-locked beam profiles.

A fiber end cap was then added to the output fiber before diverging onto a collimating lens. After collimation (3 mm beam diameter), 16 output beams were directed onto external high-power isolators for optical return protection. However, due to moderate thermal lensing and astigmatic aberrations introduced by the isolators at high powers, the beam quality of the output fiber laser beams were slightly impaired (M^2 1.2–1.3). A general schematic of the MOPA arrangement for the kilowatt scale beam combination is shown in Figure 2.13a. Sixteen lasers were configured into a tiled 4 × 4 laser array where each beam was directed onto a far-field focusing lens by turning prisms. A self-referenced LOCSET scheme with 15 phase-modulated beams and a single unmodulated reference beam was arranged. Subsequently, 16 beams were combined in the far field where a 0.1% sampling wedge was used to sample the beam and direct it into a phase locking and beam diagnostic subsystem. Here, the combined beam overlaps (interferes) onto the LOCSET photodetector and is further sampled into an imaging system. A reimaging (magnification) system is used to enlarge the focal (combining) plane and optimize spatial overlap among all 16 beams. Finally, a fast photodiode is used to monitor the output intensity and estimate the RMS phase error.

A total combined output power of 1.45 kW from 16 lasers was achieved with a residual phase fluctuation of $\lambda/25$. The resulting phase-locked beam profiles are shown in Figure 2.13b and c, respectively. Unfortunately, the fill factor is limited by the size of our turning prisms ($D = 12.5$ mm) and a low fill factor was attained ($2\omega_0/\Delta = 0.24$). The ensuing low array fill factor results in substantial power in the side lobes. This can be remedied in future experiments by inserting longer focal length collimating lenses and expanding the beams for optimal subaperture fill factors of $2\omega_0/D \sim 0.89$ [23]. Nevertheless, the phase-locked beam profile exhibits a stable interference fringe pattern with amplified intensity, thereby confirming single-frequency kilowatt-scale CBC of 16 fiber lasers.

2.4.2
Kilowatt-Scale Coherent Beam Combining of Photonic Crystal Fiber Amplifiers

Notably, kilowatt-scale beam combining was also attained with PCF amplifiers. In PCF amplifiers, micrometer-sized air holes in the cladding allow precise control of the refractive index leading to larger core diameters while maintaining single-mode operation. Another significant advantage of double-clad PCF amplifiers is the high numerical apertures that can be attained for pumping purposes through the utilization of a web of subwavelength silica bridges. Recently, novel SBS-suppressive PCF amplifiers with 494 W of single-frequency [44] and 994 W of narrow-linewidth (modulated at 300 MHz) [45] output power have been demonstrated. The PCF design is based on a segmented acoustic profile that is doped such that the core segments are optically uniform but acoustically inhomogeneous [44]. The acoustically modified PCF amplifier results in multiple Brillouin gain spectrum (BGS) peaks that help suppress SBS in single-frequency and narrow-linewidth fiber lasers.

The acoustically manipulated fiber core, illustrated in Figure 2.14 a, was designed to give two distinct Brillouin peaks. Here, the Brillouin shift in the center region is comprised of one hexagon with different acoustic velocity v_1 than the acoustic velocity of the six outer regions v_2. In particular, a combination of dopants comprised of fluorine, aluminum, and germanium was used to achieve the segmentation of the acoustic index of refraction while maintaining uniformity of the optical index. To accommodate further SBS suppression through the application of externally applied or optically induced thermal gradients, the peaks were designed to have a separation of >200 MHz. Since the Brillouin shift is approximately 2 MHz/°C [46], this design would allow introduction of a temperature variation of $\sim$100 °C without overlap in the Brillouin gain bandwidth [44].

A fiber based on the segmented acoustic design with a core diameter of 40 μm and mode field diameter (MFD) of 30 μm was fabricated by NKT Photonics. The inner cladding of the fiber was 300 μm with a nominal numerical aperture of 0.55–0.6. Furthermore, the pump absorption at 976 nm was estimated at 4 dB/m. Next, the BGS was investigated using a pump–probe technique. The resulting Brillouin spectrum, shown in Figure 2.14b, displays at a shift of approximately 16 GHz

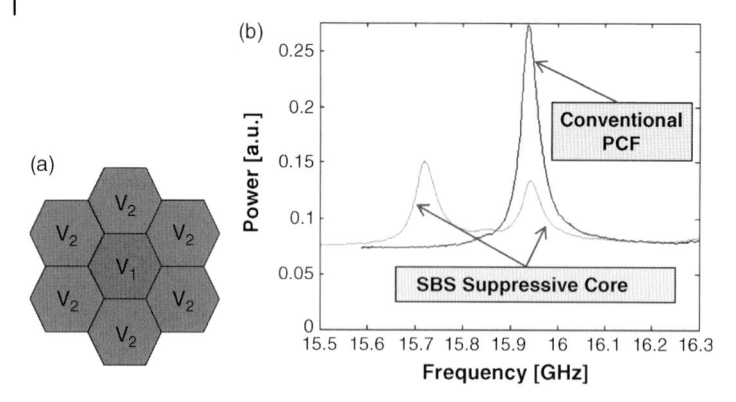

Figure 2.14 (a) The core design of the segmented acoustic fiber. The Brillouin shift in the center region comprised of one hexagon is different from that in the outer region comprised of six hexagons. (b) BGS for acoustically segmented fiber confirming existence of two primary peaks. For comparison, the BGS for the reference fiber is shown [44].

two peaks with a separation of 220 MHz as expected, thereby confirming the SBS suppression provided by the segmented acoustic PCF. Also shown is the BGS of a reference PCF, for the same pump and probe values. The reference fiber is a conventional PCF with identical core and cladding dimensions but no acoustic tailoring.

Subsequently, a high-power fiber amplifier was built around the segmented PCF amplifier. An experimental setup of the PCF amplifier arrangement with a counter-propagating pump scheme is depicted in Figure 2.15. Once again an NPRO was used as the master oscillator with the seed being amplified to 30 W using a three-stage amplifier system. This preamplifier was then free space coupled into the core of the PCF amplifier with a 10 m long gain fiber. The pump power was provided by stacks of 976 nm Laserline diodes with a maximum output of 1.5 kW. Consequently, single-frequency output powers of up to 500 W were recorded without the onset of SBS [44]. Particularly, we were prevented from fully investigating the SBS suppressing characteristic of this amplifier by the sudden onset of modal instabilities [7]. Nonetheless, we have recently designed a PCF for concurrent SBS and modal instability suppression. The design utilizing simultaneous gain and acoustic tailoring led to the development of a 994 W narrow-linewidth (modulated at 300 MHz), near-diffraction-limited ($M^2 < 1.3$) PCF amplifier [45].

Moreover, we analyzed the beam quality of the SBS suppressive amplifier. We conducted measurements using a Spiricon beam analyzer and M^2 values of less than 1.3 were obtained at all power levels. Based on this, it can be inferred that the development of an optical interface between segmented regions was minimal, if any. For the current experiments, three ~400 W segmented acoustic fiber lasers were built and arranged in a filled aperture configuration. Next, the beams were coherently combined through beam splitters and LOCSET phase electronics, as shown in Figure 2.16. Ultimately, the three-element beam combination experiments

2.4 LOCSET High-Power Beam Combining

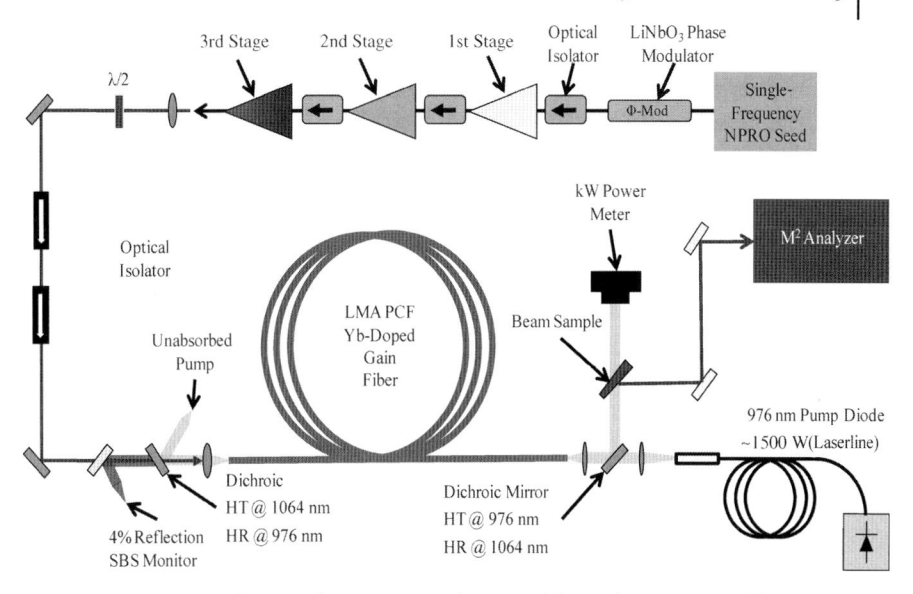

Figure 2.15 Experimental setup of counter-pumped PCF amplifier. A three-stage amplifier system was utilized to provide approximately 30 W of seed power. Preamplifier seed and pump power were free space coupled to PCF amplifier.

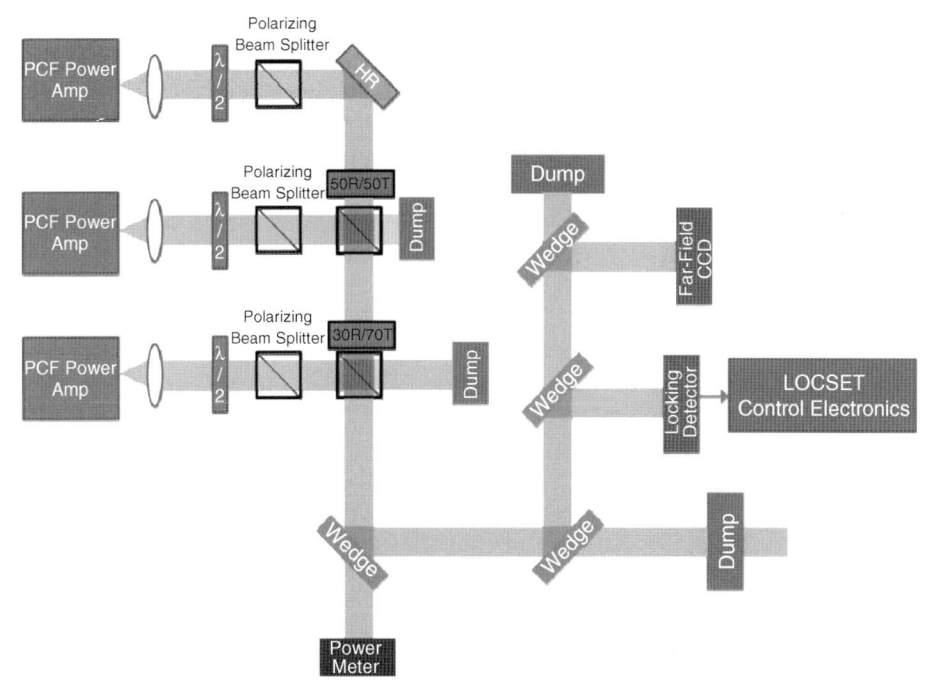

Figure 2.16 Single-frequency combination of three 400 W PCF amplifiers in a filled aperture CBC arrangement. LOCSET phase locking is used to coherently combine the lasers and produce 1 kW of power.

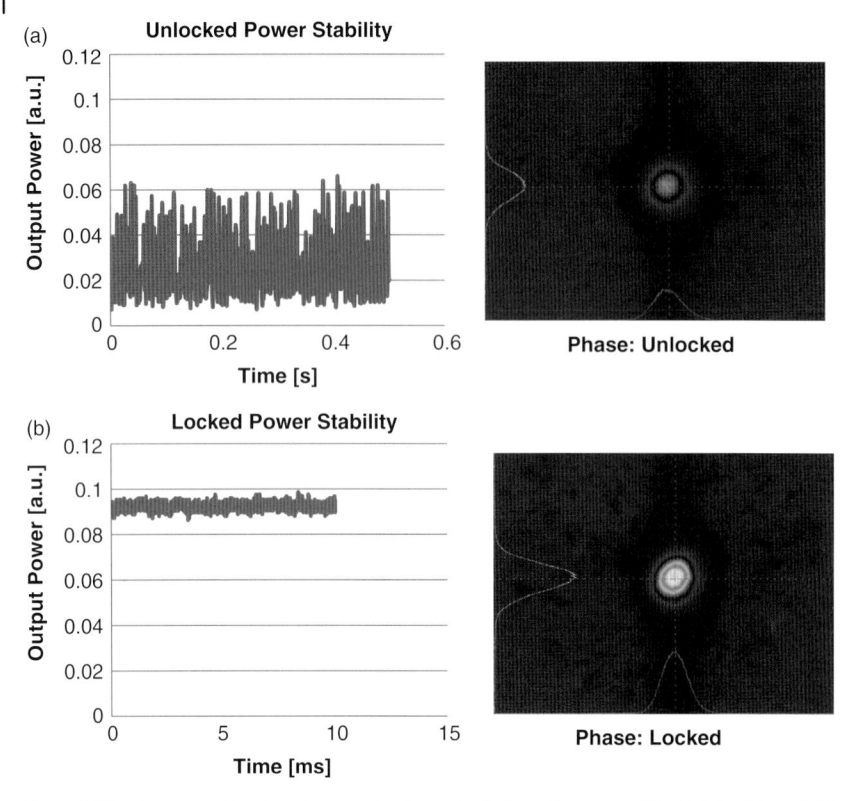

Figure 2.17 Beam combination time-varying intensity (left) and beam profile (right) of three PCF amplifiers (a) unlocked and (b) phase-locked via LOCSET.

resulted in 1.04 kW of combined power with a residual phase fluctuation of $\lambda/18$. Intensity and beam profile measurements are shown in Figure 2.17a and b, respectively. Notably, the filled aperture arrangement results in 1 kW of power in a single central lobe, with near-diffraction-limited beam quality. Hence, single-frequency kilowatt beam combining via LOCSET has been demonstrated with both conventional silica and novel SBS-suppressive PCF fiber lasers.

High-power single-frequency laser sources with good beam quality are highly desired due to their utility in CBC, gravitational wave detection [47], and nonlinear frequency conversion [48]. In terms of CBC, single-frequency beam combining circumvents path length matching techniques required to combine narrow-linewidth high-power fiber lasers. Here, single-frequency lasers denote lasers with linewidths smaller than the Brillouin linewidth (~60 MHz) and narrow-linewidth fiber lasers refer to lasers with gigahertz wide spectral widths. Regardless, LOCSET beam combining should effortlessly extend to narrow-linewidth beam combining as long as path length matching tolerances are met. Hence, we have recently demonstrated LOCSET beam combining at narrow linewidths with sinusoidal phase modulation [43].

2.5
Conclusion

Fiber lasers have superior beam quality, size, weight, and efficiency advantages over conventional solid-state and chemical lasers. Although constrained by power scaling limits of individual fibers, beam combination of multiple fiber lasers can overcome such drawbacks. Toward this end, LOCSET is an established phase locking technique that has been used for coherent fiber laser beam combination. Notably, the recent high-power experiments in both tiled (conventional silica fiber) and filled aperture (PCF) arrangements establish LOCSET's viability at kilowatt power levels, with as many as 16 high-power lasers (100 W) coherently combined. In addition, LOCSET's channel scalability and error performance were analyzed through low-power multichannel LOCSET CBC of 2, 16, and 32 channels. Here, a 32-channel LOCSET beam combining with excellent RMS phase error of $\lambda/71$ was reported [27]. More importantly, there was no phase error degradation when scaling from 2 to 32 channels, as expected from previous simulations [21]. Furthermore, due to LOCSET's high operational bandwidth and low phase error, LOCSET appears readily scalable for efficient combination of over 100 lasers.

References

1 Quintino, L., Costa, A., Miranda, R., Yapp, D., Kumar, V., and Kong, C.J. (2007) Welding with high power fiber lasers: a preliminary study. *Mater. Design*, **28**, 1231–1237.

2 Jackson, S.D. and Lauto, A. (2002) Diode-pumped fiber lasers: a new clinical tool? *Laser Surg. Med.*, **30**, 184–190.

3 Sprangle, P., Peñano, J., Hafizi, B., and Ting, A. (2007) Incoherent combining of high-power fiber lasers for long-range directed energy applications. *J. Directed Energy*, **2**, 273–284.

4 Smith, R.G. (1972) Optical power handling capacity of low loss optical fibers as determined by stimulated Raman and Brillouin scattering. *Appl. Opt.*, **11**, 2489–2494.

5 Lichtman, E., Waarts, R.G., and Friesem, A.A. (1989) Stimulated Brillouin scattering excited by a modulated pump wave in single-mode fibers. *J. Lightwave Technol.*, **7**, 171–173.

6 Ward, B., Robin, C., and Dajani, I. (2012) Origin of thermal modal instabilities in large mode area fiber amplifiers. *Opt. Express*, **20**, 11407–11422.

7 Smith, A.V. and Smith, J.J. (2011) Mode instability in high power fiber amplifiers. *Opt. Express*, **19**, 10180–10192.

8 Sprangle, P., Peñano, J., Hafizi, B., and Ting, A. (2007) Incoherent combining of high-power fiber lasers for long-range directed energy applications. *J. Directed Energy*, **2**, 273–284.

9 Augst, S.J. and Fan, T.Y. (2004) Coherent beam combining and phase noise measurements of ytterbium fiber amplifiers. *Opt. Lett.*, **29**, 474–476.

10 Sprangle, P., Ting, A., Penano, J., Fischer, R., and Hafizi, B. (2009) Incoherent combining and atmospheric propagation of high-power fiber lasers for directed-energy applications. *IEEE J. Quantum Electron.*, **45**, 138–148.

11 Wirth, C., Schmidt, O., Tsybin, I., Schreiber, T., Eberhardt, R., Limpert, J., Tünnermann, A., Ludewigt, K., Gowin, M., ten Have, E., and Jung, M. (2011) High average power spectral beam combining of four fiber amplifiers to 8.2kW. *Opt. Lett.*, **36**, 3118–3120.

12 Madasamy, P., Jander, D.R., Brooks, C.D., Loftus, T.H., Thomas, A.M., Jones, P., and

Honea, E.C. (2009) Dual-grating spectral beam combination of high-power fiber lasers. *IEEE J. Sel. Top. Quantum Electron.*, **15**, 337–343.

13 Flores, A., Shay, T.M., Lu, C.A., Robin, C.A., Pulford, B., Sanchez, A.D., Hult, D., and Rowland, K. (2011) Coherent beam combining of fiber amplifiers in a kW regime. Conference on Lasers and Electro-Optics 2011, Baltimore, MD, Paper CFE3.

14 Yu, C.X., Augst, S.J., Redmond, S.M., Goldizen, K.C., Murphy, D.V., Sanchez, A., and Fan, T.Y. (2011) Coherent combining of a 4kW, eight-element fiber amplifier array. *Opt. Lett.*, **36**, 2686–2688.

15 Redmond, S.M., Fan, T.Y., Ripin, D., Thielen, P., Rothenberg, J., and Goodno, G. (2012) Diffractive beam combining of a 2.5-kW fiber laser array. Lasers, Sources, and Related Photonic Devices, Paper AM3A1.

16 Loftus, T.H., Thomas, A.M., Norsen, M., Minelly, J., Jones, P., Honea, E., Shakir, S.A., Hendow, S., Culver, W., Nelson, B., and Fitelson, M. (2008). Four-channel, high power passively phase locked fiber array. *Advanced Solid-State Photonics*, Paper WA4.

17 Bochove, E.J. and Shakir, S.A. (2009) Analysis of a spatial-filtering passive fiber laser beam combining system. *IEEE J. Sel. Top. Quantum Electron.*, **15**, 320–327.

18 Corcoran, C.J., Durville, F., Pasch, K.A., and Bochove, E.J. (2007) Spatial filtering of large mode area fiber lasers using a self-Fourier cavity for high power applications. *J. Opt. A*, **9**, 128–133.

19 Rothenberg, J. (2008) Passive coherent phasing of fiber laser arrays. *Proc. SPIE*, **6873**, 687315.

20 Bourderionnet, J., Bellanger, C., Primot, J., and Brignon, A. (2011) Collective coherent phase combining of 64 fibers. *Opt. Express*, **19**, 17053–17058.

21 Shay, T.M. (2006) Theory of electronically phased coherent beam combination without a reference beam. *Opt. Express*, **14**, 12188–12195.

22 Jones, D.C., Scott, A.M., Clark, S., Stace, C., and Clarke, R.G. (2004) Beam steering of a fibre bundle laser output using phased array techniques. *Proc. SPIE*, **5335**, 125–131.

23 Vorontsov, M.A. and Lachinova, S.L. (2008) Laser beam projection with adaptive array of fiber collimators: I. Basic considerations for analysis. *J. Opt. Soc. Am. A*, **25**, 1949–1959.

24 Bruesselbach, H., Wang, S., Minden, M., Jones, D.C., and Mangir, M. (2005) Power-scalable phase-compensating fiber-array transceiver for laser communications through atmosphere. *J. Opt. Soc. Am. B*, **22**, 347–353.

25 Shay, T.M., Benham, V., Baker, J.T., Sanchez, A.D., Pilkington, D., and Lu, C.A. (2007) Self-synchronous and self-referenced coherent beam combination for large optical arrays. *IEEE J. Sel. Top. Quantum Electron.*, **13**, 480–486.

26 Shay, T.M., Baker, J.T., Sancheza, A.D., Robin, C.A., Vergien, C.L., Flores, A., Zerinque, C., Gallant, D., Lu, C.A., Pulford, B., Bronder, T.J., and Lucero, A. (2010) Phasing of high power fiber amplifier arrays. Advanced Solid-State Photonics, Paper AMA1.

27 Pulford, B. (2011) LOCSET phase locking: operation, diagnostics, and applications, Dissertation, University of New Mexico.

28 Goodno, G.D., Asman, C.P., Anderegg, J., Brosnan, S., Cheung, E.C., Hammons, D., Injeyan, H., Komine, H., Long, W.H., McClellan, M., McNaught, S.J., Redmond, S., Simpson, R., Sollee, J., Weber, M., Weiss, S.B., and Wickham, M. (2007). Brightness-scaling potential of actively phase-locked solid-state laser arrays. *IEEE J. Sel. Top. Quantum Electron.*, **13**, 460–472.

29 Anderegg, J., Brosnan, S., Cheung, E., Epp, P., Hammons, D., Komine, H., Weber, M., and Wickham, M. (2006) Coherently coupled high-power fiber arrays. *Proc. SPIE*, **6102**, 61020.

30 Vorontsov, M.A., Carhart, G.W., and Ricklin, J.C. (1997) Adaptive phase-distortion correction based on parallel gradient-descent optimization. *Opt. Lett.*, **22**, 907–909.

31 Liu, L., Vorontsov, M.A., Polnau, E., Weyrauch, T., and Beresnev, L.A. (2007) Adaptive phase-locked fiber array with wavefront phase tip-tilt compensation using piezoelectric fiber positioners. *Proc. SPIE*, **6708**, 67080.

32 Goodno, G., McNaught, S., Rothenberg, J., McComb, T., Thielen, P., Wickham, M., and Weber, M. (2010) Active phase and polarization locking of a 1.4kW fiber amplifier. *Opt. Lett.*, **35**, 1542–1544.

33 Yu, C.X., Kansky, J.E., Shaw, S.E.J., Murphy, D.V., and Higgs, C. (2006) Coherent beam combining of large number of PM fibers in 2-D fiber array. *Electron. Lett.*, **42**, 1024–1025.

34 Goodno, G.D., Shih, C., and Rothenberg, J.E. (2010) Perturbative analysis of coherent combining efficiency with mismatched lasers. *Opt. Express*, **18**, 25403–25414.

35 Jones, D.C., Stacey, C.D., and Scott, A.M. (2007) Phase stabilization of a large-mode-area ytterbium-doped fiber amplifier. *Opt. Lett.*, **32**, 466–468.

36 Uberna, R., Bratcher, A., Tiemann, B.G., Alley, T.G., Sanchez, A.D., Flores, A., and Pulford, B. (2010) Coherent polarization beam combination with active phase and polarization control. Solid State and Diode Laser Technology Review Technical Digest, 12–16.

37 Uberna, R., Bratcher, A., Alley, T.G., Sanchez, A.D., Flores, A., and Pulford, B. (2010) Coherent combination of high power fiber amplifiers in a two-dimensional re-imaging waveguide. *Opt. Express*, **18**, 13547–13553.

38 Jolivet, V., Bourdon, P., Bennai, B., Lombard, L., Goular, D., Pourtal, E., Canat, G., Jaoeun, Y., Moreau, B., and Vassuer, O. (2009) Beam shaping of single-mode and multimode fiber amplifier arrays for propagation through atmospheric turbulence. *IEEE J. Sel. Top. Quantum Electron.*, **15**, 257–268.

39 Wickham, M., Thielen, P., Jo, J., Goodno, G., Rice, R., Cheung, E., Rothenberg, J., Gallant, D., Baker, J., Lucero, A., Sanchez, A., Shay, T., Robin, C., Vergien, C., and Zeringue, C. (2008) High efficiency coherent fiber beam combiner. 2008 Annual Directed Energy Symposium Proceedings.

40 Thielen, P., Ho, J., Burchman, D., Goodno, G., Rothenberg, J., Wickham, M., Flores, A., Lu, C., Pulford, B., Robin, C., Sanchez, A., Hult, D., and Rowland, K. (2012) Two-dimensional diffractive coherent beam combining. Advanced Solid-State and Photonics, Paper AM3A2.

41 Jeong, Y., Nilsson, J., Sahu, J.K., Payne, D.N., Horley, R., Hickey, L.M.B., and Turner, P.W. (2007) Power scaling of single-frequency ytterbium-doped fiber master oscillator power amplifier sources up to 500W. *IEEE J. Sel. Top. Quantum Electron.*, **13**, 546–551.

42 Koplaw, J.P., Kliner, D., and Goldberg, L. (2000) Single-mode operation of a coiled multimode fiber amplifier. *Opt. Lett.*, **25**, 442–444.

43 Flores, A., Lu, C., Robin, C., and Dajani, I. (2012) Experimental and theoretical studies of phase modulation in Yb-doped fiber amplifiers. *Proc. SPIE*, **8281**, 83811.

44 Robin, C. and Dajani, I. (2011) Acoustically segmented photonic crystal fiber for single-frequency high-power laser applications. *Opt. Lett.*, **36**, 2641–2643.

45 Robin, C., Dajani, I., Zeringue, C., Ward, B., and Lanari, A. (2012) Gain-tailored SBS suppressing photonic crystal fibers for high power applications. *Proc. SPIE*, **8237**, 82371.

46 Hildebrandt, M., Büesche, S., Weßels, P., Frede, M., and Kracht, D. (2008) Brillouin scattering spectra in high-power single-frequency ytterbium doped fiber amplifiers. *Opt. Express*, **16**, 15970–15979.

47 Kracht, D., Wilhelm, R., Frede, M., Fallnich, C., Seifert, F., Willke, B., and Danzmann, K. (2006) High power single-frequency laser for gravitational wave detection. Advanced Solid-State Photonics, Paper WE1.

48 Kontur, F.J., Dajani, I., Lu, Y., and Knize, R.J. (2007) Frequency-doubling of a CW fiber laser using PPKTP, PPMgSLT, and PPMgLN. *Opt. Express*, **15**, 12882–12889.

3
Kilowatt Coherent Beam Combining of High-Power Fiber Amplifiers Using Single-Frequency Dithering Techniques

Zejin Liu, Pu Zhou, Xiaolin Wang, Yanxing Ma, and Xiaojun Xu

3.1
Introduction

High-power lasers are used in a wide range of applications in various fields such as industry, medicine, biology, and astronomy. Since the first laser demonstrated by Maiman, there is continuous need for high-power radiation. Due to undesirable effects such as laser medium damage, thermo-optical effects, and nonlinearities, it is not possible to scale a laser to arbitrary high power while simultaneously maintaining the beam quality simply by launching more pump power [1].

Coherent beam combining (CBC) of multiple laser emitters provides a constructive solution to achieve high-power laser without degrading the beam quality that is obtainable from a single laser emitter. In a CBC system, all laser elements operate with the same spectrum and phase, constructive interference could be obtained at the far field due to the vector adding the electric fields. The addition of N high-quality beams will then increase the power by a factor of N.

Here, we review the brief history and the current status of CBC technology. For the time being it can be found that CBC is able to increase the laser brightness with any types of operation regimes and radiation spectra. Therefore, CBC provides an effective power scaling solution for practical requirements.

3.1.1
Brief History of Coherent Beam Combining

Shortly after the first demonstrated solid-state laser (SSL), two He–Ne beams were effectively phase-locked to show a constructive addition at the far-field pattern in 1965 [2]. In the 1970s and afterward, due to the potential for material processing and military applications, CBC of high-power CO_2 laser has attracted the attention of worldwide laser scientists [3,4]. Subsequently, chemical laser was considered to be the most effective high-power emitter candidate in laser technology due to the unprecedented increase in output power. CBC of chemical

Coherent Laser Beam Combining, First Edition. Edited by Arnaud Brignon.
© 2013 Wiley-VCH Verlag GmbH & Co. KGaA. Published 2013 by Wiley-VCH Verlag GmbH & Co. KGaA.

lasers was also investigated in detail [5]. Then in the 1990s, CBC of a large number of semiconductor lasers was widely studied due to the revolution in semiconductor technology. CBC of 900 semiconductor laser amplifiers was thus demonstrated [6].

However, due to the poor performances of active optical component such as phase modulators, those early demonstrations were not robust against perturbations, therefore CBC was only a kind of proof-of-concept test and remained in laboratories.

3.1.2
Coherent Beam Combining: State of the Art

CBC is in its full bloom in the present century due to the great progress indeed made in scaling the output power of SSLs. Notable milestones include the demonstration of 100 kW slab lasers by CBC of seven modules [7,8] in 2009, which is the brightest SSL ever demonstrated (see Chapter 1). In addition, this architecture provides a vehicle for brightness scaling well beyond 100 kW by adding more modules.

The rapid progress in CBC technology has made their application in many research fields more acceptable and cost-effective. For example, high-power 589 nm single-line radiation laser can be used as high-brightness laser guide star in astronomical imaging. Sum frequency generation (SFG) in a solid-state system is often used for generation of 589 nm single-line radiation. Recently, it was shown that high-power radiation could be obtained by frequency doubling of an 1178 nm fiber laser system composed of three coherently combined fiber laser beams. This is considered to be the most suitable, flexible, and reliable technology for future laser guide star systems [9].

The CBC technology could also be straightforwardly extended to different types of laser emitters that operate in various regimes. CBC of two thulium-doped fiber lasers with a central wavelength of $\sim 2\,\mu m$ was demonstrated in 2010 [10], and afterward five quantum cascade lasers emitting at 4.65 µm were coherently combined successfully (see Chapter 13) [11], which undoubtedly enable the power scaling capability of mid-infrared laser systems. CBC could also be applied to ultrafast laser systems that operate in picosecond or femtosecond regimes [12] to suppress the inevitable nonlinear effects that come up in the case of high-peak-power operation. It could even be extended to the terahertz band. For example, phase locking of several terahertz oscillators by using the same driven laser beams could significantly improve the spatial resolution for active stand-off terahertz imaging [13].

It is to be noted that CBC could not only increase the output power of the laser emitter but also provide control of the spatial beam profile and a degree of beam steering and tracking. Radial vector beams that had vanished diffraction pattern in the center can be used in laser machining, particle acceleration interactions, and generating longitudinal electric fields when tightly focused. This type of laser beam is often formed in free space laser cavities by using a conical lens or axicon. Recently, it was shown that radially polarized beam can be generated effectively by coherently combining Gaussian beams, which have the advantage of simplicity and direct control over mode generation [14].

3.1.3
Key Technologies for Coherent Beam Combining

There are mainly three important factors that enabled such successful power scaling. One is the newly developed fiber-pigtailed phase modulator, which has ~200 mW power handling capability and gigahertz bandwidth, which means a sufficient control bandwidth for phase stabilization of each laser module. Then there is the incorporation of fast phase servo system based on FPGA, DSP, and industrial computers, which also gained amazing progress over time due to the development of integrated circuit technology. The third factor is the compactness and robustness of single laser module, otherwise the whole CBC system configuration would be huge and immobile. As one kind of SSL, fiber lasers are particularly amenable to CBC configuration because of its inherent modularity and ease of building tightly packed arrays. To date, the volume of a 10 kW single-mode fiber laser could be comparable to a refrigerator, and 4 kW continuous wave (CW) fiber laser had been demonstrated by CBC of eight modules in 2011 [15].

In this chapter, we will focus on the phase servo system for coherent beam combination. CBC of fiber laser/amplifiers based on master oscillator power amplifier (MOPA) configuration is believed to be a promising way to acquire high-brightness lasers. In this technique, active phasing is the key and many approaches have been proposed: heterodyne detection phase control technique, multidithering technique, and SPGD algorithm phase control technique (see Chapters 1, 2, and 4). CBC with heterodyne detection phase control technique requires a reference beam and a photodetector (PD) array with the same number of elements as the number of beamlets in the whole MOPA system. The system will become increasingly complex and sophisticated when scaled to a large number of beamlets. SPGD algorithm needs only one PD, but the implementing speed of the algorithm and control bandwidth will decrease as the number of lasers increases. Multidithering technique needs one PD and phase control can be implemented at a high speed, but every element in the array requires an individual phase modulation frequency and a corresponding phase control module. Accordingly, the modulation frequency accumulates to extremely high values and becomes difficult to be practically implemented as the number of lasers increases. In addition, when large numbers of beams are coherently combined, the difficulty and cost of the control system will also increase rapidly. In this chapter, we present single-frequency dithering (SFD) technique based on time division multiplexing technique that has been widely applied in optical communication field to alleviate above difficulties. We will show that phase control based on SFD has great potential to boost stable and robust CBC to a large number of laser beams with a simple and cost-effective phase control system.

3.2
Single-Frequency Dithering Technique

3.2.1
Theory of Single-Frequency Dithering Technique

SFD technique for CBC has been presented in Ref. [16]. SFD is based on time division multiplexing that is widely applied in communication field. The scheme of CBC using SFD is shown in Figure 3.1. The beam from a seed laser is split into N beams (four channels are shown in Figure 3.1) and coupled to N or $N-1$ optical phase modulators. The laser beams from the phase modulators are amplified by N fiber amplifiers and sent to free space via the collimators. The phase modulators are used to add phase modulation and phase control signals. The beam array exiting the collimators is focused on a PD located at the focus plane of a lens. The electronic signals delivered by the PD are used to generate appropriate phase control signals in the signal processor. Phases of all laser beams will be locked when phase control signals are applied to the proper phase modulators.

The advantage of SFD is that only one modulation frequency and one phase control module are required. The different steps i for coherent beam combination with SFD are briefly described as follows.

We assume that t represents time and that the phase locking system starts working at t_0. When $t_0 < t < t_1$, the phase modulation signal is applied to the phase modulator in channel 1. The phase error between channel 1 and the mean phase value of other channels can be derived and interpreted into a proper phase control signal, which is applied to the phase modulator in channel 1. During this first step, no phase modulation and no control signals are applied to other channels.

When $t_1 < t < t_2$, the phase modulation signal and relevant phase control signal are applied to the phase modulator in channel 2. The modulation signal of channel 1 becomes zero and the control signal is kept to its final value at $t = t_1$. The phase

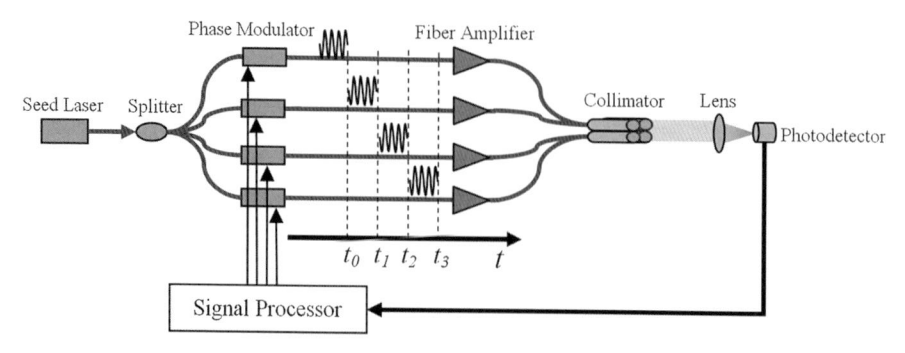

Figure 3.1 Scheme of CBC using single-frequency dithering technique.

modulation signals and relevant control signals applied to the remaining channels are still kept to zero.

When $t_{i-1} < t \leq t_i$ ($i \leq N$), the phase modulation and control signals are applied to the phase modulator in channel i. Phase control signals of other channels are kept unchanged at the value generated at $t = t_{i-1}$, while phase modulation signals become zeros. When $i > N$, let us introduce the notation $m = i \bmod N$. The phase modulation and control signals are now applied to channel m. Other channels hold the previous phase control signal levels.

If one cycle duration $T = t_{i+N} - t_i$ is short enough for the phases of all channels to experience negligible fluctuations but long enough for the dithering process to be efficient, the phase errors between different channels will be compensated after iterative operation of the control process described above.

The mathematical principle of SFD can be introduced as follows. We assume that M Gaussian beams with the same polarization are located at a source plane ($z = 0$) and propagates along the z-axis in the Cartesian coordinate system. According to the well-established Gaussian beam formulation, the optical field of the jth Gaussian beam can be expressed as

$$u_j(x, y, 0, t) = \exp\left[-\frac{(x - a_j)^2 + (y - b_j)^2}{\omega_0^2}\right] \times \exp\left[i\varphi_{j_f}(t)\right], \tag{3.1}$$

where ω_0 denotes the initial waist width, (a_j, b_j) denotes the central position of the jth beam, and $\varphi_{j_f}(t)$ is the optical phase of the jth beam at $z = 0$. $\varphi_{j_f}(t)$ continuously fluctuates with time due to environmental perturbations undergone by the fiber amplifiers.

When the modulation signal is applied to channel p, its optical field can be expressed as

$$u_p(x, y, 0, t) = \exp\left[-\frac{(x - a_p)^2 + (y - b_p)^2}{\omega_0^2}\right] \times \exp\left[i(\varphi_{p_f} + \varphi_m)\right], \tag{3.2}$$

where φ_m denotes the phase modulation signal with the frequency ω_m and the amplitude a_m that can be expressed as

$$\varphi_m = a_m \sin(\omega_m t). \tag{3.3}$$

Therefore, the optical field of the pth beam at the receiver plane ($z = L$) can be expressed as

$$u_p(x, y, L, t) = \frac{\omega_0}{\omega(L)} \exp\left[-\frac{(x - a_p)^2 + (y - b_p)^2}{\omega^2(L)}\right]$$

$$\times \exp\left\{-i\left\{k\left[\frac{(x - a_p)^2 + (y - b_p)^2}{2R(L)} + L\right] - \Psi - \varphi_{p_f} - \varphi_{p_t} - \varphi_m\right\}\right\}, \tag{3.4}$$

where φ_{p_t} denotes the phase fluctuations induced by atmosphere turbulence for the pth beam, $Z_0 = (\pi\omega_0^2)/\lambda$, $\omega(L) = \omega_0\sqrt{1 + (L/Z_0)^2}$, $R(L) = Z_0((L/Z_0) + (Z_0/L))$, and $\Psi = \tan^{-1}(L/Z_0)$. For simplicity, let $\varphi_{pL} = k\left[\frac{(x-a_p)^2+(y-b_p)^2}{2R(L)} + L\right] - \Psi$, $\varphi_{pN} = \varphi_{p_f} + \varphi_{p_t}$, and

$$A_p = \frac{\omega_0}{\omega(L)} \exp\left[-\frac{(x-a_p)^2+(y-b_p)^2}{\omega^2(L)}\right].$$

Then, Eq. (3.4) can be expressed as

$$u_p(x,y,L,t) = A_p \exp\left[-i\left(\varphi_{pL} - \varphi_{pN} - \varphi_m\right)\right]. \tag{3.5}$$

Therefore, the optical intensity of all beams at $z = L$ can be expressed as

$$I(x,y,L,t) = \sum_{j=1}^{M} u_j^2 + \sum_{j=1}^{M}\sum_{\substack{l=1 \\ l\neq j}}^{M} u_j u_l^*. \tag{3.6}$$

In Eq. (3.6), $u_j u_l^* = u_l u_j^*$, then $u_j u_l^* + u_l u_j^* = 2\text{Re}(u_j u_l^*)$. So Eq. (3.6) can be expressed as

$$I(x,y,L,t) = \sum_{j=1}^{M} u_j^2 + 2\sum_{\substack{j=1 \\ j\neq p}}^{M}\sum_{\substack{l=j+1 \\ l\neq p}}^{M} A_j A_l \cos\left(-\varphi_{jL} + \varphi_{jN} + \varphi_{lL} - \varphi_{lN}\right)$$

$$+ 2\sum_{\substack{j=1 \\ j\neq p}}^{M} A_p A_j \cos\left(-\varphi_{pL} + \varphi_{pN} + \varphi_m + \varphi_{jL} - \varphi_{jN}\right)$$

$$= \sum_{j=1}^{M} u_j^2 + 2\sum_{\substack{j=1 \\ j\neq p}}^{M}\sum_{\substack{l=j+1 \\ l\neq p}}^{M} A_j A_l \cos\left(\Phi_{jlN} + \Phi_{ljL}\right)$$

$$+ 2\sum_{\substack{j=1 \\ j\neq p}}^{M} A_p A_j \cos\left(\Phi_{pjN} + \Phi_{jpL} + \varphi_m\right), \tag{3.7}$$

where $\Phi_{jlN} = \varphi_{jN} - \varphi_{lN}$, $\Phi_{ljL} = \varphi_{lL} - \varphi_{jL}$, $\Phi_{pjN} = \varphi_{pN} - \varphi_{jN}$, and $\Phi_{jpL} = \varphi_{jL} - \varphi_{pL}$. When a PD is located at the receiver plane, the optical intensity will be transformed into electrical current signal, which can be expressed as

$$i_p(t) = R_{PD}\int_S I(x,y,L,t)ds$$

$$= R_{PD}\left[\sum_{j=1}^{M}\int_S u_j^2 ds + 2\sum_{\substack{j=1 \\ j\neq p}}^{M}\sum_{\substack{l=j+1 \\ l\neq p}}^{M}\int_S A_j A_l \cos\left(\Phi_{jlN} + \Phi_{ljL}\right)ds \right.$$

$$\left. + 2\sum_{\substack{j=1 \\ j\neq p}}^{M}\int_S A_p A_j \cos\left(\Phi_{pjN} + \Phi_{jpL} + \varphi_m\right)ds\right] \tag{3.8}$$

where R_{PD} represents the responsivity of the PD and S represents the PD area. In Eq. (3.8), the first and second term in the bracket do not include the phase modulation signal and will be neglected in later analysis. The third term is marked as $J_p(t)$ and is expanded as follows:

$$J_p(t) = 2\sum_{\substack{j=1 \\ j\neq p}}^{M} \int_S A_p A_j \cos(\Phi_{pjN} + \Phi_{jpL} + \varphi_m) ds$$

$$= 2\sum_{\substack{j=1 \\ j\neq p}}^{M} \begin{array}{l} \cos[\Phi_{pjN} + \alpha_m \sin(\omega_m t)] \int_S A_p A_j \cos\Phi_{jpL} ds \\ +\sin[\Phi_{pjN} + \alpha_m \sin(\omega_m t)] \int_S A_p A_j \sin\Phi_{jpL} ds. \end{array} \quad (3.9)$$

Fourier series expansions for the cosine and sine of $\alpha_m \sin(\omega_m t)$, $\cos[\Phi_{pjN} + \alpha_m \sin(\omega_m t)]$ and $\sin[\Phi_{pjN} + \alpha_m \sin(\omega_m t)]$ in Eq. (3.9) can be expressed as follows:

$$\cos[\Phi_{pjN} + \alpha_m \sin(\omega_m t)]$$
$$= \cos(\Phi_{pjN})\cos[\alpha_m \sin(\omega_m t)] - \sin(\Phi_{pjN})\sin[\alpha_m \sin(\omega_m t)]$$
$$= \cos(\Phi_{pjN})\left[J_0(\alpha_m) + 2\sum_{i=1}^{\infty} J_{2i}(\alpha_m)\cos(2i\omega_m t)\right] \quad (3.10)$$
$$- \sin(\Phi_{pjN})\left[2\sum_{i=1}^{\infty} J_{2i-1}(\alpha_m)\sin[(2i-1)\omega_m t]\right].$$

$$\sin[\Phi_{pjN} + \alpha_m \sin(\omega_m t)]$$
$$= \sin(\Phi_{pjN})\cos[\alpha_m \sin(\omega_m t)] + \cos(\Phi_{pjN})\sin[\alpha_m \sin(\omega_m t)]$$
$$= \sin(\Phi_{pjN})\left[J_0(\alpha_m) + 2\sum_{i=1}^{\infty} J_{2i}(\alpha_m)\cos(2i\omega_m t)\right] \quad (3.11)$$
$$+ \cos(\Phi_{pjN})\left[2\sum_{i=1}^{\infty} J_{2i-1}(\alpha_m)\sin[(2i-1)\omega_m t]\right].$$

Obviously, the phase error signal can be extracted from Eqs. (3.10) and (3.11) by demodulating. When the photocurrent is multiplied by $\alpha \sin(\omega_m t)$ and integrated over a cycle of phase modulation signal, we obtain the phase error signal of the pth beam as follows:

$$S_{np} = \int_0^{1/\omega_m} i_p(t) \times a\sin(\omega_m t) dt$$
$$= \alpha R_{PD} J_1(\alpha_m) \sum_{\substack{j=1 \\ j\neq p}}^{M} \cos(\Phi_{pjN}) \int_S A_p A_j \sin\Phi_{jpL} ds - \sin(\Phi_{pjN}) \int_S A_p A_j \cos\Phi_{jpL} ds$$
$$= \alpha R_{PD} J_1(\alpha_m) \sum_{\substack{j=1 \\ j\neq p}}^{M} \cos(\Phi_{pjN}) Q_{Cpj} - \sin(\Phi_{pjN}) Q_{Spj}.$$

$$(3.12)$$

where $Q_{Cpj} = \int_S A_p A_j \sin\Phi_{jpL} ds$ and $Q_{Spj} = \int_S A_p A_j \cos\Phi_{jpL} ds$. Then the phase control signal can be obtained based on S_{np}, which finally realizes CBC of laser beam array.

It is important to note that Q_{Cpj} and Q_{Spj} strongly affect the value of S_{np}. When all beams are recombined in the near field (filled aperture geometry), $a_p = a_j$ and $b_p = b_j$. Therefore, $Q_{Spj} = 0$ and $Q_{Cpj} = \int_S A_p A_j ds$, then Eq. (3.12) can be expressed as

$$S_{np} = -\alpha R_{PD} J_1(\alpha_m) \sum_{\substack{j=1 \\ j \neq p}}^{M} \sin(\Phi_{pjN}) \int_S A_p A_j ds. \tag{3.13}$$

In Eq. (3.13), we can see that a correct value of S_{np} is obtained only if the PD area S is large enough for proper signal integration. The PD area S has a weak effect on S_{np} in the filled aperture geometry. However, in the tiled aperture scheme, Q_{Cpj} and Q_{Spj} will strongly affect S_{np}. Unfortunately, the analytical form of Q_{Cpj} and Q_{Spj} cannot be obtained, so the numerical analysis has to be implemented. Without loss of generality, the square array of four Gaussian beams is studied and the sketch map of laser beams in near field is shown in Figure 3.2a. The laser beam radius ω_0 at the source plane is 3 cm and the diameter D of circumcircle is 30 cm. The fill factor is about 35%. Fill factor is defined as the beam width divided by the distance of the adjacent beamlets. The optical intensity pattern at $L = kD^2/8$ in the ideal case is shown in Figure 3.2b.

When the integration area S is circular and its center is located at the center of the main lobe in the far field [coordinate (0, 0) in Figure 3.2b], the changes of Q_{Cpj} and Q_{Spj} versus the radius of S are shown in Figure 3.3 where the different curves correspond to various values of p and j. In Figure 3.3, it is found that the value of Q_{Spj} is far less than Q_{Cpj} so that Q_{Spj} can be ignored and the radius of S strongly affects

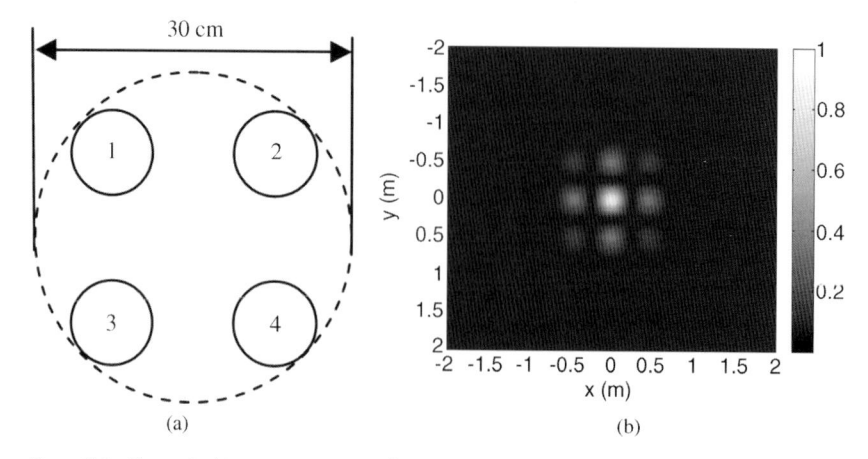

Figure 3.2 The optical intensity pattern in the (a) near and (b) far fields for four Gaussian beams.

3.2 Single-Frequency Dithering Technique | 83

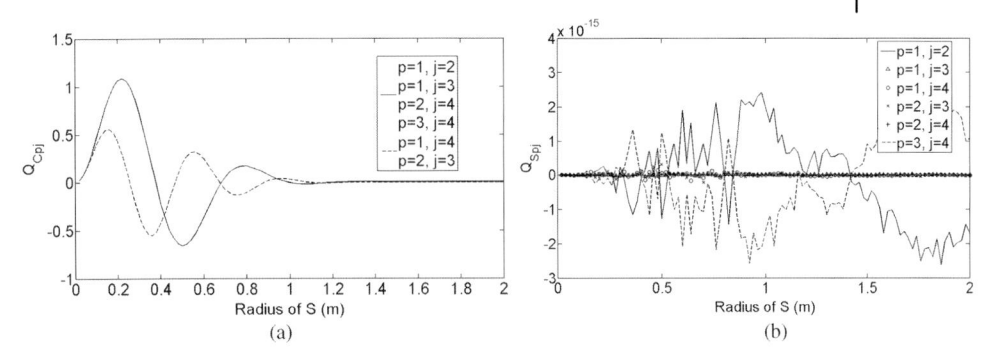

Figure 3.3 Q_{Cpj} and Q_{Spj} with the center of S located at (0, 0).

the value of Q_{Cpj}. When the radius of S is more than 1.2 m, the value of Q_{Cpj} is close to zero, which will lead to S_{np} becoming zero. This is the reason why a hard aperture is located before the PD in coherent beam combination using dithering technique. The different radius of S determines the value and sign of Q_{Cpj}, which finally determines the value and sign of S_{np}. Therefore, different aperture radii can lead to opposite results.

When the center of S is moved to other places, Q_{Spj} cannot be ignored. Figure 3.4 shows the curves of Q_{Cpj} and Q_{Spj} when the center of S is located at (0, 0.5 m) in Figure 3.2b. The results show that the location of the hard aperture in receiver plane also affects the value of S_{np}.

In CBC using dithering techniques, the phase control signal is produced based on S_{np}. Therefore, the different S_{np} will lead to different phase control signals, which finally formed the variform optical intensity patterns in far field shown in Table 3.1 where R, denotes the radius and X and Y the position of the integral area S. In calculation, the original phases of laser beams are randomly produced before phase locking and the original intensity pattern in far field is shown in Figure 3.5. From Table 3.1, we can see that the contrary phase-locking results will be obtained with the radius R increasing (i.e., R = 0.2 and R = 0.5) when the position of S remained unchanged, which is determined by the signs of P_{Cjl} and P_{Sjl}. In addition, the

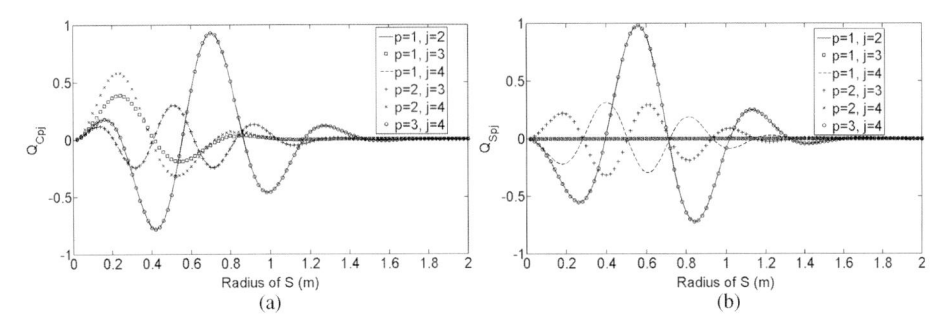

Figure 3.4 Q_{Cpj} and Q_{Spj} with the center of S located at (0, 0.5).

Table 3.1 Effects of size and position of integral area S on the optical intensity pattern in the far field.

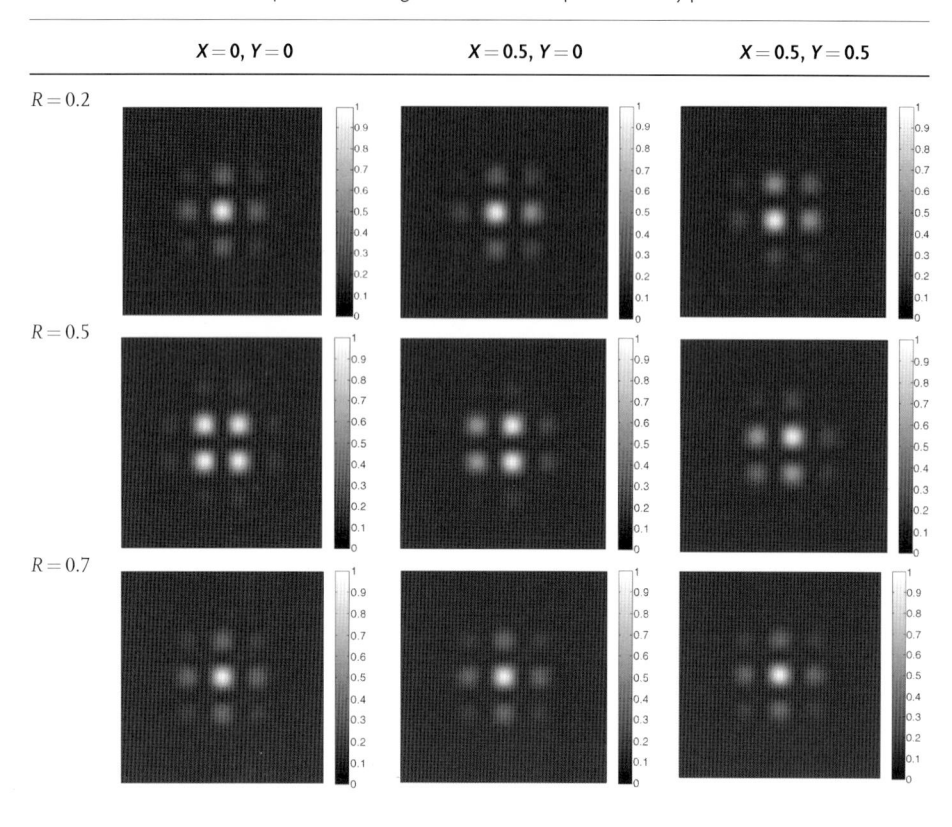

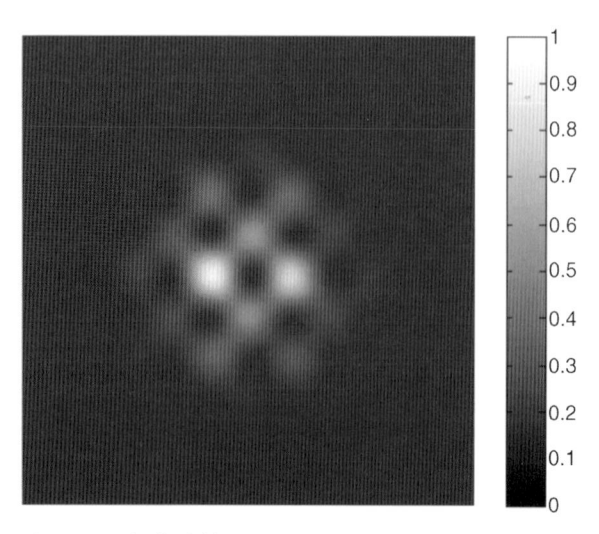

	X = 0, Y = 0	X = 0.5, Y = 0	X = 0.5, Y = 0.5
R = 0.2			
R = 0.5			
R = 0.7			

Figure 3.5 The far-field intensity pattern with the stochastic piston phase errors among laser beams.

position and size of integral area S affect the distribution of intensity among lobes. Therefore, the different far-field intensity pattern can be obtained by changing the size and position of the hard aperture in the experiment.

3.2.2
Kilowatt Coherent Beam Combining of High-Power Fiber Amplifiers Using Single-Frequency Dithering Technique

The experimental setup is shown in Figure 3.6 [17]. The seed laser is a distributed feedback (DFB) polarization-maintaining Yb-doped fiber laser with 1064 nm wavelength and 20 kHz linewidth produced by NP Photonics Corporation. The 30 mW laser power from the optically isolated seed laser is preamplified to 300 mW. The beam is then split into nine channels and coupled to nine LiNbO$_3$ phase modulators. The modulators (Photline Technologies Corporation) have 100 MHz bandwidth and operate at 1060 nm wavelength. The output power from the phase modulators is higher than 10 mW and coupled to homemade three-stage all-fiber amplifiers. The first stage is used to amplify the laser power from 10 to 100 mW level, the second stage from 100 mW to 10 W level, and the third stage from 10 to 100 W level. The maximum output powers of the third-stage amplifiers are 120.5, 119.2, 121.2, 120.6, 117.3, 121.6, 118.4, 121.2, and 122 W, respectively, with a maximum overall output power close to 1080 W. The difference in output power of fiber amplifiers can be attributed to the unequal splitting ratio of the fiber splitters, different insertion loss of phase modulators, isolators, and the different pumping diodes of the third-stage amplifiers.

The laser beams from the third-stage amplifiers are space-coupled via nine collimators. The diameter of each laser beam output from the collimator is 5 mm. The nine laser beams are tiled using free space mirrors side by side into a 3 × 3 laser array with 40% fill factor in the near field. The collimated laser beams are sampled by a mirror with 99% reflectivity and the reflected light is sent to a power meter for the total power measurement. The light passing through a sampler

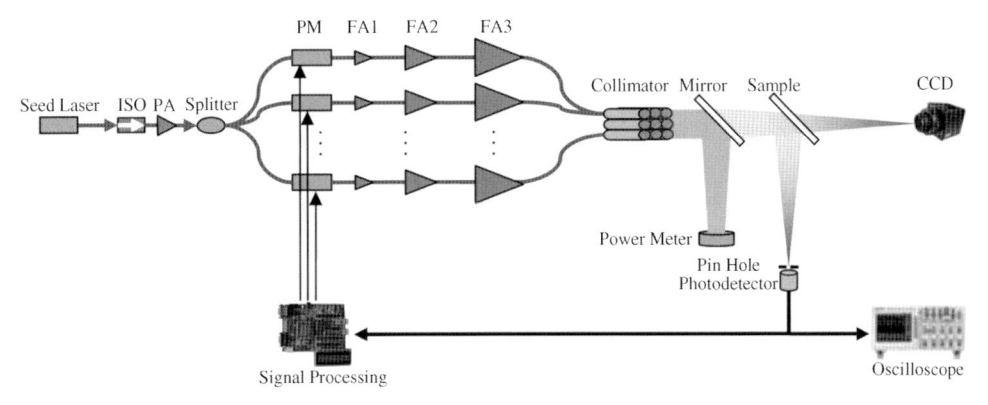

Figure 3.6 The experimental setup of coherent beam combination of nine beams. ISO, isolator; PA, preamplifier; PM, phase modulator; FA, fiber amplifier.

propagates over a distance of 10 m and reaches a CCD camera that is used to diagnose the beam quality. The light reflected by the sampler is coupled through a hard aperture, and a PD is located just behind the aperture. The hard aperture with 300 μm diameter is used to collect the power of the main lobe in the far-field interference pattern of the combined beams, which can provide phase error information for the phase control algorithm. The PD is a PDA10CS InGaAs amplifier detector (THORLABS Corporation) and has 700–1800 nm spectral response and 8.5 MHz bandwidth when the gain is at 10 dB. The electronic signals delivered by the PD are coupled into the signal processor based on field programmable gate array (Cyclone III FPGA from Altera Corporation) where the SFD algorithm is implemented. The homemade signal processor operates at 50 MHz main frequency and can provide phase control signals for 12 channels, but only 9 channels are used in the present experiment. The phase control signals produced by the signal processor are sent to the phase modulators to compensate for the phase errors between the elements.

In the experiment, when the control system is in the open loop, SFD is not performed and the phases of laser beams randomly fluctuate due to phase fluctuations in each fiber channel induced by amplifier temperature variations, fan heater, and mechanical vibrations. The intensity pattern at the observing plane keeps shifting. The 2 s long exposure far-field intensity distribution is shown in Figure 3.7a, and its fringe contrast is about zero, where the fringe contrast is defined by the formula $(I_{max} - I_{min})/(I_{max} + I_{min})$, where I_{max} and I_{min} are the maximum optical intensity and the adjacent minimum on the intensity pattern, respectively.

When the phase control system is in the closed loop, the phase control algorithm is implemented and the phase modulation and control signals are applied to the modulators of each channel by the signal processor and the phase errors are compensated. The intensity pattern at the observing plane is stable with a fringe contrast calculated to be more than 85%. The 2 s long-exposure far-field intensity distribution is shown in Figure 3.7b. The theoretical far-field pattern of the ideally phased nine-laser array, which is arranged according to the above real data, is

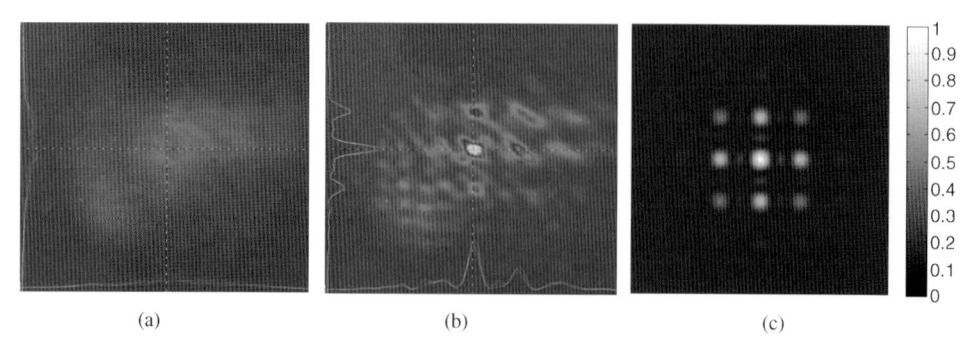

Figure 3.7 Long-exposure far-field intensity pattern of the combined laser beam. (a) Open-loop experimental result. (b) Closed-loop experimental result. (c) Closed-loop theoretical pattern.

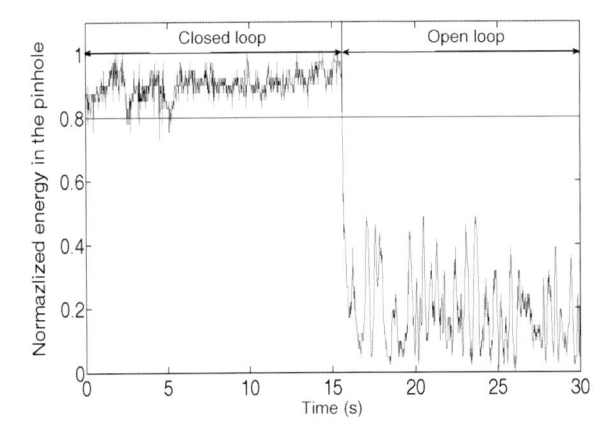

Figure 3.8 Time series signals of energy collected by the aperture in the open loop (right) and closed loop (left).

computed and presented in Figure 3.7c. It is found that the experimental results are in agreement with the theoretical ones.

The fidelity of coherent beam combination and the suppression of phase fluctuations can further be studied using the time series signals and the spectral density of energy collected by the hard aperture, as shown in Figures 3.8 and 3.9. When the control loop is open, the normalized energy collected by the hard aperture fluctuates between 0 and 0.5 randomly. When the control loop is closed, the energy collected by the aperture can be locked steadily to be more than 0.8 for most of the time and the residual phase error is less than $\lambda/15$. It can be seen in Figure 3.9 that phase noises below 400 Hz have been compensated, which denotes a remarkable increase of energy in the main lobe.

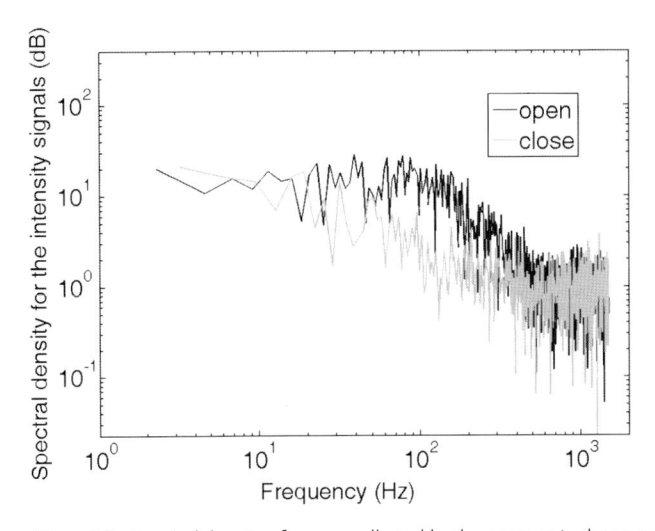

Figure 3.9 Spectral density of energy collected by the aperture in the open loop and closed loop.

3.2.3
Coherent Polarization Beam Combining of Four High-Power Fiber Amplifiers Using Single-Frequency Dithering Technique

SFD can also be used in coherent polarization beam combining (CPBC). In Ref. [18], CPBC of four polarization-maintaining high-power fiber amplifiers with a total output power of 60 W has been demonstrated. The experimental setup is shown in Figure 3.10. The seed laser is a linear polarized single-frequency fiber laser with a central wavelength of 1064.4 nm. The laser power from the seed laser through the isolator (ISO) is about 30 mW and amplified to 120 mW output power via an amplifier (AMP) before splitting into four channels and coupled to four $LiNbO_3$ phase modulators with more than 100 MHz modulating bandwidth. The output power from each phase modulators is more than 20 mW, and the loss of laser power is due to the insertion loss of the modulator. Then each fiber channel is coupled to a two-stage all-fiber homemade polarization-maintaining amplifier. The output power of the second-stage amplifiers is about 25 W. Nevertheless, due to the difference in insertion loss, after the collimators, the power of the four channels is 16.7, 18.7, 11.4, and 19.7 W, respectively, and the power ratio is 1.46 : 1.64 : 1 : 1.73. By rotating the half-wavelength plates (HWPs) and adjusting the four beams coaxially, they can be combined by the polarization beam combiners (PBCs). M_1 is an all-reflectance mirror and M_2 is a high-reflectance mirror. After M_2, a small portion of the beam is

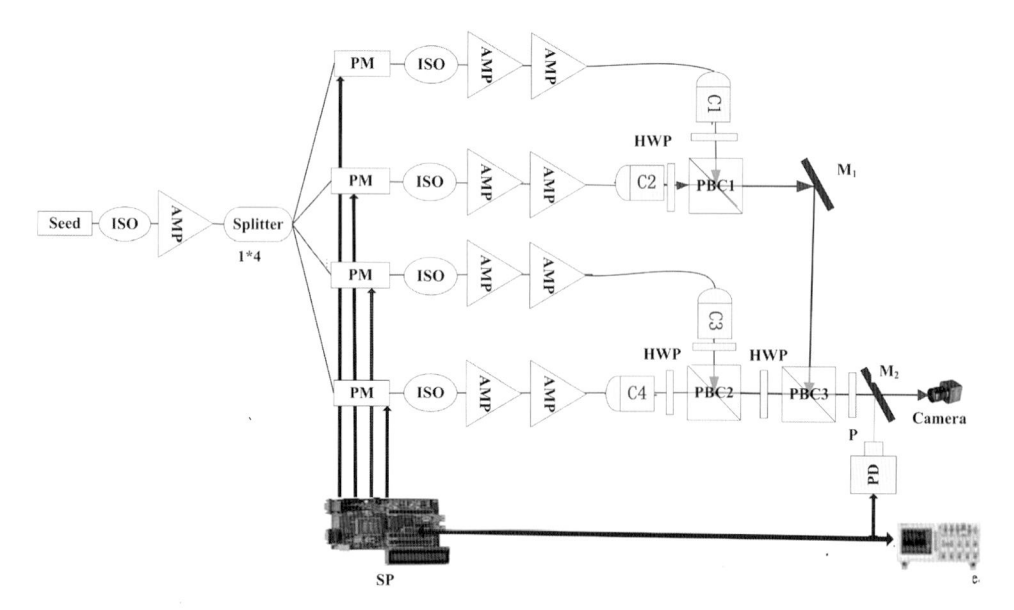

Figure 3.10 Experimental setup of CPBC of four-channel fiber amplifiers. Seed, seed lasers; ISO, isolator; AMP, amplifier; PM, phase modulator; C1–C4, collimators; HWP, half-wavelength plate; PBC_i ($i = 1$–3), polarization beam combiners; P, polarizer; M_1, all-reflectance mirrors; M_2, high-reflectance mirror; PD, photodetector; SP, signal processing.

sent to a PD. Another part of the beam after M_2 is sent to an infrared camera, which is used to diagnose the profile of the combined beam. The PD is an InGaAs amplifier detector with a 700–1800 nm spectral response and 8.5 MHz bandwidth when the gain is at 10 dB.

It is to be noted that the tip–tilt error in each laser beam should be carefully precompensated by carefully tuning the four collimators [19]. In the experiment, when the CPBC system is in the open loop, some of the laser power leaks through PBC3, the intensity profile at the camera is changing all the time, and the combined output power is unsteady due to the phase errors in each beam. Figure 3.11a shows three snapshots of the intensity pattern at moments t_1, t_2, and t_3 when the system is in the open loop. When the SFD algorithm is implemented and the whole system is in the closed loop, the intensity profile and the combined output power are steady. The intensity profile in closed loop is shown in Figure 3.11b. The combined output power measured at the output port of M_2 is 60 W. The combining efficiency η is calculated to be 90%, η being defined by the formula $\eta = P_{out}/P_{in}$, where P_{out} is the combined output power and P_{in} is the total power after the collimators. In the practical system, some factors such as overlap error, tilt error, residual phase error, radius error of the collimators, and amplitude imbalance influence the combining efficiency of the system.

The fidelity of CPBC and the phase noise suppression can further be studied by the time series signal and spectral density of energy collected by the PD, which is shown in Figure 3.12. When the control system is open, the normalized energy collected by the PD fluctuates randomly. When the control system is

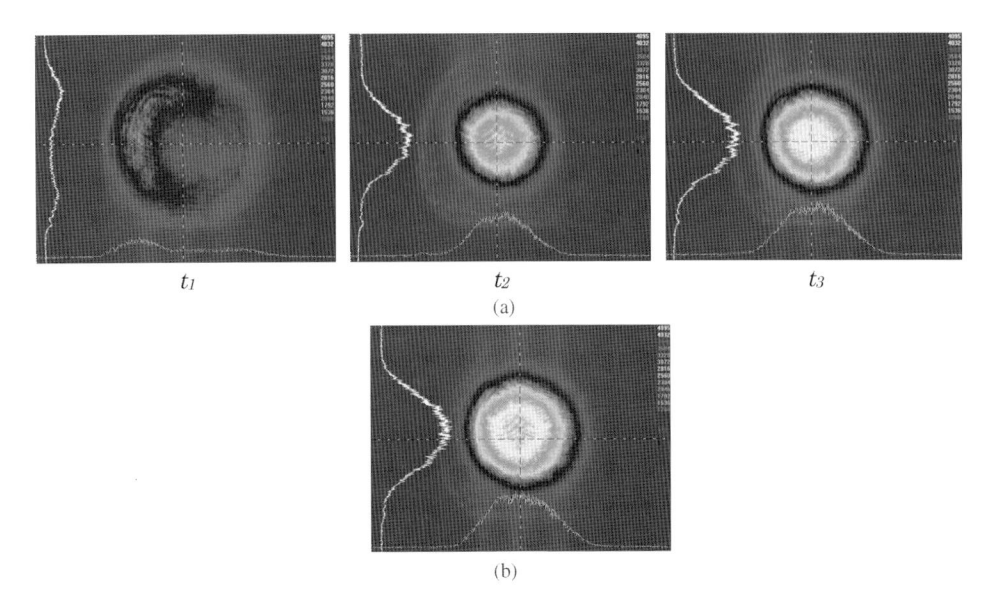

Figure 3.11 The intensity profiles with the system in the (a) open loop and (b) closed loop.

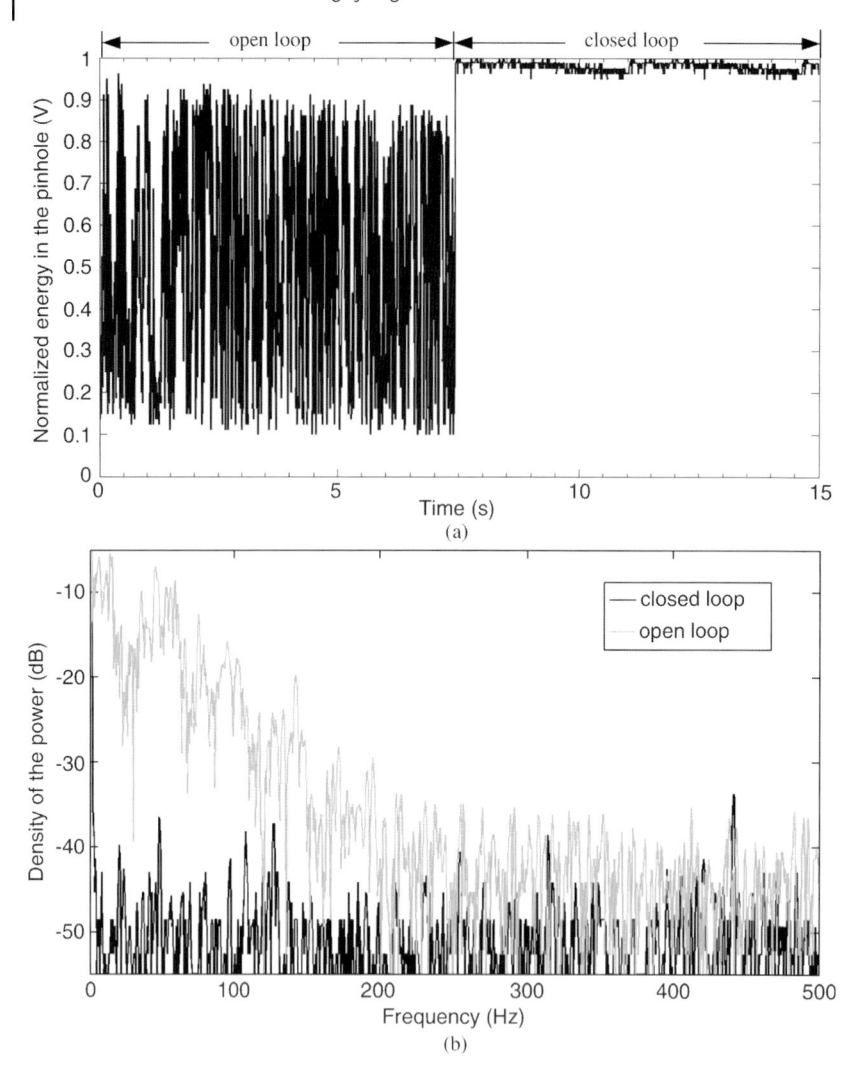

Figure 3.12 Time series signals and spectral density of energy collected by PD in the open loop and closed loop. (a) Time series signals. (b) Spectral density.

closed, the normalized power can be locked stably (see Figure 3.12a), and the spectral density of power is about 20 dB lower than that in the open loop below 200 Hz (see Figure 3.12b), which indicates an effective suppression of the phase noise.

In summary, CPBC of fiber amplifiers using SFD technique has been demonstrated. When the system is in the closed loop, the output power is 60 W and the combining efficiency of the whole system can be as high as 90%.

3.2.4
Target-in-the-Loop Coherent Beam Combination of Fiber Lasers Based on Single-Frequency Dithering Technique

CBC of fiber laser based on cooperative target has been studied by many research teams and has made great developments and the highest output power is up to 4 kW and the maximum number of laser beams is 64 [15,20–25]. However, CBC based on scattering surface is rarely studied, although this technology is very important for CBC system in real environment. Bourdon *et al.* studied CBC based on scattering surface for the first time in 2008 and multidithering technique was used for phase locking [26,27]. In their experiments, three fiber amplifiers were used for CBC and the output power of every fiber laser was less than 2 W. Recently, Weyrauch *et al.* demonstrated CBC based on noncooperative target over a 7 km propagation path with a combined optical power of 12 mW, but the target is a semicooperative corner cube retroreflector and not a noncooperative scattering surface [28]. We have also studied in detail CBC based on noncooperative target and the CBC of two beams based on scattering surface has also been demonstrated [29]. In this chapter, CBC of 9×10 W level fiber amplifiers based on scattering surface is presented and the whole output power is more than 100 W. To our best knowledge, this is the CBC based on noncooperative target with the highest output power and the largest number of fibers at the present time.

The experimental setup is shown in Figure 3.13. The seed laser is a DFB polarization-maintaining Yb-doped fiber laser with 1064 nm wavelength and 20 kHz linewidth. The 30 mW laser power from the optically isolated seed laser is preamplified to 300 mW and is further split into nine channels and coupled to nine phase modulators. These modulators are $LiNbO_3$ phase modulators with more than 100 MHz modulating bandwidth and operate on 1060 nm wavelength. The output power from the phase modulators is higher than 10 mW and coupled to homemade two-stage all-fiber amplifiers. The first stage is used to amplify the laser

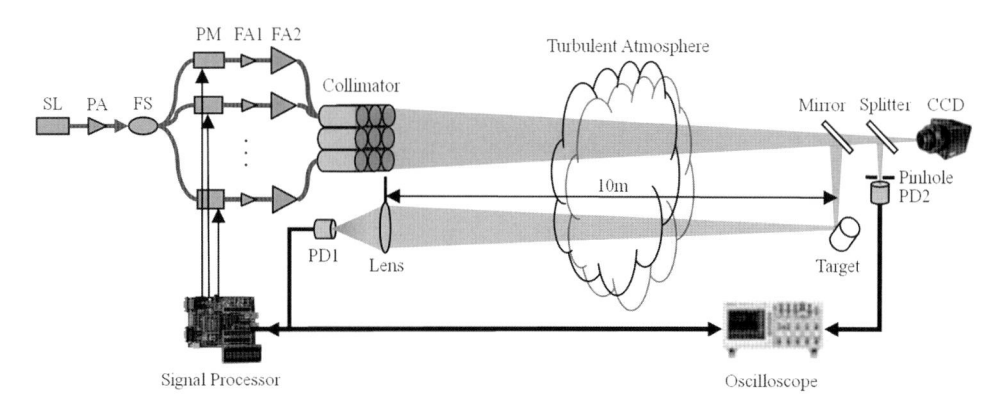

Figure 3.13 The experimental setup of CBC of nine beams based on scattering surface. SL, seed laser; PA, preamplifier; PM, phase modulator; FA, fiber amplifier; PD, photodetector.

power from 10 to 100 mW level, the second stage from 100 mW to 10 W level, and the overall output power is more than 100 W.

The laser beams from the main amplifiers are space-coupled via nine collimators. The diameter of laser beam output from the collimator is 5 mm. The nine laser beams are tiled using free space mirrors side by side into a 3 × 3 laser array with 40% fill factor in the near field. The laser beams from collimators propagate 10 m in free space and then are sampled by a mirror with 99% reflectivity. The reflected light is obtained from scattering on the target surface, which is an aluminium column with rough surface. One part of the passing light reaches a CCD camera used to diagnose the intensity pattern on the scattering surface. The other part of the passing light travels through a hard aperture before reaching a PD (PD2), which is used to diagnose the phase-locking result. Another PD (PD1) is located at the emitting plane and a lens is used to collect the optical power scattered by the target into PD1. The PD is a PDA10CS InGaAs amplifier detector. Contrary to previous CBC systems based on cooperative targets [20,21–25], the phase control signals are obtained from the optical power scattering by glint on the scattering surface and not from an aperture. The electronic signals transformed by PD1 are coupled into the signal processor where SFD algorithm is run.

In the experiment, when the control system is in the open loop, SFD is not performed and the phases of laser beams randomly fluctuate, as shown in Figure 3.14a.

When the phase control system is in the closed loop, the phase control algorithm is implemented and the phase noises are compensated. The intensity pattern at the scattering surface is clear and steady. The 2 s long-exposure far-field intensity distribution is shown in Figure 3.14b–d and its fringe contrast is calculated to be more than 85%. When the target aluminium column is circumrotated, different patterns are obtained, as shown in Figure 3.14b–d. However, the phases of laser array are steadily locked on the surface of the target independent of the intensity pattern.

The fidelity of coherent beam combination and the phase fluctuation suppression can further be studied using the voltage signals generated by PD1 and PD2 shown in Figure 3.15. When the control loop is open, the voltage signals from PD2 fluctuate between 0 and 2 V randomly, which indicates dithering of intensity pattern on the scattering surface. The voltage signals from PD1 fluctuate between 4 and 10 V

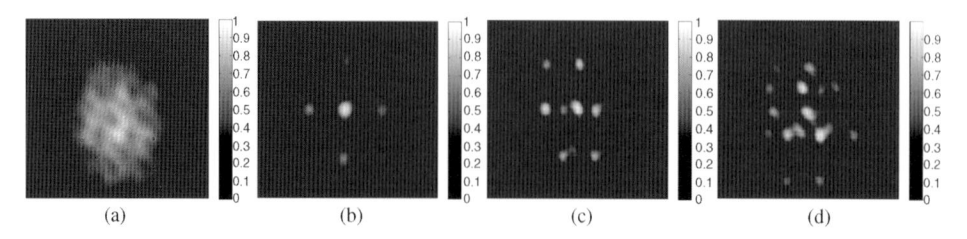

Figure 3.14 Two second long-exposure far-field intensity pattern of the combined laser beam. (a) Open-loop experimental result. (b–d) Closed-loop experimental results for different place on scattering surface.

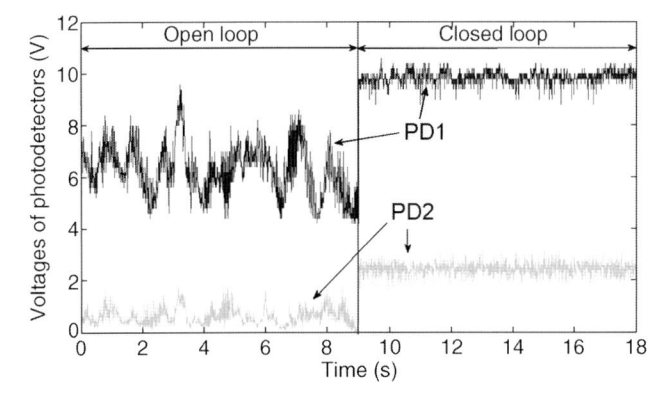

Figure 3.15 Time series signals of energy collected by the aperture in open loop (left) and closed loop (right).

randomly. The optical power collected by PD1 includes light from the glint and that reflected by the plane; the former induces fluctuations randomly and the latter results in a 4 V offset. When the control loop is closed, the voltage signals from PD1 and PD2 are locked steadily at their highest voltage, respectively. It is noted that the voltage signal from PD1 has a low-frequency slight fluctuation that is induced by turbulence during the backward propagation. However, the intensity pattern is locked steadily on the surface of the target that has not been affected by the turbulence during the backward propagation.

Another experiment has been performed to study the effect of the turbulence during the backward propagation on the phase locking of laser array on the surface of the target. In the experiment, a hot air blower has been used to introduce a strong turbulence on the onward and backward path, respectively. When the blower is located in the onward path and works at the highest power, the phases of laser array are not effectively locked because of the strong phase fluctuations induced by the hot air blower (Figure 3.16). When the same turbulence is induced on the backward

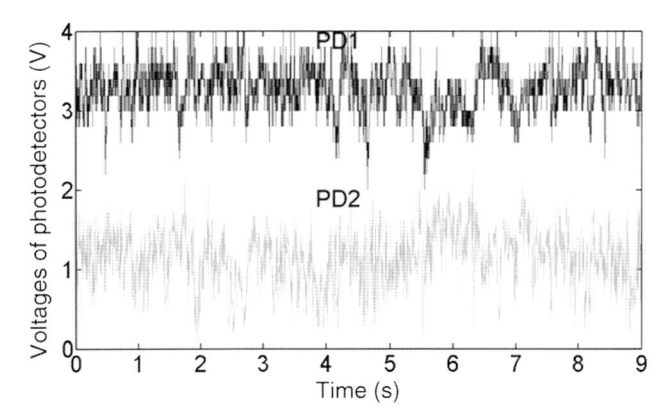

Figure 3.16 The voltage signals of photodetectors in closed loop when the strong turbulence is induced on the onward path.

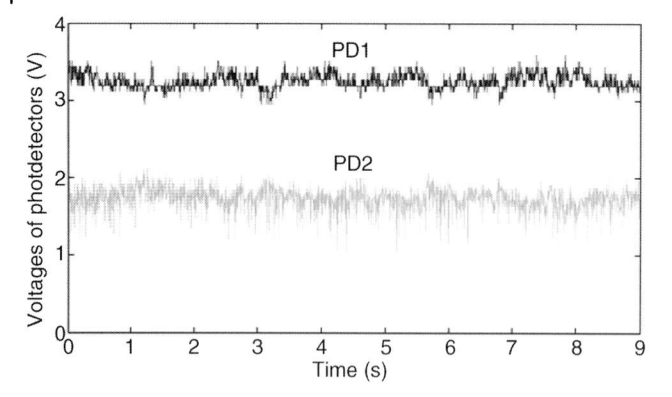

Figure 3.17 The voltage signals of photodetectors in the closed loop when the strong turbulence is induced on the backward path.

path, efficient phase locking is realized (Figure 3.17). These experimental results show that the atmospheric turbulence on the backward path has a weak effect on the phase locking of laser array and the feasibility of CBC based on noncooperative target using SFD technique has been demonstrated again.

In summary, we present the CBC of nine 10 W level fiber amplifiers based on scattering surface target using SFD. The feasibility of CBC based on noncooperative target using SFD has also been demonstrated. The fringe contrast of the long-exposure coherent combined beam profile on the target surface is improved to 85% in closed loop from 0 in open loop. In addition, the effect of atmospheric turbulence on the backward paths on CBC has been studied experimentally. The experimental results show that the turbulence on the backward paths has a weak effect on CBC.

3.3
Sine–Cosine Single-Frequency Dithering Technique

3.3.1
Theory of Sine–Cosine Single-Frequency Dithering Technique

As a result of the serial working mode, the implementing speed of SFD will decrease as the number of beamlets increases. In this section, sine–cosine single-frequency dithering (SCSFD) technique is presented and it is shown that its implementing speed is twice as fast as SFD [30].

SCSFD is presented based on the principle of quadrature modulation and demodulation. For a pair of quadrature signals, the modulation and demodulation of one signal do not affect another one, so two quadrature signals with the same frequency can be used to modulate and demodulate at the same time. In the sine–cosine technique, each pair of lasers in the array is phase modulated sequentially by a pair of sine and cosine signals with the same frequency, and then

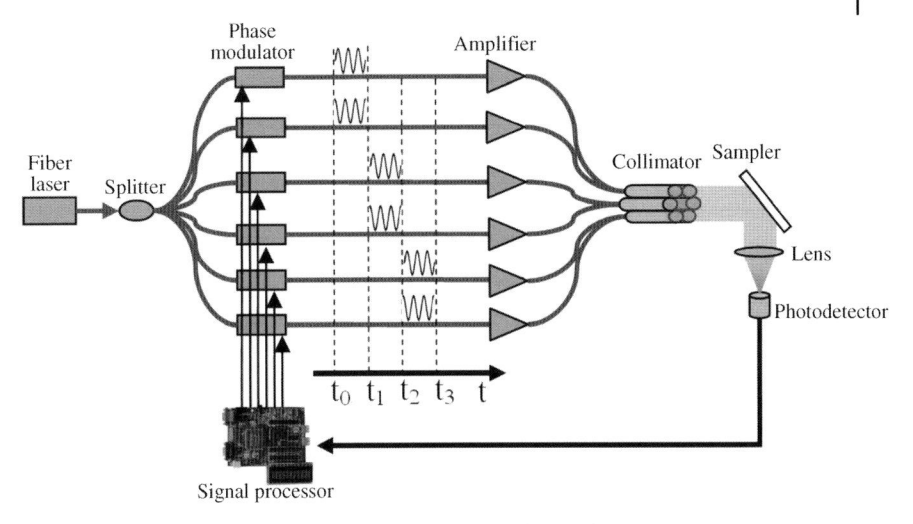

Figure 3.18 Scheme of the sine–cosine single-dithering technique for CBC.

the corresponding phase control signals are obtained by demodulating the above phase modulation signals.

The scheme of SCSFD technique is shown in Figure 3.18. The beam from seed laser is split into N beams (six beams are given in Figure 3.18) and coupled to N or $N-1$ optical phase modulators. The laser beams from the phase modulators are amplified by N fiber amplifiers and sent to free space via collimators. The phase modulator is used to implement the phase modulation and control signals. The beam array emitted from the collimators is focused by a lens and a PD is located at the focus plane. The electronic signals from the PD are used to generate the optical phase control signals in the signal processor. The whole laser array will be in phase-locked state when phase control signals are applied to the phase modulators.

The different steps for SCSFD can be briefly described as follows. Let us assume that t represents time and that the signal processor starts working at time t_0. When $t_0 < t < t_1$, the sine and cosine phase modulation signals are applied to the phase modulators in channels 1 and 2, respectively. Then the phase control signals for channels 1 and 2 can be derived and applied to the phase modulators of channels 1 and 2, respectively. During this first step, phase modulation and control signals are not applied to the remaining channels.

When $t_1 < t < t_2$, the sine and cosine phase modulation signals and relevant phase control signals are applied to the phase modulators in channels 3 and 4, respectively. The modulation signals of channels 1 and 2 are set to zero and the control signal retains the final value at $t = t_1$. The phase modulation and control signals applied to the remaining channels are still kept to zero.

When $t_{i-1} < t \leq t_i$ $(i \leq N)$, the phase modulation signals and relevant control signals are applied to the phase modulators in channels $2i-1$ and $2i$, respectively, and the phase control signals of other channels remain unchanged as the signal generated at $t = t_{i-1}$, while the phase modulation signals become zeros.

The mathematical principle of SCSFD can be briefly described as follows. As for SFD, M Gaussian beams with the same polarization are assumed to be located at a source plane ($z=0$) and propagate along the z-axis in the Cartesian coordinate system. Therefore, when the phase modulation signals are applied to channels $2p-1$ and $2p$, their optical field can be expressed as

$$u_{(2p-1)}(x,y,0,t) = \exp\left[-\frac{(x-a_{(2p-1)})^2 + (y-b_{(2p-1)})^2}{w_0^2}\right] \times \exp\left[i\left(\varphi_{(2p-1)f} + \varphi_{S_m}\right)\right], \quad (3.14)$$

$$u_{(2p)}(x,y,0,t) = \exp\left[-\frac{(x-a_{(2p)})^2 + (y-b_{(2p)})^2}{w_0^2}\right] \times \exp\left[i\left(\varphi_{(2p)f} + \varphi_{C_m}\right)\right], \quad (3.15)$$

where φ_{S_m} and φ_{C_m} are sine and cosine phase modulation signals, respectively, and can be expressed as

$$\varphi_{S_m} = a_m \sin(\omega_m t), \quad (3.16)$$

$$\varphi_{C_m} = a_m \cos(\omega_m t). \quad (3.17)$$

With the same steps as for SFD, the phase control signals of channels $2p-1$ and $2p$ can be obtained as follows:

$$\begin{aligned}S_{np} &= \int_0^{1/\omega_m} i_p(t) \times a\sin(\omega_m t)dt \\ &= \alpha R_{PD}J_1(a_m)\sum_{\substack{j=1\\j\neq p}}^{M}\cos(\Phi_{pjN})\int_S A_p A_j \sin\Phi_{jpL}ds - \sin(\Phi_{pjN})\int_S A_p A_j \cos\Phi_{jpL}ds \\ &= \alpha R_{PD}J_1(a_m)\sum_{\substack{j=1\\j\neq p}}^{M}\cos(\Phi_{pjN})Q_{Cpj} - \sin(\Phi_{pjN})Q_{Spj}.\end{aligned} \quad (3.18)$$

$$S_{n(2p-1)} = \alpha R_{PD}J_1(a_m)\sum_{\substack{j=1\\j\neq 2p-1}}^{M}\cos(\Phi_{(2p-1)jN})Q_{C(2p-1)j} - \sin(\Phi_{(2p-1)jN})Q_{S(2p-1)j}. \quad (3.19)$$

$$S_{n(2p)} = \alpha R_{PD}J_1(a_m)\sum_{\substack{j=2\\j\neq 2p}}^{M}\cos(\Phi_{(2p)jN})Q_{C(2p)j} - \sin(\Phi_{(2p)jN})Q_{S(2p)j}. \quad (3.20)$$

For SCSFD algorithm, two phase control signals are used to compensate for phase errors between beams at the same time, which makes this algorithm two times faster than the SFD.

3.3.2
Coherent Beam Combining of Nine Beams Using Sine–Cosine Single-Frequency Dithering Technique

To demonstrate the feasibility of SCSFD, we have done the active phase locking experiment of nine beams. The experimental setup is shown in Figure 3.19. The seed laser is a DFB polarization-maintaining Yb-doped fiber laser with 1064 nm wavelength and 20 kHz linewidth. The laser from the seed laser through the optical isolator is split into nine beams and coupled to nine $LiNbO_3$ phase modulators, but eight phase modulators have been used to compensate the phase errors. The laser beams from the phase modulators are coupled to optical isolators before being sent to nine fiber amplifiers whose output power can be tuned to be more than 10 W. The output beams from amplifiers are sent to free space via nine collimators in a square array arrangement. The radius of each beam is 2.5 mm, and the distance between the beams is about 15 mm (see Figure 3.19 insert). The Rayleigh distance of the collimators is more than 10 m. The output beams from collimators travel 10 m and are sampled by a cubic beam splitter. After the splitter, one part of the beam passes through a focusing lens with 1 m focal length and is sent to a homemade pinhole with 30 μm radius. A PD is located just behind the pinhole. Another part of the beam after the splitter is also focused by a lens and an infrared CCD camera is placed on the focal plane in order to diagnose the far-field beam profile of the combined beams.

In the experiment, when the phase control system is in the open loop, the SCSFD does not operate and the phases of the laser beams randomly fluctuate due to phase fluctuations in each fiber amplifiers. The 2 s long-exposure far-field intensity distribution is shown in Figure 3.20a, and the fringe contrast is about 21%. When the phase control system is in the closed loop, the phase control algorithm operates and the phase modulation and control signals are applied to the modulators of each channel by the signal processor and the phase errors are compensated efficiently. The intensity pattern at the observing plane is steady. The 2 s long-

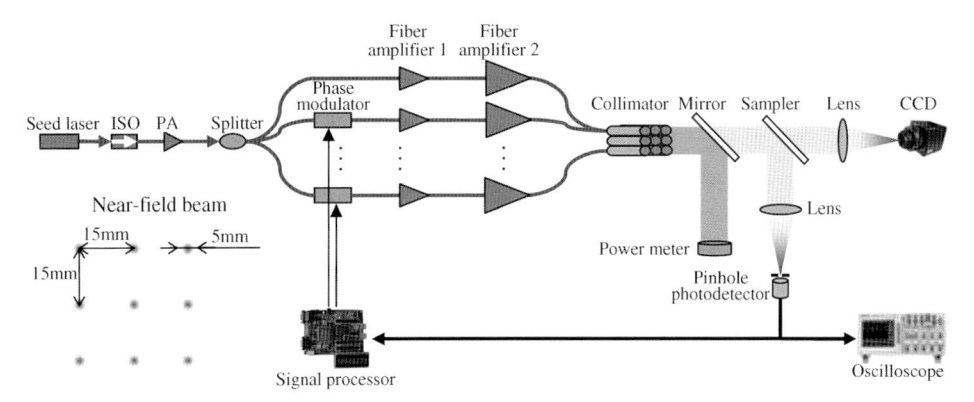

Figure 3.19 The experimental setup of coherent beam combination of nine beams. ISO, isolator; PA, preamplifier.

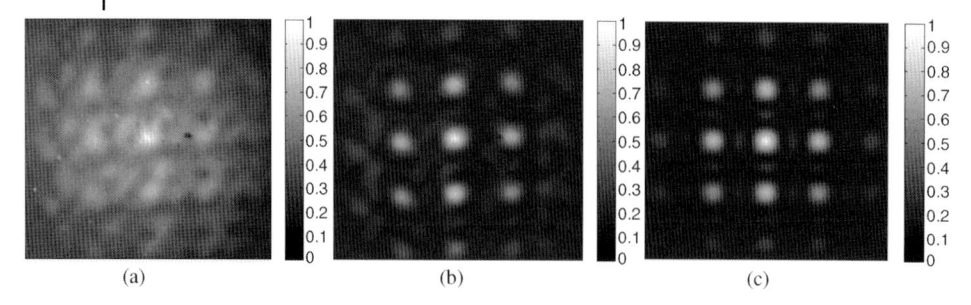

98 | *3 Kilowatt Coherent Beam Combining of High-Power*

(a) (b) (c)

Figure 3.20 Two second long-exposure far-field intensity patterns of the combined laser beam. (a) Open-loop pattern. (b) Closed-loop pattern. (c) Theoretical pattern.

exposure far-field intensity distribution is shown in Figure 3.20b and the fringe contrast is calculated to be more than 90%. The theoretical far-field pattern of the ideally phased nine-laser array, which are arranged according to the above real data, is computed and presented in Figure 3.20c. It is found that the experimental and theoretical results are in good agreement.

The fidelity of phase locking and phase fluctuation suppression results can further be studied using the time series signals of energy collected through the aperture, as shown in Figure 3.21. When the control loop is open, the normalized energy collected by the aperture fluctuates between 0 and 1 randomly. When the control loop is closed, the energy collected through the aperture can be locked steadily to be more than 90% for most of the time. The residual phase error is less than $\lambda/20$, which denotes a remarkable increase of energy in the main lobe. The phase errors that fluctuate less than $\lambda/20$ will not be compensated further because they go beyond the noise floor of the used phase control system.

To compare SCSFD and SFD techniques, the evolution in time from when the algorithm is turned on until the amplifiers are phase locked is shown in Figure 3.22. For ease of discussion, the time from when the algorithm is turned on is defined T_t.

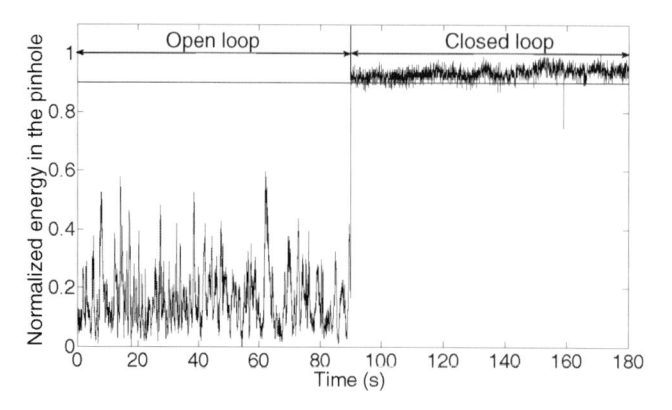

Figure 3.21 Time series signals of energy collected by the aperture in the open loop and closed loop.

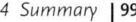

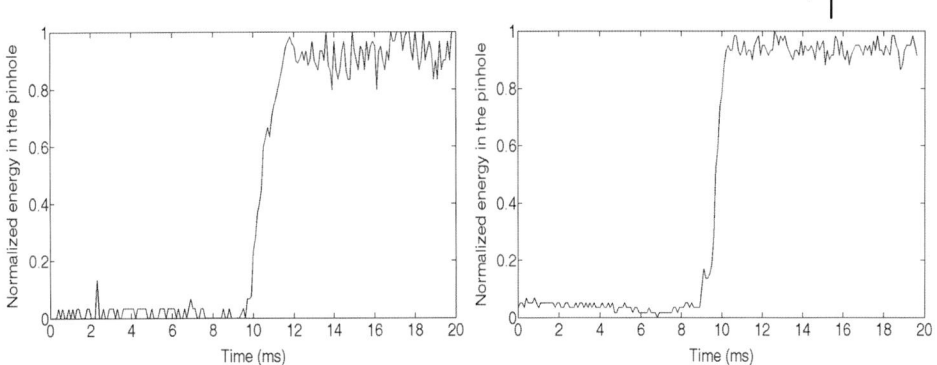

Figure 3.22 The evolution in time from when the algorithm is turned on until the amplifiers are phase locked. (a) Single-frequency dithering technique. (b) Sine–cosine single-frequency dithering technique.

Under same experimental conditions, T_t is about 2 ms for SFD (shown in Figure 3.22a) and about 1 ms for SCSFD techniques (shown in Figure 3.22b). In addition, it is found that the residual phase errors in Figure 3.22a are larger than those in Figure 3.22b. These results show that the implementing speed of SCSFD algorithm is higher than that of SFD algorithm. However, the control bandwidth of the SCSFD algorithm will decrease with an increase in the number of lasers. Hence, increasing the frequency of the modulation signal is still an effective approach to alleviate this difficulty.

3.4 Summary

In this chapter, SFD and SCSFD techniques have been systematically studied. For each, theoretical models have been developed based on Gaussian beam arrays. The results show that the size and position of the hard aperture in front of the PD have a strong effect on CBC when the laser beams are transmitted by different subapertures (tiled geometry). In the experiment, first, CBC of nine 100 W level fiber amplifiers using SFD has been demonstrated and the overall output power is more than 1 kW when the laser beams are transmitted by different subapertures (tiled geometry). Second, CPBC of four fiber amplifiers using SFD is implemented and favorable results have been obtained. Third, target-in-the-loop coherent beam combination of fiber lasers based on SFD has been demonstrated. Finally, the feasibility of CBC using SCSFD has been experimentally demonstrated with CBC of nine 10 W fiber amplifiers. The experimental results show that the phase control speed of sine–cosine technique is twice faster than the single dithering technique. This chapter demonstrates that SFD and SCSFD have the potential to boost dithering-based phase control CBC to a large number of fiber lasers with the advantages of using a simple-to-implement, compact, and cost-effective phase control system.

References

1 Fan, T.Y. (2005) Laser beam combining for high-power, high-radiance sources. *IEEE J. Sel. Top. Quantum Electron.*, **11**, 567–577.

2 Enloe, L.H. and Rodda, J.L. (1965) Laser phase-locked loop. *Proc. IEEE*, **54**, 165–166.

3 Buczek, C.J. and Freiberg, R.J. (1972) Hybrid injection locking of high power CO_2 laser. *IEEE J. Quantum Electron.*, **8**, 641–650.

4 Lebedev, F.V. and Napartovich, A.P. (eds) (1993) *High-Power Multibeam Lasers and Their Phase Locking*, SPIE Proceedings Series, vol. 2109, Society of Photo Optical.

5 Coffer, G., Bernard, J.M., Chodzko, R.A., Turner, E.B., Gross, R.W.F., and Warren, W.R. (1983) Experiments with active phase matching of parallel-amplified multiline HF laser beams by a phase-locked Mach–Zehnder interferometer. *Appl. Opt.*, **22**, 142–148.

6 Levy, J. and Roh, K. (1995) Coherent array of 900 semiconductor laser amplifiers. *Proc. SPIE*, **2382**, 58–69.

7 Goodno, G.D., Asman, C.P., Anderegg, J., Brosnan, S., Cheung, E.C., Hammons, D., Injeyan, H., Komine, H., Long, W., McClellan, M., McNaught, S.J., Redmond, S., Simpson, R., Sollee, J., Weber, M., Weiss, S.B., and Wickham, M. (2007) Brightness-scaling potential of actively phase-locked solid state laser arrays. *IEEE J. Sel. Top. Quantum Electron.*, **13**, 460–472.

8 McNaught, S., Komine, H., Weiss, S., Simpson, R., Johnson, A., Machan, J., Asman, C., Weber, M., Jones, G., Valley, M., Jankevics, A., Burchman, D., McClellan, M., Sollee, J., Marmo, J., and Injeyan, H. (2009) 100 kW coherently combined slab MOPAs. Conference on Lasers and Electro-Optics, Baltimore, MD, May 31, 2009, Paper CThA1.

9 Taylor, L.R., Feng, Y., and Calia, D.B. (2010) 50W CW visible laser source at 589nm obtained via frequency doubling of three coherently combined narrow-band Raman fibre amplifiers. *Opt. Express*, **18**, 8540–8555.

10 Zhou, P., Ma, Y., Wang, X., Xiao, H., Leng, J., Xu, X., and Liu, Z. (2011) Coherent beam combination of thulium-doped fiber lasers. *SPIE Newsroom*. doi: 10.1117/2.1201101.003488

11 Bloom, G., Larat, C., Lallier, E., Lehoucq, G., Bansropun, S., Lee-Bouhours, M.S.L., Loiseaux, B., Carras, M., Marcadet, X., Lucas-Leclin, G., and Georges, P. (2011) Passive coherent beam combining of quantum-cascade lasers with a Dammann grating. *Opt. Lett.*, **36**, 3810–3812.

12 Seise, E., Klenke, A., Breitkopf, S., Limpert, J., and Tünermann, A. (2011) 88W 0.5mJ femtosecond laser pulses from two coherently combined fiber amplifiers. *Opt. Lett.*, **36**, 3858–3860.

13 Preu, S., Malzer, S., Döhler, G.H., Zhao, Q.Z., Hanson, M., Zimmerman, J.D., Gossard, A.C., and Wang, L.J. (2008) Interference between two coherently driven monochromatic terahertz sources. *Appl. Phys. Lett.*, **92**, 221107.

14 Kurti, R.S., Halterman1, K., Shori, R.K., and Wardlaw, M.J. (2009) Discrete cylindrical vector beam generation from an array of optical fibers. *Opt. Express*, **17**, 13982–13998.

15 Yu, C.X., Augst, S.J., Redmond, S.M., Goldizen, K.C., Murphy, D.V., Sanchez, A., and Fan, T.Y. (2011) Coherent combining of a 4kW, eight-element fiber amplifier array. *Opt. Lett.*, **36**, 2686–2688.

16 Ma, Y., Zhou, P., Wang, X., Ma, H., Xu, X., Si, L., Liu, Z., and Zhao, Y. (2010) Coherent beam combination with single frequency dithering technique. *Opt. Lett.*, **35**, 1308–1310.

17 Ma, Y., Wang, X., Leng, J., Xiao, H., Dong, X., Zhu, J., Du, W., Zhou, P., Xu, X., Si, L., Liu, Z., and Zhao, Y. (2011) Coherent beam combination of 1.08kW fiber amplifier array using single frequency dithering technique. *Opt. Lett.*, **36**, 951–953.

18 Ma, P., Zhou, P., Ma, Y., Su, R., and Liu, Z. (2012) Coherent polarization beam combining of four high power fiber amplifiers using single-frequency dithering technique. *IEEE Photonics Technol. Lett.*, **24**, 1024–1026.

19 Vorontsov, M.A., Weyrauch, T., Beresnev, L.A., Carhart, G.W., Liu, L., and Aschenbach, K. (2009) Adaptive array of phase-locked fiber collimators: analysis and

experimental demonstration. *IEEE J. Sel. Top. Quantum Electron.*, **15**, 269–280.

20 Anderegg, J., Brosnan, S.J., Cheung, E., Epp, P., Hammons, D., Komine, H., Weber, M.E., and Wickham, M. (2006) Coherently coupled high power fiber arrays. *Proc. SPIE*, **6102**, 202–206.

21 Xiao, R., Hou, J., Liu, M., and Jiang, Z.F. (2008) Coherent combining technology of master oscillator power amplifier fiber arrays. *Opt. Express*, **16**, 2015–2022.

22 Liu, L., Vorontsov, M.A., Polnau, E., Weyrauch, T., and Beresnev, L.A. (2007) Adaptive phase-locked fiber array with wavefront phase tip–tilt compensation using piezoelectric fiber positioners. *Proc. SPIE*, **6708**, 67080K.

23 Shay, T.M., Baker, J.T., Sanchez, A.D., Robin, C.A., Vergien, C.L., Zeringue, C., Gallant, D., Chunte, L.A., Pulford, B., Bronder, T.J., and Lucero, A. (2009) High power phase locking of a fiber amplifier array. *Proc. SPIE*, **7195**, 71951M.

24 Shay, T.M. and Benham, V. (2004) First experimental demonstration of phase locking of optical fiber arrays by RF phase modulation. *Proc. SPIE*, **5550**, 313.

25 Bourderionnet, J., Bellanger, C., Primot, J., and Brignon, A. (2011) Collective coherent phase combining of 64 fibers. *Opt. Express*, **19**, 17053–17058.

26 Bourdon, P., Jolivet, V., Bennaï, B., Lombard, L., Canat, G., Pourtal, E., Jaouen, Y., and Vasseur, O. (2008) Coherent beam combining of fiber amplifier arrays and application to laser beam propagation through turbulent atmosphere. *Proc. SPIE*, **6873**, 687316.

27 Jolivet, V., Bourdon, P., Bennaï, B., Lombard, L., Goular, D., Pourtal, E., Canat, G., Jaouën, Y., Moreau, B., and Vasseur, O. (2009) Beam shaping of single-mode and multimode fiber amplifier arrays for propagation through atmospheric turbulence. *IEEE J. Sel. Top. Quantum Electron.*, **15**, 257–268.

28 Weyrauch, T., Vorontsov, M.A., Carhart, G. W., Beresnev, L.A., Rostov, A.P., Polnau, E. E., and Liu, J.J. (2011) Experimental demonstration of coherent beam combining over a 7km propagation path. *Opt. Lett.*, **36**, 4455–4457.

29 Tao, R., Ma, Y., Si, L., Dong, X., Zhou, P., and Liu, Z. (2011) Target-in-the-loop high-power adaptive phase-locked fiber laser array using single-frequency dithering technique. *Appl. Phys. B*, **105**, 285–291.

30 Ma, Y., Zhou, P., Wang, X., Ma, H., Xu, X., Si, L., Liu, Z., and Zhao, Y. (2011) Active phase locking of fiber amplifiers using sine–cosine single-frequency dithering technique. *Appl. Opt.*, **50**, 3330–3336.

4
Active Coherent Combination Using Hill Climbing-Based Algorithms for Fiber and Semiconductor Amplifiers

Shawn Redmond, Kevin Creedon, Tso Y. Fan, Antonio Sanchez-Rubio, Charles Yu, and Joseph Donnelly

4.1
Introduction to Hill Climbing Control Algorithms for Active Phase Control

Active coherent combination has been shown to be a viable way to effectively combine the output of several laser sources. Instrumental in any active coherent combination architecture is the ability to control and correct the phase of the individual lasers to be combined. Depending on the intended application, this could require combining tens to thousands of lasers, necessitating a solid understanding of the scaling of the control algorithms' performance and beam combination architecture. Active phase control algorithms can be broadly categorized into two types: phase sensing and metric sensing. Phase sensing detects the phase error directly and typically stabilizes it to a zero error state. This can be achieved with RF heterodyne approaches, which are the subject of other chapters and interferometry-based approaches [1–3]. Metric sensing, on the other hand, directly detects the metric, or signal, and works to optimize it, often at the expense of direct knowledge of actual errors being corrected.

One fairly recent metric sensing approach applied to this challenging control problem is the stochastic gradient descent algorithm (SPGD) demonstrated by Vorontsov *et al.* [4] in 1997 for use in active phase correction. It is an attractive algorithm due to its simplicity of implementation and will be the basis for the remainder of the chapter. It is a model-free algorithm, implying that it can be adapted to most optimization problems independent of the functional relationship between control variables. For implementation in active phase control, it uses a single photodetector to provide a metric for phasing of the individual lasers. This greatly simplifies system architectures, which is beneficial for reducing cost as well as size, weight, and power (SWaP). Understanding the scaling behavior and performance of the SPGD algorithm in potential dynamic environments can help determine appropriate and inappropriate applications.

In this chapter, we first describe the use of SPGD algorithm for active coherent combination of lasers and the scaling of the system convergence as the number of

Coherent Laser Beam Combining, First Edition. Edited by Arnaud Brignon.
© 2013 Wiley-VCH Verlag GmbH & Co. KGaA. Published 2013 by Wiley-VCH Verlag GmbH & Co. KGaA.

lasers, or elements, is increased. Next, two variations of the traditional SPGD control algorithm are described, which allow faster system convergence by optimization of the elements' dither relationship and introduction of multiple detectors that provide additional information, thereby reducing the "effective" number of independent elements.

The chapter then describes application of the SPGD algorithm toward active coherent combination of semiconductor amplifiers and fiber amplifiers. The semiconductor amplifiers combined are based on the slab-coupled optical waveguide (SCOW) concept developed at MIT Lincoln Laboratory by Walpole et al. [5] in 2002. With nearly circular ∼4–5 μm outputs and 1 W level output powers, they provide greater flexibility of beam combination architectures than the ∼1 μm highly astigmatic output of conventional semiconductor devices. Results of active coherent combination from tiled beam architectures as well as single-beam outputs using a diffractive optical element (DOE) are summarized. Ytterbium–doped fiber amplifiers (YDFAs) with properties appropriate for active beam combination have achieved kilowatt-level output powers. The important characteristics of beam-combinable fiber amplifiers are discussed, followed by a summary of tiled beam and single-beam active coherent combination results using commercially available 500 W fiber amplifiers.

4.1.1
Conventional SPGD-Based Control Algorithm for Active Phase Control

The SPGD algorithm is an optimization technique, which when applied to active phase control works to optimize or reduce the relative phase error of the lasers to be combined. The metric that is optimized is the combined power or on-axis intensity, depending on the beam combination architecture. The power or Strehl is maximized by minimizing the relative phase errors, thus enabling the use of a simple power meter as detector. This eliminates the need for a reference beam for heterodyne detection or other interferometry-based method, as well as individual frequency modulation and demodulation. However, this detection simplification does not come for free. Since the SPGD algorithm works to optimize a metric, which in this case is combined power, without directly measuring the phase, it cannot provide the phase errors. It is for this reason that it also has an array-size-dependent convergence time.

In order to conceptualize the operation of the SPGD algorithm, Figure 4.1 provides an illustration of a two-element example that has a maximum Strehl, or signal metric, when the relative phase error is 0. The SPGD algorithm works by applying small perturbations, or dithers, about the current state of the system, shown here as a phase error of x_0. The measured signal metric with the dithers applied along with the dither amplitudes provides an estimate of the slope used as the error signal. An iteration cycle involves applying the dithers both positive and negative along with the previous iteration's correction. The system converges to the maximum signal because it is only at this location that the measured gradient will be zero. In addition, in order to be positioned exactly at the peak for the

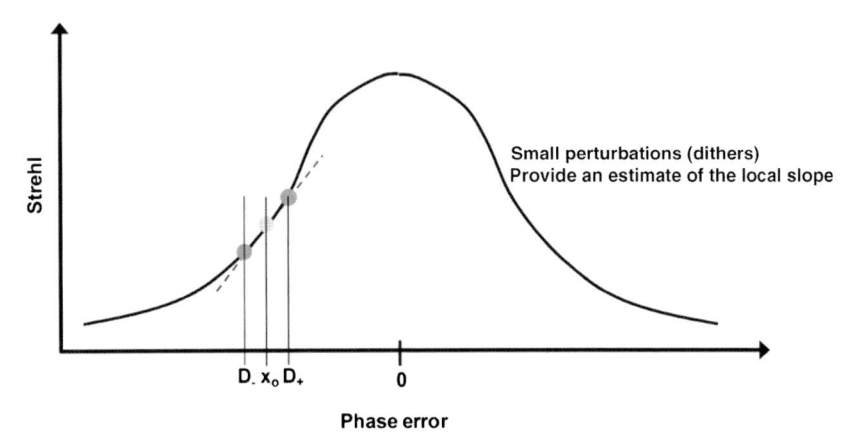

Figure 4.1 Schematic of two-element example using SPGD. With the system at a phase error x_0, a small dither is applied in both directions to estimate the local slope. This provides the error signal used for maximizing the metric.

maximum signal, the dither must be applied equally with opposite signs to ensure a zero gradient is only measured when the system is at the maximum. When scaling to N-elements, the dither becomes an N-element vector with a different dither amplitude for each element. The dither vector changes each iteration to provide the necessary information to ensure convergence to the peak in the N-dimensional space.

The standard control equation [3] commonly implemented for the applications to be later summarized is

$$A_{k+1} = g_{\text{leakage}} A_k + g \left(\frac{1}{2\sigma^2_{\text{rad,rms}}} \right) \left(\frac{S_+ - S_-}{S_+ + S_-} \right) D_k, \quad (4.1)$$

where A_k is the kth iteration of element A, g_{leakage} is a gain used to "leak" off any accumulated piston phase error typically nearly equal to 1, g is the control gain, σ is the standard deviation of the dither vector in radians, D_k is the dither amplitude for element A, and S_+ and S_- are the power detector measurements when adding or subtracting the dither, respectively. Dither arrays are made up of multiple dither vectors either as rows or columns of the array with a different dither vector used for subsequent iterations and traditionally comprise random and statistically independent values. In addition, each dither vector has an equal variance providing equal gain and an unbiased correction. For large numbers of elements, use of dither vectors that apply to all elements simultaneously is important in order to maintain a detectable response. When applying small phase dithers to all elements, the root mean square (RMS) phase change for the entire array due to the dither that can be held constant regardless of the number of elements, thereby making the dither signal always a constant fraction of the total signal. This element-independent signal-to-noise fraction is a very useful feature of this algorithm for scaling to large numbers of elements. Simulations and experiments have empirically found that the convergence time is

$$t_{\text{convergence}} = \frac{\alpha \cdot 2 \cdot N_{\text{channels}}}{f_{\text{dither}}}, \quad (4.2)$$

where α is a dither-dependent constant and f_{dither} is the frequency for each dither.

4.1.2
Orthonormal Dither-Based Control Algorithm

With the convergence time that has a linear dependence on the number of elements, it is important to keep the linear coefficient α as small as possible to reduce the convergence time. Simulations and experimental measurements have found that for conventional SPGD with random dithers, the linear coefficient is ∼4–8, depending on the definition of the convergence time. In order to assess whether this coefficient could be further reduced, one can begin with a derivation of the SPGD algorithm itself as applied to active phase control. When following the derivation reported by Vorontsov and Sivokon [6], the signal metric is expanded as a Taylor's series approximation about the different laser elements' phase. Vorontsov showed that for statistically independent dither sets, the algorithm will generally converge with each update, but with a small error due to the use of statistically independent dithers. However, upon closer inspection, the SPGD algorithm for active phase control does not require dither sets to be only statistically independent, but should rather work using orthonormal dither sets, and as it will be shown is significantly faster when doing so.

Orthonormal dither sets are arrays consisting of $(N-1)$ N-element dither vectors, where each dither vector is normalized to have a standard deviation of 1 and where all dither vectors are orthogonal to each other. Mathematically, this is expressed as $\langle D_i, D_k \rangle = \delta_{ik}$, where δ_{ik} is the Kronecker delta. Only $N-1$ dither vectors are required for a N-element active phasing system as there are only $N-1$ relative phase differences. When using orthonormal dithers, several expressions that were previously approximations for statistically independent dithers become exact. This results in eliminating any error term in the estimate of the gradient, thereby creating a faster system convergence.

This approach of orthonormal dithers was implemented in simulations and experiments. An experimental comparison is shown in Figure 4.2 using 21-semiconductor amplifiers. It was found that by using orthonormal dithers, the linear coefficient is ∼2, which is approximately two–four times faster than that when using statistically independent dithers. With a linear coefficient of ∼2, this implies that it takes two iterations through the orthonormal set to converge to steady state. Given the gradient–descent nature of the SPGD algorithm, it was puzzling that the convergence time was so short. As a point of reference, the smallest possible linear coefficient is 1, which is for the case when the exact error is detected and applied. In order to gain insight into why the convergence times are so fast, an alternative derivation offers a more intuitive understanding of system convergence. It also

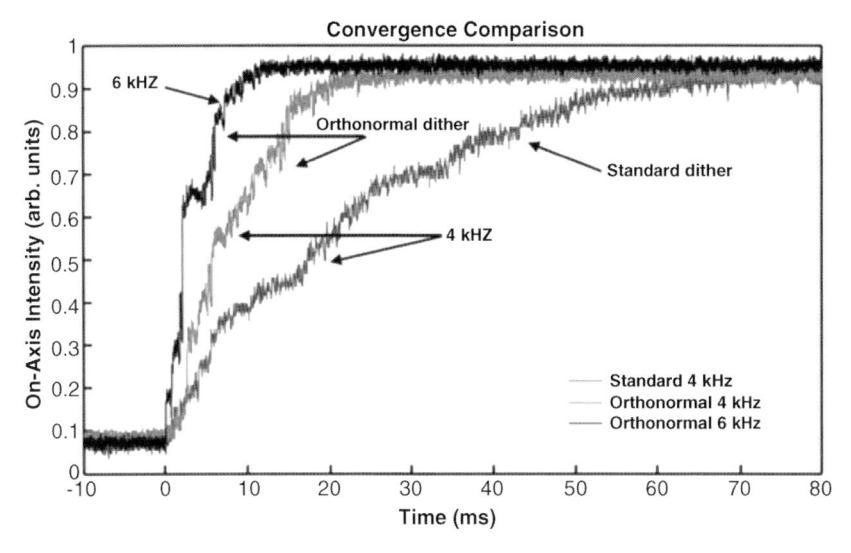

4.1 Introduction to Hill Climbing Control Algorithms for Active Phase Control | 107

Figure 4.2 Experimental comparison of orthonormal dither and standard dither approaches.

reveals that when using orthonormal dithers with the current control equation for active phase control, the algorithm is not really SPGD or gradient–descent, but rather a series-based "phase sensing" algorithm.

Imagine an N-element array of laser elements that will be phase locked to each other. The RMS phase error of the total array is

$$\varphi_{\text{rms}} = \sqrt{\frac{\sum_{N}^{i=1}(\varphi_i - \bar{\varphi})^2}{N}}, \quad \text{where } \bar{\varphi} \text{ is the mean phase.} \quad (4.3)$$

This is essentially the sum of individual element errors relative to the mean. For phase locking, the absolute value of the phase is not important, so the array phase error relative to the first element could be equivalently written as

$$\varphi_{\text{error,rms}} = \sqrt{\frac{\sum_{N}^{i=2}(\varphi_i)^2}{N-1}}, \quad \text{where } \varphi_1 \text{ is set to 0.} \quad (4.4)$$

This expression implies that in order to minimize the phase error, every element's phase error needs to be minimized, which makes intuitive sense. In active phase control applications, the metric used is the Strehl ratio, which is related to the RMS of the phase error according to the Maréchal approximation:

$$S \approx e^{-\varphi_{\text{rad,rms}}^2}. \quad (4.5)$$

Substituting the expression for the phase array error leads to

$$S \approx \left(e^{-\left(\varphi_2^2/N-1\right)}\right)\left(e^{-\left(\varphi_3^2/N-1\right)}\right) \cdots \left(e^{-\left(\varphi_N^2/N-1\right)}\right), \quad (4.6)$$

which is a product of independent Gaussians. In order to maximize the Strehl or minimize the phase error, this simplifies to minimizing $N-1$ independent errors with a Gaussian dependence on the metric. For a Gaussian hill $e^{-a(x-x_0)^2}$, when evaluating the normalization factor used in the SPGD control equation, the metric reduces to the actual error term itself, which explains the rapid convergence:

$$\left(\frac{S_+ - S_-}{S_+ + S_-}\right)\left(\frac{1}{2\Delta x}\right) = \left(\frac{e^{-a(x+\Delta x-x_0)^2} - e^{-a(x-\Delta x-x_0)^2}}{e^{-a(x+\Delta x-x_0)^2} + e^{-a(x-\Delta x-x_0)^2}}\right)\left(\frac{1}{2\Delta x}\right)$$

$$= \frac{-\sinh(2a\Delta x(x-x_0))}{2\cdot \Delta x \cdot \cosh(2a\Delta x(x-x_0))}, \quad (4.7)$$

where Δx is the dither amplitude. For small Δx,

$$\left(\frac{S_+ - S_-}{S_+ + S_-}\right)\left(\frac{1}{2\Delta x}\right) \approx -a(x-x_0). \quad (4.8)$$

This only applies to a Gaussian hill, and it can be shown that $a \approx 1$ for a random set of uncorrelated laser elements. This is the key to explaining the rapid convergence because each dither provides a nearly exact measurement of the error. In other words, if the Strehl ratio were exactly Gaussian for all phase errors, then the system should converge in $N-1$ steps when dithering one element at a time. This can be generalized to creating complete orthonormal basis sets consisting of all elements such that each dither vector actually consists of some fraction of all laser elements. This can be a source of confusion as all elements are now simultaneously dithered, but this does not imply it can converge faster than dithering one element at a time.

Combining these results allows an alternative derivation of the control equation:

$$A_{k+1} = g_{\text{leakage}} A_k + g\left(\frac{1}{2D_{k,\text{amp}}}\right)\left(\frac{S_+ - S_-}{S_+ + S_-}\right)\left(\frac{D_k}{D_{k,\text{amp}}}\right), \quad (4.9)$$

where $D_{k,\text{amp}}$ is the RMS amplitude of the dither vector and the correction term is the exact error for a Gaussian hill multiplied by the normalized dither vector. For the Strehl-based metric with equal variance dithers, this becomes

$$D_{k,\text{amp}} = \sigma_{\text{rms}}, \quad (4.10)$$

$$A_{k+1} = g_{\text{leakage}} A_k + g\left(\frac{1}{2\sigma_{\text{rms}}^2}\right)\left(\frac{S_+ - S_-}{S_+ + S_-}\right)D_k. \quad (4.11)$$

This derivation also implies that for small phase errors, the convergence should be close to one time through the dither set, $a = 1$, which is the fastest convergence possible.

As stated earlier, uncorrelated laser elements have a Strehl ratio that is well approximated by a Gaussian. This characteristic is the key reason why, for active phase control, the SPGD algorithm with orthonormal dithers converges so quickly. To provide some examples of how close the approximation is, Figures 4.3–4.6 plot the Strehl ratio of various sized linear-phased arrays with random phase errors along with a fit to a Gaussian ($e^{-\sigma^2}$). The individual data points are simulation outputs for

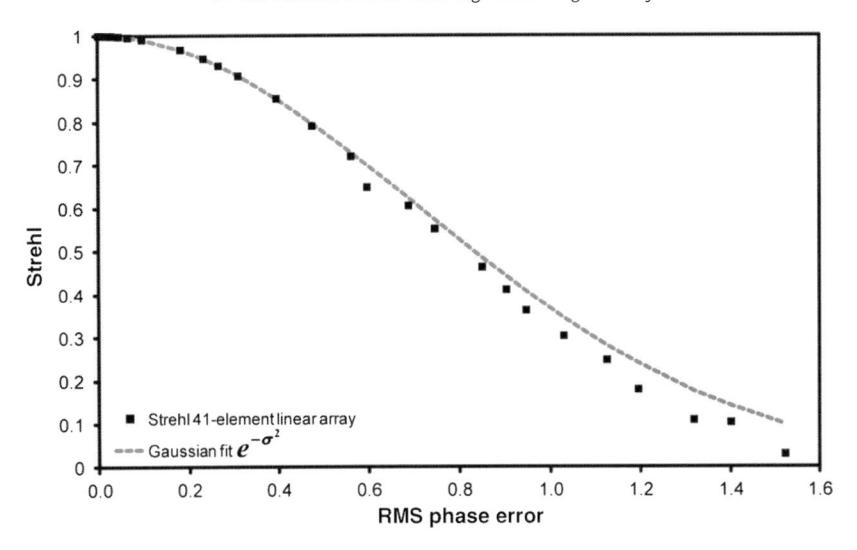

Figure 4.3 Simulated Strehl versus RMS phase error of a 41-element linear array with randomly generated individual element phase.

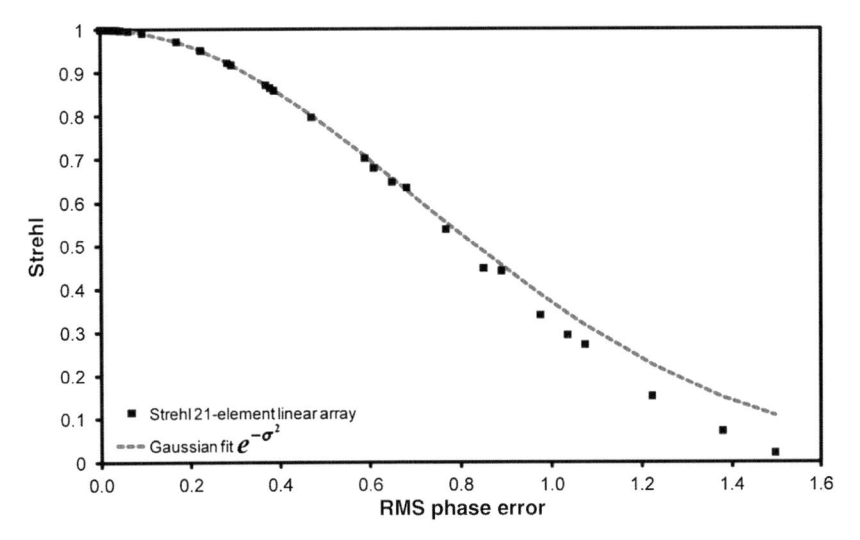

Figure 4.4 Simulated Strehl versus RMS phase error of a 21-element linear array with randomly generated individual element phase.

linear arrays when the individual element phases are randomly generated. The RMS phase error and Strehl ratio are computed for each random realization and plotted. These indicate how closely the Strehl ratio is approximated by a Gaussian function, especially for small phase errors when the number of elements to be combined is greater than 2. This implies that the control equation (4.11) does provide nearly an exact measure of the phase error. Figure 4.7 plots the case for a two-element array

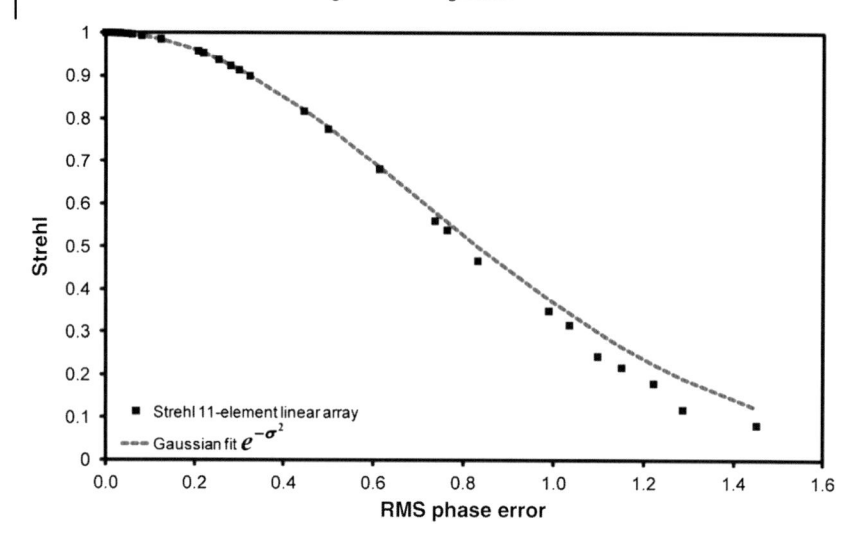

Figure 4.5 Simulated Strehl versus RMS phase error of an 11-element linear array with randomly generated individual element phase.

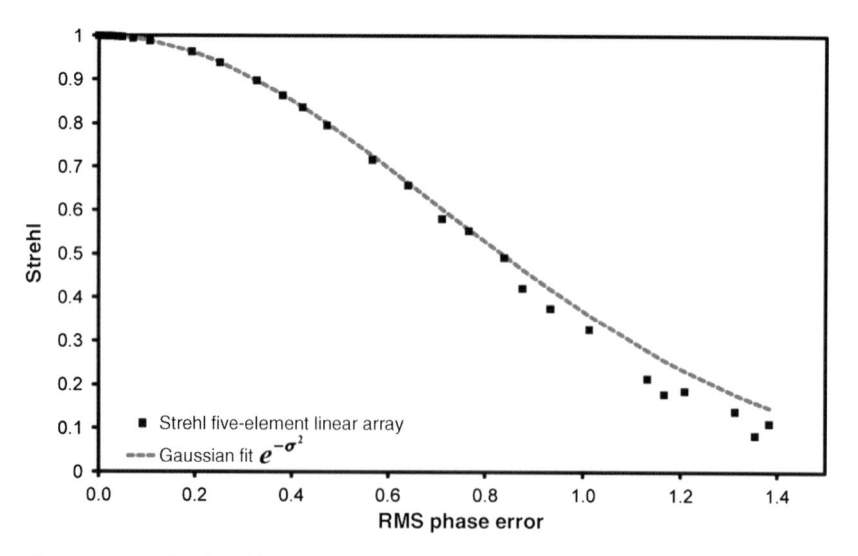

Figure 4.6 Simulated Strehl versus RMS phase error of a five-element linear array with randomly generated individual element phase.

that is exactly a $\cos^2$ functional dependence along with a Gaussian fit. This will be used later to show how changing the shape of the hill impacts the convergence time.

To illustrate it differently, Figures 4.8 and 4.9 plot the Strehl and error term assuming that the Strehl had a Gaussian functional dependence or a $\cos^2$ function

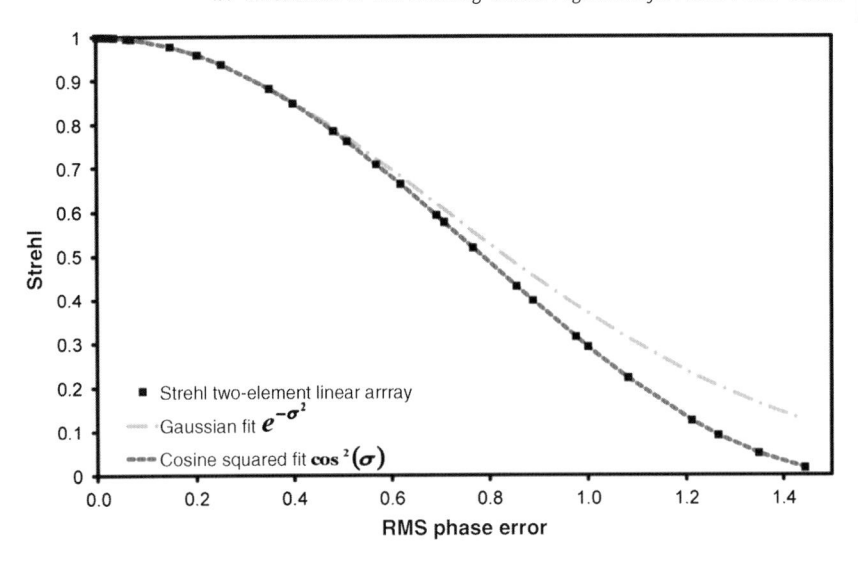

Figure 4.7 Simulated Strehl versus RMS phase error of a two-element linear array with randomly generated individual element phase.

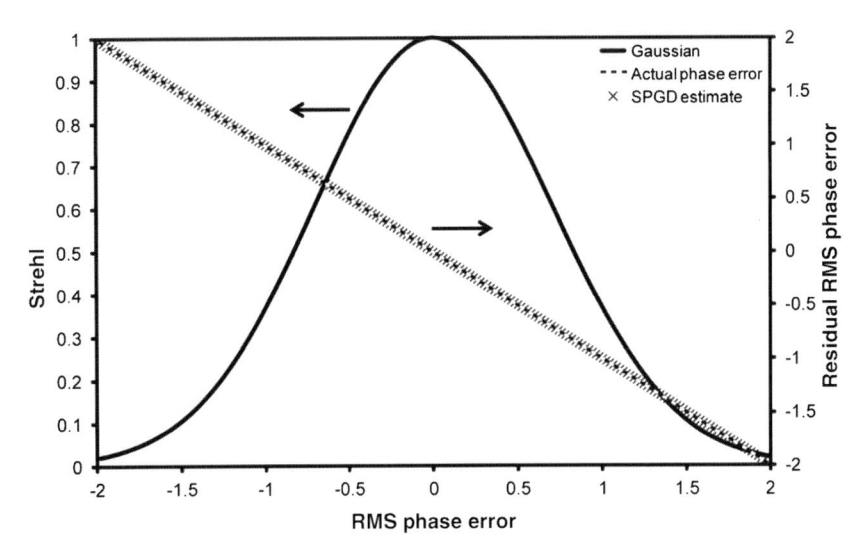

Figure 4.8 SPGD estimate compared with actual error for a Gaussian functional dependence.

dependence. It is observed that for small dither amplitudes, the error is estimated nearly exactly, regardless of the error for a Gaussian case, but degrades for the $\cos^2$ case further away from the peak. Figures 4.10–4.12 plot how this impacts the convergence of a phased array, as a function of the number of iterations through the

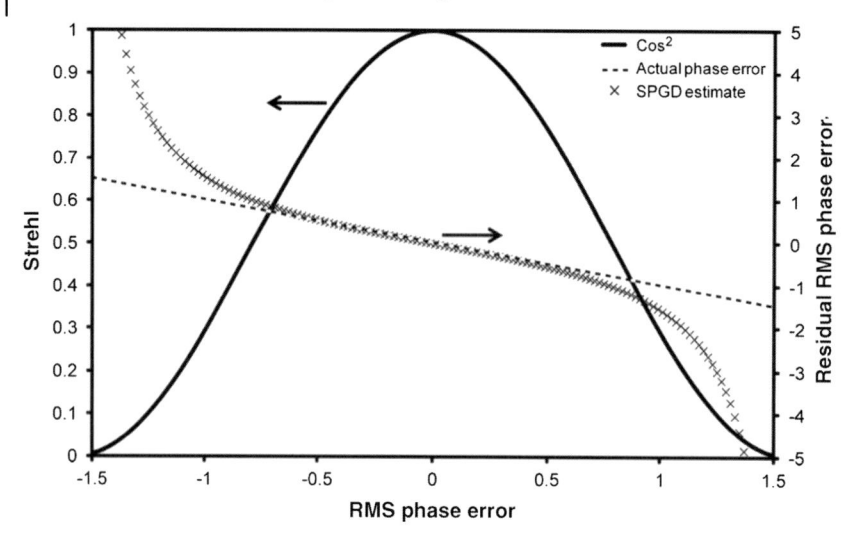

Figure 4.9 SPGD estimate compared with actual error for a $\cos^2$ functional dependence encountered when phasing two elements.

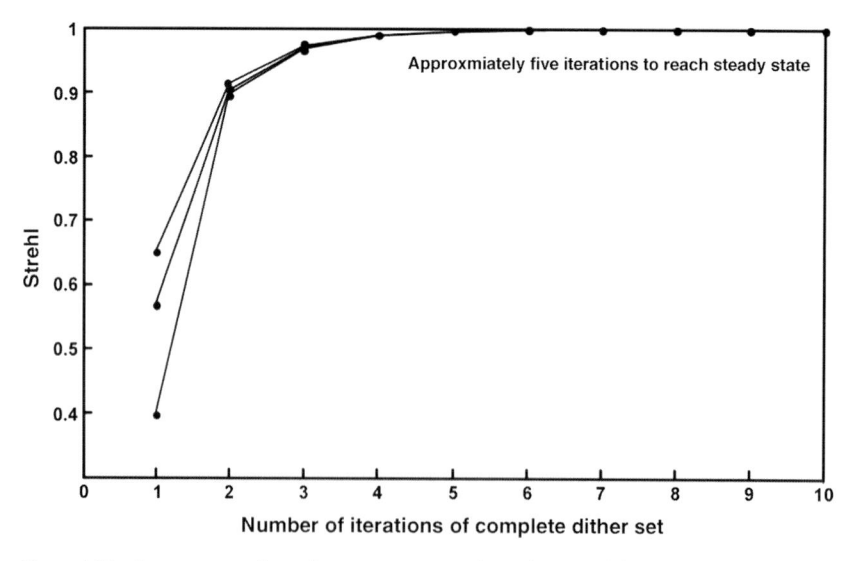

Figure 4.10 Convergence of two-element array using the orthonormal dither SPGD algorithm with three different random initial phase errors. A two-element array has a $\cos^2$ functional dependence on phase error.

complete dither set, using the standard control equation (4.11). Overall both converge very fast, but faster for arrays with greater than two elements due to the near exact estimate of the phase error caused by the nearly Gaussian functional dependence.

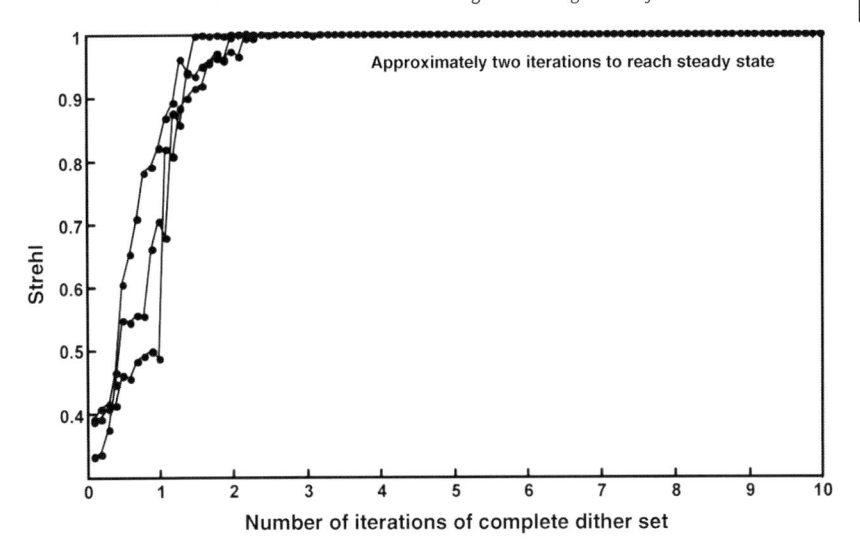

Figure 4.11 Convergence of 11-element array using the orthonormal dither SPGD algorithm with three different random initial phase errors. An 11-element array has nearly a Gaussian functional dependence on phase error. This also illustrates the ~2 linear coefficient when using orthonormal dithers.

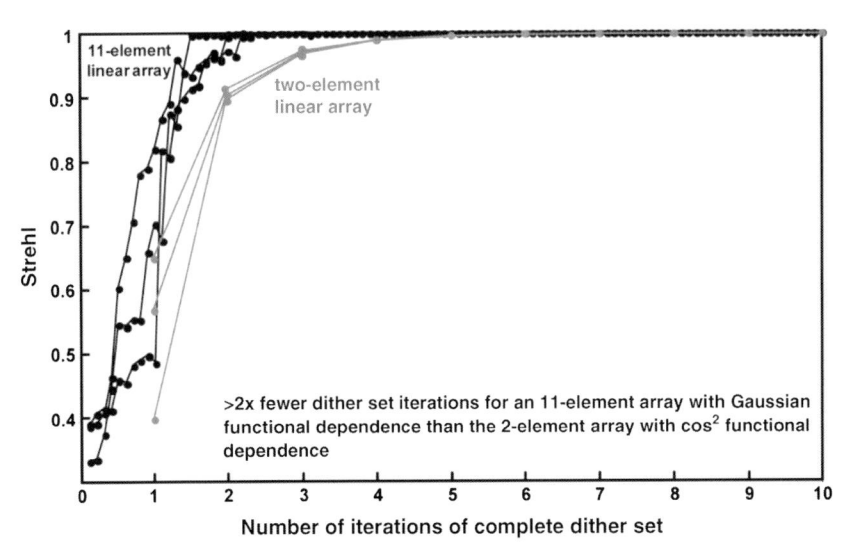

Figure 4.12 SPGD convergence comparison of 2- and 11-element arrays illustrating the slower convergence when the functional dependence is not Gaussian.

4.1.3
Multiple Detector-Based Control Algorithm

Conventional SPGD for active phase control to date has relied on a single metric or single power meter. Due to the linear dependence of the convergence time with the number of elements, it is worthwhile to consider the impact of additional metrics. To understand the potential impact, imagine a two-dimensional array to be actively combined, as illustrated in Figure 4.13. If one considers the approach of first phasing along one dimension and then the second, the convergence time would then be proportional to the sum of the elements for each dimension instead of the product. This should result in a speed increase of the total number of elements divided by the sum of the elements in each dimension. In effect, this would reduce the number of "effective" elements of the system. In order to achieve this, a detector would be required for each subgroup and the entire array. This concept was tested experimentally; a schematic is shown in Figure 4.14.

Although Figure 4.13 shows a sequential nature of the convergence, the actual implementation was simultaneous. The control equation used was

$$A_{k+1} = g_{leakage} A_k$$
$$+ g \left(\frac{1}{2\sigma_{rms}^2} \right) \left(\frac{S_{global+} - S_{global-}}{S_{global+} + S_{global-}} \cdot D_{piston} + \frac{S_{group\ 1+} - S_{group\ 1-}}{S_{group\ 1+} + S_{group\ 1-}} \right.$$
$$\left. \cdot D_{group} + \cdots + \frac{S_{group\ N+} - S_{group\ N-}}{S_{group\ N+} + S_{group\ N-}} \cdot D_{group} \right), \quad (4.12)$$

where all power detector measurements are used every iteration to compute the correction for the corresponding subgroup and the entire array. Each element now has a dither consisting of two components:

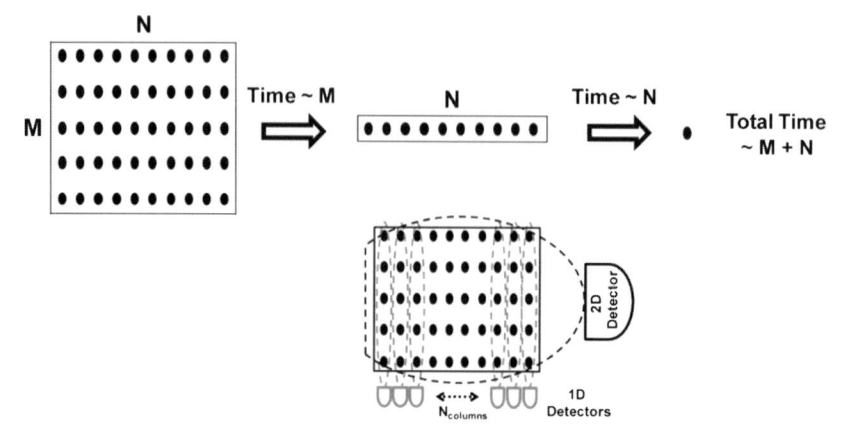

$$D_{mult} = D_{piston} + D_{group}, \quad (4.13)$$

Figure 4.13 Illustration of two-dimensional array to be actively combined.

4.1 Introduction to Hill Climbing Control Algorithms for Active Phase Control | 115

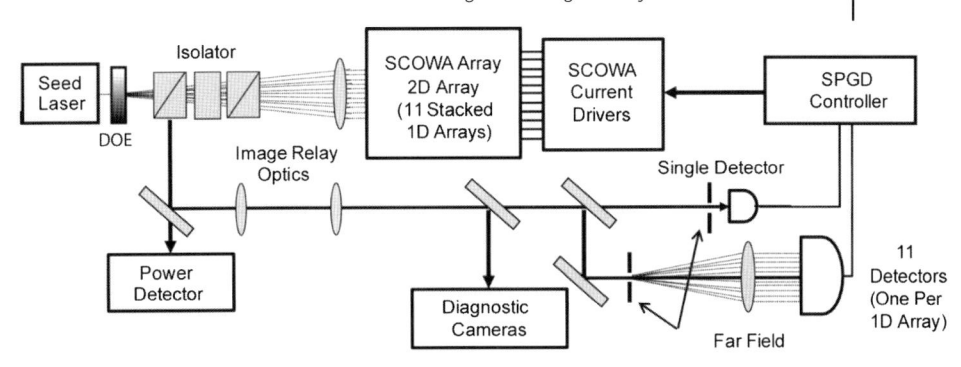

Figure 4.14 Illustration of multiple detector-based approach.

where the D_{group} is a zero mean dither set for the subgroup and D_{piston} is a dither set for all subgroups. In order to avoid coupled loops, creating instabilities, it is essential that the dither set for each subgroup has a zero mean so that it does not create a "piston" error corrupting the intentional D_{piston} term.

Experimental convergence time comparisons shown in Figure 4.15 for an 11×21 array, described in further detail in Section 4.2.1.2, highlight the drastic improvement by over an order of magnitude when using $11 + 1$ (subgroups and entire array) detectors. A summary of several configurations is presented in Table 4.1, along with simulation results for comparison. Several different starting conditions were simulated providing some statistics for the variation in convergence times. Also, included in Table 4.1 are the estimates of the speed enhancement using the previous simple approximation. Due to the simultaneous implementation, the convergence time on average is faster than the simple approximation, which agrees fairly well with the slowest observed convergence times.

Before transitioning to applications of the SPGD algorithm for active phasing, it is worthwhile to briefly discuss how the information discussed in this section can be used toward designing an active coherent combination system. Often the first questions to ask are what is the control bandwidth of the active control system and how does it compare with the phase noise spectrum that need to be corrected? In the previous sections, the speed metric discussed has been the convergence time as this is easy to measure. Although it does not provide the control bandwidth directly, it is inversely proportional to the control bandwidth. The proportionality constant depends on the control bandwidth definition used. If one uses a definition of the frequency at which the phase noise amplitude is reduced by 3 dB, then a proportionality constant of ~ 4 is empirically found in simulations and experiments described in this chapter ($B \sim 4/t_{\text{convergence}}$). Depending on how this compares with the noise spectrum, one can then determine if a single detector would be sufficient or whether multiple detectors are required. Not all optical configurations are readily amenable to multiple detectors, thus requiring initial design decisions as to whether this capability will be required. Finally, how the algorithm gets implemented in hardware is also important as this sets the dither frequency that, as previously

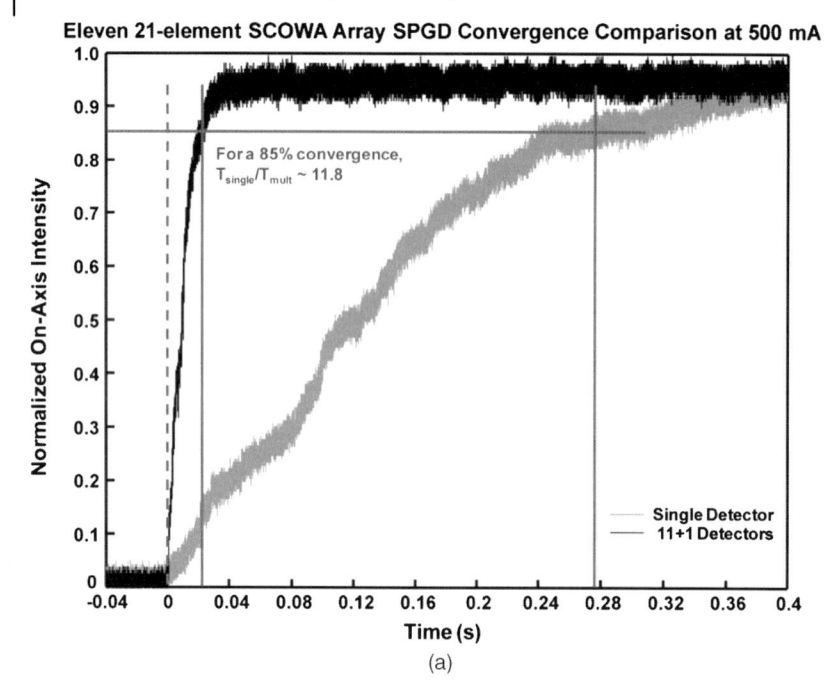

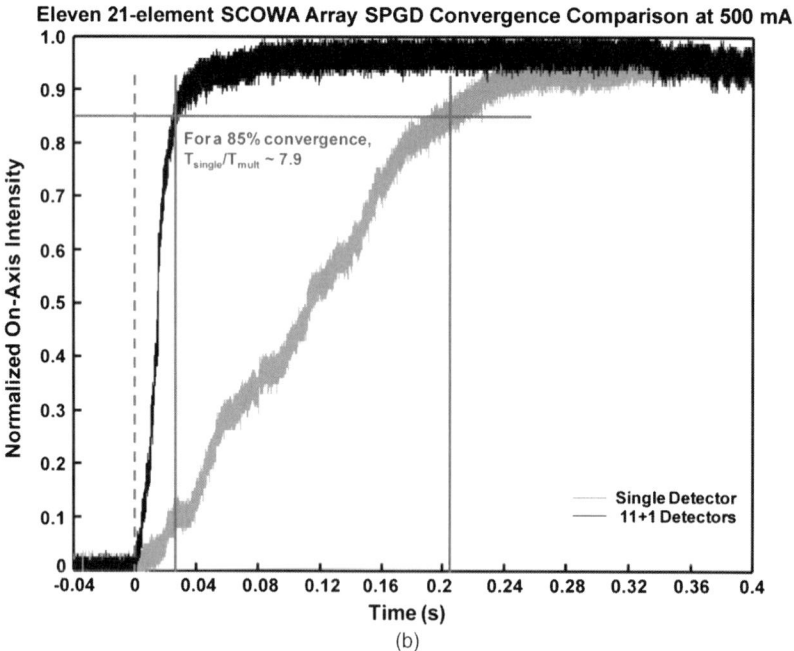

Figure 4.15 Convergence time comparison using 11 + 1 detectors over 1 detector. Shown are two convergence time measurements with different initial phase conditions capturing some of the variations observed when using a SPGD algorithm. Speed increases of 7.9–11.8 are shown.

Table 4.1 Summary of several configurations with simulation results.

Number of 21-element arrays	Number of detectors	"Simple" approximation	Simulated speed increase	Experimentally measured speed increase
6	7	4.7	4–6	4.3–6
10	11	6.8	9–12	8.4–10.5
11	12	7.2	7–12	7.9–11.8

discussed, is linearly proportional to the convergence time or likewise inversely proportional to the control bandwidth. Description of some hardware implementations is the subject of the next section with maximum dither frequencies of 6 kHz demonstrated for semiconductors and 300 kHz for fiber amplifiers. With this information, one can determine how many laser elements could be combined with a preselected number of detectors compatible with the combining architecture to be used, as well as set the speed requirements for the control algorithm hardware.

4.2
Applications of Active Phase Control Using Hill Climbing Control Algorithms

This section describes hardware implementation of the SPGD algorithm along with the orthonormal and multiple detector variations. These were used to coherently combine both semiconductor amplifiers and fiber amplifiers in tiled beam and single-beam configurations. Over 200 semiconductor amplifiers were coherently combined with combined output powers up to $\sim$40 W, and up to 8 fiber amplifiers were coherently combined with kilowatt-level output powers. The algorithm was implemented differently for the two types of amplifiers as the control bandwidth requirements vary dramatically (approximately hertz for semiconductor amplifiers and kilohertz for fiber amplifiers).

4.2.1
Semiconductor Amplifier Active Coherent Combination

Coherent combination of semiconductors is an attractive idea due to the potential efficiency gains and size reductions over alternative laser media often pumped by semiconductors. Large-scale efforts [7] have sought to realize these potential benefits; however, because of the low-power output of single-mode semiconductors and thereby requiring large numbers of devices, it has still not been practically realized. The design of multiwatt-level, near-diffraction-limited semiconductor diode lasers poses significant challenges. These challenges include designing suitable semiconductor laser structures with good beam quality in the slow and fast axes and reasonable electrical efficiency such that thermal considerations do not limit the output power. Numerous approaches have been pursued in the past. Ridge waveguide (RWG) lasers are commonly employed as high-power, single-spatial-mode lasers. Low-threshold, low-modal-loss RWG lasers have been demonstrated to the 1 W level [8]. Limitations on the output power of RWG lasers include dissipation of the heat along a narrow ridge, higher order lateral mode instability, and catastrophic optical mirror damage (COMD). Tapered lasers [9] have the advantage of a large lateral single mode size, which helps to mitigate the problems associated with heat removal and COMD. However, the large lateral mode leads to issues of astigmatic output and single-mode stability at high power levels.

4.2.1.1 Introduction to SCOWA Semiconductor Waveguide and Phase Control

The slab-coupled optical waveguide laser (SCOWL) is a new class of in-plane laser that is fundamentally different in design from the RWG or the tapered laser, which has epitaxial layer design similar to the RWG laser. The fundamental difference is that the SCOWL design is based on a multimode waveguide that converts a highly multimode beam into a single-mode beam by coupling higher order waveguide modes into the slab creating a higher propagation loss for these modes. RWG lasers, on the other hand, incorporate waveguides that are inherently single mode in the transverse direction. The advantage of the slab-coupled waveguide is that the mode size can be made relatively large (approximately four–five times the operating wavelength λ) compared to that of RWG lasers ($1 \times \lambda$). This is a significant benefit from the point of view of COMD since the mode can be spread out at the device facet, reducing the optical intensity at the facet. Another benefit is that the SCOWL allows nearly symmetric modes, which greatly improve the coupling to single-mode fiber.

The 980 nm SCOWL device has been described in Refs [10–12]. A typical 980 nm SCOWL structure is shown in Figure 4.16. The structure consists of $Al_{0.3}Ga_{0.7}As$

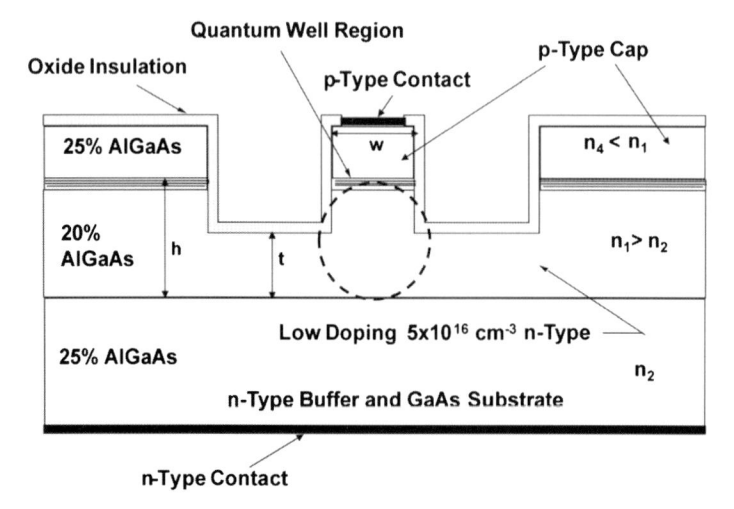

Figure 4.16 SCOWL structure of 980 nm and device schematic. In this structure, $w = 5.7\,\mu m$, $h = 5.0\,\mu m$, and $t = 4.5\,\mu m$ (w: ridge width, h: waveguide height, t: slab thickness). Reprinted with permission of IEEE [10].

cladding layers, a 5 μm thick, n-$Al_{0.25}Ga_{0.75}As$ waveguide with a low doping level ($n = 5 \times 10^{16}$ cm^{-3}), and an active region that consists of three compressively strained 70 Å thick $In_{0.17}Ga_{0.83}As$ quantum wells (QWs) with 70 Å thick $GaAs_{0.92}P_{0.08}$ tensile strained barrier and bounding layers.

A key point in the design is that the multi-QW active region is offset from the peak of the optical mode, which itself is peaked in the waveguide region of the structure, as indicated by the dashed circle in Figure 4.16. This design allows the large, approximately 5 μm, mode to be defined primarily by the low-loss waveguide. It also leads to a low confinement factor ($\Gamma \approx 0.003$–0.005), which reduces the modal gain of the laser and consequently requires that the waveguides have low loss. Typical SCOWL designs have losses of $\alpha \approx 0.5$ cm^{-1}. The large waveguide allows the optical mode to be scaled to large sizes, typically to 5 μm × 5 μm, and the larger mode size is beneficial for mitigating COMD. In addition, the low-loss, low-gain design enables the construction of extremely long cavity lengths (up to 1 cm), which helps in spreading the thermal load of the device along the length.

Waveguide designs are accomplished with the aid of a complex-index-mode solver. The entire structure includes the RWG, slab region, and unetched region, which is retained in order to help with higher order mode discrimination. The physics of slab coupling in passive waveguides is described by Marcatili [13], and more recently incorporating a gain region as in the SCOWL [5,10]. The essential principle is that an arbitrarily large single-mode waveguide can be constructed by coupling a multimode waveguide with a slab, and the slab serves as a mode filter against higher order modes. This slab coupling concept translates into RWG structures like the SCOWL in which critical dimensions are ridge width (w), waveguide height (h), and slab thickness (t); see Figure 4.16 for these dimensions with respect to the SCOWL structure. According to Marcatili's first-order coupled mode analysis, the single-mode nature of the composite slab waveguide system is determined by the ratios T/H and T/W (T, H, and W are related to t, h, and w by increasing the dimensions of the latter by the evanescent field lengths). Therefore, the values of T, H, and W independently can be large relative to λ, while their ratios still satisfy the single-mode criteria.

There are certainly practical limits to the basic concept. For instance, in practice, it becomes rather difficult to scale beyond the waveguide height $h \approx 10 \times \lambda$, since the number of "bound" higher order modes increases with h, and it is necessary to filter out all the higher order modes. To date, robust SCOWL operation has been demonstrated up to $\sim 6 \times \lambda$, for $\lambda = 980$ nm. A two-dimensional complex-mode solver was used to analyze the structure shown in Figure 4.16. The result of the mode solver is shown in Figure 4.17, in which it is clear that only the lowest order SCOWL mode has gain, and all other modes that are discretized due to the choice of artificial boundaries in the model are lossy.

The 980 nm SCOWL devices have demonstrated >1 W continuous wave (CW) output power. A typical L–I characteristic of a junction-side-down mounted device is shown in Figure 4.18 along with the near-field and far-field profiles at various operating points. The beam quality M^2 in both the horizontal and vertical directions are nearly ideal, $M^2_{x,y} = 1.1$, over the entire range of operation. The measured mode

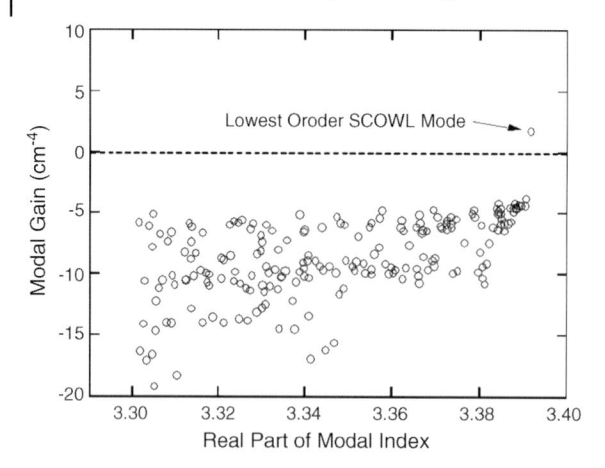

Figure 4.17 Complex-index plot of discretized modes of the 980 nm SCOWL. Only the lowest order SCOWL mode has gain. All the higher order modes are lossy due to slab coupling. Reprinted with permission of IEEE [10].

size is typically $4 \times 5\,\mu m$ (measured at the $1/e^2$ intensity). This mode size is a good match to common single-mode fiber with butt-coupling efficiencies of approximately 84% demonstrated [10].

Newer SCOWL designs do not require etching through the QWs and have exhibited higher power and higher reliability [14–18], and operating wavelengths have been extended to 1.06 μm.

4.2.1.2 Tiled Array Beam Combination

Arrays of slab-coupled optical waveguide amplifiers (SCOWAs) were actively coherently combined in a tiled beam configuration [19]. Figure 4.19 is a schematic

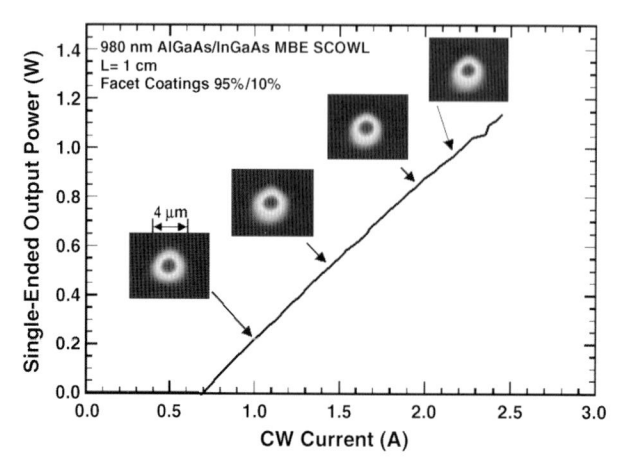

Figure 4.18 L–I characteristic of 980 nm SCOWL device. Near-field profiles are also shown at various current levels. Reprinted with permission of IEEE [10].

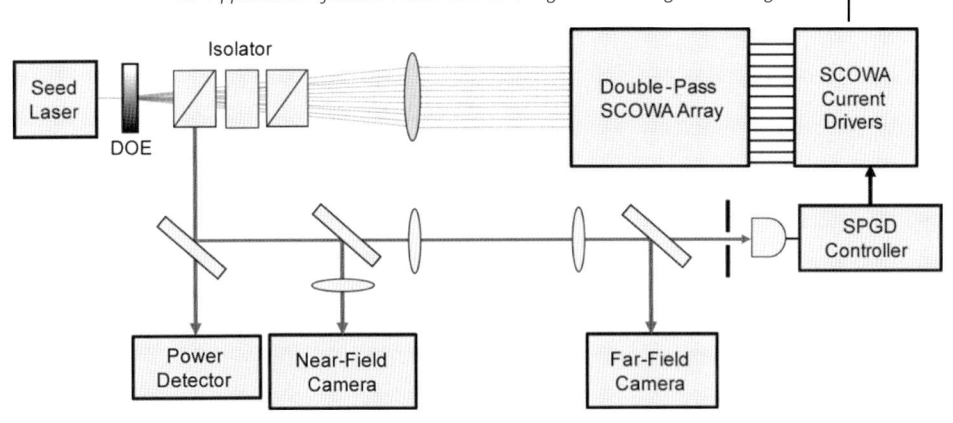

Figure 4.19 Schematic of the active CBC architecture concept.

of the architecture based on a standard master-oscillator (MO) power amplifier configuration. A narrow-linewidth seed is divided into multiple beams using a DOE designed to match the number of elements in the amplifier array. The DOE is placed at the focal plane of a transform lens that determines the array element pitch and mode matching for optimal coupling into the double-passed SCOWA. The input and output beams are separated using an isolator. GaAs-based 960 nm SCOWA arrays, comprised of 21 elements on a 200 μm pitch, were assembled on low-profile single-bar microchannel coolers and mounted junction-side down on patterned AlN carriers to enable individual electrical addressability of array elements. The SCOWA arrays were high reflectivity (HR) coated on the back facet and antireflection (AR) coated on the front facet to prevent laser oscillation and collimated with 500 μm focal length GaP microlens arrays.

Phase actuation was achieved using the drive current to the SCOWA. The physical mechanism is largely the temperature-dependent refractive index at the update frequencies ($<$10 kHz) of the SPGD controller. The micrometer-level SCOWA dimensions allow the drive current to be modulated at frequencies up to hundreds of kilohertz before the amplitude of the induced phase response decreases. The induced phase change for the given current change was characterized by interfering two SCOWA elements, as shown in Figure 4.20.

The phase response is measured by first adjusting the currents such that the far field is that of an in-phase mode. The drive current of one element is then fixed while the other is varied, and the on-axis intensity is measured. This process is repeated for several initial drive currents to generate a phase-to-current response as a function of drive current. Due to the nonlinear amplifier extraction efficiency, this response is not linear with current. Typical values are $\sim$100 mA to induce a 2π phase change. While changing the drive current to impart a phase change leads to slight differences in individual beam power, the effect on combining efficiency with amplitude variation [20] is small. The phase-noise dynamics, which set the active control system bandwidth requirement, also need to be understood. The phase noise of a single-passed SCOWA element was measured with a heterodyne detection approach

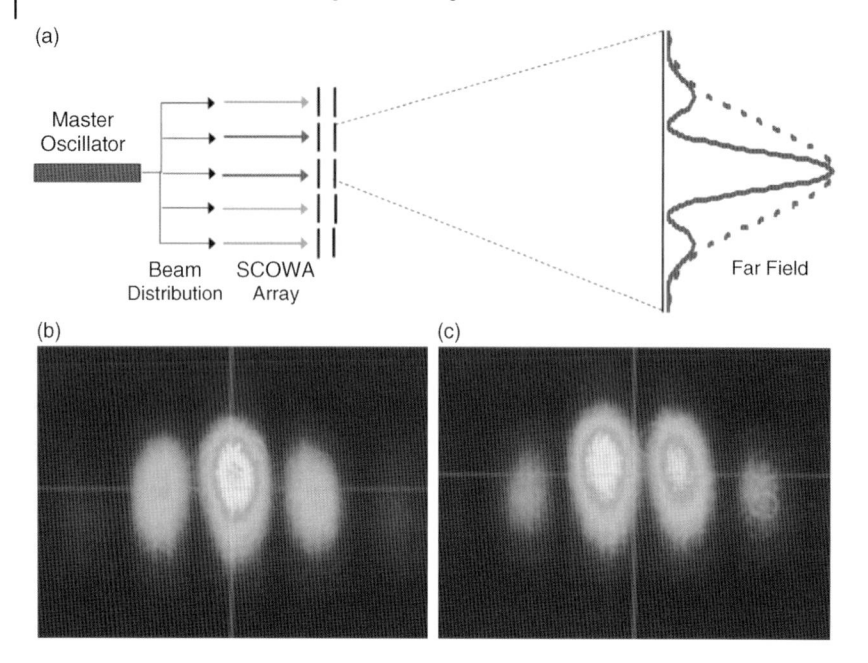

Figure 4.20 (a) Schematic of phase-to-current characterization setup. (b) Experimental far field of two in-phase elements. (c) Experimental far field of two out-of-phase elements.

used in Ref. [21]; the phase noise spectral density and integrated phase noise are shown in Figure 4.21, which demonstrates that controller bandwidths of only several hertz are needed to stabilize the phase.

The SPGD controller was a personal computer with a Linux real-time operating system capable of controlling 231 elements at a dither frequency up to 6 kHz. The

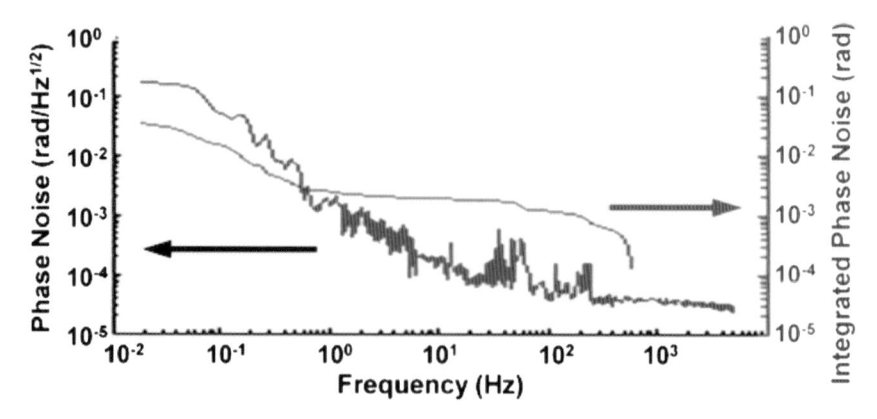

Figure 4.21 Phase-noise spectral density and integrated phase noise of a single-element single-passed SCOWA.

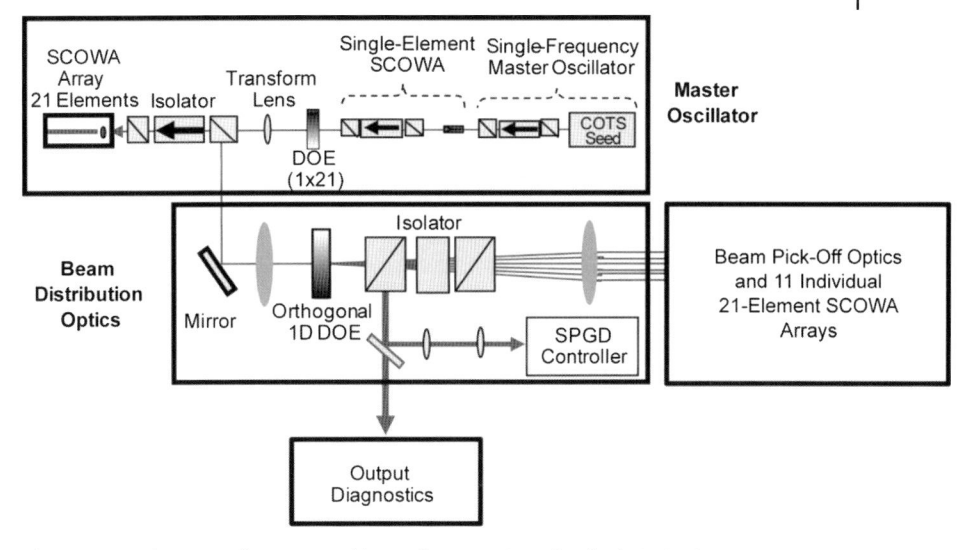

Figure 4.22 Schematic of experimental layout for active CBC of multiple SCOWA arrays.

dithers used were orthonormal with typical dither amplitudes corresponding to $\sim\lambda/50$ phase change, which contributes <2% combining loss.

Figure 4.22 shows a schematic of the experiment. The seed was amplified by a single SCOWA before being split by a one-dimensional 21-channel DOE, which served to seed a 21-element SCOWA array. The outputs of the 21-element SCOWA array were further divided, for use in up to eleven 21-element SCOWA arrays, by image relaying to the final stage with a second one-dimensional 11-channel DOE aligned orthogonal to the array axis placed at the Fourier plane. The final-stage 21-element SCOWA arrays were individually mounted to allow accurate positioning and pointing. A series of HR-coated prisms spatially separated the beams to the 11 arrays. The final pitch of the composite two-dimensional array was $\sim$2.6 mm $\times$ 200 µm. The far field of the two-dimensional SCOWA array was image relayed through the output isolator to various diagnostic cameras, power meters, and the SPGD controller. The input signal to the SPGD controller was generated by sampling only the on-axis portion of the far field using a multimode optical fiber connected to a silicon detector.

The far field from two, six, and eleven 21-element SCOWA arrays coherently combined are shown in Figures 4.23–4.25, respectively. They were recorded using a 10 bit resolution CMOS camera with peak intensities near saturation. The low intensity between the high-intensity lobes indicates high coherence as well as good combining efficiency. The combined output power for the two, six, and eleven 21-element array configurations were $\sim$8.2, 20.7, and 38.5 W CW when operated at an average current of 500 mA. Multiple lobes are present in the far field due to the low fill factor, $\sim$40 and 6% in the 21-element axis and orthogonal axis, respectively, in the near field. The scalability of this approach is evident when comparing the far fields, which maintained very high coherence when combining 2–11 arrays.

124 | *4 Active Coherent Combination Using Hill Climbing-Based*

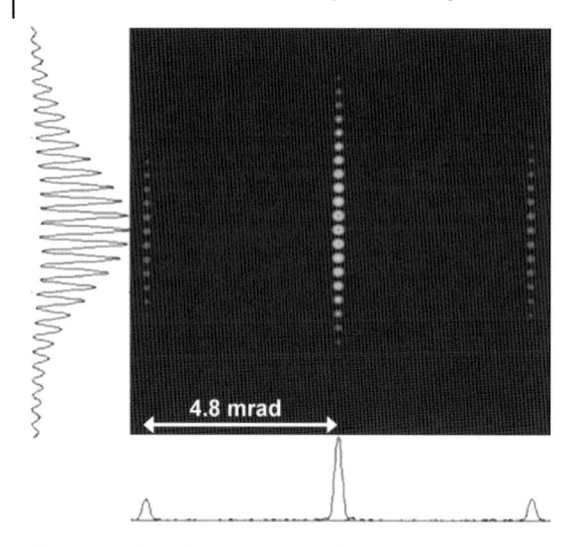

Figure 4.23 Experimental far field of two 21-element SCOWA arrays coherently combined.

To quantify the brightness increase, two techniques were used. The first was camera based, which calculated the ratio of the power contained in the central far-field lobe to the total integrated power and compared it to an ideal, no-phase-error case simulated numerically. This technique gave experimental values ~60% of the ideal value for the 6 and 11 SCOWA array combination. However, camera background noise limited the accuracy, biasing it smaller due to large differences in size

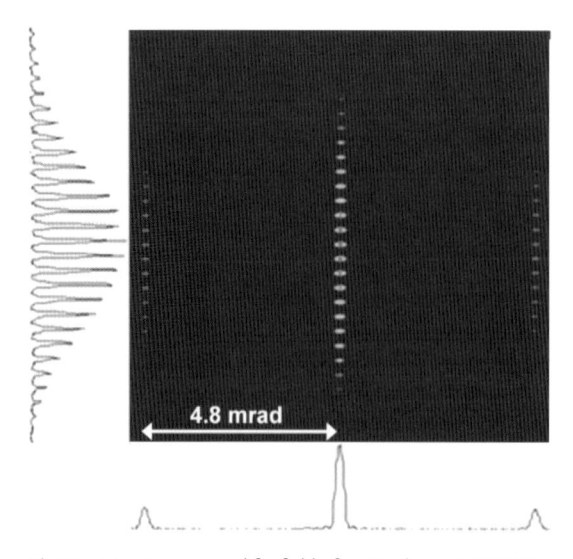

Figure 4.24 Experimental far field of six 21-element SCOWA arrays coherently combined.

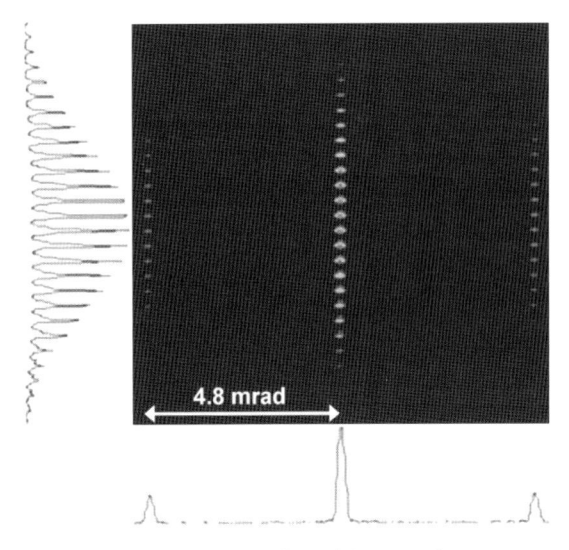

Figure 4.25 Experimental far field of eleven 21-element SCOWA arrays coherently combined.

of illuminated areas to dark areas. The second method used the on-axis intensity signal input to the SPGD controller to determine the ratio of the signal with all elements unphased and phased. The increase for ideal phasing should be N, where N is the number of active elements. This technique gave experimental values of ∼80% of ideal for the 6 and 11 SCOWA array combination. The average of several runs was used to estimate the unphased signal, which fluctuated due to the coherent interference of the SCOWAs. Both techniques provided similar estimates of the quality of the combination. SCOWA array positional variation and microlens collimation variation are believed to be the biggest contributors to the reduction from ideal. With 11 arrays, however, an additional smaller degradation in the array-to-array dimension occurred due to high-frequency mechanical vibrations outside the SPGD control bandwidth, which is observable by the blurring of the lobes in one dimension.

4.2.1.3 Single-Beam Active Coherent Combination Using Diffractive Optical Elements

As an alternative to tiled beam architectures, coherent combination architectures can be configured to produce a single-beam output using a DOE first demonstrated by Leger et al. [22]. This approach was used with semiconductor amplifiers whose outputs were actively coherently combined into a single beam using a DOE [23]. Previous double-passed architectures using SCOWAs [19] suffered from parasitic on-chip lasing limiting the output power at which the semiconductors could be coherently combined as well as power-handling limitations of the outcoupling isolator. To address these issues, a single-pass architecture, shown in Figure 4.26, was used. To eliminate on-chip modes, the 5 mm long waveguides across the 47-element, 1.16 cm wide bar were angled in the plane of the array by 3.4° to the wafer cleave plane. With the addition of an AR coating to both facets, this suppressed

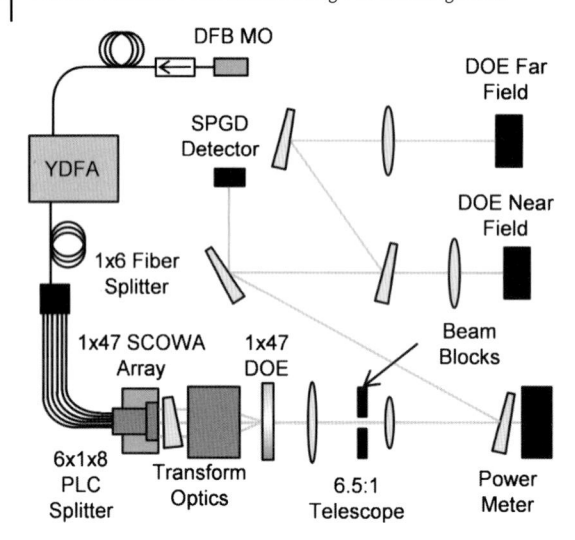

Figure 4.26 Coherent combining setup. A single-frequency source is amplified and split into 48 beams, 47 of which are coupled to an array of 47 amplifiers. The beams are combined using a single lens and DOE, and the output sent to diagnostics. Only three elements are shown for clarity. DFB, distributed feedback laser.

on-chip lasing up to at least 2.3 A drive current per element, creating an array of high-fidelity single-pass power amplifiers. To make use of commercial fiber components and silicon facet passivation, the SCOWAs were designed for use at 1064 nm. The arrays were soldered junction-side down to a patterned multilevel AlN carrier for individual addressability and then bump bonded to a microimpingement cooler.

The fiber-coupled MO was amplified by a single-frequency YDFA, split using custom high-power 1×6 fused fiber splitters, and further split using a $6 \times 1 \times 8$ planar lightwave circuit (PLC) splitter array that was butt-coupled to the SCOWA array. Each SCOWA produced 1 W output power with 37% electrical-to-optical efficiency at 1.4 A. The nearly Gaussian output modes from all SCOWAs were then collimated and angularly mapped onto a 1×47 DOE combiner by a common, all-fused-silica optical system consisting of a prism and an $f = 100$ mm four-element scanning lens. In addition to matching the 2.5 mrad interbeam splitting angle of the DOE to the nominal 252 μm pitch of the array, the optics were designed to minimize aberrations and correct for the tilted field imparted by the angled SCOWA facet without using microlenses. This was enabled by the relatively large ~4 μm × 6 μm mode field of the SCOWA to form a compact coherent beam combining (CBC) module.

Due to some small residual distortion of the optical system and nonlinear interbeam splitting angle dependence, the SCOWA spacing of the array was adjusted in the lithographic mask according to the lens ray-tracing model. The positional accuracy of each emitter was confirmed by turning on each element individually and measuring the deviation of the on-axis beam position on the DOE

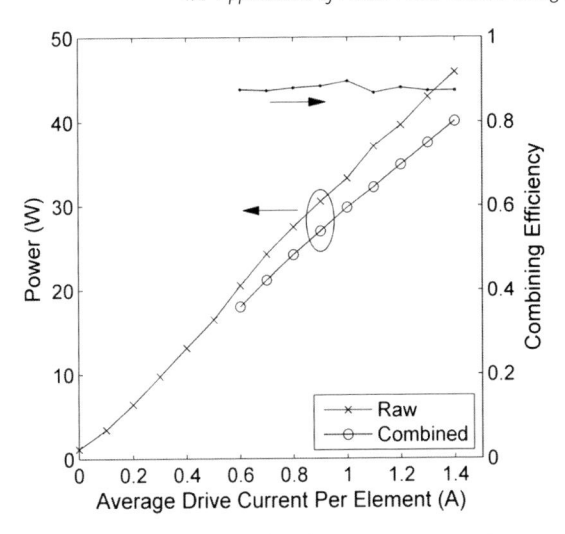

Figure 4.27 Coherently combined power was measured on a power meter. Raw power refers to that measured at the DOE position with control loops open. Drive current here is the average for all the emitters in the array.

far-field camera. The array achieved 40 W with a combining efficiency of nearly 90% at 1.4 A per amplifier, as shown in Figure 4.27. The combining efficiency was determined by comparing the power in the signal beam to the total power at the DOE. Positional errors for this array were 0.25 μm (slow axis) by 0.36 μm (fast axis) RMS, contributing an estimated 4% loss in combining efficiency. The 47-channel DOE fabricated using grayscale lithography achieved ~99% combining efficiency with greater than 99% splitting efficiency defined as the power in the 47 orders divided by the total power. The polarization extinction and ASE content of the SCOWAs were measured at 2 and <1% loss, respectively. Phase fluctuations inherent in the SPGD dither scheme similarly contributed <1%. The remaining few percent loss is believed to be due to optical path differences imparted by the transform optics.

The SPGD controller was implemented on a real-time PC with a dither rate of 6 kHz. For the single-pass SCOWAs, a full wave of phase delay corresponded to ~100–300 mA of current change, depending on incident seed powers and DC current level. When implementing the SPGD algorithm for semiconductors using the drive current to impart a phase change, it is required to impose a constraint of keeping the entire array average drive current constant. This is to prevent the system from continuing to increase the drive currents, which also increases the power.

The combined beam quality M^2, shown in Figure 4.28, was ~1.2 × 1.3, which was better than the beam quality of ~1.3 × 1.7 for an individual beam. This improvement is possible for DOE CBC due to the constructive interference of only common wavefronts, with loss being filtered out to the higher DOE orders.

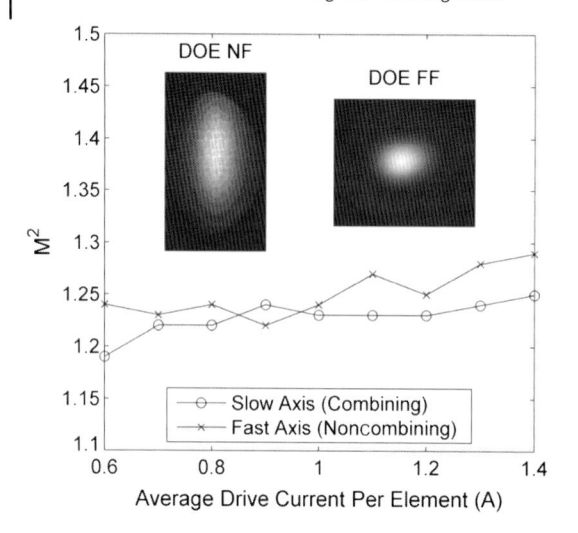

Figure 4.28 Beam quality evolution of the combined beam as a function of average drive current. The M^2 of a single element was measured at 1.3 (slow axis and combining direction) by 1.7 (fast axis). Mode profiles (inset) were recorded at 600 mA. NF, near field; FF, far field.

4.2.2
Fiber Amplifier Active Coherent Combination

Cladding-pumped fiber amplifiers were demonstrated in the late 1990s and opened the way for low-brightness pumping of such amplifiers [24–27]. Because of the availability of high-power, low-brightness pump diodes, the output power from fiber amplifiers increased quickly. YDFAs are now available with multikilowatt diffraction-limited output. Because of their guided nature, output from a fiber laser can be inherently diffraction limited. Thus, fiber lasers are excellent candidates for beam combining to increase their overall brightness.

4.2.2.1 Introduction to Fiber Amplifier Active Beam Combination Architectures

Although multikilowatt output powers have been demonstrated from Yb fiber lasers [28], it is not with the appropriate specifications required for CBC. The key specifications required for CBC include beam quality, preservation of spectrum, polarization purity, coherence, and phase noise. Typical performance characteristics of beam-combinable Yb-fiber amplifiers are kilowatt-level powers, polarization extinction ratios of $\sim$20 dB, M^2 of $\sim$1.1, and spectral linewidths of approximately tens of gigahertz. Recently, the beam combinability of a 1.4 kW Yb-doped non-polarization-maintaining fiber amplifier was demonstrated in which 25 GHz linewidth was used to mitigate stimulated Brillouin scattering (SBS) effects [29].

It is important to point out that, although convenient, single-frequency amplifiers are not necessary for CBC. Because output powers in beam-combinable fibers are limited by SBS and increasing linewidth can increase SBS threshold, using broad-

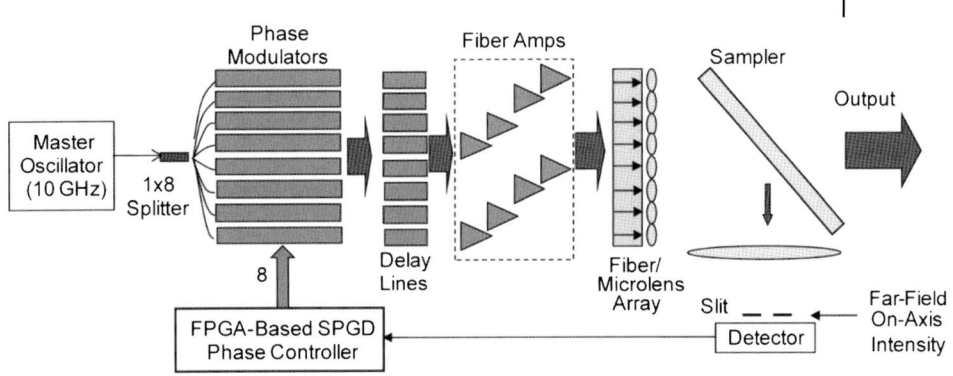

4.2 Applications of Active Phase Control Using Hill Climbing Control Algorithms | 129

Figure 4.29 Linear array tiled beam configuration.

linewidth fiber amplifiers has enabled higher power per fiber. This does however require that the path lengths from the fiber amplifiers be matched to the coherence length [30].

Although the total fiber laser beam combined power has not yet exceeded those coming out of a single-fiber laser, this will not likely be the case in the next few years.

4.2.2.2 Tiled Array Beam Combination

Eight commercial 500 W polarization-maintaining (PM) fiber amplifiers were beam combined in a linear–array tiled beam configuration, shown schematically in Figure 4.29 [31]. The MO output is split and sent through eight phase modulators used for phase control and adjustable delay lines used for path length matching. The outputs from the delay lines are used as inputs to the high-power fiber amplifiers. The outputs from the fiber amplifiers are collimated by a microlens array to increase the fill factor. The output from the microlens array is sampled, with the sample going through a lens to transform to the far field. A slit is used to look at the on-axis intensity with a photodiode. The electrical signal from the photodiode is sent to the SPGD control system implemented in an FPGA. The individual fiber outputs had free space, 1 mm diameter, 6 mm long endcap terminations fused to the output pigtails. These endcaps had a 3° angle polish and an AR coating to reduce back reflection to the amplifier. The eight endcapped outputs were aggregated in a silicon V-groove array. This V-groove array had a pitch of 1.5 mm. A microlens array does aperture filling to increase the fraction of power in the far-field central lobe. This microlens array has a focal length of 17.5 mm, and an individual beam has a Gaussian beam radius of 0.5 mm. The path lengths of the fiber amplifiers were matched to better than 1 mm to accommodate the 10 GHz linewidth of the system.

To demonstrate the amplifiers' applicability to CBC, we measure their coherence and phase noise. Figure 4.30a shows the coherence measurement results for one-fiber amplifier. A 5 mW, 10 GHz phase-modulated optical source is used to seed the 500 W amplifier. The input and the output amplifier taps are interfered with a fiber coupler to ensure mode matching. Oscillations are observed because of mechanical and thermal drifts. The 96% visibility shows that the amplifier keeps the same

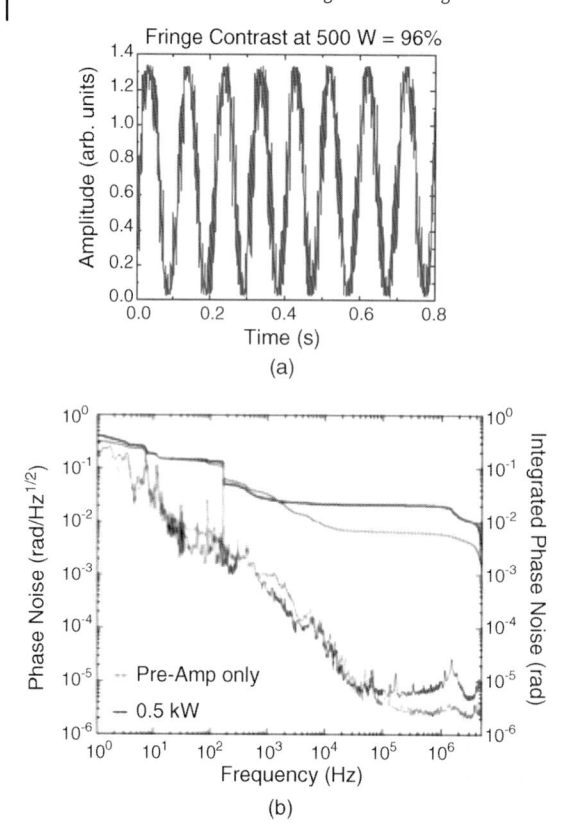

Figure 4.30 (a) Fringe visibility from a fiber amplifier and (b) phase noise power spectral density and integrated phase noise at low power (15 W) and at full power for one of the 0.5 kW fiber amplifiers.

frequency content and preserves the signal coherence between its input and output. The remaining 4% is largely due to imperfect power equalization.

To ascertain that such fiber amplifiers are coherently beam combinable, one must also measure their phase noise to determine the required bandwidth of the control electronics. The amplifier phase noise was characterized using a heterodyne measurement [21]. Figure 4.30b shows the measured phase noise at 15 and 500 W. The integrated phase noise falls off very sharply around ∼200 Hz. Thus, phase locking of such amplifiers requires feedback electronics with nominal kilohertz bandwidth. Furthermore, there is virtually no difference in the phase noise between the two powers below 200 Hz. Therefore, the phase noise is not dominated by fiber amplifier nonlinearity or by thermal effects for noise frequencies above a few hertz, but rather by mechanical vibrations.

Figure 4.31a shows the far-field intensity patterns for combining the eight fibers at different power levels per fiber. The less-than-unity fill factor leads to side lobes. At the highest power, 500 W per fiber, the fraction of power in the central lobe is 58%, while at 125 W per fiber this fraction is 57%. This compares with a calculated

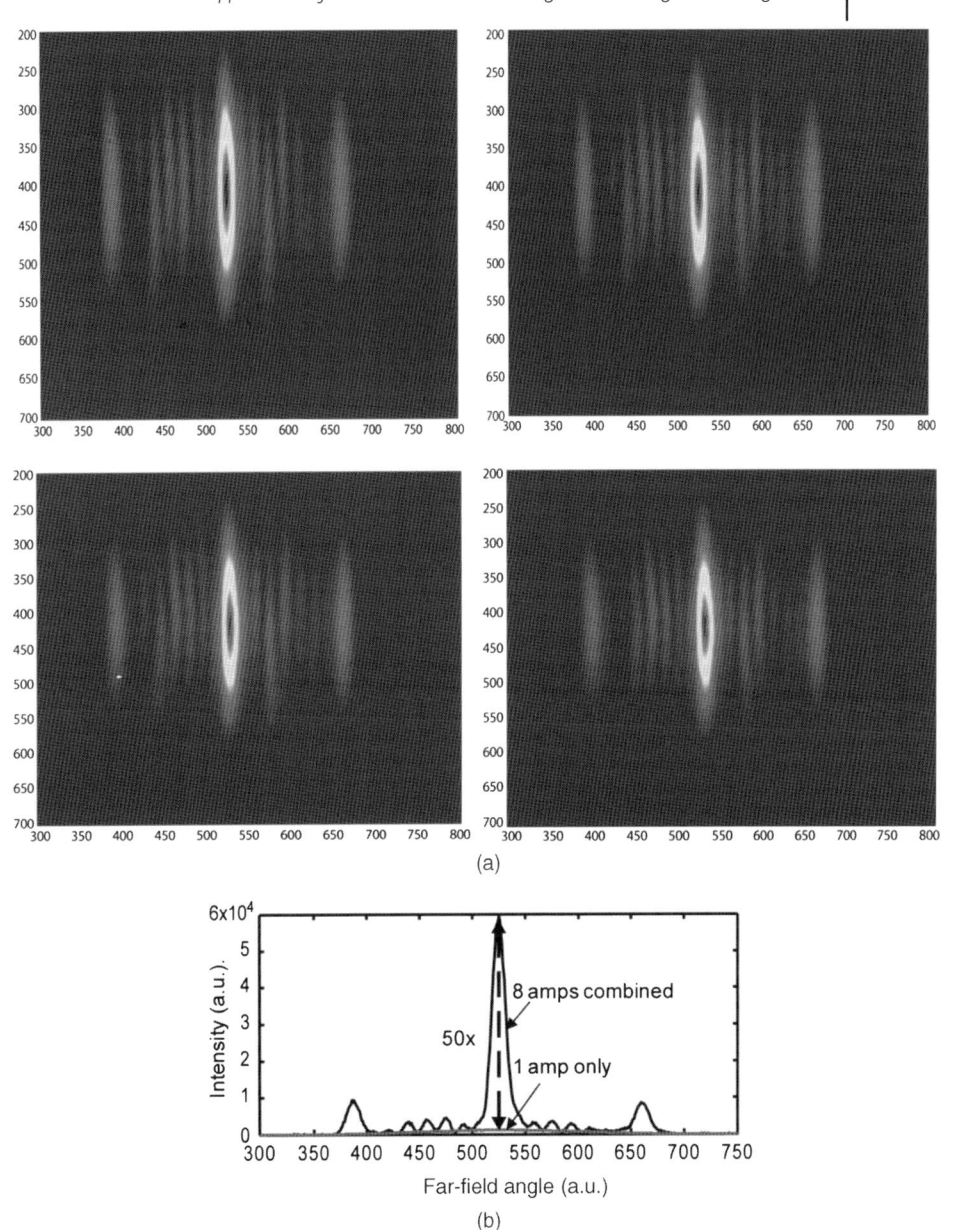

Figure 4.31 Far-field intensity patterns at different output powers for the eight-amplifier combined array and (b) far-field intensity for a single-amplifier array element and for all eight-amplifier elements combined at 4 kW. On-axis intensity with all eight elements is 50× that of single element.

fractional power in the central lobe of 68% for an output with uniform phase and the experimental near-field intensity profile. Figure 4.31b shows the far fields across the beam combining direction with only a single fiber operating and with all eight fibers operating. Ideally, the on-axis far-field intensity should increase as N^2, where N is the number of elements. The experimentally observed increase is 50, leading to a beam combining efficiency of 78%. Nonidealities leading to reduced beam combining efficiencies included imperfect polarization purity, errors in the position of the fibers in the output array, differences in collimation caused by differences in focal length within the microlens array and in endcap length, residual phase error, and higher order mode content.

Another measure of the ideality of the beam, the fraction of the power as a function of the far-field diffraction angle, is shown in Figure 4.32a. The three curves represent the far fields for a near field of an ideal top hat in a square aperture

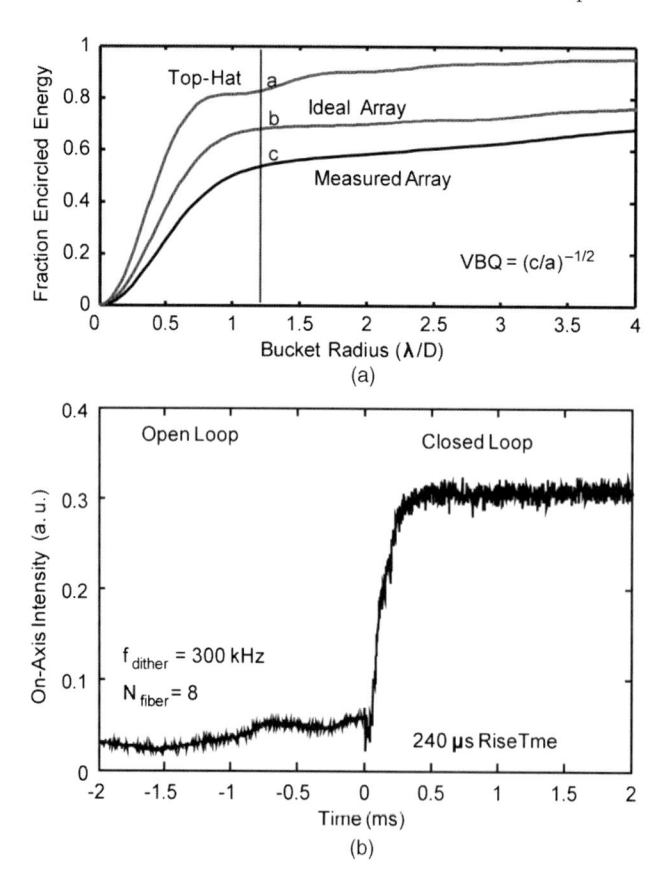

Figure 4.32 (a) Encircled far-field power as a function of far-field angle, with the vertical line at $1.22\lambda = D$. The top hat curve is for a uniform phase and intensity beam, the ideal curve is for a uniform phase with the experimental near-field pattern, and the measured array is the experimental data. (b) The on-axis intensity as a function of time when the phase controller is engaged.

(uniform intensity and phase), a near field same as our experiment but with uniform phase, and the experimental data. A common measure of beam quality in high-energy laser systems is the power-in-the-bucket (PIB) vertical beam quality (VBQ) [32], which is related to the fraction of the power within a given far-field angle compared with a reference ideal beam (the ideal top hat in this case). The PIB VBQ is defined as $(c/a)^{-1/2}$ (Figure 4.32a) at a far-field angle of $1.22\lambda/D$, and for our experiment the VBQ was determined to be 1.25. The best VBQ possible given the fill factor of the near field is 1.10 times the diffraction limit.

The final measure of performance is the dynamic response of the phase control system. This can be characterized by the convergence time of the far-field on-axis intensity when the phase control system is activated, which is shown in Figure 4.32b. In this experiment, the SPGD dither frequency is 300 kHz using random dithers. For this eight-fiber system, the convergence time is $\sim$240 μs.

4.2.2.3 Single-Beam Active Coherent Combination Using Diffractive Optical Elements

In this section, we extend diffractive beam combining to high-power fiber amplifier arrays by integrating a reflective five-beam DOE with a portion of the compact phased fiber amplifier array described in the previous section. This configuration demonstrated a 1.93 kW output beam that contained 79% of the input power and exhibited improved beam quality compared to the input beams [33].

Five adjacent beams from the array were directed onto the DOE using the geometry shown in Figure 4.33. Each fiber beam was partially collimated by one element of a monolithic microlens array that allowed an adjustable beam size on the DOE. A close-coupled pair of lenses served as a composite Fourier optic to fully collimate each beam and direct them to overlap on the DOE at angles matching the diffractive orders. The collimated Gaussian beam diameter on the DOE was approximately 3.3 mm.

The DOE combined most of the incident power into the $m=0$ diffractive order, with the uncombined power diffracted into higher $|m|>0$ orders. The DOE was tilted slightly so that the combined beam was reflected out of the plane of diffraction for geometric separation from the input beams. Following the DOE, a loose aperture terminated any residual power left in the $|m|>0$ orders. A low-power sample of the combined beam was picked off by a beam sampler and sent to diagnostics and a detector to lock the beams using the FPGA SPGD controller. This automatically locks each beam at the DOE input to the conjugate of the phase imposed upon diffraction rather than in-phase that occurs when using a tiled beam configuration.

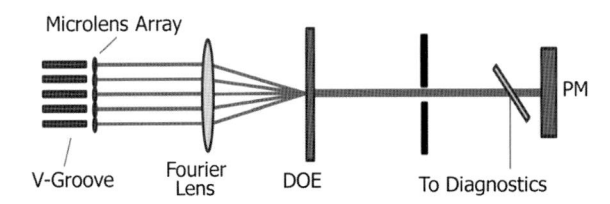

Figure 4.33 Optical schematic of the DOE fiber beam combiner.

The SPGD dither frequency was ~300 kHz using random dithers that led to typical convergence times of ~130 μs.

The combined power in the central $m = 0$ output order is shown in Figure 1.26a. At low powers, the overall combining efficiency was 90%. This was less than the ideal 96% DOE combining efficiency, implying an additional efficiency loss of ~6%. Measurements indicated that this additional loss was almost entirely due to the fiber positional errors of the fiber array, fiber higher order modes, and polarization mismatch coming from the rotational orientation errors of the fibers in the fiber array [30]. As input power P increased to 2.5 kW, the combining efficiency dropped to 79% due to the fiber array expanding at higher powers, which is described in Chapter 1. Both the combined output and the input beams were very close to diffraction limited (Figure 1.26b), although M^2 values increased slightly with power. At all powers, the combined beam exhibited slightly improved M^2 over the inputs. As stated earlier, the reason for the improvement in the combined beam M^2 is that uncommon aberrations that differ from beam to beam are effectively spatially filtered by coherent combination in the plane of the DOE [30,34].

4.3
Summary

The use of the SPGD algorithm for active coherent combination has been described. Its linear scaling behavior as the number of elements is varied was shown, and we discussed why it works so effectively for active phase control by providing nearly an exact estimate of the phase error when combining more than two elements. Orthonormal and multiple detector variations of the algorithm were introduced, which were found in both simulation and experiment to provide >2× speed increases and >10× speed increases with $11 + 1$ detectors, respectively. These were subsequently used to coherently combine both semiconductor amplifiers and high-power fiber amplifiers, demonstrating over 200 combined semiconductor amplifiers with combined output powers of up to ~40 W and eight-fiber amplifiers coherently combined with kilowatt-level output powers.

The use of an SPGD-based control algorithm has been shown to be a viable way to coherently combine lasers in a relatively simple implementation. Although the basic nature of its scaling is fairly well understood, there still exist opportunities to improve the understanding of its spectral phase noise rejection capability relative to the convergence time. In addition, for applications with a long time-of-flight between when a dither is applied and its effect is detected, further adaptations can provide opportunities to increase the "effective" dither frequency.

Disclaimer

The views expressed are those of the authors and do not reflect the official policy or position of the Department of Defense or the US Government. This is in accordance with DoDI 5230.29, January 8, 2009.

References

1 Primot, J. (1993) Three-wave lateral shearing interferometer. *Appl. Opt.*, **32**, 6242–6249.

2 Barchers, J.D., Fried, D.L., Link, D.J., Tyler, G.A., Moretti, W., Brennan, T.J., and Fugate, R.Q. (2003) Performance of wavefront sensors in strong scintillation. *Proc. SPIE*, **4839**, 217–227.

3 Kansky, J.E., Yu, C.X., Murphy, D.V., Shaw, S.E.J., Lawrence, R.C., and Higgs, C. (2006) Beam control of a 2D polarization maintaining fiber optic phased array with high-fiber count. *Proc. SPIE*, **6306**, 63060.

4 Vorontsov, M.A., Carhart, G.W., and Ricklin, J.C. (1997) Adaptive phase-distortion correction based on parallel gradient–descent optimization. *Opt. Lett.*, **22**, 907–909.

5 Walpole, J.N., Donnelly, J.P., Taylor, P.J., Missaggia, L.J., Harris, C.T., Bailey, R.J., Napoleone, A., Groves, S.H., Chinn, S.R., Huang, R., and Plant, J. (2002) Slab-coupled 1.3-μm semiconductor laser with single-spatial large-diameter mode. *IEEE Photon. Technol. Lett.*, **14**, 756–758.

6 Vorontsov, M. and Sivokon, V. (1998) Stochastic parallel-gradient–descent technique for high-resolution wave-front phase distortion correction. *J. Opt. Soc. Am. A*, **15**, 2745–2758.

7 Levy, J. and Roh, K. (1995) Coherent array of 900 semiconductor laser amplifiers. *Proc. SPIE*, **2382**, 58–69.

8 Schmidt, B., Lichtenstein, N., Sverdlov, B., Matuschek, N., Mohrdiek, S., Pliska, T., Müller, J., Pawlik, S., Arlt, S., Pfeiffer, H.-U., Fily, A., and Harder, C. (2003) Further development of high power pump laser diodes. *Proc. SPIE*, **5248**, 42–54.

9 Walpole, J.N. (1996) Semiconductor amplifiers and lasers with tapered gain regions. *Opt. Quantum Electron.*, **28**, 623–645.

10 Donnelly, J.P., Huang, R.K., Walpole, J.N., Missaggia, L.J., Harris, C.T., Plant, J., Bailey, R.J., Mull, D.E., Goodhue, W.D., and Turner, G.W. (2003) AlGaAs/InGaAs slab-coupled optical waveguide lasers. *IEEE J. Quantum Electron.*, **39**, 289–298.

11 Huang, R.K., Donnelly, J.P., Missaggia, L.J., Harris, C.T., Plant, J., Bailey, R.J., Mull, D.E., and Goodhue, W.D. (2003) High power, nearly diffraction limited AlGaAs–InGaAs semiconductor slab-coupled optical waveguide laser. *IEEE Photon. Technol. Lett.*, **15**, 900–902.

12 Huang, R.K., Missaggia, L.J., Donnelly, J.P., Harris, C.T., and Turner, G.W. (2005) High-brightness slab-coupled optical laser arrays. *IEEE Photon. Technol. Lett.*, **17**, 959–961.

13 Marcatili, E.A.J. (1974) Slab-coupled waveguides. *Bell Syst. Tech. J.*, **53**, 645–674.

14 Huang, R.K., Donnelly, J.P., Missaggia, L.J., Harris, C.T., Chann, B., Goyal, A.K., Sanchez-Rubio, A., Fan, T.Y., and Turner, G.W. (2007) High-brightness slab-coupled optical waveguide lasers. *Proc. SPIE*, **6485**, 64850F-1-9.

15 Huang, R.K., Chann, B., Missagia, L.J., Augst, S.J., Connors, M.K., Turner, G.W., Sanchez-Rubio, A., Donnelly, J.P., Hostetler, J.L., Miester, C., and Dorsch, F. (2009) Coherent combination of slab-coupled optical waveguide lasers. *Proc. SPIE*, **7230**, 72301G-1-12.

16 Smith, G.M., Huang, R.K., Donnelly, J.P., Missaggia, L.J., Connors, M.K., Turner, G.W., and Juodawlkis, P.W. (2010) High-power slab-coupled optical waveguide lasers. 23rd Annual Meeting of the IEEE Photonics Society, pp. 479–480.

17 Smith, G.M., Duerr, E.K., Siegel, A.M., Donnelly, J.P., Missaggia, L.J., Connors, M.K., Mathewson, D.C., Turner, G.W., and Juodawlkis, P.W. (2011) Directly-modulated high-power slab-coupled optical waveguide lasers. 24th Annual Meeting of the IEEE Photonics Society, pp. 288–289.

18 Donnelly, J.P., Juodawlkis, P.W., Huang, R., Plant, J.J., Smith, G.M., Missaggia, L.J., Loh, W., Redmond, S.M., Chann, B., Connors, M.K., Swint, R.B., and Turner, G.W. (2012) High-power slab-coupled optical waveguide lasers and amplifiers, in *Advances in Semiconductor Lasers, Semiconductor and Semimetals Series*, vol. 86 (eds J.J. Coleman, A.C. Bryce, and C. Jagadish), Academic Press, Amsterdam, pp. 1–47.

19 Redmond, S.M., Creedon, K.J., Kansky, J.E., Augst, S.J., Missaggia, L.J., Connors, M.K., Huang, R.K., Chann, B., Fan, T.Y., Turner, G.W., and Sanchez-Rubio, A. (2011) Active coherent beam combining of diode lasers. *Opt. Lett.*, **36**, 999–1001.

20 Fan, T.Y. (2009) The effect of amplitude (power) variations on beam-combining efficiency for phased arrays. *IEEE J. Sel. Top. Quantum Electron.*, **15**, 291–293.

21 Augst, S.J., Fan, T.Y., and Sanchez-Rubio, A. (2004) Coherent beam combining and phase noise measurements of ytterbium fiber amplifiers. *Opt. Lett.*, **29**, 474–476.

22 Leger, J.R., Swanson, G.J., and Veldkamp, W.B. (1986) Coherent beam addition of GaAlAs laser by binary phase gratings. *Appl. Phys. Lett.*, **48**, 888–890.

23 Creedon, K.J., Redmond, S.M., Smith, G.M., Missaggia, L.J., Connors, M.K., Kansky, J.E., Fan, T.Y., Turner, G.W., and Sanchez-Rubio, A. (2012) High efficiency coherent beam combining of semiconductor optical amplifiers. *Opt. Lett.*, **37**, 5006–5008.

24 Dominic, V., MacCormack, S., Waarts, R., Sanders, S., Bicknese, S., Dohle, R., Wolak, E., Yeh, P.S., and Zucker, E. (1999) 110 W fiber laser. Conference on Lasers and Electro-Optics, Baltimore, MD, Paper CPD11.

25 Limpert, J., Liem, A., Höfer, S., Zellmer, H., and Tünnermann, A. 150W Nd/Yb codoped fiber laser at 1.1 μm. *Conf. Lasers Electro Optics*, **73**, 590–591.

26 Platonov, N.S., Gapontsev, D.V., Gapontsev, V.P., and Shumilin, V. (2002) 135 W cw fiber laser with perfect single mode output. Conference on Lasers and Electro-Optics, Long Beach, USA, Paper CPDC3.

27 Ueda, K.I., Sekiguchi, H., and Kan, H. (2002) 1 kW cw output from fiber embedded lasers. Conference on Lasers and Electro-Optics, Long Beach, CA, Paper CPDC4.

28 Yeong, Y., Sahu, J.K., Payne, D., and Nilsson, J. (2004) Ytterbium-doped large-core fiber laser with 1.36kW cw output power. *Opt. Exp.*, **12**, 6088–6092.

29 Goodno, G.D., McNaught, S.J., Rothenberg, J.E., McComb, T.S., Thielen, P.A., Wickham, M.G., and Weber, M.E. (2010) Active phase and polarization locking of a 1.4 kW fiber amplifier. *Opt. Lett.*, **35**, 1542–1544.

30 Goodno, G.D., Shih, C.-C., and Rothenberg, J.E. (2010) Perturbative analysis of coherent combining efficiency with mismatched lasers. *Opt. Express*, **18**, 25403–25414.

31 Yu, C.X., Augst, S.J., Redmond, S.M., Goldizen, K.C., Murphy, D.V., Sanchez, A., and Fan, T.Y. (2011) Coherent combining of a 4kW, eight-element fiber amplifier array. *Opt. Lett.*, **36**, 2686–2688.

32 Slater, J.M. and Edwards, B. (2010) Characterization of high-power lasers. *Proc. SPIE*, **7686**, 76860.

33 Redmond, S.M., Ripin, D.J., Yu, C.X., Augst, S.J., Fan, T.Y., Thielen, P.A., Rothenberg, J.E., and Goodno, G.D. (2012) Diffractive coherent combining of a 2.5kW fiber laser array into a 1.9kW Gaussian beam. *Opt. Lett.*, **37**, 2832–2834.

34 Cheung, E.C., Ho, J.G., Goodno, G.D., Rice, R.R., Rothenberg, J., Thielen, P., Weber, M., and Wickham, M. (2008) Diffractive-optics-based combination of a phase-locked fiber laser array. *Opt. Lett.*, **33**, 354–356.

5
Collective Techniques for Coherent Beam Combining of Fiber Amplifiers
Arnaud Brignon, Jérome Bourderionnet, Cindy Bellanger, and Jérome Primot

5.1
Introduction

Fiber lasers are increasingly a preferred choice for high-power solid-state laser because of their advantages in terms of compactness, reliability, efficiency, and beam quality delivery. The development of high-brightness semiconductor diodes as pump sources and the emergence of large-mode-area fibers have allowed a rapid increase in the output power of fiber laser systems. Even given these advantages, it is desirable to increase the system power or energy levels beyond what is possible with a single-mode fiber laser. This becomes of particular interest for applications that require a narrow-linewidth source with polarized emission. Indeed, in this case the performances of a single conventional fiber are often spoiled by parasitic nonlinear effects such as stimulated Brillouin scattering (SBS). Coherent beam combining (CBC) of several fiber amplifiers is a straightforward approach to circumvent the limitations of single-fiber laser master oscillator power amplifier (MOPA) channel.

The different methods for achieving CBC fall under two main categories: passive (Part Two of this book) and active (Part One) phase locking. As examples of passive approaches, an all-optical feedback loop in a single-ring cavity has been demonstrated and techniques based on SBS phase conjugation mirror have also been proposed for compensation of phase differences between fiber emitters. Alternatively, active phase locking involves phase detection and active compensation of phase errors [1–7]. After appropriate phase locking of the different fiber amplifier outputs, the combination of the beams can be performed using filled or tiled aperture configurations. In the filled aperture geometry, the beams overlap in the near field. This can be performed using a set of beam splitters or a set of polarizers, or a single diffractive optical element (DOE). In the tiled aperture geometry, the output beams are placed side by side in the near field. After propagation and proper near-field intensity distribution, a great part of the total power can be concentrated in a central lobe in the far field. Although a fraction of the total power is lost in side lobes, the tiled aperture configuration brings additional functionalities such as beam steering and beam shaping. This can be useful to correct a wavefront that is distorted

Coherent Laser Beam Combining, First Edition. Edited by Arnaud Brignon.
© 2013 Wiley-VCH Verlag GmbH & Co. KGaA. Published 2013 by Wiley-VCH Verlag GmbH & Co. KGaA.

after propagation through atmosphere or to track a target or a photoreceiver in free space communication [8–10].

In previous chapters, CBC of fiber amplifiers has been demonstrated with relatively moderate number of fibers. The present chapter is dedicated to CBC of a large number of fibers and the associated technologies and system configurations. Phase locking of a large number of fibers (typically larger than 50) can be useful for different practical situations. For high-power applications for instance, in order to avoid the difficulties of manipulating very high-power fiber lasers, it can be interesting to work with a large number of relatively low-power fiber lasers instead of a few high-power fibers. For applications that require beam agility, the use of a large number of fibers provides the possibility to realize beam steering and shaping. The precision of these functionalities increases with the number of emitters. A set of a large number of fibers having individual phase control can thus behave like optical phased arrays antennas.

When combining a large number of fibers, the difficulty is the increase of the complexity and cost of the system and the concomitant reduction of the phase correction loop bandwidth. In order to overcome these limitations, the measurement of the phase errors between the elementary sources can be advantageously obtained by using a wavefront sensor based on an array of detectors, or, in other words, based on a camera. This way, it is possible to retrieve the phase errors with only one acquisition.

The aim of this chapter is to present some examples of collective techniques that offer practical solutions for beam combining of a large number of fibers. These techniques include wavefront sensor-based camera, arrays of phase modulators, and collimating fiber arrays. Using such key technologies, experimental proof of concepts are presented in the continuous-wave (CW) regime.

5.2
The Tiled Arrangement

As presented elsewhere in this book (see, for instance, Figures 8.1 and 8.2), two principal arrangements can be chosen in order to recombine the multiple beams of a CBC system: the filled and the tiled aperture schemes. In the filled aperture geometry, the beams are recombined in the near field by using polarizers, beam splitters, or a DOE. The latter is of particular interest when dealing with a large number of fibers, as shown in Chapters 1 and 4. For a very large number of fibers (>100), the design complexity and efficiency of the DOE may be a concern. In this particular case, the tiled aperture configuration may be the most simple solution. The beams coming from the fiber outputs are placed side by side in the near field. The recombination is then obtained in the central lobe in the far field. In order to maximize the energy in the central lobe, stringent conditions must be satisfied. The aim of this section is to introduce modeling of the tiled aperture recombination in order to derive the influences of the main parameters. This will allow specifying and designing collimated fiber arrays for CBC.

5.2.1
Calculation of the Far-Field Intensity Pattern

In order to calculate the far field from a fiber array, we start with the propagation of a Gaussian distribution of the optical field u at the end of a polarization-maintaining fiber in a cylindrical coordinate system (r, z):

$$u(\mathbf{r}, z) = E_0 \frac{w_0}{w(z)} \cdot \exp\left(\frac{-\mathbf{r}^2}{w^2(z)}\right) \cdot \exp\left(ik\frac{\mathbf{r}^2}{2R(z)}\right) \cdot \exp(i(kz - \varphi(z))), \quad (5.1)$$

where E_0 is the maximum amplitude of the optical field, w_0 is the waist of the Gaussian beam, and

$$\begin{cases} w(z) = w_0 \sqrt{1 + \frac{z^2}{z_R^2}}, \\ R(z) = z\left(1 + \frac{z_R^2}{z^2}\right), \\ z_R = \frac{\pi w_0^2}{\lambda}, \\ \varphi(z) = \arctan\left(\frac{z}{z_R}\right), \\ k = \frac{2\pi}{\lambda}, \end{cases} \quad (5.2)$$

where λ is the wavelength and z_R is the Rayleigh range. Each beam is then collimated by a lens of diameter $2R$. The end facet of the fiber is placed at $z = 0$ and at a distance f from the lens, where f is the focal length, as shown in Figure 5.1.

For one fiber, the optical field becomes

$$u_l(\mathbf{r}, f) = u(\mathbf{r}, f) \cdot \exp\left(i\pi \frac{\mathbf{r}^2}{\lambda f}\right) \cdot \text{disk}(|\mathbf{r}|), \quad (5.3)$$

where $\text{disk}(x) = 1$ for $x \leq R$ and $\text{disk}(x) = 0$ for $x > R$.

The fibers and their associated collimating lenses are disposed side by side in a matrix arrangement. For the sake of clarity, we consider in this case a $N_x \times N_y$ rectangular lattice arrangement, with N_x, d_x, N_y, and d_y, respectively, being the

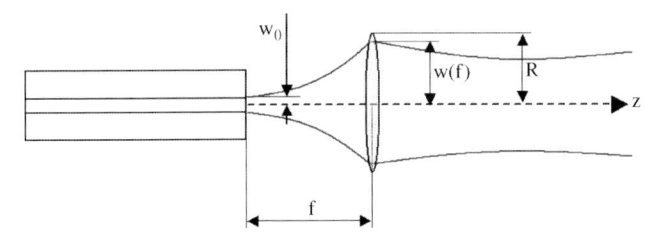

Figure 5.1 Optical field at the output of one collimated fiber.

number of fibers and the array pitches along the x and y directions. The resulting total optical field U in the plane of the lens array is given by

$$U(x,y) = \text{Rect}\left(\frac{x}{N_x d_x}, \frac{y}{N_y d_y}\right) \times \left(u_l(x,y) \otimes \text{Comb}\left(\frac{x}{d_x}, \frac{y}{d_y}\right)\right), \quad (5.4)$$

where $\otimes$ is the convolution operator,

$$\begin{cases} x = |\mathbf{r}|\cos(\arg(\mathbf{r})) \\ y = |\mathbf{r}|\sin(\arg(\mathbf{r})) \end{cases}, \quad (5.5)$$

and Comb denotes the Dirac comb function:

$$\text{Comb}\left(\frac{x}{d_x}, \frac{y}{d_y}\right) = \sum_j \sum_k \delta(x - jd_x, y - kd_y). \quad (5.6)$$

"Rect" is the aperture function, defined by $\text{Rect}(x,y) = 1$ if $|x| \leq 1/2$ and $|y| \leq 1/2$, otherwise $\text{Rect}(x,y) = 0$.

The Fourier transform of the near-field U distribution gives the far-field $\tilde{U}$. The final intensity in the far field is given by

$$I(x',y') = \tilde{U}(x',y') \cdot \tilde{U}(x',y')^*, \quad (5.7)$$

with $\tilde{U}(x',y') = \tilde{u}_l(x',y') \times \left[\text{sin}\,c(x'N_x d_x, y'N_y d_y) \otimes \text{Comb}(x'd_x, y'd_y)\right]$.

Figure 5.2 shows typical intensity patterns of near and far fields. In the near field, the pitch between the lenses is d_x and the total pupil is given by $D_x = N_x \times d_x$, where N_x is the number of emitters along the x-axis. The far field is composed of a series of grating lobes with an angular spacing of λ/d_x and a divergence of λ/D_x. The spatial

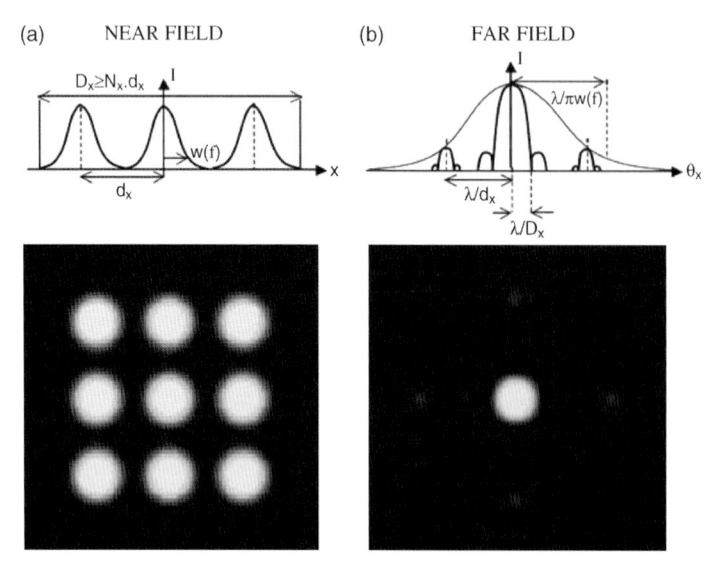

Figure 5.2 Schematic of the intensity profiles in the near field and after propagation in the far field, example with nine fibers.

profile of each lobe in the far field exhibits the *sinc* shape that is given by the Fourier transform of the overall aperture of the near-field profile according to Eqs. (5.4)–(5.7). Then the grating lobes comb in the far field is modulated by an envelope function, which is given by the Fourier transform of the elementary pupil optical field u_l. In the present case of a Gaussian illumination of the microlenses, one obtains a Gaussian envelope having a divergence at $1/e^2$ of $\lambda/\pi w(f)$, where $w(f)$ is the beam waist of one individual beam in the near field.

On the other hand, in the ideal situation of a uniform illumination that perfectly matches the microlenses, the envelope function then becomes a sinc function, whose zeros fall exactly on the positions of the far-field grating lobes. This last remark emphasizes the importance in the system designing of an optimized filling of the near-field elementary pupil by each of the optical beamlets. The fill factor of the near-field pupil is then defined by

$$\text{FF} = \frac{\pi w^2(f)}{\iint_{\text{Lattice cell}} dS} \qquad (5.8)$$

and corresponds to the ratio between the equivalent surface occupied by one optical beamlet in the plane of the microlenses and the surface of one elementary cell of the fiber lattice.

To evaluate the quality of the coherent combining, we choose in this chapter to calculate the energy in the main far-field lobe encircled in the area defined by the first zero of the sinc lobe function. This energy is then normalized to the total near-field energy taken before the microlenses array, which is before the filtering by the disk function in Eq. (5.3). This combining efficiency η is then given by

$$\eta = \frac{\iint_{r' \leq \lambda/D} I(x', y') dx' dy'}{N_x \times N_y \times \iint u \cdot u \cdot (x, y) dx dy}. \qquad (5.9)$$

5.2.2
Influence of Design Parameters on the Combining Efficiency

The number of fibers of the CBC system will drive the achievable output power, and is thus an important parameter for the design of the system. In addition, for a given pitch in the near field, more fibers also lead to a larger synthetic pupil and hence a lower divergence source. The number of fibers however does not affect the combining efficiency as defined in Eq. (5.9). The envelope function is indeed independent of the number of fibers, and so is the ratio between the main and side grating lobes in the far field. The results presented hereafter, which are calculated with limited number of fibers (typically 100), can thus be scaled to any number of fibers.

5.2.2.1 Impact of the Near Field Arrangement
The collimated fibers can be organized in different patterns. Basically, the impact of this pattern is dual: first, Eqs. (5.8) and (5.9) show that the most closely packed pattern will provide the best filling of the near-field pupil by the optical beamlets and hence the highest combining efficiency. Second, the symmetries and periodicities of near-field fiber pattern determine the way the energy will be distributed in the far

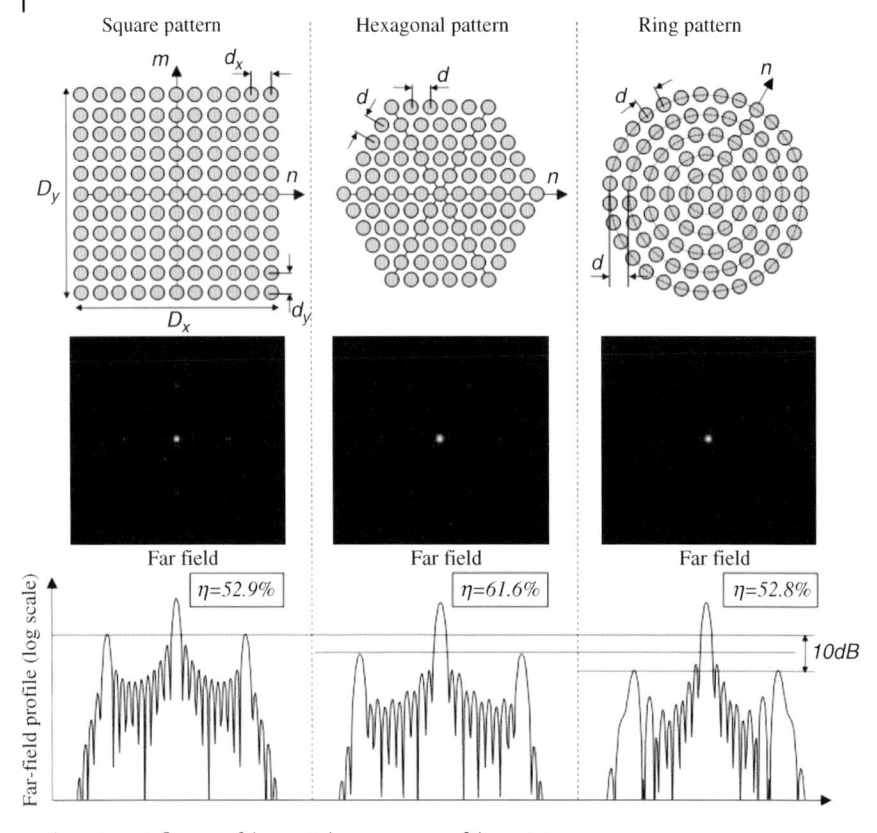

Figure 5.3 Influence of the spatial arrangement of the emitters.

field. Figure 5.3 shows the calculated far field obtained with three closed-packed patterns: square, hexagonal, and ring.

In all cases, calculations were carried out considering a ratio between beamlets and microlens radii, $\omega(f)/R$ of 0.9. The far-field profiles presented in log scale are normalized in each geometrical configuration with respect to the total near-field energy as in Eq. (5.8).

According to Eq. (5.8), the square pattern has a near-field fill factor of 64% $[=\pi/4\,(\omega/R)^2]$. The four side lobes are 10 dB lower than the central lobe, and the far-field pattern calculation gives a combining efficiency of 52.9%. The main advantage of the square lattice pattern is an easier manufacturability compared to more complex patterns.

The fill factor for the hexagonal pattern is $\pi/2\sqrt{3} \times (\omega/R)^2$ and equals 73% for ω/R of 0.9. This time, the far field presents six side lobes at -14 dB from the main lobe. The combining efficiency is almost 62%. Hexagonal stacking provides the closest possible packing and consequently provides the highest combining efficiency. Although less convenient than square lattice, the manufacturing of such a pattern is perfectly feasible.

The ring pattern (i.e. circular arrangement) is more original because its structure is no longer periodic. As a consequence, the side lobes are severely reduced in the far field, with a diffuse ring surrounding the main lobe at −20 dB. As the structure is no longer periodic, the definition of the fill factor of Eq. (5.8) cannot be applied. In particular, the fill factor of the pattern varies with the number of rings considered: for the first ring, the fill factor is one of the hexagonal patterns, whereas for distant rings, the fill factor tends to be one of the square lattices. Defining a fill factor for the whole near-field pupil, instead of considering only the elementary pupil, one obtains a fill factor of 63% for the structure of Figure 5.3, very close to the value obtained for the square lattice. We calculate a combining efficiency of 52.8%, again very close to the one obtained for the square lattice. The advantage of the ring pattern is the quasi-absence of side lobes, but the nonperiodicity of the structure makes more difficult the manufacturing of fiber array and microlenses array with identical patterns.

5.2.2.2 Impact of Collimation System Design and Errors

The influence of the size of each beam compared with the aperture size of its associated lens is shown in Figure 5.4. In this example, we considered disk-shape lenses in square, hexagonal, and circular arrangements, as described in Figure 5.3. For small beam diameters, the beams do not cover the entire lenses. The total fill factor is then reduced and the combining efficiency is low. Numerical calculation shows that for all near-field geometries, a ratio ρ of about 90% between beam diameter and lens diameter maximizes the combining efficiency. Above this value, a fraction of the power is clipped by the lens aperture and the combining efficiency decreases. With a usual fiber numerical aperture of 0.1, this condition leads to choose a lens array with a f-number of 4.5.

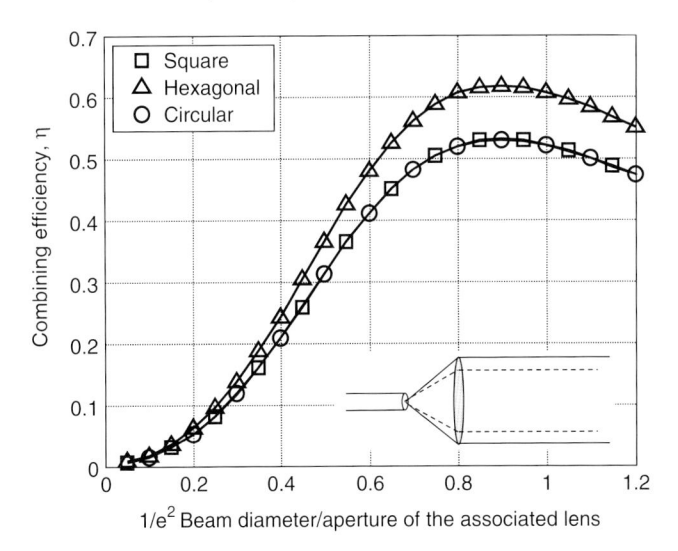

Figure 5.4 Calculated combining efficiency as a function of the waist of each beam in the aperture of its associated lens. Results for square, hexagonal, and circular geometrical configurations are shown in squares, triangles, and circles, respectively.

In the following simulations, we considered a 1.5 mm pitch for the various near-field patterns. A 90% ratio between beam and lens diameters then implies a 6.8 mm focal length for the collimating microlenses.

Figure 5.5 presents the influences of two other key parameters for the combining efficiency. An important degradation of the combining efficiency can be caused by an offset between the fiber axis and the axis of its associated lens. The off-axis beam is then

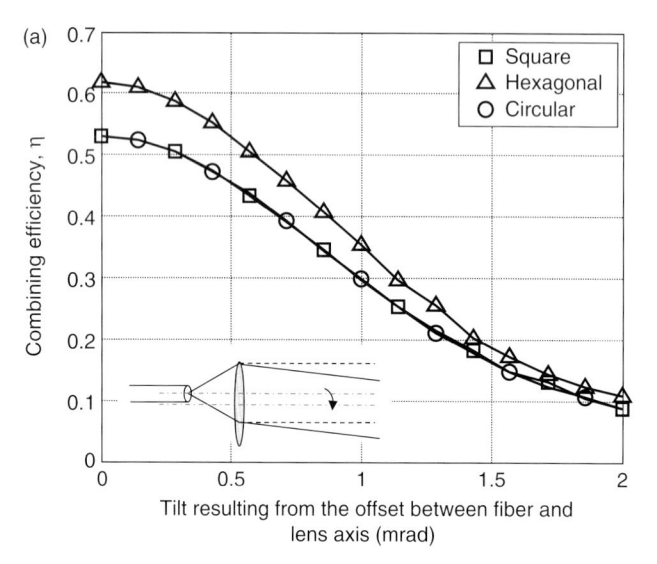

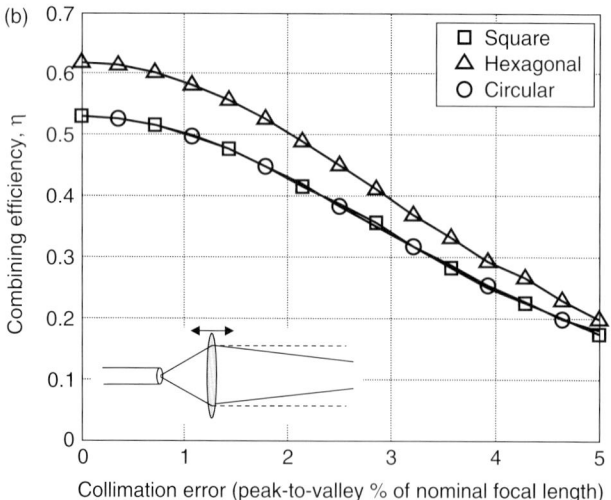

Figure 5.5 (a) Combining efficiency as a function of the tilt error produced by a misalignment between fiber and collimation lens axis. (b) Combining efficiency as a function of the relative focal length variation of the collimating lens. For both figures, square, hexagonal, and circular geometries are respectively represented by squares, triangles, and circles.

tilted after passing through the lens, which degrades the overlapping of all the beamlets in the far field. To obtain the results presented in Figure 5.5a, we considered a Gaussian random tilt distribution across the fiber pattern. The abscissa coordinate is then the peak-to-valley amplitude of the random distribution. To maintain a combining efficiency penalty at less than 5%, the tilt angle distribution amplitude must be kept lower than about 0.3 mrad. For a lens having a focal length of 6.8 mm, this gives an offset of less than 2 µm. The combining efficiency penalty reaches 50% for a tilt angle of about 1.1 mrad, which corresponds to an offset of 7.5 µm with the same lenses. On the other hand, a fiber tilt translates into an offset after the lens. This effect does not contribute in first order to the degradation of the far-field combination.

The precision of the focal lengths of the collimating lenses is another important parameter. We saw that tilt errors degrade the far-field combination by spreading the positions in the far field of all the beamlets. Similarly, collimation errors spread the size distribution of the beamlets in the far field and thereby degrade their combination (by degrading the overlapping of the far-field spots). It can be seen in Figure 5.5 that a precision of at least 1% on the focal length is required to keep the combining efficiency penalty below 5%. The combining efficiency drops by 50% for a focal length variation of 3.7% peak to valley.

As shown in Figures 5.4 and 5.5, the parametric analysis of the combining efficiency seems identical for all the studied geometries. It can then be concluded that no geometry is more robust than another with respect to the misalignment errors.

5.2.2.3 Impact of Phase Error

Figure 5.6 represents the combining efficiency calculated for different peak-to-valley amplitudes of a random phase error distribution (with a Gaussian statistic). The calculations show that the combining efficiency is obviously very sensitive to the

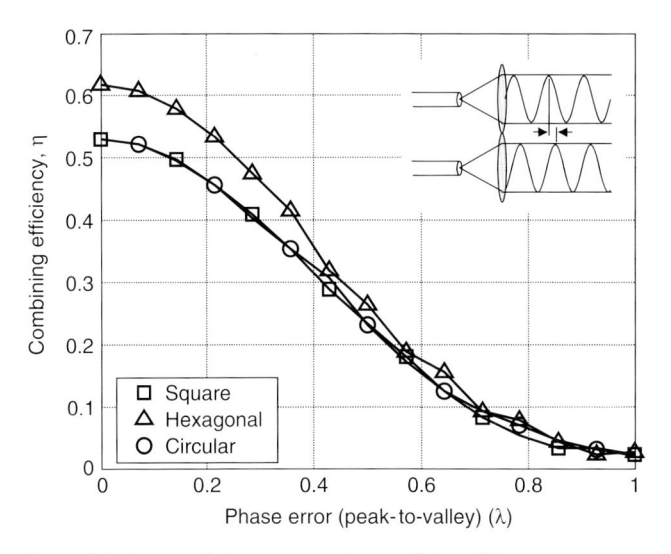

Figure 5.6 Impact of phase error on the combining efficiency.

quality of the fibers cophasing, that is, to the residual phase error between the fibers. It can be seen that the phase of each fiber channel has to be controlled with a precision of at least $\lambda/10$ to keep a combining efficiency above 95% of its maximum value. The combining efficiency decreases by 50% for a phase error of about $\lambda/2$. Typical fiber amplifier noise measurement shows that a kilohertz correction is enough to ensure beam combining with high efficiency.

5.2.2.4 Impact of Power Dispersion

The combining efficiency is not very sensitive to power dispersion on the fiber array. Indeed, calculations show that even with a power dispersion of 60%, the combining efficiency remains up to 90% of its maximal value.

5.2.3
Beam Steering

As demonstrated above, the tiled aperture or the far-field recombination approach can provide a combination efficiency of 60% for a hexagonal. This approach provides an additional feature, which is the ability to control the phase profile of the total output pupil. This allows, for instance, applying a phase profile that would precompensate propagation distortions (propagation in the atmosphere for instance) or performing beam steering by applying a phase ramp. A steering system can be defined by its resolving power or the number of points (or directions) that can be resolved in the scanning region. The resolving power is then given by the total scan angle divided by the FWHM divergence of the output beam. Figure 5.7 illustrates the case of the synthetic pupil obtained with a square lattice of beamlets, as described in Figure 5.3. The divergence of the combined beam is $\lambda/(N_{x,y} \times d_{x,y})$, where N_x is the number of fibers per row (N_y per column) and $d_{x,y}$ are the lattice

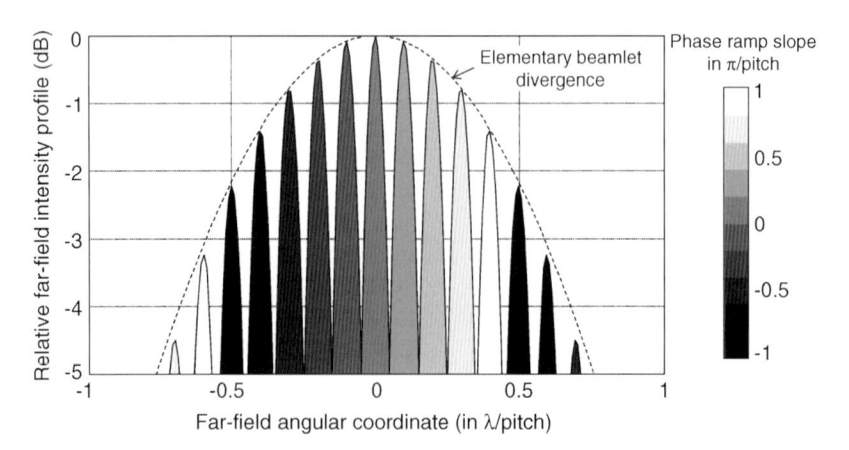

Figure 5.7 Beam steering ability of the tiled aperture CBC system. Far-field profiles are plotted for various values of the applied phase ramp slope, covering the angular space between two adjacent lobes.

pitches in the x- and y-directions, respectively. The scanning angle is given by the angular distance between two adjacent grating lobes and is given by $\lambda/d_{x,y}$. The resolving power of the CBC setup as a steering system is then directly given by $N_x \times N_y$, the total number of fibers that are combined. Figure 5.7 shows in gray scale the various far-field profiles when a phase ramp is applied with a slope varying from $-\pi/d_{x,y}$ to $+\pi/d_{x,y}$, therefore covering the entire angular space between two adjacent lobes.

It must however be noticed that only a step-like phase ramp can be applied onto the synthetic pupil, because only phase pistons can be applied to the fibers. As a consequence, only the grating lobes comb will be steered in the far field, while the overall energy envelope (corresponding to the divergence of a elementary beam) will not be affected. Steering the combined beam will therefore systematically be accompanied by a power reduction. In the specific case of Figure 5.7, the scanning angle (from $-\lambda/2d_{x,y}$ to $+\lambda/2d_{x,y}$) is entirely covered with a power variation of more than 2 dB. Solutions have been proposed to get always the maximum of energy in the steered direction. In this case, the microlens array has to be translated with respect to the fiber array in order to steer the energy envelope and the grating lobes comb simultaneously [11].

5.3
Key Elements for Active Coherent Beam Combining of a Large Number of Fibers

A generic architecture for fiber CBC is presented in Figure 5.8. A master oscillator provides a signal that is distributed to N single-mode polarization-maintaining (PM) fiber amplifiers. One or several preamplifiers could be required for proper power

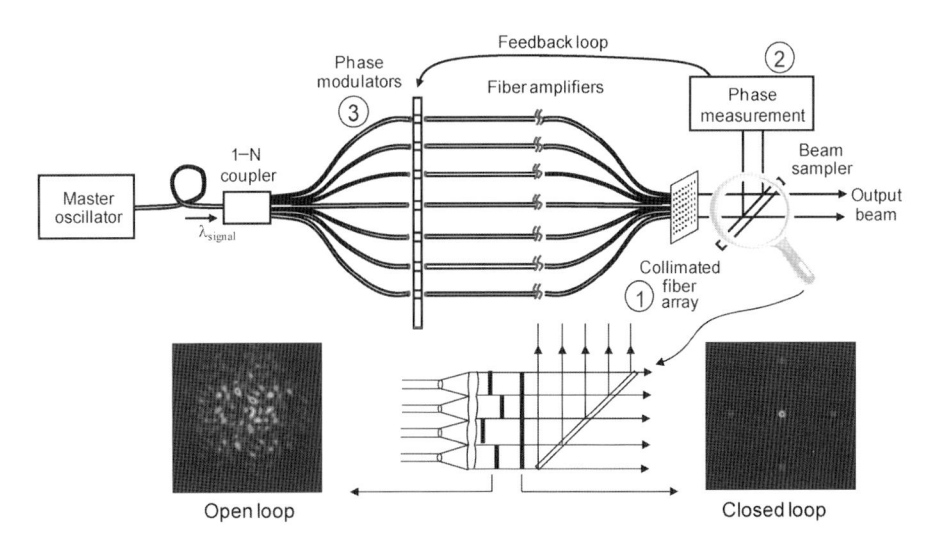

Figure 5.8 Active CBC of a large number of fibers. The specific key elements are as follows: 1 – the collimated fiber array, 2 – the phase measurement device, and 3 – the phase modulators.

extraction, depending on the power or energy level required for the application. The N fiber outputs are disposed in a matrix arrangement and collimated by a lens array. A small fraction of the output beams is sampled and launched to a phase sensor device in order to measure the piston-phase errors between the N fibers. The knowledge of phases then allows driving the modulators placed before the amplifiers for compensation of the phase errors. When dealing with a large number of arms (typically $N > 50$), the key elements for CBC are (i) the collimated fiber array, (ii) the phase measurement device, and (iii) the phase modulators. In the following sections, these three elements will be analyzed, potential technologies proposed, and some practical realizations will be presented.

5.3.1
Collimated Fiber Array

In order to obtain maximum power or energy in the main central lobe in the far field, the collimated fiber array must have very stringent specifications, as shown in Section 5.2.2. Typical requirements are summarized in Table 5.1 [12].

Let us suppose a collimated fiber array with 64 fibers in a square pattern arrangement (8×8 array). Fibers used in the device are standard PM fibers having a total divergence of 0.1 rad. Requirement toward individual lens fill factor is established around 90% and leads to choose an individual microlens with an f-number of 4.5 in order to match with the fiber divergence. Silicon microlens arrays, with 1500 μm pitch and with an individual focal length of 5.77 mm–1550 μm leading to an f-number of 3.8, are chosen to match as close as possible this criterion. For operation around 1 μm, microlens array in silica should be chosen to minimize material absorption.

A typical microlens array for fiber collimation is shown in Figure 5.9. A photolithographic manufacturing process can be used for high precision [12]. Sixty-four planospherical circular lenslets are placed in a 8×8 square arrangement. Measures performed on these lenses show a perfect pitch and focal regularity and the main aberration measured is a spherical aberration with a value of 0.7 rad ($\lambda/9$). These values are in agreement with what we could expect from a planospherical lens and match the requirements for beam combining.

Table 5.1 Summary of the requirements of the collimated fiber array to realize beam combining with a combining efficiency above 80% of its higher value.

Parameter	Required value
Phase error (peak-to-valley)	$<\lambda/10$
Collimating lens f-number	$\rho/(2 \times ON) = 4.5$ (for standard single-mode fibers)
Pointing error (peak-to-valley)	<0.6 mrad
Focal length error (peak-to-valley)	$<1\%$

ρ is the ratio between individual beam diameter and the lens diameter. ON is the numerical aperture of the fiber (ON = 0.1 for standard single-mode fiber).

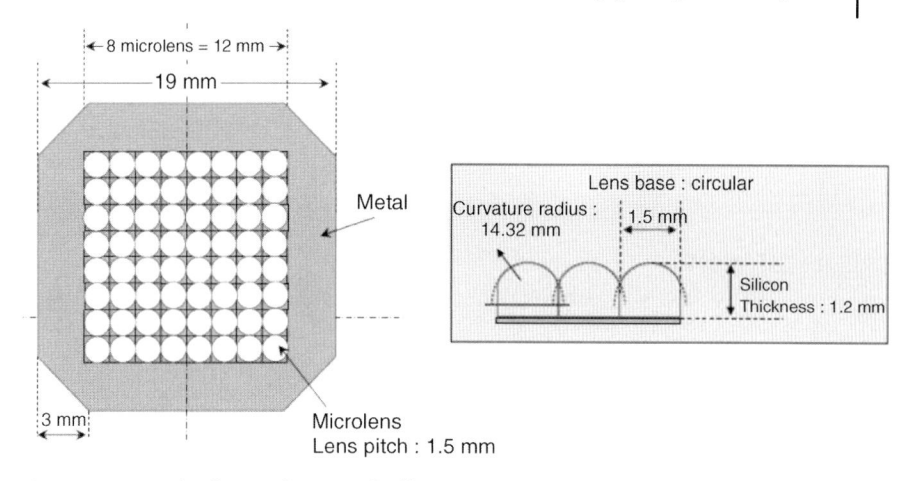

5.3 Key Elements for Active Coherent Beam Combining of a Large Number of Fibers

Figure 5.9 Example of a microlens array for fiber collimation.

To support the 64 polarization-maintaining fibers with an accurate distance between the fibers, different techniques can be used. For instance, with multiple horizontal V-grooves, it is possible to ensure proper distance between the fibers along a single line. Then, the stacking of several lines allows obtaining a two-dimensional (2D) arrangement. Accurate position of different lines is the most critical issue with this technique. Another solution is the manufacturing of a dedicated positioning plate with a 2D array of holes. This component can be made of a plate drilled with 64 holes of 125 μm diameter and a pitch of 1500 μm. In each hole, a fiber is inserted, polarization aligned, glued, and polished. To limit the fiber offset errors that lead to a strong combining efficiency degradation, particular attention must be paid to pitch and hole diameters regularity. For this reason, lithographic manufacturing techniques must be employed.

Fabrication of high-precision optical fiber holder was achieved by means of deep X-ray lithography (DXRL) technology, which can handle the requested high precision, that is, 1 μm precision for the fiber hole diameter and for the positioning between two fibers. The theoretical precision of DXRL is mainly related to the vertical and horizontal divergences of the synchrotron source, which are less than 0.1 mrad, well beyond the required tolerance. The material chosen for the holder is poly(methyl methacrylate) (PMMA), which is a standard DXRL resist that offers appropriate mechanical qualities to tightly maintain a matrix of optical fibers. To manufacture the DXRL mask, duplication is done using UV optical lithography. A standard chromium mask was used, which guarantees an accuracy of better than 1 μm for the dimensions and relative positions of the holes. The UV lithography was performed on a 300 μm thick graphite substrate (SFG-2.3 from Poco Graphite) on top of which a 25 μm thick SU-8 layer was spun up. After development, gold was electroplated up to a thickness of 20 μm. The duplication of the chromium mask to manufacture the DXRL mask maintains the precision of the hole diameter and of the positioning between holes within the tolerance of 1 μm. To manufacture the high-precision optical fiber holder, a 1 mm thick PMMA sheet was baked for 1 h at 110 °C in order to remove accumulated stress due to former treatments such as cutting and machining. The PMMA resist was exposed to synchrotron light at ANKA, Forschungszentrum Karlsruhe (FZK), on litho II LIGA station. This LIGA beamline is equipped with a silicon mirror covered with 200 nm thick nickel film in order to remove high-energy photons (higher than 12 keV) and a cutoff angle of 4.85 mrad was chosen. The used radiation dose at the bottom of the resist was 3 kJ/cm^3. The PMMA sheet and the mask were kept at 20 °C during exposure in order to avoid heating due to the absorbed high photon flux. After exposure, the self-standing PMMA sheet was then developed in a standard GG bath [13] for 24 h, rinsed in water for 20 min, and then dried at room temperature. We can note that our manufacturing process, based on masking techniques, allows the realization of a holder compatible with a much higher number of fibers with the same accuracy for high-power beam combining application, as mentioned in Ref. [14]. Also, the use of other materials than PMMA, such as metallic plates, can be envisaged for holding a higher power fiber amplifier and for better thermal management.

Figure 5.10 presents some pictures at different porcessing steps of a fiber array designed to hold 64 fibers. Figure 5.10a shows some details of the PMMA manufactured plate. The 64 PM fibers were then inserted into the plate (Figure 5.10b) and the slow axes of the PM fibers were all aligned parallel. The fibers were then glued with epoxy glue on the PMMA plate and the whole array was fixed into a tube holder for polishing process. The resulting fiber array is shown in Figure 5.10c and d.

The complete collimating fiber array is obtained by aligning the fiber array with the microlens array. First, the array of microlens is adjusted so that the beam issued

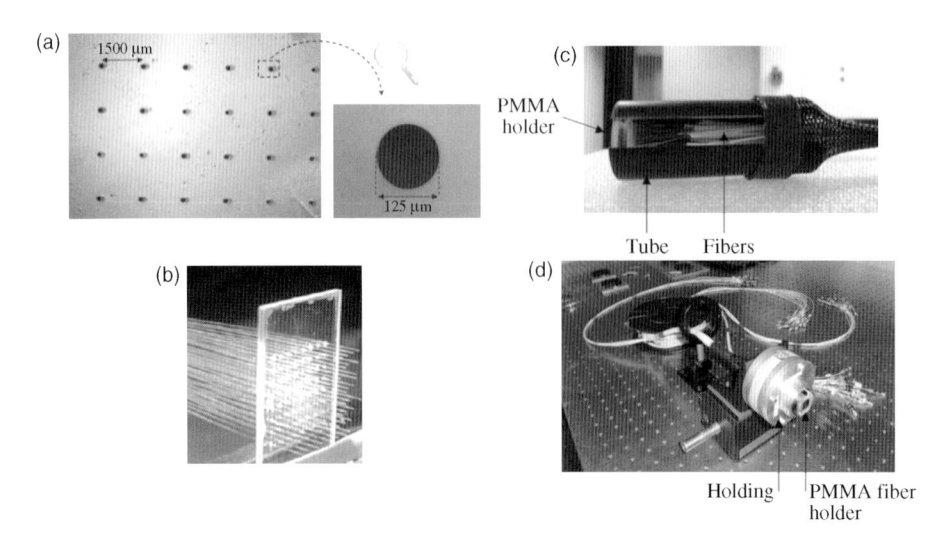

Figure 5.10 Photographs of the fiber holder. (a) Details of the PMMA plate drilled with 1500 μm pitch 125 μm diameter holes. (b) Insertion of the 64 PM fibers in the holder. (c) Final component with the fibers aligned, glued, and polished. (d) General view of the fiber array.

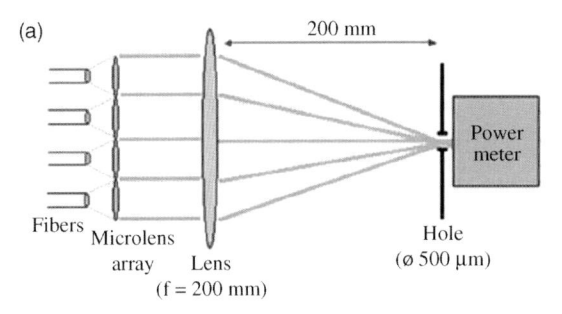

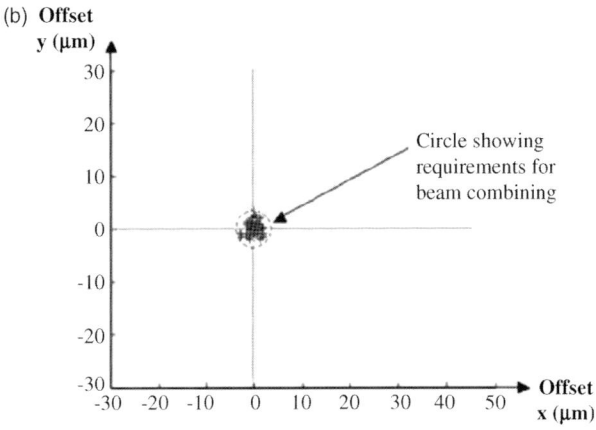

Figure 5.11 (a) Experimental setup for collimation alignment tests. (b) Impact of each beam on the focal plane of the lens.

from each fiber is centered on the associated lens. To measure fiber position accuracy, the position of the centroid of each beam was plotted in the far-field plane. As shown in Figure 5.11, this has been done through a 200 mm focal length lens and a camera placed exactly at its focal point. These data allowed us to retrieve the offsets between each fiber and its associated lenslet axis. A dashed circle is plotted in the figure to represent the offset limit to achieve beam combining with high combining efficiency. The maximum pointing error we measured on our collimation system was 0.52 mrad, and the average error was 0.2 mrad. Thus, such a system meets requirements of beam combining.

5.3.2
Collective Phase Measurement Technique

As previously explained, CBC techniques allow minimization of the phase differences between the fiber amplifiers in order to maximize the power or the energy in the central lobe in the far field. Different techniques have been presented in previous chapters. These techniques are mainly based on indirect phase locking techniques. In this case, the control loop optimizes the intensity through a pinhole in a power-in-the-bucket geometry. Two main phase locking methods are commonly used: the stochastic parallel gradient descent (SPGD) (see Chapters 4 and 6) [15] and the locking of optical coherence by single-detector electronic-frequency tagging (LOCSET) (see Chapters 2 and 3) [3,4]. Such techniques are quite simple to implement because only one single detector is required. However, the complexity of these techniques rises with the number of fibers to combine (see Section 8.2). For combining of a very large number of fibers, direct measurement of the phase is an interesting approach. In this case, instead of a single detector, one can use an array of detectors, that is, a camera. The idea is to record with the camera an interferometric pattern formed between the beamlets with each other or with a plane phase reference beam. With only one acquisition, it is then possible to collectively calculate the phase errors between all the fibers. Today, high-speed cameras with bandwidth greater than 1 kHz either around 1 μm with silicon technology or around 1.5 μm with InGaAs technology are available and ready to use for such applications.

In this section, a self-referenced collective method is presented to achieve the measurement of phase errors on large periodic beam arrays. A square pattern has been chosen for the following descriptions. Results could be readily extrapolated to hexagonal patterns.

5.3.2.1 Principle of the Measurement

In the near field, the phase retrieval of the fiber array can be seen as the analysis of a complex segmented wavefront, which stems from the juxtaposition of subwavefronts. The quadriwave lateral shearing interferometer (QWLSI), based on a diffraction grating [16,17], has proved to be an efficient self-referenced way to analyze a complex stepped wavefront [18–20]. This concept has been adapted to the measurement of a periodic fiber array [21].

The fundamental principle is to make interferences between four replicas of the wavefront from the square-patterned beam array and to analyze the interference pattern. No additional external reference wave is required. Thus, as shown in Figure 5.12, the

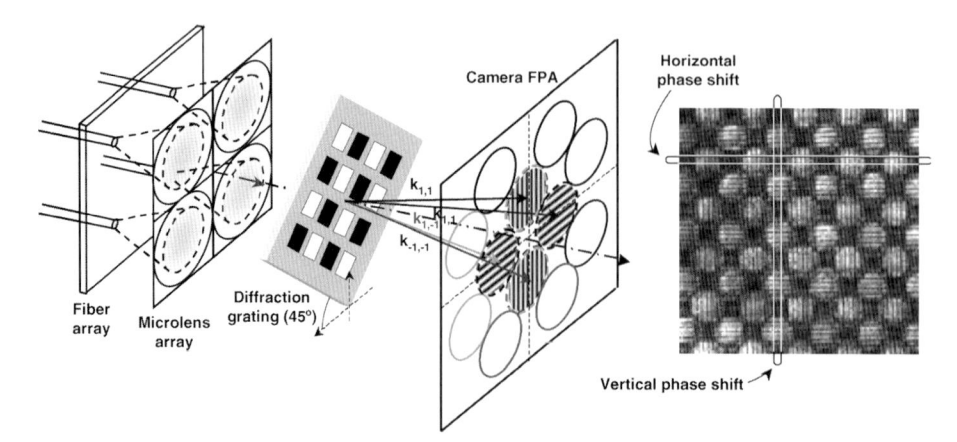

Figure 5.12 Principle of the self-referenced wavefront analysis technique based on the QWLSI (on four beams in a square pattern). *Right*: Part of an experimentally recorded interference pattern. FPA, focal plane array.

incident wavefront from the collimated fibers is split into four tilted and laterally sheared replicas, thanks to a two-dimensional diffraction grating. The grating orientation is rotated by 45° with regard to the beam pattern axis, and the lateral shearing is chosen to ensure a perfect overlap between replicas of adjacent beams.

On each overlap area, we obtain a set of two-wave sinusoidal fringes, whose phase shift is in direct relation to the phase step between the considered subwavefronts. Areas with horizontal fringes lead to a measurement of the phase step between each beam and its nearest neighbor in the vertical direction, and vertical fringes give the same information in the horizontal direction. Hence, the camera records a pattern that contains sets of vertical and horizontal fringes representative of the step between the fibers in two directions without correlation, allowing the reconstruction of the whole segmented wavefront. This operation can be very fast, allowing the combination of high-power fiber amplifiers. To address a hexagonal pattern, three replicas instead of four will be necessary, and the grating orientation toward pattern axis should be equal to 30° instead of 45°. This method can also be applied for pulsed regime. A two-color version of this technique can be implemented to retrieve phase shifts higher than 2π.

5.3.2.2 Implementation in the Experimental Setup

A QWLSI was used in an experimental setup to measure the phase relationship of 64-fiber array. The experimental QWLSI is composed of an InGaAs camera and a 2D diffraction grating with a 240 μm period. Experimentally, to perform the measure, a small amount of output beams is imaged on the camera with a magnification ratio of 2:1 to avoid parasitic overlap due to Gaussian beam divergence. This ratio has been chosen because the size of the fiber array had to be adapted to the size of the camera FPA. We hence obtain a beam array with a pitch of 750 μm. In order to obtain a perfect overlap of the replicas of the wavefront, the diffraction grating was placed at 40 mm from the front of the FPA and rotated by 45° from the axis of the laser head.

For the 64 beams in a square pattern, we obtained a pattern with 7×8 horizontal fringe sets and 8×7 vertical fringe sets, as presented in Figure 5.13. As the phase steps of the fibers evolve during the experiment in open loop, we could observe the scrolling of the fringes on the camera. After calibration, errors on phase retrieval were measured to be 0.11 rad ($\lambda/60$) RMS with our device.

5.3.2.3 Phase Retrieval Techniques

Various methods can be applied to reconstruct the phase map of the array from the interference pattern. In this section, we have chosen to describe two of these methods.

5.3.2.3.1 Fourier Transform Method
The intensity pattern of one line of vertical fringes set can be approximately written as

$$I(x) = \sum_n \text{rect}\left(\frac{x - nd}{a}\right) \times \cos(\Omega x + \Delta\varphi_n), \tag{5.10}$$

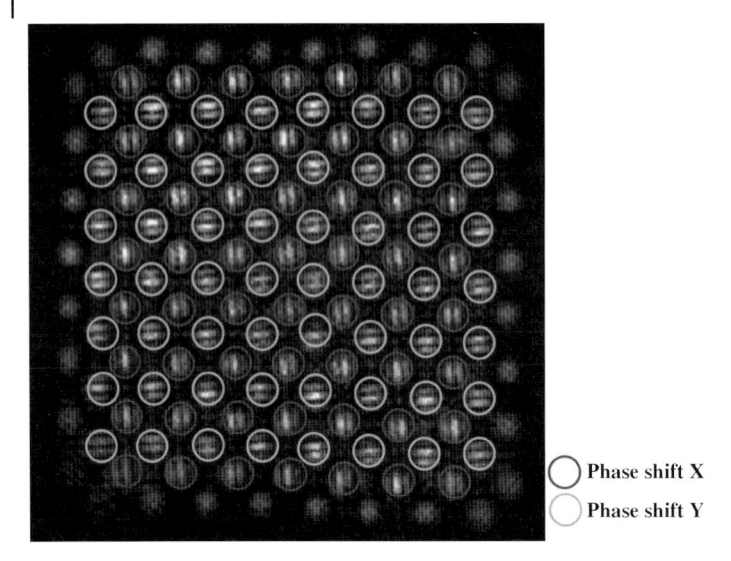

Figure 5.13 Experimental interference pattern obtained with QWSLI with the laser head. Sets of fringes relative to horizontal (X) and vertical (Y) phase shifts are differentiated in the figure.

where a is the pupil width, d is the pitch, Ω is the fringes frequency, and $\Delta\varphi_n$ denotes the phase shift between fibers n and $n+1$. The first calculation step is a Fourier transform of $I(x)$:

$$\tilde{I}(u) = \frac{1}{2}\left[\sum_n \mathrm{FT}\left(\mathrm{rect}\left(\frac{x-nd}{a}\right)\right) \cdot \delta(u-\Omega) \times e^{i\Delta\varphi_n}\right.$$

$$\left. + \sum_n \mathrm{FT}\left(\mathrm{rect}\left(\frac{x-nd}{a}\right)\right) \cdot \delta(u+\Omega) \times e^{-i\Delta\varphi_n}\right], \tag{5.11}$$

where $\sim$ and FT denote Fourier transform operator and δ is the Dirac function. Then, $\tilde{I}(u)$ is numerically filtered by selecting the frequency content around Ω and centered at zero frequency. This means to consider only the left-hand summation in the above equation and to turn $\delta(u-\Omega)$ into $\delta(u)$. Finally, a reverse Fourier transform is applied:

$$\tilde{I}_{\mathrm{filtered}}(u) = \frac{1}{2}\left[\sum_n \mathrm{FT}\left(\mathrm{rect}\left(\frac{x-nd}{a}\right)\right) \cdot \delta(u) \times e^{i\Delta\varphi_n}\right],$$

$$\mathrm{TF}^{-1}\left(\tilde{I}_{\mathrm{filtered}}\right)(x) = \frac{1}{2}\left[\sum_n \mathrm{rect}\left(\frac{x-nd}{a}\right) \times e^{i\Delta\varphi_n}\right]. \tag{5.12}$$

The phase map is finally obtained by taking the argument of the above expression.

5.3.2.3.2 Demodulation Method
As the interference pattern is only composed of sinusoidal fringes, a simple demodulation process can be applied to retrieve the

phase difference between one fiber and its neighbor. This technique brings an advantage in terms of calculation speed as no Fourier transform is involved.

The intensity pattern of one line of vertical fringes set can be approximately given by Eq. (5.10). By multiplying $I(x)$ by carriers C_p and C_q at the frequency Ω:

$$C_p(x) = \cos(\Omega x),$$
$$C_q(x) = \sin(\Omega x),$$
(5.13)

we obtain the following expressions:

$$I(x) \cdot C_p(x) = \sum_n \text{rect}\left(\frac{x - nd}{a}\right) \times \left(\frac{1}{2}\cos(2\Omega x + \Delta\varphi_n) + \frac{1}{2}\cos(\Delta\varphi_n)\right),$$

$$I(x) \cdot C_q(x) = \sum_n \text{rect}\left(\frac{x - nd}{a}\right) \times \left(\frac{1}{2}\sin(2\Omega x + \Delta\varphi_n) + \frac{1}{2}\sin(\Delta\varphi_n)\right).$$
(5.14)

With a numerical low-pass filter, it is possible to extract the terms $\cos(\Delta\varphi_n)$ and $\sin(\Delta\varphi_n)$. The phase shift between one fiber and its neighbors is then given by

$$\Delta\varphi_n = \arctan\left(\frac{\sin(\Delta\varphi_n)}{\cos(\Delta\varphi_n)}\right).$$
(5.15)

The whole phase map can then be deduced by propagating phase shifts from one fiber to another in vertical and horizontal directions. The use of the redundancy of the data will give some robustness to the results.

5.3.3
Phase Modulators

Phase modulators are key elements for the control of the phase of the beam from each fiber. Different solutions can be employed for this purpose. The most common technology is based on fiber-coupled electro-optic modulators (EOMs) placed before each fiber amplifier. $LiNbO_3$ is the most widely known material for EOMs. Typical CW power handling of such a device is in the range of $\sim$100 mW. The bandwidth is larger than 100 MHz and is thus perfectly suitable for CBC. The main disadvantage of discrete EOM on each fiber arm is the overall cost and size when dealing with a large number of fibers to phase lock. Alternative solutions may include piezoelectric fiber stretchers [5] and direct pump current modulation. Piezoelectric fiber stretchers present the advantage of controlling the phase outside the fiber amplifier and exhibit a phase range control much greater than 10π, but a bandwidth limited to about 1–10 kHz. The advantage of direct pump current modulation is that no other devices are required for phase control, but additional relative intensity noise (RIN) may be an issue.

For a large number of fibers, an attractive technology may rely on integrated photonics such as silicon photonics for 1.5 μm operation. Another possibility is the use of compact array of modulators using electro-optic ceramics. For this purpose, four-channel PLZT-based phase modulators have been recently developed [22].

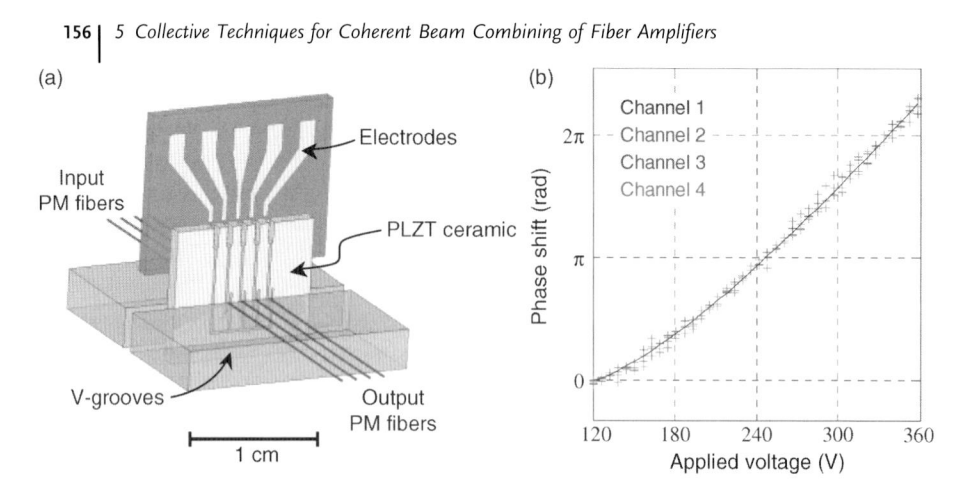

Figure 5.14 Schematic of a phase modulator array.

A schematic of the device is shown in Figure 5.14a. A pattern of interdigited electrodes is etched in an electro-optic ceramic (PLZT) using ultrasonic machining. The polarization axes of the input and output PM fibers are aligned parallel to the applied electric field so as to provide the voltage-controlled phase shift. As seen in Figure 5.2a, a 2π phase shift at 1.55 µm is obtained for a 240 V excursion for all the four channels. The response time of the device is in the 1–10 µs range. PLZT ceramic is transparent over a large spectral range and could be also designed for 1 µm operation.

5.4
Beam Combining of 64 Fibers with Active Phase Control

In this section, we present a beam combining experiment involving 64 fibers based on active phase control [22]. The setup involves a laser head composed of an array of collimated fibers, arrays of electro-optic ceramic modulators, and a QWLSI for phase measurement.

Figure 5.15 shows the experimental setup. A continuous 1.55 µm distributed feedback (DFB) laser is preamplified to 1 W by the first polarization-maintaining erbium-doped fiber amplifier (PM EDFA). The output is then split into four beams to feed four additional 1 W PM EDFAs. Each of the 4 outputs is further split into 16, leading to 64 amplified fibered outputs. Sixteen modules of compact, four-channel PLZT-based electro-optic phase modulators (as the one shown in Figure 5.14) are then inserted for phase feedback control before connecting to the output laser head. The output beams are first sampled for near-field imaging onto the QWLSI for phase measurement and correction through the phase modulators. The main laser output is then sampled again for far-field imaging onto a control camera and beam combination diagnosis.

5.4 Beam Combining of 64 Fibers with Active Phase Control

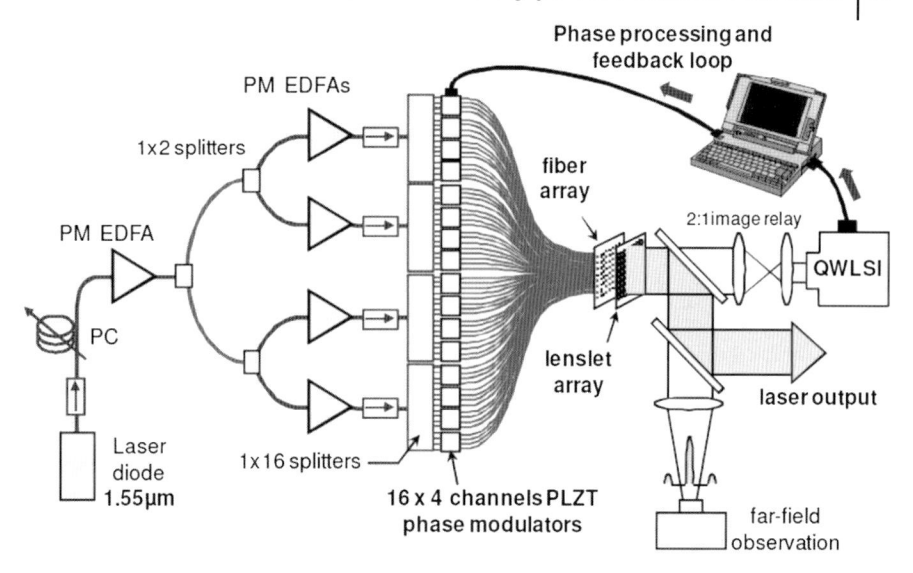

Figure 5.15 Experimental setup. PC, Polarization controller; PM EDFA, polarization-maintaining erbium-doped fiber amplifier.

Figure 5.16 shows the experimental far-field pattern obtained when the control loop is enabled. In the figure, the dashed line shows the theoretical profile corresponding to an ideal in-phase coherent summation of the 64 beamlets in a square lattice arrangement. Both are very close, with a slight deviation corresponding to a combining efficiency degradation of 0.64. This means that the peak energy of the central lobe of the experimental far-field pattern is 0.64 times the peak energy of the theoretical profile. This degradation is consistent with a measured phase error of $\lambda/10$ rms or $\lambda/3$

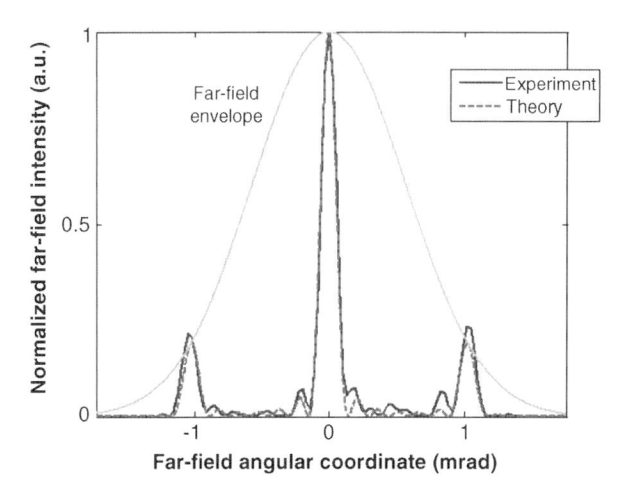

Figure 5.16 Experimental (solid line) and theoretical (dashed line) far-field intensity profiles. Gray curve shows the Fourier transform of an individual output pupil.

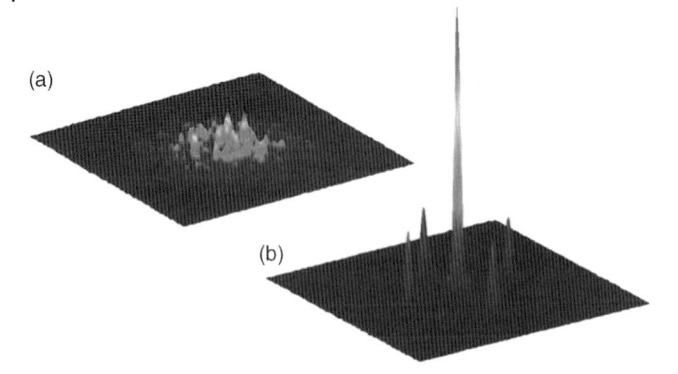

Figure 5.17 Far-field intensity pattern of 64 fibers, when the phase-locked loop is open (a) and closed (b).

peak-to-valley ratio, which theoretically drops the SR by 0.8. Another ×0.8 reduction is due to additional pointing and collimating errors of the output beamlets.

To further quantify the beam combining efficiency of the system, we evaluated the ratio between the energy contained in the central lobe of the far-field pattern and the total emitted energy. The central lobe width is determined by the first zeros of the theoretical far-field pattern in Figure 5.16. Experimentally, we measured 34% of the total emitted energy contained in the central lobe compared to 44% for the perfect fibers combination in a square lattice arrangement.

Figure 5.17 shows the far-field intensity pattern recorded on the IR camera used for far-field monitoring. In the open-loop configuration, when no phase correction is applied, the far field results from the interference of all the 64 beamlets with fluctuating phases. This leads to a speckle-like pattern, with an overall envelope divergence corresponding to the Fourier transform of an individual near-field beamlet, while the speckle grain divergence corresponds to the Fourier transform of the total near-field pupil. On the other hand, when the phase loop is closed, we observe constructive interference of all the 64 beams, leading to the presence of an intense central lobe, whose divergence is given by the inverse of the total output pupil size. The system operated at 20 Hz. Since the computation time is below 100 µs, and the modulators response time is in microsecond range, the system bandwidth is only limited by the QWLSI camera acquisition rate. This bandwidth was however appropriate to cophase the 64 fibers and to compensate the relative phase fluctuations of our four EDFAs in CW regime. More recent investigations demonstrate phase locking at 1 kHz bandwidth using the same technique.

5.5
Beam Combining by Digital Holography

In this section, we propose an alternative method for beam combining, based on self-adaptive digital holography [23–26]. This method brings a "passive" approach where the feedback loop is reduced to an electric connection only and can also be easily applied to a large number of fibers.

5.5.1
Principle

The principle of fiber beam combining with digital holography is shown in Figure 5.18. It uses a two-dimensional array of fiber amplifiers, each one being collimated by a lens array with parallel axis at its extremity. A low-power reference plane wave Φ_{R1}, at a wavelength λ, propagates through the fiber array. We record on a CCD camera the interference between the resulting wavefront $\Phi(x,y)$ and a reference plane wave Φ_{R2}. The two waves are coherent with each other and tilted by a small angle α. For each fiber i, the recorded interference pattern consists of a series of linear fringes whose position is given by the relative dephasing φ_i between fiber i of the array and the reference Φ_{R2}. As this dephasing evolves in time, the recorded fringes scroll on the CCD. The angle α is chosen small enough to maintain the fringe spacing d times larger than several CCD pixel size to ensure enough phase correction precision. The CCD camera signal is then used to drive a phase-only spatial light modulator (SLM), which displays the recorded fringe pattern and acts as a dynamic digital hologram. A reference plane wave Φ_{R3} is then used to diffract on the SLM in the Raman–Nath regime. On the -1 diffraction order, propagating at an angle $-\alpha$ with respect to Φ_{R3}, the phase-conjugated segmented wavefront $-\Phi(x,y)$ is generated.

By reinjecting this wavefront into the fiber array, all the dephasing terms are compensated and the resulting output beams are phase locked with a common plane phase Φ_{out}. In this scheme, coherent combining is obtained by a process of conjugate wavefront generation.

In this technique, precision toward phase correction is given by the sampling of the fringe pattern by the pixels of the camera and the SLM. As commercially available SLMs have up to 1920×1080 pixels for 1.06 µm operation, a precision of $\lambda/10$ can be

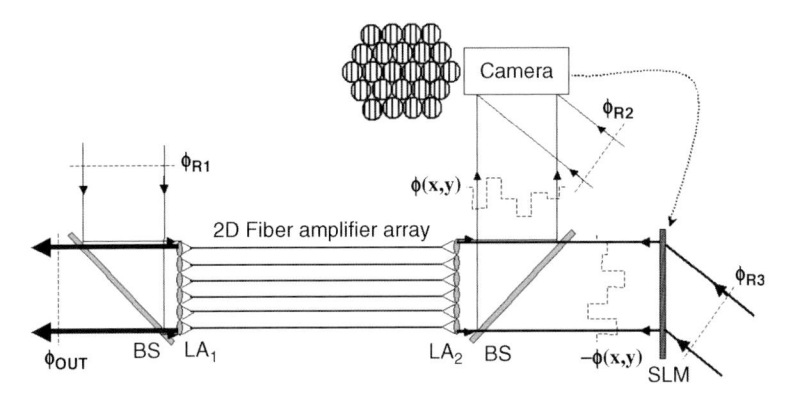

Figure 5.18 Schematic of beam combining by digital holography. BS, low-reflectivity beam splitter; LA, lens array; SLM, spatial light modulator [27].

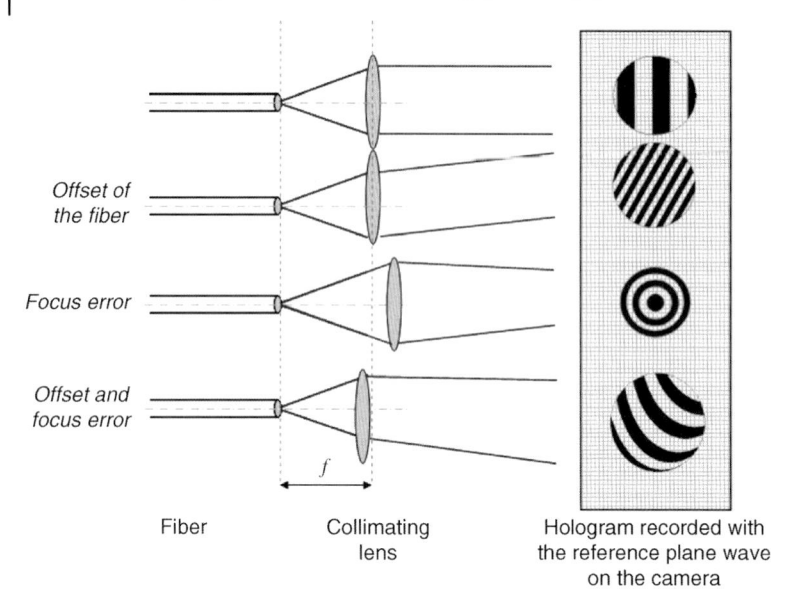

Figure 5.19 Schematic effect of collimation errors on the shape of the fringe in the interference pattern.

expected with up to ~200 fibers in the array. On the output side of the collimated fiber array (LA$_1$ in Figure 5.18), collimation alignment must respect the requirements described in Section 5.2.2. This will grant coupling of the reference plane wave Φ_{R1} into the fiber array and then ensure beam combining with high efficiency on the output corrected beams. As we realize phase conjugation, the collimation of the fibers on the sensor side (LA$_2$ in Figure 5.18) does not need so strict requirement in terms of alignment precision. Indeed, as described in Figure 5.19, a small misalignment or defocus will turn into a deformation of the fringe pattern recorded. When reading the hologram written on the SLM, the diffracted wave will then correct the collimation error by phase conjugation. The only requirement to grant a proper correction is to have enough pixels per fringes on the CCD camera in order to reach the required phase correction accuracy. The magnification ratio between the image recorded by the camera and the image displayed by the SLM should be equal to 1 in order to permit the coupling of the -1 diffracted order $-\Phi(x,y)$ into the fiber array.

The recorded interferogram results from the interference of the beams from each collimated fibers and a common large reference wave. Thus, for each fiber, we obtain the interference between a Gaussian and a plane field distribution and the corresponding modulated area on the interference pattern presents naturally a contrast with a Gaussian-like distribution (see the left hand-side image in Figure 5.20). This does not optimize the diffraction efficiency when reading, as the diffraction efficiency is in direct proportion to the contrast of the fringes.

To overcome this limitation, a simple improvement of the contrast can be operated while recording the interferogram, as described in Figure 5.20. This operation

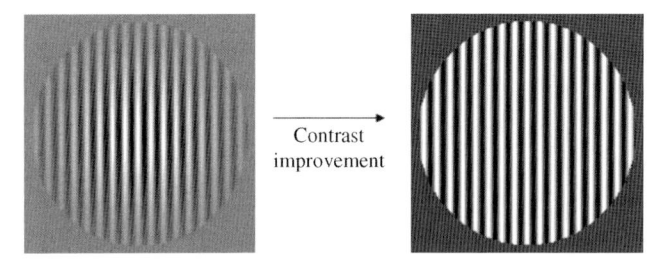

Figure 5.20 Contrast improvement operated to increase the diffraction efficiency of the hologram.

increases greatly the diffraction efficiency (from ~10 to up to 33% on each modulated area) and has no consequence on the phase correction of the beams because the phase relationship between the fiber and the reference wave is coded by the relative position of the fringes and not by the contrast of the interferogram. In fact, a binary image is enough to ensure the hologram functionality and a fast ferroelectric binary SLM could be used to compensate the typical phase fluctuations of fiber amplifiers.

5.5.2
Experimental Demonstration

The experimental proof-of-concept of CBC by digital holography is shown in Figure 5.21 [27]. The master oscillator is a nonplanar ring oscillator (NPRO) Nd: YAG laser having a narrow bandwidth of 10 kHz that ensures long-enough coherence length. An isolator protects the NPRO from a possible back-propagating beam. The beam from the laser is expanded to form a 15 mm diameter plane wave

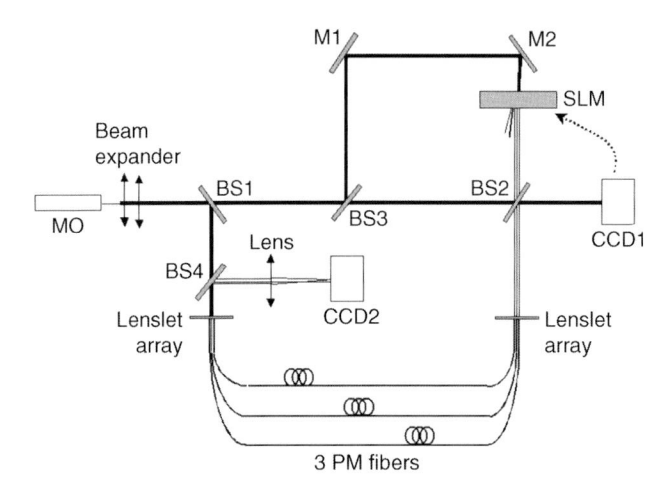

Figure 5.21 Experimental setup of beam combining by digital holography. MO, master oscillator; BS, beam splitter; M, mirror; CCD, charge-coupled device; SLM, spatial light modulator; PM, polarization maintaining.

reference. With the first low-reflectivity beam splitter BS1, we take a small amount of the beam to generate plane wave Φ_{R1} and feed the PM fibers. Thanks to a second beam splitter BS2, the beams from the fibers with segmented wavefront $\Phi(x,y)$ can interfere with the transmitted reference beam with plane phase Φ_{R2}. The angle α between the beams was 5 mrad, leading to a grating period of 200 μm. We recorded the interference pattern with a CCD monochrome camera with a 720×576 resolution (CCD1 in Figure 5.21). The video signal is then transferred to a computer via an acquisition card and displayed on both the screen of the computer and the SLM with the help of the clone mode of the graphic card. The computer is optional and is only used in the experiment to observe the interference pattern.

The SLM is an amplitude-only twisted liquid crystal (LC) array, with a resolution of 800×600 and the same pixel size as CCD1. In this configuration, one pixel from CCD1 is directly converted into one pixel on the SLM. The pixel lines of the SLM that are not addressed by the CCD are not used. With another beam splitter BS3, we take a part of the reference beam to generate a plane wave Φ_{R3} and read the digital hologram displayed on the LC SLM. By adjusting mirror M2, the -1 order wavefront diffracted from the SLM is coupled into fibers. A low-diffraction efficiency on the order of 1% was measured in the -1 order. This is due to our twisted nematic LC SLM, which acts as an amplitude hologram. This efficiency could be greatly improved by the use of a phase-only SLM, thus reaching a possible 33% diffraction efficiency in the -1 order. Finally, a second camera CCD2 allows us to observe the phase-locked beams in the far field at the focal plane of a lens.

The experimental results of beam combining by digital holography are shown in Figures 5.22 and 5.23. A time-varying dephasing of many waves with a bandwidth of about 2 Hz was obtained by heating the fibers with the help of a lamp. The resulting moving fringes were recorded on CCD1 and then displayed on the SLM for the generation of the digital hologram. A typical interference pattern recorded on CCD1 is shown in Figure 5.22a. Three distinct areas can be observed corresponding to the interferences between the output of the three fibers $\Phi(x,y)$ and the plane reference wave Φ_{R2}. Owing to nonperfect collimation conditions of the fibers, the grating periods and orientations are different, but will be fully compensated after the phase conjugation process. The phase-locked output beams with wavefront Φ_{out} are observed in the far field on CCD2. A stable time intensity diffraction profile is shown in Figure 5.22b. The precision of phase locking is given by the number of pixels on each fringe recorded on the CCD1 and displayed on the SLM. In our experiment, about 30 pixels per fringe were used, giving a precision of phase locking above $\lambda/10$, thus ensuring suitable phase error compensation. For comparison, Figure 5.22c shows the blurred far-field intensity pattern obtained when the phase correction is disabled. In this case, a static interference pattern was displayed on the SLM, avoiding dynamic compensation of the phase errors. Figure 5.22b and c were obtained after 15 s time integration.

The dynamic performance of the phase locking system was also measured by placing a photodiode on the main central lobe of the far-field pattern. Figure 5.23 shows the intensity stability of the main lobe versus time. When the phase locking is stopped, strong intensity fluctuations are observed. The present experimental setup

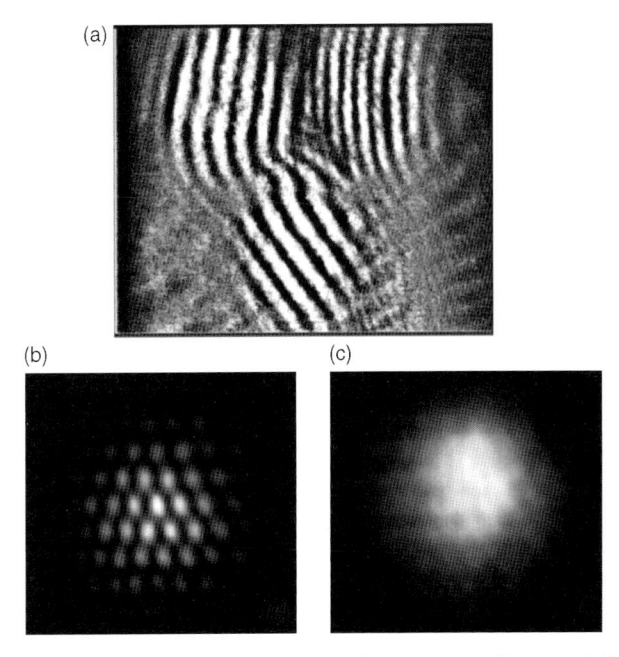

Figure 5.22 (a) Recorded hologram with three passive fibers. Far-field intensity patterns of the (b) three phase-locked beams and the (c) three unlocked beams obtained after 15 s time integration.

works at video frame rate, but as already mentioned, higher speed operation could be achieved by using a kilohertz camera to record the hologram and a fast SLM.

5.6
Conclusion

In this chapter, various technologies have been presented for CBC of a large number of fiber amplifiers. Among the different key elements, we have emphasized the

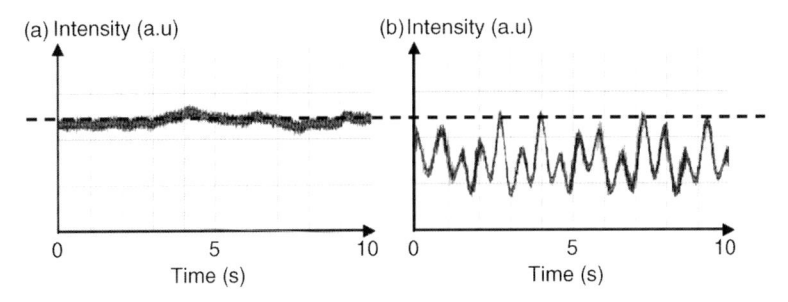

Figure 5.23 Intensity versus time of the central part of the far-field pattern when the three fibers are (a) phase locked and (b) unlocked.

necessity of developing arrays of collimating fibers allowing accurate positioning of fibers. A collective phase measurement technique based on interferometry has been proposed and demonstrated. The technique allows measurement of the phase errors between all fibers with only one acquisition.

Coherent phase combining of 64 independent amplified fibers is demonstrated. The presented concept also involves that compact array arrangement of the modulators is intrinsically scalable to a much larger number of fibers. It is extrapolated that phase locking of thousands of fibers would be possible with commercially available high-speed and high-resolution cameras.

Acknowledgments

The authors acknowledge the French National Research Agency (ANR) for partial funding of this work within the Coherent Amplifying Network (CAN) Project led by Gérard Mourou. The authors also thank Fayçal Bouamrane, Thierry Bouvet, and Sephan Megtert from Unité Mixte de Recherche CNRS/Thales.

References

1 Fan, T.Y. (2005) Laser beam combining for high-power, high-radiance sources. *IEEE J. Sel. Top. Quantum Electron.*, **11**, 567–577.

2 Labaune, C., Hulin, D., Galvanauskas, A., and Mourou, G. (2008) On the feasibility of a fiber-based inertial fusion laser driver. *Opt. Commun.*, **281**, 4075–4080.

3 Shay, T.M., Benham, V., Baker, J.T., Ward, C.B., Sanchez, A.D., Culpepper, M.A., Pilkington, D., Spring, J., Nelson, D.J., and Lu, C.A. (2006) First experimental demonstration of self synchronous phase locking of an optical array. *Opt. Express*, **14**, 12015–12021.

4 Bennai, B., Lombard, L., Jolivet, V., Delezoide, C., Pourtal, E., Bourdon, P., Canat, G., Vasseur, O., and Jaouen, Y. (2008) Brightness scaling based on 1.55 μm fiber amplifiers coherent combining. *Fiber Intergr. Opt.*, **27**, 355–369.

5 Yu, C.X., Klansky, J.E., Shaw, S.E., Murphy, D.V., and Higgs, C. (2006) Coherent beam combining of a large number of PM fibers in 2-D fiber array. *Electron. Lett.*, **42**, 1024–1025.

6 Demoustier, S., Bellanger, C., Brignon, A., and Huignard, J.P. (2008) Coherent beam combining of 1.5 μm Er-Yb doped fiber amplifiers. *Fiber Integr. Opt.*, **27**, 392–406.

7 Bellanger, C., Brignon, A., Colineau, J., and Huignard, J.P. (2008) Coherent fiber combining by digital holography. *Opt. Lett.*, **33**, 2937–2939.

8 Brusselback, H., Wang, S., Minden, M., Jones, D.C., and Mangir, M. (2005) Power-scalable phase-compensating fiber array transceiver for laser communications through the atmosphere. *J. Opt. Soc. Am. B*, **22**, 347–353.

9 Neubert, W.M., Kudielka, K.H., Leeb, W.R., and Scholtz, A.L. (1994) Experimental demonstration of an optical phased array antenna for laser space communications. *Appl. Opt.*, **33**, 3820–3830.

10 Stace, C., Harisson, C.J.C., Clarke, R.G., Jones, D.C., and Scott, A.M. (2004) Fiber bundle lasers and their applications. 1st Electro Magnetic Remote Sensing Defence Technical Conference, Edinburgh, UK, Paper B21.

11 Bourderionnet, J., Rungenhagen, M., Dolfi, D., and Tholl, H.D. (2008) Continuous laser beam steering with micro-optical arrays: experimental results. *Proc. SPIE*, **7113**, 71130Z.

12 Bellanger, C., Brignon, A., Toulon, B., Primot, J., Bouamrane, F., Bouvet, T., Megtert, S., Quétel, L., and Allain, T. (2011) Design of a fiber-collimated array for beam combining. *Opt. Eng.*, **50**, 025005.

13 Glashauser, W. and Ghica, G.V. (1980) Verfahren für die spannungsfreie Entwicklung von bestrahlten Polymethylmetacrylatschichten, Siemens Patent, Germany, German Patent No. 3039110.

14 Jones, D., Turner, A., Scott, A., Stone, S., Clark, R., Stace, C., and Stacey, C. (2010) A multi-channel phase locked fibre bundle laser. *Proc. SPIE*, **7580**, 75801V-1.

15 Liu, L., Vorontsov, M.A., Polnau, E., Weyrauch, T., and Beresnev, L.A. (2007) Adaptive phase-locked fiber array with wavefront tip–tilt compensation. *Proc. SPIE*, **6708**, 67080.

16 Primot, J. and Sogno, L. (1995) Achromatic three-wave (or more) lateral shearing interferometer. *J. Opt. Soc. Am. A*, **12**, 2679–2685.

17 Velghe, S., Primot, J., Guerineau, N., Cohen, M., and Wattellier, B. (2005) Wavefront reconstruction from multidirectional phase derivatives generated by multilateral shearing interferometers. *Opt. Lett.*, **30**, 245–247.

18 Toulon, B., Primot, J., Guérineau, N., Haïdar, R., Velghe, S., and Mercier, R. (2007) Step-selective measurement by grating-based lateral shearing interferometry for segmented telescopes. *Opt. Commun.*, **279**, 240.

19 Toulon, B., Vincent, G., Haïdar, R., Guérineau, N., Collin, S., Pelouard, J.L., and Primot, J. (2008) Holistic characterization of complex transmittances generated by infrared sub-wavelength gratings. *Opt. Express*, **16**, 7060.

20 Mousset, S., Rouyer, C., Marre, G., Blanchot, N., Montant, S., and Wattellier, B. (2006) Piston measurement by quadriwave lateral shearing interferometry. *Opt. Lett.*, **31**, 2634.

21 Bellanger, C., Toulon, B., Primot, J., Lombard, L., Bourderionnet, J., and Brignon, A. (2010) Collective phase measurement of an array of fiber lasers by quadriwave lateral shearing interferometry for coherent beam combining. *Opt. Lett.*, **35**, 3931–3933.

22 Bourderionnet, J., Bellanger, C., Primot, J., and Brignon, A. (2011) Collective coherent phase combining of 64 fibers. *Opt. Express*, **19**, 17053–17058.

23 Stappaerts, E.A. (1995) Holographic system for interactive target acquisition and tracking. U.S. Patent No. 5,378,888, January 3.

24 Schnars, U. and Jueptner, W. (2005) *Digital Holography: Digital Hologram Recording, Numerical Reconstruction, and Related Techniques*, Springer.

25 Cuche, E., Marquet, P., and Depeursinge, C. (1999) Simultaneous amplitude-contrast and quantitative phase-contrast microscopy by numerical reconstruction of Fresnel off-axis holograms. *Appl. Opt.*, **38**, 6994–7001.

26 Gross, M. and Atlan, M. (2007) Digital holography with ultimate sensitivity. *Opt. Lett.*, **32**, 909–911.

27 Bellanger, C., Brignon, A., Colineau, J., and Huignard, J.P. (2008) Coherent fiber combining by digital holography. *Opt. Lett.*, **33**, 2937–2939.

6
Coherent Beam Combining and Atmospheric Compensation with Adaptive Fiber Array Systems

Mikhail Vorontsov, Thomas Weyrauch, Svetlana Lachinova, Thomas Ryan, Andrew Deck, Micah Gatz, Vladimir Paramonov, and Gary Carhart

6.1
Introduction

One of the most promising technologies that may dramatically change traditional laser beam transmitter (beam director) systems is related with recent advances in *coherent combining* of beams (beamlets) that are generated in a multichannel optical power amplifier (MOPA) fiber system and transmitted through a coupled MOPA array of fiber collimators [1–7]. In the directed energy (beam projection) applications considered here, coherent beam combining implies phasing of the outgoing beamlets at a remotely located target in the atmosphere by controlling their piston phases at the fiber collimator subapertures. Ideal target plane phasing of N_{sub} outgoing beamlets results in their constructive interference leading potentially to an N_{sub}-fold increase of combined beam target plane peak intensity, compared to incoherent beam combining in the absence of beamlets' phasing [3–6]. Note that coherent beam combining can only be achieved if the outgoing beamlets are quasi-monochromatic and have identical (or nearly identical) polarization states and spatial mode structure [1,4]. These requirements are fulfilled in a so-called *coherent fiber array system* based on a narrow-line seed laser and polarization-maintaining (PM), single-mode fiber elements and subsystems. A notional schematic of a coherent fiber array-based laser beam projection system is shown in Figure 6.1. In this system, control of piston phases at the fiber array pupil plane is performed using an optoelectronic target-in-the-loop (TIL) feedback system [7–9]. This control system includes a bistatic optical receiver that transforms the captured target return wave power into an electric signal J_{PIB} known as the power-in-the-bucket (PIB) metric, a metric processor that computes a phase-locking control metric J, and a phase-locking controller that utilizes the metric signal for computation of control voltages (controls) $\{u_j\}, j = 1, 2, \ldots, N_{sub}$. Phase shifters that are integrated into the MOPA system transform the control voltages $\{u_j\}$ into optical wave time delays, resulting in controllable changes of the outgoing beam phases averaged over the fiber collimator subaperture areas, which are referred to as *piston phases*. The

Coherent Laser Beam Combining, First Edition. Edited by Arnaud Brignon.
© 2013 Wiley-VCH Verlag GmbH & Co. KGaA. Published 2013 by Wiley-VCH Verlag GmbH & Co. KGaA.

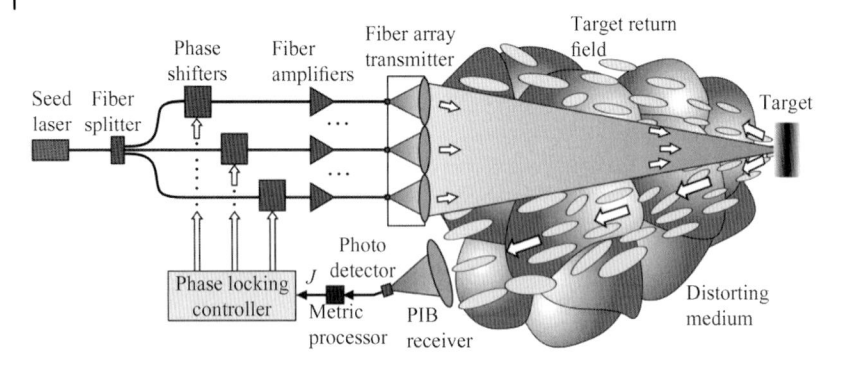

Figure 6.1 Notional schematic of a laser beam projection system based on coherent phased fiber array technology.

transmitted beamlets form a *combined beam* that propagates to the target. Atmospheric turbulence-induced refractive index inhomogeneities result in scintillations of the projected combined beam intensity inside the illuminated target area (target hit spot) and a decrease in the average projected beam power density (target hit spot brightness).

Achievement of the smallest (ideally, diffraction-limited) target hit spot in the vicinity of an assigned target aim point – the ultimate goal in the directed energy applications – requires precise overlapping of the transmitted beamlets, also referred to here as *combined beam focusing*, fine combined beam pointing and stabilization, and real-time compensation of both the MOPA system-induced random phase shifts and atmospheric turbulence-induced phase aberrations. In this chapter, we describe recent results of basic research in the area of fiber array system development with major focus on fiber array architectures, integration of adaptive optics (AO) and beam control capabilities, as well as coherent beam combining and atmospheric compensation techniques with the common goal of the projected beam power density increase at either unresolved (point source) or resolved (extended or speckle) targets.

6.2
Fiber Array Engineering

There are several considerations that impact the design of fiber collimator array-based beam directors. First, since in most applications the position of the target in space can be dynamically changing due to either target or beam director platform (or both) movement, the fiber array beam director should provide capabilities for the combined beam pointing and target hit spot stabilization in the conditions of changing beam propagation direction and distance. This requires integration of tracking, beam pointing, and beam focusing functions into the fiber array-based beam director. Similar to conventional beam directors based on laser transceiver telescopes with monolithic mirrors, target tracking and coarse beam pointing can be performed by integrating the fiber array system into a gimbaled platform. Nevertheless, both fine beam pointing and hit spot stabilization require significantly higher accuracy than gimbals can provide. In conventional beam directors, the required accuracy is achieved by controlling the outgoing beam tip and tilt phase with relatively small size beam steering mirrors located in the beam director optical train prior to the final beam expansion with a transmitter telescope. In the case of fiber array-based systems, fine beam pointing can only be achieved with integration of tip and tilt wavefront phase control into each fiber collimator.

Second, in conventional systems, beam focusing is performed via displacement of the transmitter telescope secondary mirror, which leads to formation of a parabolic (spherical) wavefront phase of the outgoing beam. Since fiber array-based systems do not have external beam-forming optics, the combined beam focusing can be only performed using phase-shaping elements that are directly integrated into fiber collimators.

Consider the following two options for the integration of wavefront-shaping capabilities into the fiber collimator array, which can be utilized for the fine steering and focusing of the combined beam. The simplest option is related with approximation of tip–tilt and parabolic phase functions using subaperture-averaged (piston) phases, as illustrated in Figure 6.2a. This wavefront approximation, referred to as the *stair-mode approximation*, can be achieved using the fiber-integrated phase shifters of the MOPA system. A more accurate approximation of the combined beam phase can be obtained using, in addition to pistons, control of tip and tilt phase components at each fiber collimator subaperture (Figure 6.2b) [4]. As shown in Figure 6.2c, such tip–tilt control of the outgoing beamlet phase can be achieved using x and y displacements of the fiber tip that is located in the collimating lens focus. In the fiber-tip positioner devices that are specially developed for this purpose, displacements of the fiber tips are performed using piezoelectric actuators [7,10].

Examples of fiber array-based laser transmitters with integrated capabilities for piston and tip–tilt phase control at each fiber collimator aperture, developed by the authors, are shown in Figure 6.3. These fiber arrays have identical subaperture diameter d, subaperture fill factor $f_{sub} = d_0/d$, and the fiber array aperture fill factor f_c that is defined by the ratio $f_c = l/d$, where d_0 is the diameter of the Gaussian beam at the fiber collimator exit and l is the distance between the beamlets' optical axes [4]. The fiber array system, shown in Figure 6.3a and referred to here as the *fiber array cluster*, is composed of seven densely packed fiber collimators [11]. This fiber array cluster is envisioned as a building block (module) that can be used for increasing (scaling) the number of subapertures in the array by assembling together fiber array systems composed of several clusters, as illustrated in Figure 6.3b. A different approach to a fiber array system scaling is illustrated in Figure 6.3c. The increase in the number of subapertures from 7 to 19 is achieved here by incorporating the external chain of 12 additional fiber collimators into the fiber array cluster.

To estimate the potential benefits of integrating the wavefront phase tip–tilt control into individual fiber collimators, consider ideal target plane phasing in vacuum of the coherent fiber arrays in Figure 6.3 that utilize either solely piston

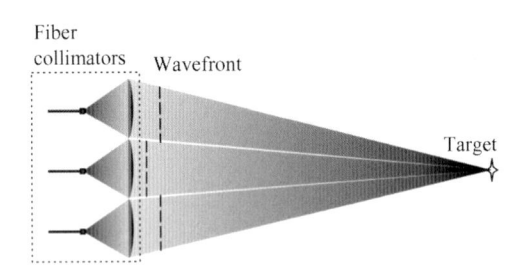

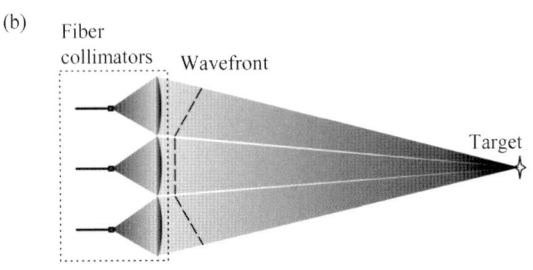

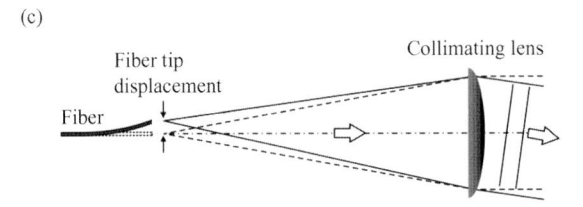

Figure 6.2 Combined beam focusing using (a) stair-mode and (b) piston and tip–tilt approximations of the parabolic wavefront. Tip and tilt control in each fiber collimator subaperture in (b) can be performed using displacement of the fiber tip, as shown in (c) [7, 10].

(stair mode) or both piston and tip–tilt phase control at each fiber collimator subaperture. Spatial distributions of the combined beam power projected onto the target plane over a distance of 2 and 7 km are characterized in Figure 6.4 by the dependencies of the total power J_{PIB} inside the on-axis circular area (bucket), commonly referred to as the *target plane* PIB metric, on the bucket diameter d_T and by the corresponding target plane intensities shown as grayscale images on the right. For comparison, the PIB metrics for the seven-subaperture incoherent system are shown in Figure 6.4a by the dotted lines. Numerical calculations of the target plane intensity $I_T(\mathbf{r})$ were performed using Fresnel (parabolic) approximation of the diffraction theory [12,13]. As the results presented in Figure 6.4 suggest, the tip–tilt control allows the desired redirection of a portion of projected beam energy from the side lobes into the central (on-axis) lobe that is associated with the target hit spot. The expected benefit from the tip–tilt control is most pronounced for beam projections over relatively short distances and with increased number of subapertures N_{sub}.

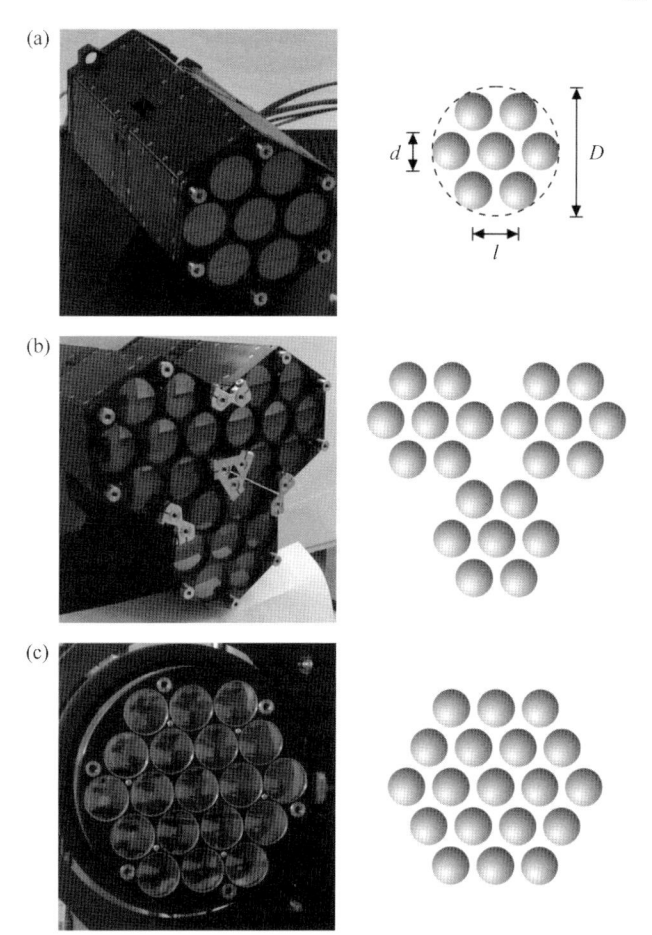

Figure 6.3 Coherent fiber array systems with (a) 7, (b) 21, and (c) 19 subapertures. In all systems, $d = 33$ mm, $l = 37$ mm, and $f_{sub} = 0.89$. The grayscale images on the right show the Gaussian-shaped intensity distributions inside the subapertures. Fiber arrays in (a) and (b) are developed by Optonicus [11], and in (c) by US Army Research Laboratory.

This conclusion is further elaborated in Figure 6.5, which represents the on-axis target plane intensity values I_T^0 as functions of the propagation distance L for the fiber array transmitter geometries in Figure 6.3 with $N_{sub} = 7, 19$, and 21 subapertures. As seen from Figure 6.5, the stair-mode approximation of the parabolic phase (dashed lines) leads to a general decrease of the target plane peak intensity value, compared to the piston and tip–tilt phase approximation (solid lines). As expected, this decrease is smaller for longer propagation distances.

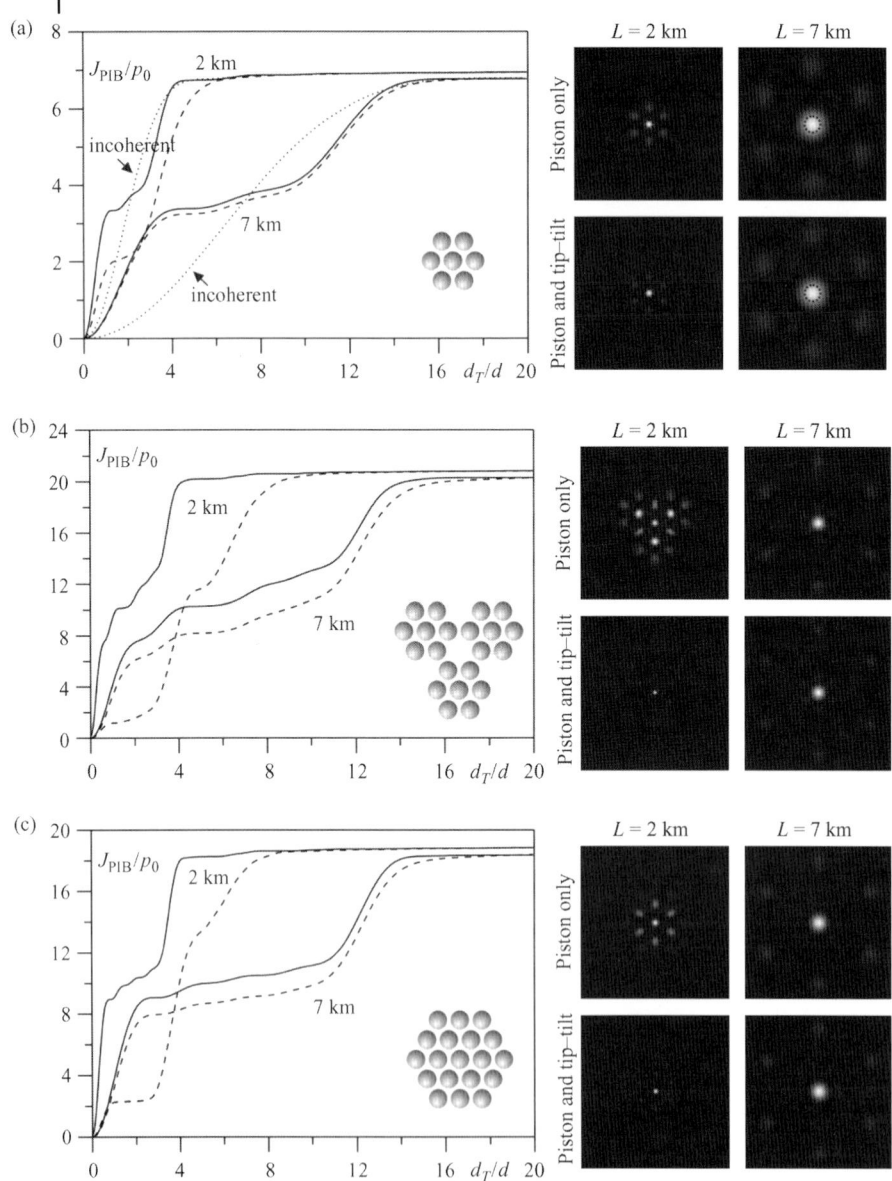

Figure 6.4 Efficiency of the target plane coherent beam combining (phase locking) over 2 and 7 km distances in vacuum for the fiber arrays in Figure 6.3 with piston and tip–tilt (solid lines) and piston-only (dashed lines) control of the outgoing beamlet phases. The PIB metric of the projected combined beam, J_{PIB}, is normalized by the power p_0 transmitted through a single-fiber array subaperture, and the on-axis bucket diameter d_T is normalized by the fiber collimator subaperture diameter d. The tip–tilt phase components are assumed to be zero with piston-control only. Corresponding target plane intensity distributions are illustrated by the grayscale images. The dashed circles in (a) indicate the target plane receivers of diameters $d_T = 2.5$ cm for $L = 2$ km and $d_T = 5$ cm for $L = 7$ km. The dotted lines in (a) correspond to the incoherent fiber array system with seven subapertures.

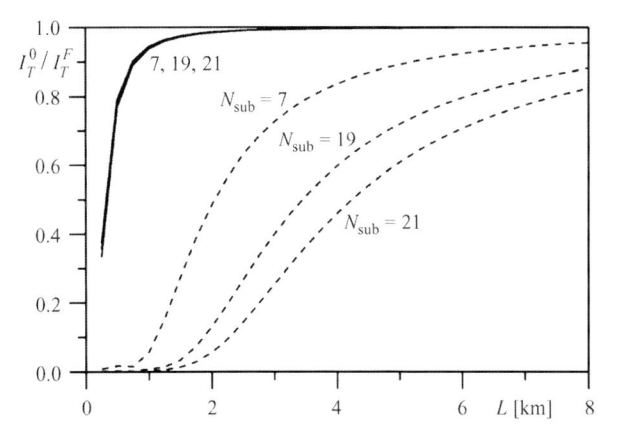

Figure 6.5 Beam projection on a remote target a distance L in vacuum for the fiber arrays in Figure 6.3 with piston and tip–tilt (solid lines) and piston-only (dashed lines) control of the outgoing beamlet phases. The on-axis target plane intensity values $I_T^0 = I_T^0(\mathbf{r} = 0)$ are normalized by the corresponding values of the on-axis target plane intensities $I_T^F = I_T^F(\mathbf{r} = 0)$ obtained using the parabolic phase inside each subaperture area.

6.3
Turbulence-Induced Phase Aberration Compensation with Fiber Array-Integrated Piston and Tip–Tilt Control

Consider now the impact of wavefront tip–tilt control at each fiber array subaperture on the mitigation of atmospheric turbulence effects. For simplicity, we assume a point source (unresolved) target at distance L in optically inhomogeneous and isotropic random medium with Kolmogorov refractive index fluctuation power spectrum (Kolmogorov turbulence model) [14]. In this commonly used model, atmospheric turbulence strength is associated with the refractive index structure parameter C_n^2. The characteristic spatial scale of atmospheric turbulence-induced phase fluctuations at the fiber array pupil plane can be described by the Fried parameter r_0 [15] that for spherical target return wave can be represented in the form $r_0 = 3.02(k^2 C_n^2 L)^{-3/5}$, where $k = 2\pi/\lambda$, λ is the optical wavelength, and C_n^2 is assumed to be a constant along the propagation path (model for homogeneous turbulence). For estimation of the potential benefit from tip–tilt control integration into fiber collimators, assume that the outgoing beamlet phase control is based on an ideal conjugation of the measured local (subaperture-averaged) piston and tip–tilt phase components – the control approach known as *phase-conjugate aberration precompensation* [5,16,17]. In numerical simulations presented in this section, the phase conjugate control of the combined beam transmitted by the fiber array was implemented by computing the phase of the target return wave originating from a monochromatic coherent small size light source (beacon) located at the target plane. The beacon optical wave propagated through a "thin" turbulent layer (Kolmogorov's phase screen) located near the fiber array system aperture. The local piston and

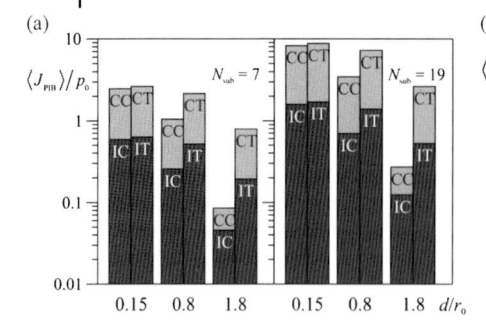

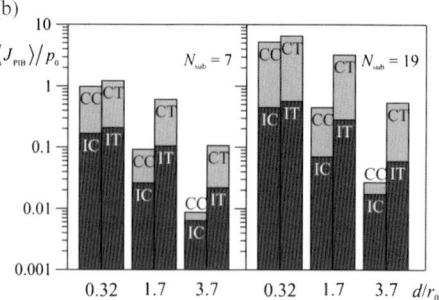

Figure 6.6 Efficiency comparison of laser beam projection using coherent (phase-locked) and incoherent fiber arrays with AO precompensation of atmospheric turbulence-induced piston and piston and tip–tilt phase aberrations using atmospheric turbulence-averaged target plane PIB metrics $\langle J_{PIB}\rangle$ for (a) 2 and (b) 7 km propagation distances and for fiber arrays with 7 and 19 subapertures. The results are obtained for the following fiber array operational modes: incoherent combining without (IC) and with (IT) tip–tilt control and coherent combining without (CC) and with (CT) tip–tilt control. The metric values $\langle J_{PIB}\rangle$ are normalized by the power p_0 transmitted through a single fiber array subaperture. The bucket size d_T equals to 1/2 of the diffraction-limited target plane intensity central lobe size for a coherent fiber array beam with seven subapertures ($d_T = 0.75d$ for $L = 2$ km and $d_T = 1.5d$ for $L = 7$ km, depicted by the dashed circles in Figure 6.4a). Each set of bars corresponds to different d/r_0 ratios.

tip–tilt phase turbulence-induced components were computed using decomposition of the phase screen-induced phase aberrations over Zernike polynomials inside each subaperture [18]. The obtained phase components were conjugated and used for generation of the outgoing combined beam that propagated through the same phase screen to the target plane. Compensation efficiency was estimated using the target plane PIB metric J_{PIB}. To obtain statistically averaged values of the PIB metric, computations were repeated with a set of 100 independent phase screen realizations and the obtained metric values were averaged.

The results of the system performance analysis are summarized in Figure 6.6 in the form of the target plane PIB metric bar diagrams presented in logarithmic scale. The atmospheric turbulence-averaged metric values $\langle J_{PIB}\rangle$ are compared for adaptive fiber arrays with coherent and incoherent beam combining and with and without integrated tip and tilt wavefront control capabilities. The results presented in Figure 6.6 were obtained for two different fiber arrays (with number of subapertures $N_{sub} = 7$ and 19) and for three different atmospheric turbulence conditions corresponding to $C_n^2 = 1 \times 10^{-15}$ m$^{-2/3}$ (weak turbulence), 1.7×10^{-14} m$^{-2/3}$ (moderate turbulence), and 6×10^{-14} m$^{-2/3}$ (strong turbulence). The associated d/r_0 values computed for $L = 2$ and 7 km range from $d/r_0 = 0.15$ ($L = 2$ km, weak turbulence) to $d/r_0 = 3.7$ ($L = 7$ km, strong turbulence).

The results in Figure 6.6 clearly demonstrate that for the examined system configurations, the achieved metric values are notably higher for the coherent phase-locked adaptive fiber arrays than that for the incoherent systems [compare the dark-shaded (IC) and light-shaded (CC) bars]. The gain of using the phase-locking control increases with an increase in N_{sub}. The gain in PIB metric increase from integration of tip–tilt control appears to be insignificant for relatively weak turbulence, but grows with d/r_0 and becomes rather substantial for $d/r_0 > 1$ – compare the corresponding CC/IC and CT/IT bars. One of the most important conclusions that can be derived from the presented analysis is that under conditions of strong turbulence, the efficiency of both coherent and incoherent beam projection systems can be significantly increased by incorporating tip–tilt wavefront aberration compensation capabilities into each fiber array subaperture [5]. This integration can lead to a decrease in the required number of subapertures as well as to more efficient compensation of turbulence effects, resulting in an increase in the target hit spot brightness with even less power transmitted through the combined beam director.

It is important to point out that similar results were obtained in numerical experiments simulating laser beam projection over distributed turbulence. In these calculations, the turbulence-induced piston and tip–tilt phase control components were obtained by considering propagation of a beacon wave through a set of 10 phase screens with Kolmogorov statistics, equidistantly distributed along the propagation path. For accurate estimation of the return wave piston and tip–tilt phase components, the return field phase was unwrapped to remove 2π-phase discontinuities. The combined beam with the updated beamlet phases was propagated through the same set of phase screens to the target, where the PIB metrics were computed. However, one should note that the laser beam propagation through a distributed turbulent medium results in both intensity scintillations of the received wave and phase singularities (branch points) [19,20]. Both effects complicate computation of local piston and tip–tilt phase components.

6.4
Target Plane Phase Locking of a Coherent Fiber Array on an Unresolved Target

6.4.1
Fiber Array Control System Engineering: Issues and Considerations

In this and the following sections, we consider control algorithms and systems that can be used in fiber array-based laser beam projection systems for coherent combining at a remote target. We assume that the distance L to the target is relatively short so that the double-pass propagation delay time $\tau_{2L} = 2L/c$ (the round-trip time, where c is the speed of light) does not exceed the characteristic time τ_{at} of the atmospheric turbulence-induced refractive index inhomogeneities update inside the propagating beam footprint. This condition is typically fulfilled for the so-called tactical range distances. At this operational range, one can apply what is known as target-in-the-loop (TIL) control techniques for target plane phasing of the outgoing beamlets and atmospheric turbulence-induced aberration compensation. In the TIL phasing control concept, the target is considered as a part of the control loop in the sense that the controls applied to reshape phases of the outgoing beamlets are dependent on measurements of the backscattered (target return) wave at the fiber array transmitter plane. Correspondingly, these measurements depend on characteristics of the target (its size, shape, surface roughness, etc.). This dependence of the backscattered wave on target characteristics significantly complicates development of TIL control techniques. To simplify the analysis, in this section we assume that the target is small in respect to the size of the diffraction-limited target plane central lobe formed by the fiber array (unresolved or point source target). In Section 6.5, we depart from this assumption and consider the more general case of an extended (resolved) target with randomly rough surface. This target type is commonly referred to as *speckle target*.

For both unresolved and resolved targets, we consider target plane fiber array phasing using the iterative control algorithm known as stochastic parallel gradient descent (SPGD) [21–23]. The SPGD control is based on optimization of a measured signal (metric) that depends on the control variables – voltages applied to either phase shifters of the MOPA system (piston phase control) or to both the phase shifters and piezoactuators of the tip–tilt control system (piston and tip–tilt control). The SPGD-based metric optimization control can result in the combined beam phasing only if the measured metric depends monotonically on the projected beam (target hit spot) quality, which is typically estimated in terms of power density or the hit spot size. In the case of a point source (unresolved) target, one can use the target return optical wave power measured inside the bistatic optical receiver aperture as the metric for the SPGD-based phase-locking control [24]. This control system is illustrated in Figure 6.1. Indeed, the measured target return wave power (PIB) monotonically depends on the transmitted beam footprint at the target plane since beamlets phasing and atmospheric turbulence effects' mitigation lead to better energy concentration at the unresolved target. Thus, the measured PIB signal can be used as a metric for SPGD-based locking of the beamlets transmitted by the fiber array. Note that optimization of the PIB metric with SPGD control automatically results in compensation of phase shifts that are introduced by both the MOPA system and the atmospheric turbulence. Since the metric optimization process requires a number of iterations N_{it} (from tens to hundreds) [25] and each SPGD iteration takes some time τ_{it}, it is critically important that the SPGD process convergence time $\tau_{SPGD} \sim N_{it}\tau_{it}$ does not exceed the atmospheric characteristic time τ_{at}. The condition $\tau_{SPGD} < \tau_{at}$ is relatively straightforward to achieve in fiber array systems with fiber-integrated phase shifters that can operate with several gigahertz bandwidth and for propagation over relatively short distances for which the double-pass time delay $\tau_{2L} < \tau_{it} \ll \tau_{at}$ and hence can be neglected. Note that the operational frequency bandwidth of the fiber-tip positioning devices is significantly lower (on the order of a few kilohertz), which makes compensation of turbulence-induced local tip–tilt phase aberration components more challenging [7].

6.4.2
SPGD-Based Coherent Beam Combining: Round-Trip Propagation Time Issue

Assume for simplicity an unresolved target at a relatively short distance ($L < c\tau_{it}/2 \ll c\tau_{at}/2$) from a fiber array-based beam projection system and consider control of the fiber array local piston and tip–tilt phase using a SPGD metric optimization technique. Control of the beamlets' piston phases with a conventional SPGD algorithm can be described as follows [7,23,24]. During each iteration cycle n ($n = 1, 2, 3, \ldots$), the SPGD controller generates a set of small-amplitude random control voltage perturbations $\{\delta u_j^{(n)}\}$ that are superimposed with the set of piston phase control signals, $\{u_j^{(n)}\}$, where $j = 1, 2, \ldots, N_{\text{sub}}$. In the simplest SPGD algorithm implementation, the perturbations represent statistically independent numbers of identical magnitude with random signs, having equal probabilities for positive and negative values. The application of signals $\{u_j^{(n)} + \delta u_j^{(n)}\}$ to the phase shifters is followed by the measurement of the performance metric value $J_+^{(n)}$ (here the PIB metric). After the metric $J_+^{(n)}$ is measured within the same nth SPGD iteration cycle, the controls with perturbations of the opposite sign $\{u_j^{(n)} - \delta u_j^{(n)}\}$ are applied to the phase shifters. This is followed by the measurement of the corresponding metric value $J_-^{(n)}$, where $J_\pm^{(n)} = J(u_1^{(n)} \pm \delta u_1^{(n)}, \ldots, u_j^{(n)} \pm \delta u_j^{(n)}, \ldots, u_{N_{\text{sub}}}^{(n)} \pm \delta u_{N_{\text{sub}}}^{(n)})$. The computed metric variation $\delta J^{(n)} = J_+^{(n)} - J_-^{(n)}$ and the applied control voltage perturbations $\{\delta u_j^{(n)}\}$ are used to generate updated control signals for the next, that is, $(n+1)$st iteration:

$$u_j^{(n+1)} = u_j^{(n)} + \gamma \delta J^{(n)} \delta u_j^{(n)}. \tag{6.1}$$

This control algorithm has two essential control parameters that need to be optimized: the update gain coefficient γ and the perturbation magnitude $|\delta u_j^{(n)}| = \xi = \text{const}$. Note that in more advanced SPGD control algorithms, both gain γ and perturbation magnitude ξ are automatically adjusted based on the current operation condition [26,27].

Besides control of piston phases, one can use an SPGD iterative procedure similar to Eq. (6.1) to control the outgoing beamlets' wavefront tips and tilts by applying voltages $\{v_{j,x}^{(n)}, v_{j,y}^{(n)}\}$ ($j = 1, 2, \ldots, N_{\text{sub}}$) to the x- and y- actuators of the fiber positioner devices in Figure 6.2c. Since piston and tip–tilt control channels have significantly different response times ($\leq 10^{-9}$ and $\lesssim 10^{-4}$ s, respectively), the piston and tip–tilt SPGD controllers operate at considerably different iteration rates and practically do not impact each other [28–30].

The SPGD process requires precise temporal synchronization between applied controls and metric measurements, which in turn demands accounting for various delays related with finite response times of metric sensor, phase shifters, and tip–tilt actuators, the time required for computation of controls, and the round-trip propagation time delay τ_{2L}. Therefore, the SPGD controller needs to postpone its operation (to pause) for a time of duration,

$$\tau_{\text{delay}} \geq \tau_{\text{sys}} + \tau_{2L}, \tag{6.2}$$

starting from the moment when the control signals are applied, before resuming operation for metric measurement. Here, the response time τ_{sys} includes all control system time delays mentioned above. Since in each SPGD cycle control voltages are changed twice, the characteristic SPGD iteration time τ_{SPGD} need to be at least twice as long as $\tau_{2L} + \tau_{sys}$.

For coherent beam combining over distances of about 1 km or longer, the propagation delay τ_{2L} becomes the major factor that precludes an increase of the SPGD iteration rate and thus limits the improvement of the phase-locking control convergence. For example, for a target at $L = 10$ km distance, the round-trip propagation delay is $\tau_{2L} = 66.7 \mu s$, which is considerably longer than the typical control system response time, $\tau_{sys} \geq 2\mu s$. Thus, in the example considered here, the round-trip propagation delay would limit the piston phase SPGD iteration rate, $f_{SPGD} = 1/\tau_{SPGD}$, to about 7 kHz – more than 30 times less than what a commercially available multichannel SPGD controllers can achieve [11].

The propagation delay problem may be overcome using a modified SPGD algorithm (referred to here as delayed-feedback SPGD or DF-SPGD) [24,31]. In this algorithm, the controller does not need to pause during the time $\tau_{delay} \geq \tau_{sys} + \tau_{2L}$ before performing the measurement of metric signals. Instead, the control parameters update is performed using the metric values $J_+^{(n)}$ and $J_-^{(n)}$ measured without delay in connection with the perturbations $\{\delta u_j^{(n-\Delta n)}\}$, which were applied Δn iterations earlier and stored in the controller's memory:

$$u_j^{(n+1)} = u_j^{(n)} + \gamma \left[J_+^{(n)} - J_-^{(n)} \right] \delta u_j^{(n-\Delta n)}. \qquad (6.3)$$

The DF-SPGD control in Eq. (6.3) requires some adjustment of parameters so that the time duration Δt between applying perturbations $\{\pm \delta u_j^{(n-\Delta n)}\}$ and the corresponding measurements of the metrics $J_\pm^{(n)}$ is at least $\tau_{sys} + \tau_{2L}$ and is approximately equal to $\tau_{DF-SPGD} \Delta n$, as illustrated in Figure 6.7. Correspondingly, the number Δn and the duration of a single iteration $\tau_{DF-SPGD}$ must be properly selected. Both parameters can be modified by changing the DF-SPGD controller iteration rate and duration of perturbations. Note that, in general, different combinations of Δn and $\tau_{DF-SPGD}$ can be chosen to fulfill the condition $\Delta t \cong \tau_{DF-SPGD} \Delta n$. Nevertheless, it is always desirable to keep the iteration rate as high as possible, which corresponds to using the highest Δn possible. In order to determine a proper set of parameters Δn and $\tau_{DF-SPGD}$, one needs to know the distance to the target, L, which could be determined, for example, by a target ranging system. Alternatively, a supervisory control loop could continuously adjust Δn and $\tau_{DF-SPGD}$ so that the system performance is optimized – an approach that may be applied especially in application scenarios with changing target distance.

6.4.3
Coherent Beam Combining at an Unresolved Target over 7 km Distance

In this section, we discuss practical issues related with the experimental implementation of the TIL coherent beam combining over tactical-range atmospheric paths.

6.4 Target Plane Phase Locking of a Coherent Fiber Array on an Unresolved Target

(a) Conventional SPGD

(b) Delayed-Feedback SPGD

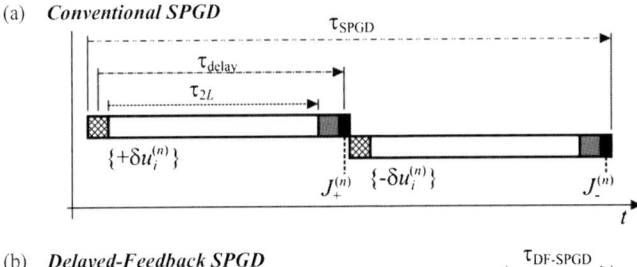

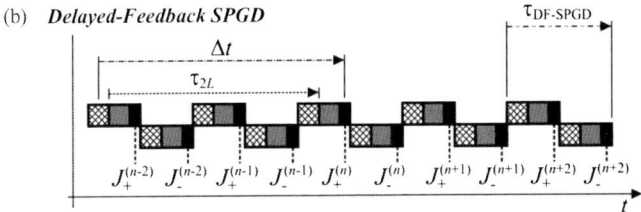

Figure 6.7 Timing diagrams for the conventional SPGD control (a) and the delayed-feedback SPGD control with $\Delta n = 2$. A single iteration indicating positive and negative perturbations is shown for conventional SPGD. In this example, utilization of the DF-SPGD control provides a fivefold increase in the iteration rate.

Experimental studies were performed using the University of Dayton's outdoor test range. The transmitter fiber array as in Figure 6.3a and the PIB receiver were located close to a window at the Intelligent Optics Laboratory (IOL). An unresolved target (corner cube retroreflector with 50 mm diameter) was installed inside a shed on the rooftop of the Dayton VA Medical Center $L = 7$ km away from the fiber array transmitter. The propagation path profile is shown in Figure 6.8. The test range was equipped with a boundary layer scintillometer, which continuously recorded the path-averaged refractive index structure constant C_n^2.

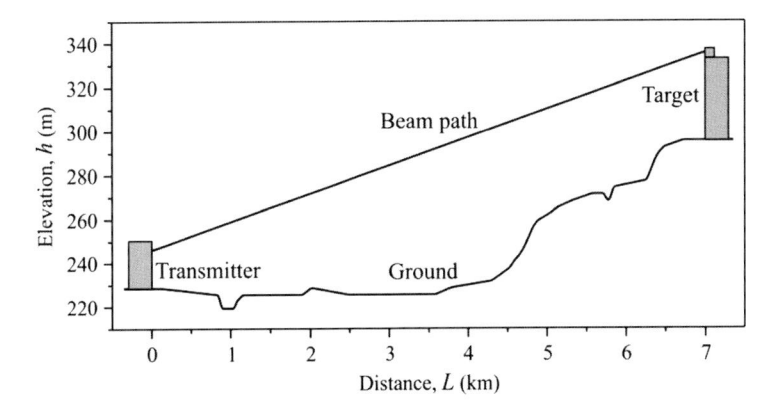

Figure 6.8 Combined laser beam 7 km long propagation path between fiber array transmitter and receiver located at the University of Dayton IOL and the retroreflector target on the rooftop of the Medical Center.

6 Coherent Beam Combining and Atmospheric Compensation with Adaptive Fiber Array Systems

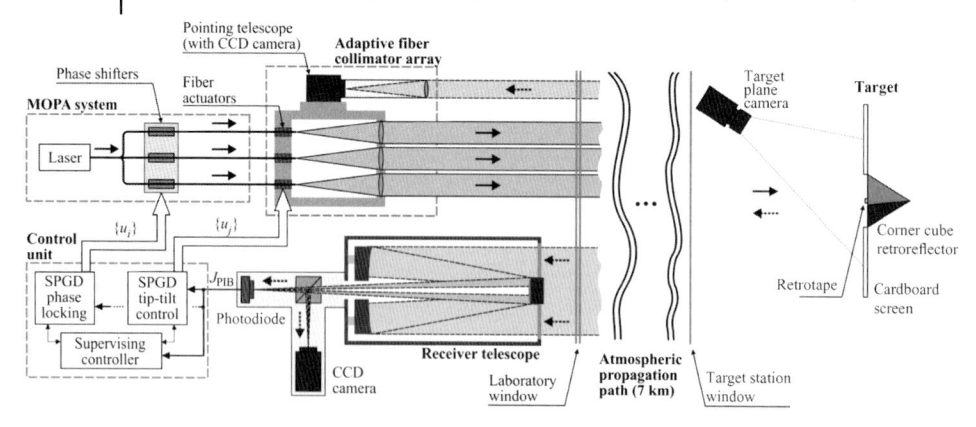

Figure 6.9 Schematic of the experimental setup used for TIL phase locking with an unresolved target over a 7 km atmospheric propagation path.

A schematic of the transmitter setup is shown in Figure 6.9. The light from a fiber-coupled laser with wavelength $\lambda = 1064$ nm and bandwidth $\Delta \nu = 5$ kHz was split into eight channels by a 1×8 fiber splitter with integrated phase shifters. Seven of the polarization-maintaining output fibers were connected to a fiber collimator array, as shown in Figure 6.3a. Each collimator was equipped with a piezoactuated fiber positioning system for control of the lateral fiber-tip position within a range of about $\pm 35 \mu$m. This corresponds to a tilt range of about ± 0.2 mrad for the aspheric collimation lenses with focal length of $f = 174$ mm [7] and results in ± 1.4 m lateral beam displacements at the target plane. The collimator array was mounted onto a gimbal together with a coboresighted small telescope, which was used to point the beams toward the target. Note that the window glass at the transmitter side introduced wavefront aberrations of about one wavelength peak-to-valley (PV) over the array aperture and $\lambda/4$ PV over a single subaperture. These aberrations were partially compensated by the subaperture tip–tilt control system.

A part of the light was reflected by the target retroreflector and propagated back to the PIB receiver based on a Schmidt–Cassegrain telescope with 20 cm aperture. A CCD camera with narrow field of view was used for the receiver telescope alignment. The receiver telescope was placed at the minimum possible distance to the collimator array. The received light power was measured by a photodetector and its output signal was used as performance metric (PIB metric, J_{PIB}) for the SPGD-based piston and tip–tilt control (Figure 6.9). The SPGD multichannel optimization controller (Optonicus, LLC) [11] was used for piston phase control (phase locking). Because the response bandwidth of the fiber actuators limits the tip–tilt control to a few kilohertz iteration rate, the corresponding control was implemented using a personal computer (PC). The two controllers operated in parallel without iteration cycles synchronization. The supervising controller (also a PC) was used to trigger the piston and tip–tilt controllers to begin or stop operation during experimental trial as well as to digitize and record metric data.

For coherent beam combining efficiency evaluation, the fiber array control system repeatedly performed sequential trials comprised of the following three stages:

Stage 1 – Feedback off: Randomized, but static, control voltages were applied to the phase shifters, tip–tilt control voltages were set to average values from previous control cycles. On average, this operational condition corresponds to incoherent beam combining.

Stage 2 – Piston control only: The SPGD controller optimized the received PIB signal by applying voltages to fiber-integrated phase shifters. For an unresolved target, this corresponds to maximization of the power within the target retroreflector. Tip–tilt control remained off.

Stage 3 – Piston and tip–tilt phase control: Both piston and tip–tilt SPGD control systems operated in parallel.

The duration of each control phase was about 1.75 s and the trials were repeated 50 times. Values for the performance metric (received PIB), J_{PIB}, were acquired for all cycles at a sampling rate of about 10 kHz. From the measured metric data of all recorded trials, the average values $\langle J_{PIB} \rangle$ and probability distributions $\rho(J_{PIB})$ of the PIB metric values J_{PIB} were calculated for each stage separately.

Experiments were performed using either the conventional SPGD (with $\tau_{SPGD} = 130\mu s$) or the DF-SPGD control algorithms (with $\Delta n = 7$ and $\tau_{SPGD} = 7\mu s$). In Figure 6.10, the dependences $\rho(J_{PIB})$ for piston control (stage 2) with both SPGD and DF-SPGD controllers are compared with the corresponding

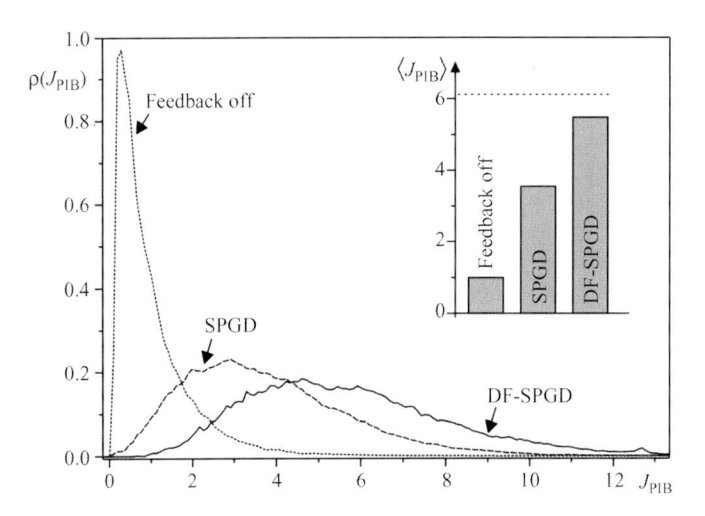

Figure 6.10 Experimental results of the coherent beam combining experiments over 7 km propagation path with an unresolved retroreflector target using the fiber array in Figure 6.3a. Measured probability densities $\rho(J_{PIB})$ for the PIB metric J_{PIB} without and with piston phase control using either the conventional SPGD or the DF-SPGD algorithms. The inset at the right compares the corresponding average metric values $\langle J_{PIB} \rangle$. The experimental results were obtained in atmospheric turbulence conditions corresponding to $C_n^2 = 6 \times 10^{-16}$ m$^{-2/3}$.

dependences obtained with feedback control system off (stage 1). Coherent beam combining (phase locking) using nondelayed (conventional) SPGD control resulted in a significant increase in the observed metric values; nevertheless, as shown in Figure 6.10, utilization of the delayed SPGD control (DF-SPGD) resulted in a noticeable improvement of system performance. A comparison of average metric values, $\langle J_{PIB} \rangle$, is shown as inset in Figure 6.10. Here, all metric values were normalized to the average value measured during the feedback-off stage, that is, $\langle J_{PIB} \rangle_{(\text{Feedback off})} \equiv 1$. The average metric improvement in comparison to the uncontrolled state was 3.7 for piston phase control with the conventional SPGD and 5.6 with the DF-SPGD; the latter value corresponds to about 90% of the value 6.1 expected for vacuum propagation (indicated by a dashed line).

As seen from the probability densities $\rho(J_{PIB})$ in Figure 6.10, there were considerable fluctuations of the PIB metric J_{PIB}. However, the fluctuation level was reduced by utilizing piston phase control, which also resulted in the average metric value increase.

In the discussed phase-locking experiments, we found a negligible difference between the values of $\langle J_{PIB} \rangle$ obtained with and without subaperture tip–tilt control. This is in accordance with the discussions presented in Section 6.3, where it is shown that the impact of tip–tilt control is expected to be low for the measured relatively weak turbulence conditions (see Figure 6.6b). Moreover, the faster iteration rate of piston phase control (nearly 50-fold in case of DF-SPGD) allows a fast compensation of the overall tilt through a stair-mode approximation.

In order to verify that maximization of the PIB metric J_{PIB} indeed corresponded to a higher peak irradiance at the target, the irradiance distribution at the target plane was directly monitored. As shown in Figure 6.9, the 50 mm diameter target retroreflector was placed behind a hole of the same size in a cardboard screen. In addition, a small patch of retroreflecting tape (6 mm diameter) was attached to the center of the target retroreflector's cover glass. A camera with a wide-angle objective, placed about 1 m in front of the screen and 20 cm to the side of the line-of-sight, recorded the beam footprint on the screen at 30 frames/s. The beams at the target plane showed a considerable level of scintillations – as expected for atmospheric turbulence conditions with a Rytov variance on the order of or near unity – so a considerable frame averaging was necessary to evaluate the irradiance at the target. Figure 6.11a and b shows the center part of the target plane irradiance distributions obtained with averaging of 270 frames that were recorded while the piston phase control was off and on, respectively. The distributions clearly demonstrate the significantly higher irradiance level at the retro-reflector with TIL phase locking.

6.5
Target Plane Phase Locking for Resolved Targets

In this section we consider the case of laser beam projection onto an extended (resolved) target with randomly rough surface. The coherent beam scattering off the target's rough surface results in a strong speckle modulation at the transceiver

(a) (b)

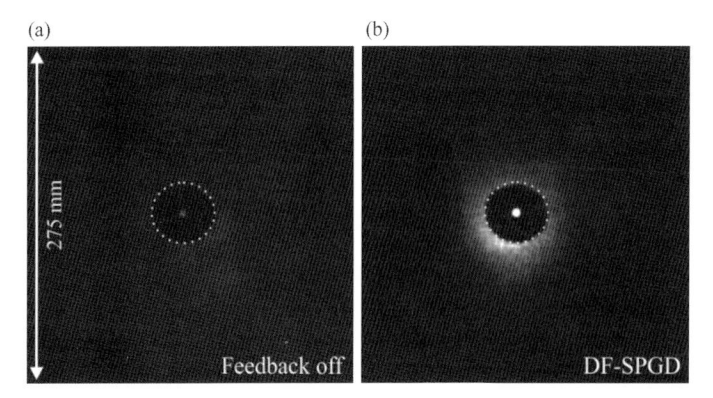

275 mm

Feedback off | DF-SPGD

Figure 6.11 Experimental long-exposure target plane irradiance distributions in the plane of the retroreflector (a) without piston phase control, that is, with feedback off, and (b) with DF-SPGD control. The brighter spot in the center corresponds to a small patch of retroreflecting tape, which was attached to the retroreflector's cover glass. The dotted circle indicates the retro-reflector.

plane, which represents a long-standing major challenge (known from the late 1970s as the speckle problem in AO) [32–34]. We address this problem by utilizing a speckle metric optimization-based phase locking (SMPL) technique, which enabled, to our knowledge, the first successful demonstration of TIL laser beam projection onto an extended target with randomly rough surface.

6.5.1
Speckle Metric Optimization-Based Phase Locking

In the SMPL technique described here, control of the outgoing laser beam phase is performed using the optimization of speckle-averaged characteristics of the target return speckle field that are referred to here as speckle metrics [35–37]. The term "speckle averaging" implies that the return wave characteristic $J(t)$ is averaged over a time period τ_J, which exceeds significantly the characteristic time τ_{sp} needed for a speckle field realization update inside the receiver aperture. The measured characteristic $J_{sp} = \langle J \rangle_{sp}$, where $\langle \cdots \rangle_{sp}$ denotes speckle averaging, can be utilized for phase-locking control as a performance measure (speckle metric) if the following conditions are fulfilled: (i) J_{sp} depends monotonically on a target plane beam quality metric J_T, which characterizes the power density distribution inside the target hit spot, and (ii) J_{sp} can be measured over a time τ_J that is considerably shorter than the characteristic times τ_{at} and τ_{AO} of turbulence and closed-loop phase control, respectively. From condition (ii) follows a hierarchy of characteristic time scales that is required for SMPL control implementation:

$$\tau_{sp} \ll \tau_J \ll \tau_{AO} \leq \tau_{at}. \tag{6.4}$$

To estimate the upper limit for the characteristic time for speckle realization updates, τ_{sp}, assume in Eq. (6.4) that $\tau_{sp} \approx 10^{-2} \tau_J \approx 10^{-4} \tau_{at}$. With a common

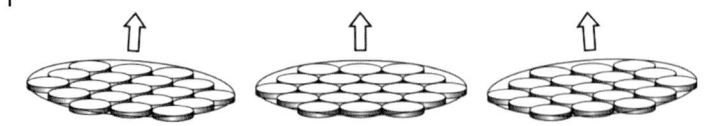

Figure 6.12 Illustration of wavefront phase tilt control using subaperture piston phases (stair-mode approximation of wavefront tilt aberration).

estimate for the characteristic atmospheric time $\tau_{at} = 1$ ms, we obtain $\tau_{sp} = 0.1$ µs. Note that this condition can be naturally fulfilled only for extremely fast spinning targets. Therefore, in the SMPL technique described here, the fast speckle field realization update that is required for speckle metric measurements is generated by artificially induced hit spot dithering, achieved by modulating the outgoing combined beam's wavefront tip and tilt. Because this tip–tilt phase modulation with hit spot dithering frequencies $\omega_{dith} \sim 1/\tau_{sp}$ in the 10 MHz range cannot be achieved using conventional optomechanical beam-steering mirrors, in the SMPL approach the required high-frequency hit spot dithering is obtained using a piston-wise (stair-mode) approximation of the outgoing beam wavefront tilts, as illustrated in Figure 6.12. Since this dithering can be performed using fiber-integrated phase shifters with bandwidths in the gigahertz range, the condition for τ_{sp} in Eq. (6.4) can easily be fulfilled.

Note that dithering of the outgoing beam also results in an undesired overall increase of the projected beam's long-exposure hit spot footprint and the corresponding decrease of the time-averaged power density. For this reason, the stair-mode dithering amplitude should be small, but still large enough to provide a statistically representative ensemble of uncorrelated (or at least weakly correlated) speckle field realizations that can be used for speckle metric evaluation. A small dithering amplitude is also important for mitigation of anisoplanatic effects [38]. As analysis and experiments show, the hit spot dithering with amplitudes of 75–100% of the diffraction-limited beam size represents an acceptable compromise between the factors already mentioned [38].

6.5.2
Speckle Metrics

In this section we show that processing of the PIB signal $J_{PIB}(t)$ measured with a receiver telescope (PIB receiver) allows obtaining speckle metrics J_{sp} that can be utilized for SMPL in the beam projection system depicted in Figure 6.1. The speckle metrics considered here are derived from an analysis of the temporal correlation function of the time-varying (AC) component $\delta J_{PIB}(t)$ of the measured PIB signal $J_{PIB}(t)$: $\Gamma_{PIB}(\tau) \equiv \langle \delta J_{PIB}(t) \delta J_{PIB}(t+\tau) \rangle_{sp}$.

Consider laser beam projection in an optically homogeneous medium onto a flat randomly rough target surface and assume that the characteristic roughness correlation distance l_s and roughness root-mean-square (rms) amplitude σ_s are significantly smaller than the hit spot size b_s, but larger than the transmitted beam wavelength λ. In the case of the hit spot dithering with velocity $\mathbf{v}_s$ at the target, one can obtain the following relationship between correlation function $\Gamma_{\text{PIB}}(\tau)$ and the target plane intensity distribution $I_T(\mathbf{r})$ [35,37]:

$$\Gamma_{\text{PIB}}(\tau) = C \int I_T(\mathbf{r}) I_T(\mathbf{r} + \mathbf{v}_s \tau) d^2\mathbf{r}, \tag{6.5}$$

where C is a constant. We assumed here that $\sigma_S \geq l_S$ (very rough surfaces) and that the receiver aperture D_R exceeds the characteristic speckle size a_{sp}. The dependence described by Eq. (6.5) can be utilized for derivation of a set of different speckle metrics. Consider first the PIB signal fluctuation variance that can be obtained by substituting $\tau = 0$ into Eq. (6.5):

$$\sigma_{\text{PIB}}^2 = \Gamma_{\text{PIB}}(0) = \langle \delta J_{\text{PIB}}^2 \rangle = C \int I_T^2(\mathbf{r}) d^2\mathbf{r}. \tag{6.6}$$

From Eq. (6.6) follows that σ_{PIB}^2 is proportional to the sharpness function $J_2 = \int I_T^2(\mathbf{r}) d^2\mathbf{r}$ – the target plane metric that is widely used for characterization of image and hit spot quality [39]. The relationship (6.6) shows that σ_{PIB}^2 can be considered as a speckle metric whose maximization results in an increase of the J_2 metric.

The PIB fluctuation power spectrum $G_{\text{PIB}}(\omega)$ offers another possibility for defining speckle metrics [35,40]. Using the Wiener–Khinchin theorem, from Eq. (6.5) we get

$$G_{\text{PIB}}(\omega) = \frac{C}{\pi} \int_0^\infty \int \cos(\omega \tau) I_T(\mathbf{r}) I_T(\mathbf{r} + \mathbf{v}_s \tau) d^2\mathbf{r} d\tau. \tag{6.7}$$

For a Gaussian beam $I_T(\mathbf{r})$ of width b_s, one can obtain from Eq. (6.7) the following analytical expression for $G_{\text{PIB}}(\omega)$:

$$G_{\text{PIB}}(\omega) = G_{\text{PIB}}(0) \exp(-\omega^2/\omega_{\text{PIB}}^2), \tag{6.8}$$

where $\omega_{\text{PIB}} = |\mathbf{v}_s|/b_s$ is the characteristic frequency bandwidth of PIB signal fluctuations [35,40,41]. The bandwidth monotonically increases with decreasing hit spot size b_s. This dependence of the PIB signal power spectrum on the hit spot size suggests that changes in the hit spot size impact the power spectrum components and can be evaluated by band-pass filtering of the PIB signal. The corresponding signals

$$P(\omega_j, \Delta_j) = \int_{\omega_j - \Delta_j/2}^{\omega_j + \Delta_j/2} G_{\text{PIB}}(\omega) d\omega \tag{6.9}$$

or their various combinations can be used to define speckle metrics of the type

$$J_{\text{sp}} = \sum_{j=1}^{N} \beta_j P(\omega_j, \Delta_j), \tag{6.10}$$

where $j = 1, \ldots, N$ denotes a number of band-pass filters with central frequencies $\{\omega_j\}$ and bandwidths $\{\Delta_j\}$, and $\{\beta_j\}$ are weighting coefficients [35,37,40]. In contrast to the speckle metric σ^2_{PIB} from Eq. (6.6), the power spectrum frequency components below $\omega_1 - \Delta_1/2$ and higher than $\omega_N + \Delta_N/2$ do not contribute to the spectral speckle metric, as defined by Eqs. (6.9) and (6.10). Control of the parameters $\{\omega_j\}$, $\{\Delta_j\}$, and $\{\beta_j\}$ in Eq. (6.10) allows optimization of the speckle metric's dependence on the target hit spot intensity distribution.

Contrary to the speckle metric σ^2_{PIB}, whose value is directly associated with the target plane metric J_2, a similar type of analytical expression linking the speckle metric defined in Eq. (6.6) with a physically meaningful target plane metric is not available. Nevertheless, both experiments and numerical simulations show that with a correct selection of parameters in Eq. (6.10), the obtained PIB signal characteristic can be used as a speckle metric, where its global maximum corresponds to the undistorted hit spot beam intensity distribution [42,43].

Note that even though the speckle metrics defined in Eqs. (6.6) and (6.10) are obtained for speckle field propagation in vacuum, it was shown that, at least in weak and medium-strength atmospheric turbulence conditions, turbulence has a relatively small impact on the speckle field statistical characteristics and the dependence of the speckle metrics on the target hit spot size is practically unchanged [43]. This property of the speckle field forms the physical basis for the use of the SMPL for beam projection systems operating in atmospheric turbulence conditions.

6.5.3
Experimental Evaluation of Speckle Metric-Based Phase Locking

For experimental validation of speckle metric sensing and speckle metric-based coherent beam combining, a series of laboratory benchtop experiments were performed using a fiber collimator array with seven subapertures. A notional schematic of the experimental setup is shown in Figure 6.13. The transmitted collimated beamlets emerging from the fiber array system were focused by a lens.

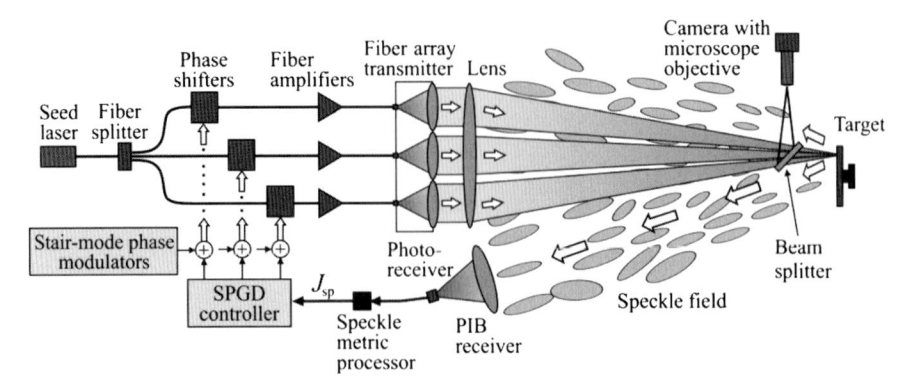

Figure 6.13 Notional schematic of the experimental setup for evaluation of speckle metric-based coherent beam combining at an extended target.

The converging combined beam was split by a beam splitter into two legs. The extended target was located in the focal plane of the first leg, while in the second leg a CCD camera (with an attached microscope objective) recorded the irradiance distribution at a plane conjugate to the target surface plane. Part of the scattered light (speckle field) was picked up by the PIB receiver, which comprised a lens with a photodetector placed at the location of the target hit spot's image formed by the receiver lens. The photodetector's output signal, the PIB metric $J_{PIB}(t)$, was proportional to the scattered wave power incident to the PIB receiver aperture. The speckle metric processor computed the PIB metric's standard deviation σ_{PIB} using an analogue circuit with an integration time $\tau_J \approx 1\mu s$. The signal $J_{sp} = \sigma_{PIB}$ was used by the SPGD controller as the speckle metric [44].

As shown in Figure 6.13, in each control channel, the output signal of the SPGD controller was mixed with a stair-mode modulation signal of 50 MHz frequency. The amplitudes and phases of the modulation signals were set as to provide a linear tilt dithering (stair-mode steering) of the outgoing combined beam. The amplitude of the linear displacement of the target hit spot was about $20\,\mu m$, which approximately corresponds to the diffraction-limited size of the central lobe of the fiber collimator array's far-field irradiance pattern.

Figure 6.14 shows the target plane irradiance distributions, which were recorded with the CCD camera under different operational conditions of the coherent beam combining system. Random piston phases with both SPGD controller and stair-mode dithering off resulted in the irradiance distribution of Figure 6.14a. SPGD

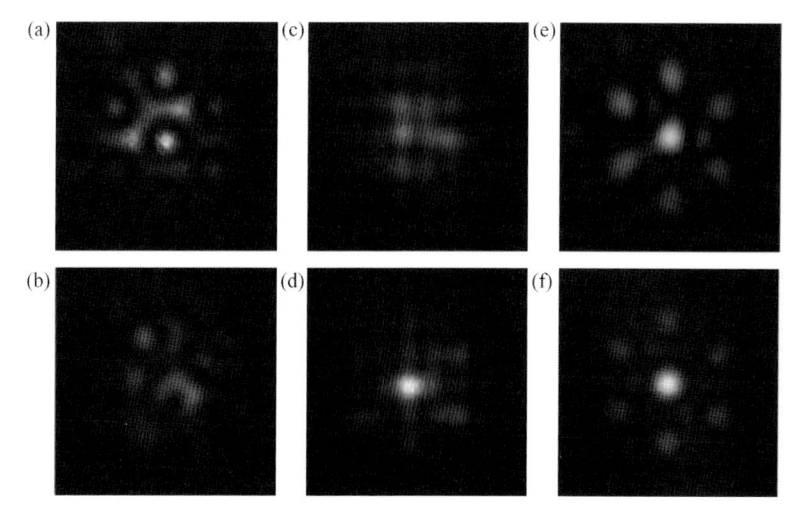

Figure 6.14 Intensity patterns at the randomly rough target surface (in the experimental setup shown in Fig. 6). (a) No phase control. (b) TIL SPGD phase control using the PIB metric with a resolved target. (c) Uncontrolled phase with stair-mode beam dithering on. (d) TIL SMPL with resolved target using hit spot dithering. (e) TIL SMPL with a rotating resolved target without dithering. (f) SPGD phase locking with an unresolved target. The side length of each panel corresponds to about $170\,\mu m$.

optimization of the PIB metric J_{PIB} – a setting conventionally used for beam combining on an unresolved target – resulted in the irradiance pattern of Figure 6.14b, which clearly indicates random relative phases of the beamlets. Figure 6.14c depicts the target plane when the beamlet phases were not controlled, while the stair-mode modulation signals were applied to the phase shifters. The dithering caused a reduction of the contrast of the interference pattern. The irradiance pattern resulting from phase locking using SPGD-based maximization of the speckle metric, $J_{sp} = \sigma_{PIB}$, with stair-mode beam steering is shown in Figure 6.14d. In contrast to PIB metric optimization, SMPL with SPGD optimization resulted in an increase of the projected beam's power density at the extended target with about twofold increase of the average hit spot peak irradiance.

The experimental setup was also used to evaluate SMPL without the hit spot dithering, but with a target rotating at about 100 revolutions per second. In this case the dynamics of the speckle pattern was much slower and to enable phase control, the integration time τ_J for speckle metric measurements and the SPGD controller's iteration time τ_{iter} had to be increased by more than three orders of magnitude. This was still sufficient to compensate for the random piston phases inherent to the MOPA and fiber optical systems. The resulting target plane intensity pattern, which is shown in Figure 6.14e, was close to the pattern observed with a PIB metric and an unresolved target (shown for comparison in Figure 6.14f). These results clearly indicate that the SMPL technique indeed offers a path toward TIL phase locking for laser beam projection on an extended target with randomly rough surface.

6.6
Conclusion

In this chapter, we considered coherent beam combining using different configurations of fiber array-based laser transmitters with integrated capabilities for piston and tip–tilt phase control at each fiber collimator aperture. In the TIL experiments over a 7 km atmospheric propagation path, it was demonstrated that SPGD-based adaptive control of the piston and the tip and tilt phases at each fiber collimator results in automatic focusing of the combined beam onto an unresolved target with precompensation of atmospheric turbulence-induced phase aberrations. The system performance was significantly increased by using an SPGD control that accounts for the round-trip propagation delay (delayed-feedback SPGD).

A new adaptive optics control technique that allows coherent beam combining at an extended (speckle) target was described. This control technique is based on SPGD optimization of the target return speckle field's statistical characteristics – speckle metrics. A characteristic feature of speckle metrics is their monotonic dependence on the high-energy laser beam power density inside the target hit spot. In the experiments, the speckle metric sensing was achieved by utilizing a megahertz-rate beam dithering with a stair-mode approximation of the outgoing combined beam's wavefront tip and tilt with subaperture piston phases. Fiber-integrated phase shifters were used for both the stair-mode beam dithering and speckle metric optimization with the SPGD control technique.

Acknowledgments

This work was supported in part through cooperative agreements between the U.S. Army Research Laboratory (ARL) and the University of Dayton as well as between ARL and Optonicus, LLC.

References

1 Leger, J.R., Nilsson, J., Huignard, J.-P., Napartovich, A.P., Shay, T.M., and Shirakawa, A. (2009) Special issue on laser beam combining and fiber laser systems. *IEEE J. Sel. Top. Quantum Electron.*, **15**, 237–470.

2 Wagner, T.J. (2012) Fiber laser beam combining and power scaling progress: Air Force Research Laboratory Laser Division. *Proc. SPIE*, **8237**, 823718-1–1823718-9.

3 Fan, T.Y. (2005) Laser beam combining for high-power, high-radiance sources. *IEEE J. Sel. Top. Quantum Electron.*, **11**, 567–577.

4 Vorontsov, M.A. and Lachinova, S.L. (2008) Laser beam projection with adaptive array of fiber collimators: I. Basic considerations for analysis. *J. Opt. Soc. Am. A*, **25**, 1949–1959.

5 Lachinova, S.L. and Vorontsov, M.A. (2008) Laser beam projection with adaptive array of fiber collimators: II. Analysis of atmospheric compensation efficiency. *J. Opt. Soc. Am. A*, **25**, 1960–1973.

6 Sprangle, P., Ting, A., Penano, J., Fischer, R., and Hafizi, B. (2009) Incoherent combining and atmospheric propagation of high-power fiber lasers for directed-energy applications. *IEEE J. Quantum Electron.*, **45**, 138–148.

7 Vorontsov, M.A., Weyrauch, T., Beresnev, L.A., Carhart, G.W., Liu, L., and Aschenbach, K. (2009) Adaptive array of phase-locked fiber collimators: analysis and experimental demonstration. *IEEE J. Sel. Top. Quantum Electron.*, **15**, 269–280.

8 Vorontsov, M.A. and Kolosov, V.V. (2005) Target-in-the-loop beam control: basic considerations for analysis and wave-front sensing. *J. Opt. Soc. Am. A*, **22**, 126–141.

9 Valley, M.T. and Vorontsov, M.A. (eds) (2004) *Target-in-the-Loop: Atmospheric Tracking, Imaging, and Compensation*, Proceedings of SPIE, vol. **5552**, Society of Photo Optical, Bellingham, WA.

10 Beresnev, L.A. and Vorontsov, M.A. (2005) Design of adaptive fiber optics collimator for free-space communication laser transceiver. *Proc. SPIE*, **5895**, 58950R-1–58950R-7.

11 Optonicus (2012) http://www.optonicus.com, August 17.

12 Born, M. and Wolf, E. (1980) *Principles of Optics*, 6th edn, Pergamon Press, New York.

13 Goodman, J.W. (1996) *Introduction to Fourier Optics*, 2nd edn, McGraw-Hill, New York.

14 Kolmogorov, A.N. (1941) The local structure of turbulence in incompressible viscous fluid for very large Reynolds numbers. *Dokl. Akad. Nauk SSSR*, **30**, 299–303 (English translation in *Turbulence: Classic Papers on Statistical Theory*, eds S.K. Friedlander and L. Topper, Interscience, New York, 1961, pp. 151–155).

15 Fried, D.L. (1966) Optical resolution through a randomly inhomogeneous medium for very long and very short exposures. *J. Opt. Soc. Am.*, **56**, 1372–1379.

16 Tyson, R.K. (1997) *Principles of Adaptive Optics*, 2nd edn, Academic Press, Boston, MA.

17 Vorontsov, M.A., Kolosov, V.V., and Kohnle, A. (2007) Adaptive laser beam projection on an extended target: phase- and field-conjugate precompensation. *J. Opt. Soc. Am. A*, **24**, 1975–1993.

18 Noll, R.J. (1976) Zernike polynomials and atmospheric turbulence. *J. Opt. Soc. Am.*, **66**, 207–211.

19 Fried, D.L. and Vaughn, J.L. (1992) Branch cuts in the phase function. *Appl. Opt.*, **31**, 2865–2882.

20 Fried, D.L. (1998) Branch point problem in adaptive optics. *J. Opt. Soc. Am. A*, **15**, 2759–2768.

21 Vorontsov, M.A. and Sivokon, V.P. (1998) Stochastic parallel-gradient-descent technique for high-resolution wave-front phase-distortion correction. *J. Opt. Soc. Am. A*, **15**, 2745–2758.

22 Vorontsov, M.A., Carhart, G.W., and Ricklin, J.C. (1997) Adaptive phase-distortion correction based on parallel gradient-descent optimization. *Opt. Lett.*, **22**, 907–909.

23 Vorontsov, M.A., Carhart, G.W., Cohen, M., and Cauwenberghs, G. (2000) Adaptive optics based on analog parallel stochastic optimization: analysis and experimental demonstration. *J. Opt. Soc. Am. A*, **17**, 1440–1453.

24 Weyrauch, T., Vorontsov, M.A., Carhart, G.W., Beresnev, L.A., Rostov, A.P., Polnau, E.E., and Liu, J.J. (2011) Experimental demonstration of coherent beam combining over a 7km propagation path. *Opt. Lett.*, **36**, 4455–4457.

25 Liu, L., Vorontsov, M.A., Polnau, E., Weyrauch, T., and Beresnev, L.A. (2007) Adaptive phase-locked fiber array with wavefront phase tip–tilt compensation using piezoelectric fiber positioners. *Proc. SPIE*, **6708**, 67080K-1–67080K-12.

26 Vorontsov, M.A., Riker, J., Carhart, G.W., Gudimetla, V.S.R., Beresnev, L.A., Weyrauch, T., and Roberts, L.C. (2009) Deep turbulence effects compensation experiments with a cascaded adaptive optics system using a 3.63m telescope. *Appl. Opt.*, **48**, A47–A57.

27 Weyrauch, T. and Vorontsov, M.A. (2005) Atmospheric compensation with a speckle beacon in strong scintillation conditions: directed energy and laser communication applications. *Appl. Opt.*, **44**, 6388–6401.

28 Vorontsov, M.A. and Carhart, G.W. (2006) Adaptive wavefront control with asynchronous stochastic parallel gradient descent clusters. *J. Opt. Soc. Am. A*, **23**, 2613–2622.

29 Weyrauch, T. and Vorontsov, M.A. (2004) Free-space laser communications with adaptive optics: atmospheric compensation experiments. *J. Opt. Fiber Commun. Rep.*, **1**, 355–379.

30 Weyrauch, T., Liu, L., Vorontsov, M.A., and Beresnev, L.A. (2004) Atmospheric compensation over a 2.3 km propagation path with a multi-conjugate (piston-MEMS/modal DM) adaptive system. *Proc. SPIE*, **5552**, 73–84.

31 Vorontsov, M.A. and Carhart, G.W. (2011) *Iteration Rate Improvement for SPGD Procedure, Invention Disclosure*, U.S. Army Research Laboratory.

32 Pearson, J.E., Kokorowski, S.A., and Pedinoff, M.E. (1976) Effects of speckle in adaptive optical systems. *J. Opt. Soc. Am.*, **66**, 1261–1267.

33 Vorontsov, M.A., Karnaukhov, V.N., Kuz'minskii, A.L., and Shmal'gauzen, V.I. (1984) Speckle effects in adaptive optical systems. *Kvant. Elektron.*, **11**, 1128–1137 (English translation in *Sov. J. Quantum Electron.* (1984) **14**, 761–766).

34 Piatrou, P. and Roggemann, M. (2007) Beaconless stochastic parallel gradient descent laser beam control: numerical experiments. *Appl. Opt.*, **46**, 6831–6842.

35 Vorontsov, M.A. and Shmal'hauzen, V.I. (1985) *Principles of Adaptive Optics*, Nauka, Moscow.

36 Vorontsov, M.A. and Carhart, G.W. (2002) Adaptive phase distortion correction in strong speckle-modulation conditions. *Opt. Lett.*, **27**, 2155–2157.

37 Vorontsov, M.A. (2004) Target in the loop propagation in random media. Technical Report 2004/22, FGAN FOM, Germany.

38 Vorontsov, M.A., Kolosov, V.V., and Polnau, E. (2009) Target-in-the-loop wavefront sensing and control with a Collett–Wolf

beacon: speckle-average phase conjugation. *Appl. Opt.*, **48**, A13–A19.

39 Muller, R.A. and Buffington, A. (1974) Real-time correction of atmospherically degraded telescope images through image sharpening. *J. Opt. Soc. Am.*, **64**, 1200–1210.

40 Vorontsov, M.A., Carhart, G.W., Pruidze, D.V., Ricklin, J.C., and Voelz, D.G. (1996) Image quality criteria for an adaptive imaging system based on statistical analysis of the speckle field. *J. Opt. Soc. Am. A*, **13**, 1456–1466.

41 Goldfischer, L.I. (1965) Autocorrelation function and power spectral density of laser-produced speckle patterns. *J. Opt. Soc. Am.*, **55**, 247–253.

42 Deng, Y., Cauwenberghs, G., Polnau, E., Carhart, G., and Vorontsov, M. (2005) Integrated analog filter bank for adaptive optics speckle field statistical analysis. *Proc. SPIE*, **5895**, 58950L-1–58950L-9.

43 Dudorov, V.V., Vorontsov, M.A., and Kolosov, V.V. (2006) Speckle-field propagation in "frozen" turbulence: brightness function approach. *J. Opt. Soc. Am. A*, **23**, 1924–1936.

44 Vorontsov, M., Weyrauch, T., Lachinova, S., Gatz, M., and Carhart, G. (2012) Speckle-metric-optimization-based adaptive optics for laser beam projection and coherent beam combining. *Opt. Lett.*, **37**, 2802–2804.

7
Refractive Index Changes in Rare Earth-Doped Optical Fibers and Their Applications in All-Fiber Coherent Beam Combining

Andrei Fotiadi, Oleg Antipov, Maxim Kuznetsov, and Patrice Mégret

7.1
Introduction

Fiber lasers have become ubiquitous devices equally important in fundamental science, engineering technologies, and in a wide range of practical high-power laser applications [1]. Nonlinear effects significantly reduce power levels available with fiber lasers. One of the ways to overcome this limitation is to use coherent combining of single-mode fiber amplifiers. The general idea of this approach is to split a coherent beam into many beams that are then amplified by a parallel array of similar power amplifiers and finally recombined to a single diffraction-limited beam. To combine the power of all the channels in one single-mode output, the fiber amplifiers must be phase locked together. In past years, nonlinear optical phase conjugation mirrors have been widely investigated for this purpose [2,3]. Now other coherent beam combining techniques have become the most prominent means of brightness scaling. Generally, a major drawback of the bulk solid-state systems is the increased complexity due to the multiple paths, required control loops, and actuators. The most obvious step toward system simplification is moving to an all-fiber configuration. This eliminates the need for mode matching of the output beams to each other and potentially reduces the sensitivity to environmental noise. Naturally, this introduces new challenges as well, as the power handling capabilities of fiber components are often not as good as for their free space components. For example, mechanically mounted mirrors can be replaced with piezofiber stretchers, but these require comparatively long fiber lengths, which can cause problems due to fiber nonlinearity. The effect of refractive index changes (RICs) induced in rare earth-doped optical fibers by resonance pumping (that is essentially a side effect of the population inversion) offers a simple solution for all-fiber coherent combining of intense laser beams at any IR wavelength far from rare earth-doped ion resonances. For example, ytterbium fiber amplifiers could be used as phase actuators at the erbium wavelength, and vice versa. This method is based on the refractive index changes due to heat load and electronic effects (which can be described by Kramers–Kronig

Coherent Laser Beam Combining, First Edition. Edited by Arnaud Brignon.
© 2013 Wiley-VCH Verlag GmbH & Co. KGaA. Published 2013 by Wiley-VCH Verlag GmbH & Co. KGaA.

relations) in the additional amplifier or directly in the working fiber amplifier and does not rely on any mechanical parts nor require any high-voltage sources. It is also capable of handling high-power levels, because it is a true all-fiber method with low insertion loss.

In this chapter, by the example of Yb-doped fibers, we describe the RIC effect and discuss the application of this effect on coherent beam combining in Er-doped fiber laser systems. In particular, we consider a coherent combining of two erbium amplifiers operating at 1550 nm with a dynamical phase control provided either by one low-power laser diode at 980 nm alone or in combination with a laser at 1060 nm, both stimulating electron transitions in an Yb-doped optical fiber. Two wavelengths approach is shown to improve the response time of the phase control loop. The chapter includes remarkable amount of original experimental studies, theoretical modeling, and demonstrations. Section 7.2 contains background material; here, we give a clear physical insight into the electronic and thermal RIC mechanisms and present important expressions helping the reader to clearly understand the scientific context and practical perspectives of these effects. The main body of the work is centered around experimental studies [4,5] significantly extended by new experimental observations reproduced in Sections 7.3 and 7.4. Section 7.3 deals with the index sensitivity of the YDFs to the diode pumping at 980 nm and signal amplification at 1060 nm. Section 7.4 focuses on potential applications of the RIC effect for all-fiber coherent beam combining.

7.2
Theoretical Description of the RIC Effect in Yb-Doped Optical Fibers

7.2.1
Introduction: Thermal and Electronic RIC Mechanisms

The propagation of an optical beam at resonance frequencies through the rare earth-doped optical fibers induces changes in the core refractive index. Two main mechanisms of the pump- or signal-induced RIC in the rare earth-doped fibers have been discussed recently: the well-known thermal index change Δn^T caused by heating owing to the matrix absorption and quantum defects of the exited rare earth ions [1,6–8], and the athermal index change Δn^e (also called electronic or Kramers–Kronig effect) related to the population change of the active ion levels with different polarizabilities [9]. We have to distinguish the electronic RIC effect from the regular Kerr nonlinearity that does not rely on the population inversion and requires much higher optical beam intensity. In many experiments, it is not straightforward to separate electronic and thermal contributions to refractive index change. In the next section, we present an analytical treatment and comparative estimation of the thermal and electronic contributions of the index changes in the fiber amplifiers used for coherent beam combining.

7.2.2
Description of the Spectroscopic Properties of Yb-Doped Optical Fibers

Silica glass, the most common material for the production of fibers, is a good host for Yb ions. The spectroscopy of the Yb ion is simply compared with other rare earth ions (Figure 7.1) [11]. The useful transitions for optical amplification are those inside the 4f shell manifold and generally a simplified two-level model (arising from four-level system) can be used to correctly describe the phenomena as shown in Figure 7.1a. To be complete, 5d shell and the charge transfer band are also included as UV transitions to these levels are allowed. Indeed, for amplification in the optical spectrum range, only two level manifolds are important: the ground-state manifold ($^2F_{7/2}$) and the excited-state manifold ($^2F_{5/2}$). They consist of four and three sublevels, respectively. The transitions between the sublevels are smoothed by strong homogeneous and inhomogeneous broadening. As a result, ytterbium fibers are able to provide optical gain over the very broad wavelength range from 975 to 1200 nm with peaks at 975 and 1030 nm. For the pumping wavelengths, Figure 7.1b reveals that the pumping overcomes the emission in the range of 850–1000 nm with best values at 915 nm. In the infrared spectrum range outside these resonances (>1200 nm), ytterbium fibers (Figure 7.1c) are optically transparent and are thus subject to pure RIC effects.

7.2.3
Description of the Electronic RIC Mechanism

Population change of the energy levels δN due to transitions in materials is accompanied by an index change δn^e, which can be described by well-known Kramers–Kronig relation between real and imaginary parts of electric susceptibility.

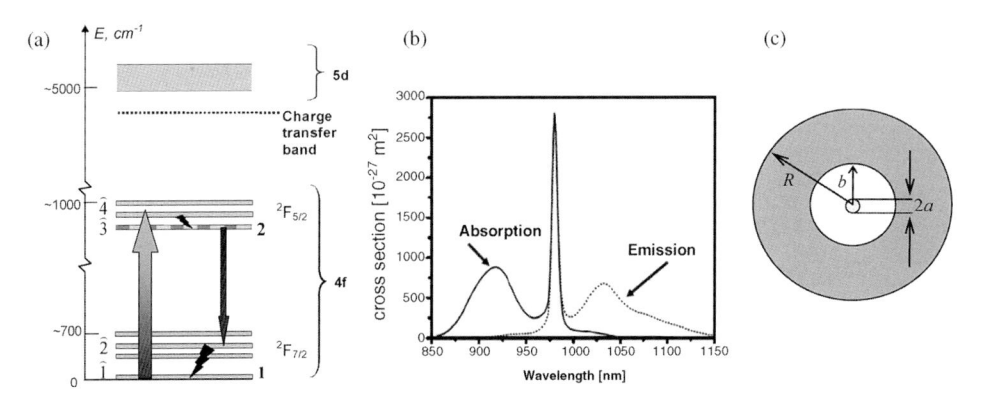

Figure 7.1 (a) Atomic manifold level system [10]. (b) Emission and absorption cross sections of ytterbium in silica host [11]. (c) Typical cross section profile of an optical fiber. 1, 2 and $\bar{1}, \bar{2}, \bar{3}, \bar{4}$ indicate the levels for two- and four-level laser models, respectively; a is the Yb^{3+}-doped core radius, b is the glass clad radius, and gray denotes the layer of plastic with radius R.

For the solid-state laser materials based on the doped glasses or crystals, the relation can be rewritten in the following form:

$$\delta n^e(\lambda) = \frac{\delta N}{2\pi^2} P \int_0^\infty \frac{\lambda^2 \Delta\sigma(\lambda')}{(\lambda'^2 - \lambda^2)} d\lambda', \quad (7.1)$$

where $P \int_0^\infty$ stands for the Cauchy principal part of the integral and:

$$\Delta\sigma(\lambda) = \sigma_{gsa}(\lambda) - \sigma_{esa}(\lambda) + \sigma_e(\lambda), \quad (7.2)$$

where $\sigma_{gsa}(\lambda)$ is the ground-state absorption cross section at wavelength λ, $\sigma_{esa}(\lambda)$ is the exited-state absorption cross section, and $\sigma_e(\lambda)$ is the emission cross section.

The electronic population index change can be described by the term of polarizability difference. Let us remind that the electronic polarizability is defined as the ratio of the dipole moment induced in an ion to the electric field that produces this dipole moment. So, the polarizability of the Yb ion on some electron level q at the given testing optical frequency ν_T is determined by the probabilities of all possible transitions from this level to other levels $i \neq q$ and can be expressed as the discrete sum of the transitions:

$$p_q(\nu_T) = \frac{e^2}{4\pi^2 m} \sum_i \frac{f_{qi}(\nu_{qi}^2 - \nu_T^2)}{(\nu_{qi}^2 - \nu_T^2)^2 + (\nu_T \Delta\nu_{qi})^2}, \quad (7.3)$$

where e and m are the electron charge and mass, respectively, f_{qi} is the oscillator force of the transition between levels q and i, and ν_{qi} and $\Delta\nu_{qi}$ are the resonance frequency and the linewidth, respectively.

Remaining in the frame of the two-level laser approximation, we have only to use $p_1(\nu)$ and $p_2(\nu)$, that is, the Yb ion polarizabilities associated with the ground ($^2F_{7/2}$) and excited ($^2F_{5/2}$) states, respectively. According to Eq. (7.3), the dominating contributions to these polarizabilities are expected from the transitions whose resonant frequencies are closer to the testing frequency ν_T and/or from the non-resonance transitions with the strongest oscillator forces.

In Yb-doped materials, the well-allowed UV transitions to the 5d electron shell and the charge transfer transition are characterized by the oscillator forces that are several orders of magnitude higher than the forces of optical transitions inside the 4f electron shell. As a consequence, in the IR spectrum band, the polarizability difference $\Delta p(\nu) = p_2 - p_1$ is expressed from Eq. (7.3) as a sum of contributions from the near-resonance transitions (between the ground and excited states) and nonresonance UV transitions. But far from the optical resonances, that is, within the YDF IR transparency band ($\lambda_T > 1.2$ μm), only the UV transitions play a major role and the polarizability difference can be represented by two main terms with different dependencies on the testing wavelength λ_T:

$$\Delta p(\lambda_T) \approx \frac{e^2}{4\pi^2 c^2 m} \left\{ A \left[\frac{1}{1 - \lambda_R^2/\lambda_T^2} + o\left(\frac{\lambda_R^2}{\lambda_T^2}\right) \right] + B \left[1 + o\left(\frac{\lambda_{UR}^2}{\lambda_T^2}\right) \right] \right\}, \quad (7.4)$$

where $A \equiv \sum_{l \in R} f_{2l} \lambda_{2l}^2 - f_{1l} \lambda_{1l}^2$, $B \equiv (f_{2U} \lambda_{2U}^2 - f_{1U} \lambda_{1U}^2)$, $\lambda_R \sim 1\,\mu m$ is a typical wavelength of the resonance transitions, f_{1U}, λ_{1U}, f_{2U}, λ_{2U} are oscillation forces and wavelengths of well-allowed charge transfer transitions to the 5d electron shell from the ground and excited states, respectively, and o is a little-o notation. The free space wavelengths λ_{1U}, λ_{2U} are located in the UV spectrum band around $\sim 0.1\,\mu m$ and obviously $\lambda_{1U} \neq \lambda_{2U}$. So, the origin of nonresonance contribution to the polarizability difference (i.e., the second term, $\sim B$) is simply the difference in the probabilities of the transitions at the testing wavelength λ_T from the ground and excited states to the 5d electron shell or the charge transfer transition.

In the two-level approximation, the Kramers–Kronig relation (7.1) for the RIC in the doped fibers can be rewritten through the polarizability difference of the energy levels (for the pumped and unpumped medium):

$$\delta n^e = \frac{2\pi F_L^2}{n_0} \Delta p \delta N_2, \quad (7.5)$$

where δN_2 is the change of the excited-state population, n_0 is the refractive index of host glass, and $F_L = (n_0^2 + 2)/3$ is the Lorentz factor.

The phase shift corresponding to the electronic RIC detected at the test wavelength λ_T in the fiber of length L is evaluated from Eq. (7.5) by its integration over the fiber volume with $\rho_T(r)$ as the weight function:

$$\delta \varphi = \frac{4\pi^2}{\lambda_T} \int_0^L \int_0^\infty \delta n^e(z,r) \rho_T(r) r \, dr \, dz \approx \frac{\bar{\eta} \rho_T(0)}{\lambda_T} \left[4\pi^2 \frac{F_L^2}{n_0} \Delta p \right] \delta N_2^\Sigma, \quad (7.6)$$

where $\delta N_2^\Sigma = 2\pi \int_0^L \int_0^\infty \delta N_2(z,r) r \, dr \, dz$ is the pump-induced change in the number of the excited Yb^{3+} ions in the whole fiber volume, $\rho_T(r)$ is the normalized power radial distribution of the probe light, r is the polar coordinate describing the fiber cross section, and z is the linear coordinate along the fiber.

The parameter $\bar{\eta} \rho_T(0)$ approximates the efficiency of the probe mode interaction with the population changes $\delta N_2(r)$ induced in the doped fiber area. Here, we intentionally isolate the factors $\rho_T(0)$ and $\bar{\eta}$ that represent the major dependencies of $\bar{\eta} \rho_T(0)$ on λ_T and Yb ion concentration profile, respectively.

The factors $\rho_T(0)$ and $\bar{\eta}$ can be evaluated from the step-index fiber approach [12]. Assuming the testing wavelength λ_T to be within the IR fiber transparency band, one could express the power mode distribution $\rho_T(r)$ as

$$\rho_T(r) = \begin{cases} \dfrac{1}{2\pi a^2} \left(\dfrac{J_0(ur/a)}{J_0(u)} \right)^2, & \text{for } r < a, \\[2mm] \dfrac{1}{2\pi a^2} \left(\dfrac{K_0(vr/a)}{K_0(v)} \right)^2, & \text{for } r > a, \end{cases} \quad (7.7)$$

where a is a step-index fiber core radius and J_n and K_n denote Bessel and modified Bessel functions, respectively. The mode parameters u and v are defined by

$$u^2 = a^2(n_0^2 k_0^2 - \beta^2),$$
$$v^2 = a^2(\beta^2 - n_0^2 k_0^2), \qquad (7.8)$$
$$V^2 = u^2 + v^2 = k_0^2 a^2(n_0^2 - n_1^2),$$

and are connected through the characteristic equation:

$$u \frac{J_1(u)}{J_0(u)} = v \frac{K_1(v)}{K_0(v)}. \qquad (7.9)$$

If the index profile of the fiber is not a step one, the above relation can still be used provided that the real fiber core radius is replaced by an effective one defined as [13]

$$\bar{a} = w \left[1.3 + 0.864 (\lambda_S/\lambda_c)^{3/2} + 0.0298 (\lambda_S/\lambda_c)^6 \right]^{-1}, \qquad (7.10)$$

where $V = 2.05 \lambda_c / \lambda_T$ is the fiber dimensionless modal parameter, w is the modal field diameter usually specified at $\lambda_S \approx 1.06\,\mu m$, and λ_c is the fiber cutoff wavelength.

Figure 7.2 explains the dependencies of the parameter $\rho_T(0)$ on λ_T and $\bar{\eta}$ on Yb ion concentration profile. One can see that the longer testing wavelength λ_T corresponds to the wider distribution $\rho_T(r)$ providing the smaller overlap between the signal power and the doped fiber core area. As a result, the parameter $\rho_T(0)$ decreases with increasing the testing wavelength. As far as all changes $\delta N_2(r)$ are assumed to occur close to the fiber axis, the factor $\bar{\eta} \to 1$. However, the wider Yb ion distribution $n_{Yb}(r)$ decreases the overlap between the signal power and the doped fiber core area resulting in $\bar{\eta} < 1$.

The factor $\bar{\eta}$ can be estimated from Eq. (7.6) to be blocked within two bounds that correspond to the limit cases of high and low pump power levels, that is, overpumped and underpumped ions within the fiber cross section:

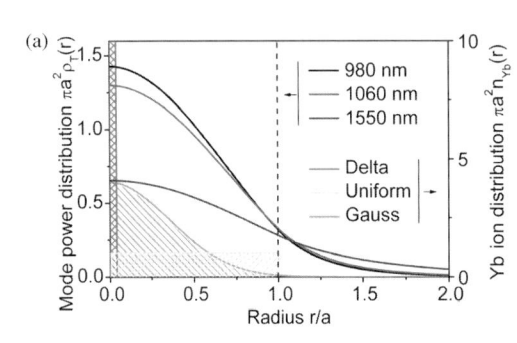

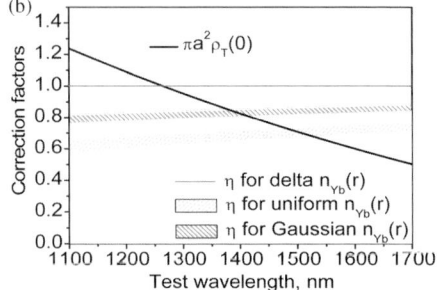

Figure 7.2 The normalized mode power distributions at different wavelengths in comparison with examples of Yb ion distributions (a) Delta-like $n_{Yb}(r) = \delta(r)/\pi a^2$, uniform $n_{Yb}(r) = 1/\pi a^2$, and Gaussian $n_{Yb}(r) = 4\exp[-(2r/a)^2]/\pi a^2$. (b) The coefficient $\rho_T(0)$ and correction factors $\bar{\eta}$ in Eq. (7.6) calculated for different Yb ion distributions.

$$\int_0^\infty n_{Yb}(r)\rho_T(r)r dr < \rho_T(0)\bar{\eta} < \frac{\int_0^\infty n_{Yb}(r)\rho_P(r)\rho_T(r)r dr}{\int_0^\infty n_{Yb}(r)\rho_P(r)r dr}, \quad (7.11)$$

where $\rho_P(r)$ is the normalized mode radial power distribution at the pump wavelength λ_P and $n_{Yb}(r) = N(r)/2\pi \int_0^\infty N(r)r dr$ is the normalized distribution of Yb ions over the fiber core.

The typical values of the correction factor calculated for uniform and Gaussian distributions of the Yb ion dopants are $\bar{\eta} \approx 0.7$ and $\bar{\eta} \approx 0.85$, respectively. The factors $\bar{\eta}$ determined from inequalities (7.11) perform a typical error of ~10% (Figure 7.2b).

Equation (7.6) predicts that the phase shift is proportional to the pump- or signal-induced changes in the whole number of the excited ions in the fiber. Its dynamics is governed by rate equations that (in the absence of the strong pump and signal absorption by silica matrix) can be rewritten in the following form:

$$\frac{d\delta N_2^\Sigma}{dt} = \frac{P_P^{in} - P_P^{out}}{h\nu_P} - \frac{P_S^{in} - P_S^{out}}{h\nu_S} - \frac{P_{ASE}}{h\nu_{ASE}} - \frac{\delta N_2^\Sigma}{\tau_{sp}},$$
$$\frac{d\delta\varphi}{dt} = K\left[P_{in} - P_{out} - \frac{\nu_P}{\nu_S}(P_S^{in} - P_S^{out}) - \frac{\nu_P}{\nu_{ASE}}P_{ASE}\right] - \frac{\delta\varphi}{\tau_{sp}}, \quad (7.12)$$

where $K \equiv (\lambda_P/\lambda_T)[\bar{\eta}\rho_T(0)/hc][4\pi^2(F_L^2/n_0)\Delta p]$; h is the Planck's constant; ν_P, ν_S, and ν_{ASE} are the average frequencies of the pump, amplified signal, and amplified spontaneous emission (ASE), respectively; τ_{sp} is the excited state lifetime, P_P^{in} and P_P^{out} are the input and output (residual) pump powers, respectively; P_S^{in} and P_S^{out} are the input and output signal powers, respectively; and P_{ASE} is the ASE emitted power. For self-consisting description of the RIC dynamics, Eq. (7.12) should be complemented by equations for the powers of the pump, signal, and ASE beams in the fiber that can be given as follows (neglecting the difference in distribution of the excited and total Yb^{3+} ions in the core):

$$\frac{dP_P(z)}{dz} = \eta_P\rho_P(0)\left[\left(\sigma_{21}^{(p)} + \sigma_{12}^{(p)}\right)N_2(z) - \sigma_{12}^{(p)}N\right]P_P(z), \quad (7.13)$$

$$\frac{dP_S(z)}{dz} = \eta_S\rho_S(0)\left[\left(\sigma_{21}^{(s)} + \sigma_{12}^{(s)}\right)N_2(z) - \sigma_{12}^{(s)}N\right]P_S(z), \quad (7.14)$$

$$\frac{dP_{ASE}(z)}{dz} = \eta_{ASE}\rho_{ASE}(0)\left[\left(\sigma_{21}^{(ASE)} + \sigma_{12}^{(ASE)}\right)N_2(z) - \sigma_{12}^{(ASE)}N\right]P_{ASE}(z) + N_2\zeta, \quad (7.15)$$

where N_2 is the number of excited Yb^{3+} ions per unit length, N is the total number of Yb^{3+} ions per unit length, $\sigma_{21}^{()}$ and $\sigma_{12}^{()}$ are emission and absorption cross sections at ν_P, ν_S, and ν_{ASE}, and

$$\zeta = \frac{h\nu_{ASE}}{4\pi\tau_{sp}}\left(\frac{2a}{L}\right)^2 \quad (7.16)$$

is the effective Langevin source describing the full-line luminescence.

7.2.4
Description of the Thermal RIC Mechanism

The described electronic RIC is not the only mechanism responsible for the induced phase shift in YDFs and governed by the population inversion. In fact, thermalization of the pump power inside the fiber, leading to a temperature rise, may result in thermally induced index changes in the fiber media. Moreover, thermal expansions of the fiber core along its length and its radius may also cause phase modulation. However, one can show that the transverse and longitudinal expansions of the single-mode fiber core make a small contribution to the thermally induced phase shift in comparison to the volume thermal index gradient.

The refractive index change due to temperature rise δT is then expressed as

$$\delta n^T = \left(\frac{\partial n}{\partial T}\right)\delta T + \delta n_{ph} = \left[\left(\frac{\partial n}{\partial T}\right) + 2n_0^3 \alpha^T C'\right]\delta T, \qquad (7.17)$$

where $\partial n/\partial T$ is the thermo-optic coefficient, δn_{ph} accounts for the photoelastic effect, C' denotes a photoelastic constant averaged over polarizations, and α^T is the thermal longitudinal expansion coefficient.

In the simplest approximation, a four-level system is used to describe the thermal RIC mechanisms inside the YDF (Figure 7.1a). The thermal load is thus achieved through nonradiative transitions between the sublevels of the $^2F_{5/2}$ state, transitions between the sublevels of the $^2F_{7/2}$ ground state, and thermalization of the silica host absorption. The corresponding heat conductivity equation expressing the evolution of the temperature $T(r, z, t)$ within the fiber is

$$\frac{\partial T}{\partial t} - d_i^2 \Delta T = Q(z, r, t), \qquad (7.18)$$

where $d_i^2 = \kappa_i/(\rho_i C_{pi})$ is thermal diffusivity; κ_i is the thermal conductivity, ρ_i is the density and C_{pi} is the thermal capacity, all for silica glass ($i = 1$) and plastic ($i = 2$), and $Q(z, r, t)$ is a heat source.

In the case of pumping at ~980 nm (in the lowest sublevel of $^2F_{5/2}$ state) and negligible small silica host absorption, the heat source can be described by the sum

$$Q(z,r,t) = \frac{h\nu_{B1}\delta N_2(r,z)}{\rho_1 C_{p1} \tau_{sp}} + \frac{\nu_{B3}\delta N_2(r,z) P_S \rho_S(r) \sigma_e(\nu_A)}{\rho_1 C_{p1} \nu_S}$$
$$+ \frac{\nu_{BL}\delta N_2(r,z) P_{ASE} \rho_{ASE}(r) \sigma_e(\nu_{ASE})}{\rho_1 C_{p1} \nu_{ASE}}, \qquad (7.19)$$

where ν_{B1} is the frequency of the nonradiative transition between sublevels of the $^2F_{7/2}$ ground state, $\nu_{B3} \equiv \nu_P - \nu_S$, and $\nu_{BL} \equiv \nu_P - \nu_{ASE}$. Three terms in Eq. (7.19) correspond to thermalization of the pump, amplified signal, and luminescence, respectively.

The heat dissipation occurs from the fiber core to the lateral surface of the clad with a sink into the air (natural convection). The temperature distribution in the plastic-air interface is expressed by the Newton formula:

$$\left.\kappa\frac{\partial T}{\partial r}\right|_{r=R} + H(T-T_0)|_{r=R} = 0, \tag{7.20}$$

where T_0 is the temperature on the external surface of the fiber and H is the heat transfer coefficient accounted independently for core, cladding, and plastic.

To evaluate the thermally induced RIC in the fiber, one has to solve numerically the nonstationary heat [Eqs. (7.18) and (7.19)] in combination with the boundary condition [Eq. (7.20)] on the cooled surfaces. The temperature of the fiber (inner cladding and core regions) can be found analytically from Eq. (7.18), and written by the following expression [28]:

$$T(z,r,t) = \sum_{n=1}^{\infty} \frac{1}{Z_n} \frac{\kappa_1}{d_1^2} \int_0^a J_0\left(\frac{\mu_n r'}{d_1}\right) \int_0^t \exp\left[-\mu_n^2(t-t')\right] Q(z,r',t') dt' r' dr' J_0\left(\frac{\mu_n r}{d_1}\right), \tag{7.21}$$

where J_0 denotes the Bessel function of the zeroth order and Z_n is the following expression:

$$Z_n = \kappa_1 b^2 / 2 d_1^2 \left[J_0^2(\psi_{n,1,1}) + J_1^2(\psi_{n,1,1})\right] + \kappa_2 / d_2^2 \{0.5 R^2 C_2^2(\mu_n) \left[J_0^2(\psi_{n,2,2}) + J_1^2(\psi_{n,2,2})\right]$$
$$+ R^2 C_2(\mu_n) D_2(\mu_n) \left[J_0(\psi_{n,2,2}) Y_0(\psi_{n,2,2}) + J_1(\psi_{n,2,2}) Y_1(\psi_{n,2,2})\right]$$
$$+ 0.5 R^2 D_2^2(\mu_n) \left[Y_0^2(\psi_{n,2,2}) + Y_1^2(\psi_{n,2,2})\right] - 0.5 b^2 C_2^2(\mu_n) \left[J_0^2(\psi_{n,1,2}) + J_1^2(\psi_{n,1,2})\right]$$
$$- 0.5 b^2 C_2(\mu_n) D_2(\mu_n) \left[J_0(\psi_{n,1,2}) Y_0(\psi_{n,1,2}) + J_1(\psi_{n,1,2}) Y_1(\psi_{n,1,2})\right]$$
$$- 0.5 b^2 D_2^2(\mu_n) \left[Y_0^2(\psi_{n,1,2}) + Y_1^2(\psi_{n,1,2})\right]\}, \tag{7.22}$$

where $\psi_{n,1,j} = \mu_{n,1,j} b / d_j$, $\psi_{n,2,j} = \mu_{n,2,j} R / d_j$,

$$D_2(\mu_n) = \frac{(\kappa_1 d_2 / \kappa_2 d_1) J_0(\psi_{n,1,2}) J_1(\psi_{n,1,1}) - J_1(\psi_{n,1,2}) J_0(\psi_{n,1,1})}{J_0(\psi_{n,1,2}) Y_1(\psi_{n,1,2}) - J_1(\psi_{n,1,2}) Y_0(\psi_{n,1,2})}, \tag{7.23}$$

$$C_2(\mu_n) = \left[J_0(\psi_{n,1,1}) - D_2(\mu_n) Y_0(\psi_{n,1,2})\right] / J_0(\psi_{n,1,2}). \tag{7.24}$$

μ_n is the nth positive root of the equation

$$C_2(\mu) \left[J_0(\psi_{n,2,2}) - \frac{\mu \kappa_2}{d_2 H} J_1(\psi_{n,2,2})\right] + D_2(\mu) \left[Y_0(\psi_{n,2,2}) - \frac{\mu \kappa_2}{d_2 H} Y_1(\psi_{n,2,2})\right] = 0, \tag{7.25}$$

and Y_0 and Y_1 are Neumann functions of the zero and first order, respectively.

7.2.5
Comparison of Electronic and Thermal Contributions to the Pump-Induced Phase Shift

The fundamental relations [Eqs. ((7.12)–(7.20)] reported in the previous sections allow us to compare contributions of the electronic and thermal mechanisms to the RIC effects induced in YDFs under diode pumping. To be more concrete in our calculations, we use typical aluminum silicate fiber parameters and pumping conditions shown in Table 7.1.

Table 7.1 Parameters of the fibers used for calculations.

Parameter	Phosphate silicate	Aluminum silicate	References and note
Glass density ρ_1 (g/cm^3)	2.2		[14]
Specific heat capacity of glass C_{p1} [cal/(g K)] (1 cal = 41 868 J)]	0.188		[14]
Thermal conductivity of glass κ_1 [W/(cm·K)]	0.014		[1,14]
Plastic density ρ_2 (g/cm^3)	1.19		[14]
Specific heat capacity of plastic C_{p2} [cal/(g K)]	0.3		[15]
Thermal conductivity of plastic κ_2 [W/(cm·K)]	0.002		[15]
Energy allocated to heat from the upper therm $^2F_{5/2}$ (cm^{-1})	0	0	Pumping in the low level
Energy allocated to heat from the low therm $^2F_{7/2}$ (cm^{-1})	740	1210	[16]
Energy of the pumping quantum (cm^{-1})	10 260	10 245	[16], Pumping in the low level
Effective energy of luminescence quantum (cm^{-1})	990	990	[16]
Doping concentration (cm^{-3})	8.56×10^{19}		[16]
Lifetime of the level $^2F_{5/2}\tau_{sp}$ (ms)	1.276	0.83	[16]
Pump absorption cross section $\sigma_{gsa}(\nu_p)$ (pm^2)	1.38	2.69	[16]
Pump emission cross section $\sigma_e(\nu_p)$ (pm^2)	1.46	2.97	[16]
Luminescence cross section $\sigma_e(\nu_L)$ (pm^2)	0.485	0.495	[16]
Absorption cross section of luminescence $\sigma_{gsa}(\nu_L)$ (pm^2)	0.068	0.086	[16]
Core radius, a (μm)	1.8		
Glass radius b (μm)	62.5		
External plastic radius R (μm)	125		
Fiber length l (m)	2		
Heat transfer coefficient H [cal/(cm^2 sK)]	0.000118		[17]
Polarizability difference at 1550 nm × 10^{-26}, Δp (cm^3)	0.9 ± 0.2	1.2 ± 0.3 2.6 ± 0.4	[18] [19]
Glass refractive index n_0	1.5		[14]
Derivative of refractive index with respect to temperature, $\partial n/\partial T$ (K^{-1})	1.2×10^{-5}		[20]
Wavelength of the probe beam λ (μm)	1.55		[18]
Effective luminescence wavelength, λ_L (μm)	1.01	1.02	[16]
Pumping beam power I_p (mW)	145		
Thermal expansion coefficient of silica glass d_T (K^{-1})	5.1×10^{-7}		[14]
Poisson's coefficient ν	0.164		[14]
Wavelength of the amplified beam (μm)	1.064		
Absorption cross section of the amplified beam $\sigma_{gsa}(\nu_A)$ (pm^2)	0.0016	0.0046	[16]
Cross section of the amplified beam $\sigma_e(\nu_A)$ (pm^2)	0.13	0.3	[16]

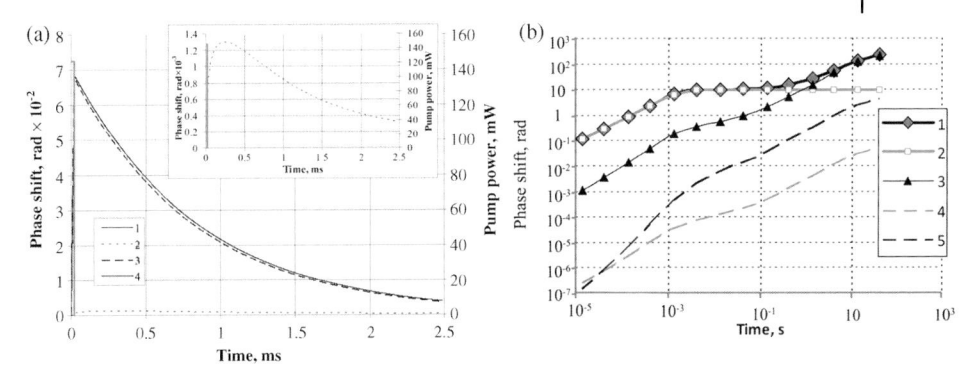

Figure 7.3 Electronic and thermal contributions to pump-induced phase shifts. (a) Temporal trace of the phase shift induced by 20 μs pump pulse of 145 mW amplitude (the pulse energy is ~3 μJ): the total phase shift (1), thermal contribution (2), electronic contribution (3), and pump pulse (4). (b) Dependence of the phase shifts on the pump pulse duration: electronic contribution (2), thermal contribution (3), total phase shift (1), part of the thermal contribution associated with the elongation of the fiber (4), and the part of thermal contribution associated with fiber expansion (5).

The results of simulations are presented in Figure 7.3. One can see that at short pump pulse excitation, the electronic contribution to the phase shift is an order of magnitude higher than the thermal contribution. With the increase of pulse duration, the situation changes: at some threshold duration, the contribution of the electronic part decelerates, while that of the thermal part continues to rise. At pulse durations over hundreds of milliseconds, the thermal contribution becomes dominating.

7.2.6
Phase Shifts in the Case of Periodic Pulse Pumping and in the Presence of Amplified Signal

The system with periodic pulse pumping has also been simulated. As an example, Figure 7.4 demonstrates dynamics of the phase shift induced in highly doped Yb-fiber by power meander with a period of 50 μs and maximum and minimum values of 15 and 105 mW, respectively. One can see that both electronic and thermal parts of the phase shift grow in the same manner as under long continuous wave (CW) pump pulse excitation. The rise time and steady-state level are determined by average pump power level. Thermal contribution to the phase shift is lower than the electronic contribution all over the simulation.

Figure 7.5 demonstrates the effect of short signal pulses at 1060 nm on the phase shift induced in the fiber by a long pump pulse at 980 nm. The signal pulse of 20 μs duration converts Yb ions from the excited state to the ground state consuming energy on signal amplification and heating. It causes a decrease of the electronic contribution to the RIC at the expense of the thermal one. The total phase shift also decreases, although it could still continue to rise during a short

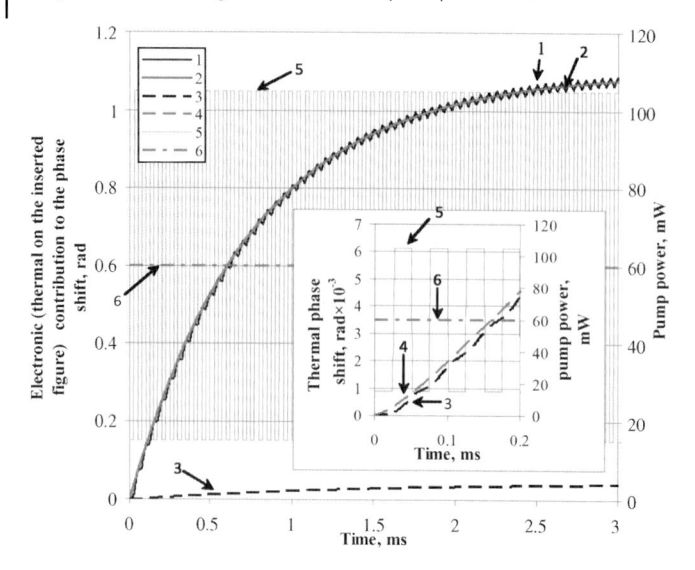

7 Refractive Index Changes in Rare Earth-Doped Optical Fibers

Figure 7.4 Comparison of electronic and thermal phase shift dynamics under periodic (5) and long CW pulse pumping (6): electronic contribution (1, 2) and thermal contribution (3, 4).

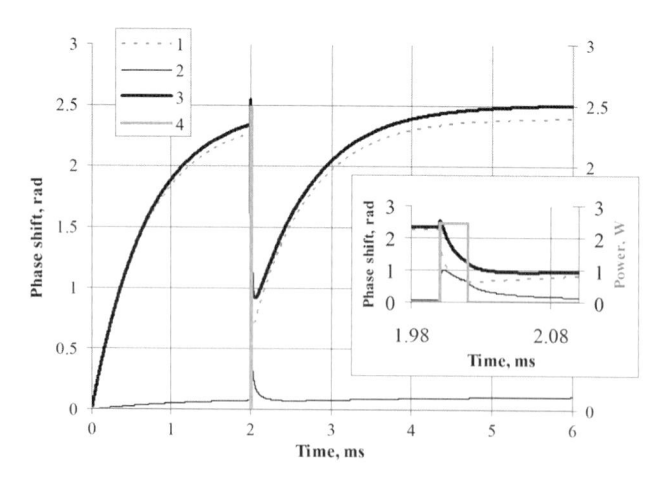

Figure 7.5 Dynamics of phase shifts induced by long pump and a short signal pulse at 980 and 1060 nm, respectively. The pump is switched-on at $t = 0$, and then 20 μs rectangular signal pulse of 2.48 W amplitude is switched-on at $t = 2$ ms: electronic contribution (1), thermal contribution (2), total phase shift (3), and signal power at 1060 nm (4).

time interval. When the signal is switched off, the population starts growing again together with the electronic contribution to the RIC. At that moment, the thermal contribution decreases (while the emitted heat is relaxing) and then starts to increase again.

7.2.7
Conclusion

In conclusion, we have presented theoretical analysis of the electronic and thermal RIC effects taking place in Yb-doped optical fibers. We have clarified their joined population inversion nature and explained the principal difference between the two mechanisms. We derived key equations of the dynamical effects that will be fruitfully employed in the next sections for detailed analysis of numerous experimental observations. Importantly, the thermal contribution to the RICs observed in YDFs is shown to be negligible under conditions of the all-fiber low-power experiments considered in the following sections.

7.3
Experimental Studies of the RIC Effect in Yb-Doped Optical Fibers

7.3.1
Previous Observations of the RIC Effect in Laser Fibers

Correct understanding of these electronic phenomena in YDFs [21] is very important for numerous fiber applications. The pump-induced RICs could significantly affect the fiber laser behavior. The enhanced nonlinear phase shift could be employed for coherent beam combining discussed in this chapter, optical switching [22], and all-fiber adaptive interferometry [23]. The pump-induced thermal and electronic RICs affecting waveguide changes become significant for very large-mode-area (LMA) fibers, and result in a reduction of the mode field diameter, but simultaneously in an improvement of the beam quality [8]. The electronic and thermal long-period index gratings induced in LMA fibers can affect the mode structure and mode instability of the powerful LMA fiber lasers [24].

The nature of the polarizability difference in rare earth ions has been also widely discussed. Some authors believe that the main contribution to the polarizability difference expressed by Eq. (7.4) comes from the first term responsible for near-resonance IR transitions [25–28]. An alternative model suggests the predominant contribution from the second term responsible for strong UV transitions located far from the resonance [9], similar to those observed in laser crystals [10,29–31].

In the following sections, we discuss our original RIC observations in commercial single-mode Yb-doped optical fibers under laser diode pumping at $\sim$980 nm and signal amplification at 1060 nm. The experiment is performed with probe wavelength far from the absorption and emission Yb^{3+} ion resonances. The main objectives are to characterize the RIC dynamics and to report the values of the polarizability difference in standard aluminum silicate and phosphate silicate YDFs in spectral range of 1460–1620 nm.

7.3.2
Methodology of Pump/Signal-Induced RIC Measurements

The experimental setup is shown in Figure 7.6. The all-fiber spliced Mach–Zehnder interferometer comprises a tested YDF in one arm and a balanced Corning HI 1060 fiber of comparable length in the second arm. The CW radiation emitted by a diode laser "Tunics" with a coherence length of about 10 m is used as a probe signal. The signal wavelength λ_T is continuously adjustable from 1450 to 1620 nm. The probe signal is detected at the interferometer output by a fast photodiode. The tested fiber is pumped from a standard pump laser diode operating CW or pulsed mode at $\lambda_P \approx$ 980 nm with a power up to ~150 mW. Signal pulses at $\lambda_S \approx$ 1060 nm are introduced to the fiber from another 50 mW laser diode (Frankfurt Lasers Company, Germany).

The RIC signal is measured as the photodiode response to a single rectangular pump pulse or a signal pulse (alone or in combination with CW 980 nm pumping). The pulse duration and amplitude are controlled by the electrical generator in the ranges of 10 μs–10 ms and 0–145 mW, respectively.

The induced phase shift $\delta\varphi(t)$ is recovered from the oscilloscope trace amplitude $U(t)$ as

$$\delta\varphi(t) = \varphi(t) - \varphi(0),$$
$$\varphi(t) = (-1)^k \arcsin\left(\frac{2U(t) - U_{max} - U_{min}}{U_{max} - U_{min}}\right) + \pi k, \tag{7.26}$$

where $k = 0, 1, 2, \ldots$ provides continuity of $\delta\varphi(t)$.

A typical example of the recorded oscilloscope trace and the corresponding phase shift is shown in Figure 7.7.

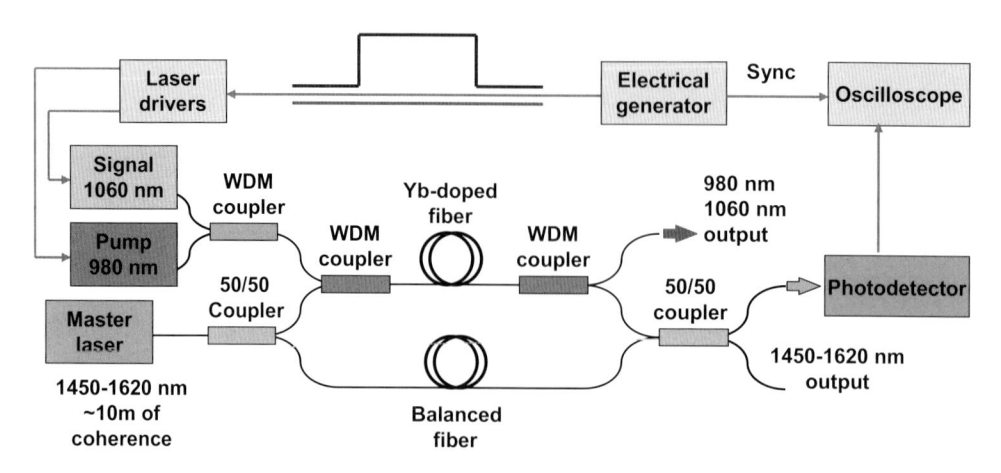

Figure 7.6 Experimental setup for testing of single-mode Yb-doped optical fibers.

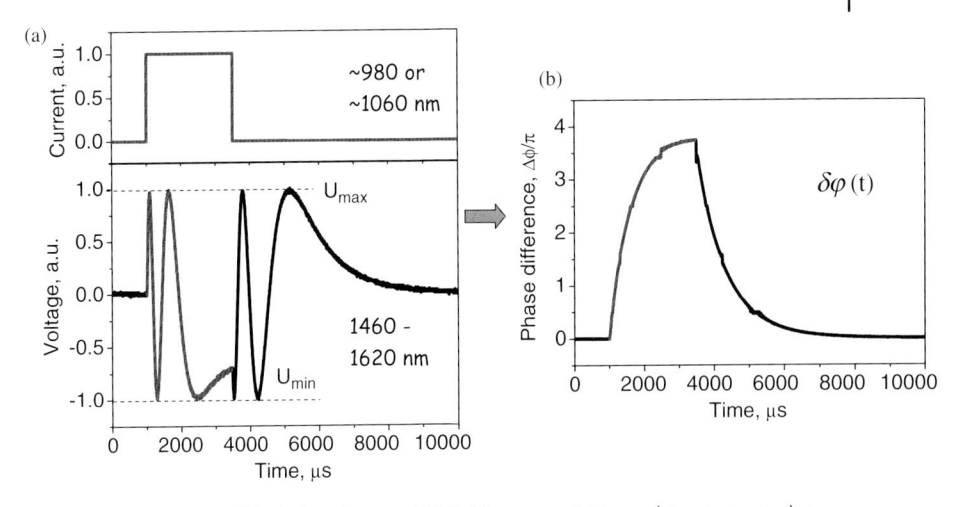

Figure 7.7 An example of the induced phase shift $\delta\varphi(t)$ recovery. (a) Laser driver current pulse profile and the recorded oscilloscope trace. (b) Reconstructed phase trace.

7.3.3
Characterization of RIC in Different Fiber Samples

This section describes RIC effects observed in samples of aluminum silicate YDFs excited by 980 nm pump pulses alone, that is, without a signal at 1060 nm. Four fiber samples of different lengths, index profile geometries, and Yb^{3+} ion concentrations have been examined. All fibers are single mode at the pump and test wavelengths. Fiber A is fabricated in Fiber Optics Research Center (Russia) with rectangular $\sim 125 \times 125\ \mu m^2$ cladding profile. Fibers B–D are Yb-198, Yb-118, and Yb-103 CorActive (Canada) products, respectively. They have circular cladding geometry with the diameter of $\sim 125\ \mu m$. All fibers possess a Gaussian distribution of Yb^{3+} ion concentration in the core. Other fiber parameters are presented in Table 7.2.

The recorded phase shifts up to $\sim 4\pi$ shown in Figure 7.8 highlight strong RIC effect induced in fiber A by pump pulses of different amplitudes. For different pump powers, the effect exhibits smooth saturation to different steady-state levels.

Table 7.2 Single-mode aluminum silicate Yb-doped fibers tested in the experiment.

Parameter	Fiber A	Fiber B	Fiber C	Fiber D
Peak absorption at 980 nm (α_P) (dB/m)	~ 900	~ 1073	~ 245	~ 35
Mode-field diameter at λ_S (w) (μm)	~ 4.5	~ 3.6	~ 4.5	~ 3.6
Cutoff wavelength (λ_c) (nm)	~ 810	~ 870	~ 680	~ 816
Equivalent core radius (a) (μm)	~ 1.6	~ 1.4	~ 1.3	~ 1.3
Coefficient K at 1550 nm [rad/(ms mW)]	$\sim 0.043\pi$	$\sim 0.056\pi$	$\sim 0.067\pi$	$\sim 0.067\pi$

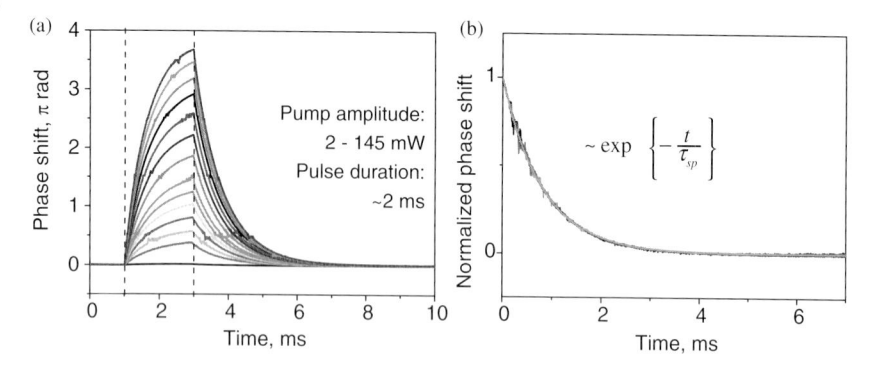

Figure 7.8 (a) Phase shifts induced in fiber A by pulses of different pulse amplitude. (b) Relaxing parts of the same curves normalized to the maximal values are shown in the delayed scale. The test wavelength is ~1550 nm and the fiber length is ~2 m.

Generally, saturation depends on the pulse amplitude and on the pulse duration or, more precisely, on the pulse energy. Decaying parts of the phase traces are due to relaxation of the refractive index at the end of the pulse excitation. They are perfectly fitted by an exponential decay function $\varphi(t) \sim \exp(-t/\tau_{sp})$, with the relaxation time constant equal to the Yb ion excited-state lifetime $\tau_{sp} \approx 850\,\mu s$, almost the same for all fiber samples. The normalized curves shown in Figure 7.8b in the delayed timescale are similar, although they relate to different pumping conditions. Such relaxation behavior corresponds to the electronic mechanism of RIC predicted by Eq. (7.12). No other features that might be attributed to the thermally induced RICs have been observed in the experiment.

Under conditions of rectangular pump pulse excitation in highly doped fibers, that is, when the pump power is totally absorbed, neglecting the low ASE we can solve Eq. (7.12) to find:

$$\delta\varphi(t) = K\tau_{sp}\left[1 - \exp\left(-\frac{t}{\tau_{sp}}\right)\right]P_0, \tag{7.27}$$

where P_0 is the pump pulse amplitude. In highly doped fiber samples A–C, the pump power is completely absorbed by the fiber length. At low pump pulse energies, the experimental traces shown in Figure 7.9 perfectly reproduce the predictions of Eq. (7.27). One can verify the exponential character of the phase growth to its steady-state level (Figure 7.9a) and the linear dependence on the pump pulse amplitude (Figure 7.9b). As the pass gain and ASE level remain low, the slopes $\delta\varphi/\delta t|_{t \to 0}$ can be measured for different pulse amplitudes (Figure 7.9c). Their linear fit gives us the factor $K = 0.056\pi$ rad/(ms mW), which is the only unknown material parameter in Eq. (7.12). At higher excitation levels, an error between the recorded traces and Eq. (7.27) appears and increases as the ASE power increases.

Factor K in Eqs. (7.12) and (7.27) is the only model parameter relating the induced phase shift and the pump pulse parameters at a given test wavelength. Phase dynamics observed in two different lengths of fiber C shows the factor K to be

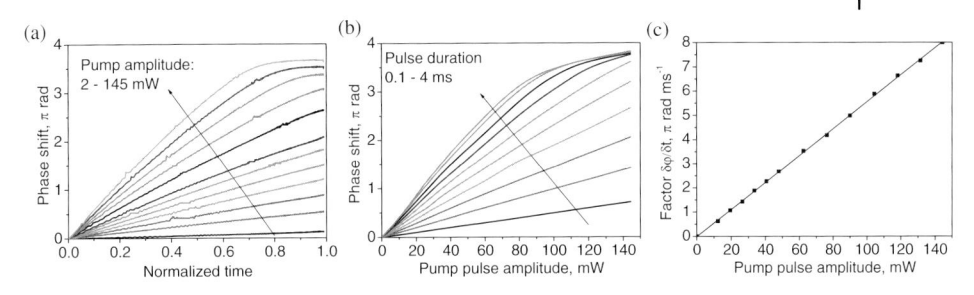

Figure 7.9 Phase shifts induced in fiber B by ~4 ms rectangular pulses as a dependence on the normalized time $\tau = 1 - \exp(-t/\tau_{sp})$ (a) and pulse amplitude (b). The slope $\delta\varphi/\delta t|_{t\to 0}$ of the phase characteristics as a dependence on the pulse amplitude (c). The test wavelength is ~1550 nm and the fiber length is ~2 m.

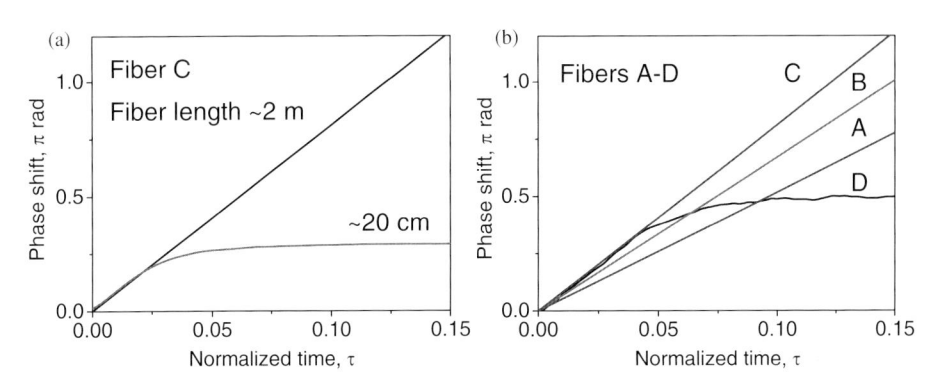

Figure 7.10 Phase shift induced by 145 mW pulses in different fiber samples: in the same fiber of different lengths (a) and different fibers of ~2 m length.

independent of the fiber length (Figure 7.10a). Reduction in the fiber length causes reduction of the saturation energy, but does not affect the slope in phase changes at low pump pulse energies. Under same experimental conditions, the phase shift dependencies observed with different fiber samples reveal different slopes and other distinct features (Figure 7.10b). Fiber D with lowest Yb ion concentration exhibits lower saturation power than other fibers, but the slope in this fiber is nearly the same as the slope relating to fiber C and is larger than the slope relating to fibers B and A. Approximation by Eq. (7.27) gives different factors K for different fibers presented in Table 7.2. However, the mutual ratios of the factors are in good agreement with our RIC model that predicts the factor K independent of Yb ion concentration and inversely proportional to the square of the fiber effective core radius $\bar{a}$ determined by Eq. (7.10).

7.3.4
Phase Shifts Induced by Signal Pulses

RIC effects in the YDF could also be induced by signal pulses at 1060 nm. The phase shifts recorded at 1550 nm and shown in Figure 7.11 highlight positive and negative RICs induced in fiber A by 4 ms signal pulses of different amplitudes alone [part (a)] and in the presence of CW pumping at 980 nm [part (b)]. In both cases, the effect exhibits smooth saturation at some steady-state level and relaxation after the end of the excitation pulse. However, the phase dynamics is qualitatively different for these two cases. In the first case (Figure 7.11a), the induced phase shifts are caused by absorption of the signal power at $\lambda_S \approx 1060$ nm in a noninverted fiber. The process is similar to RIC reported in the previous section. The only difference is that only small part of the signal power is used in 2 m of the fiber to excite Yb ions from the ground state due to relatively low absorption cross section of Yb ion at $\sim$1060 nm in comparison to the absorption cross section at 980 nm. Therefore, the achieved population inversion and the saturate level of the positive phase shifts are much lower than with 980 nm pumping. Dynamics of phase shifts involves dynamics of the rectangular 1060 nm pulse absorption in the fiber. Similar to 980 nm excitation, the relaxing parts of the phase shifts are perfectly fitted by an exponential decay function with the time constant equal to the Yb ion excited-state lifetime.

In the second case (Figure 7.11b), the induced phase shifts are due to amplification of the signal power at $\lambda_S \approx 1060$ nm in the fiber already pumped by CW 980 nm radiation. During amplification, the signal power grows at the expense of pump power providing transition of the Yb ions already excited by CW pump to the ground state. Therefore, the phase shifts are negative and their saturation level are determined by a steady-state equilibrium of populations in ground and excited states governed by CW pump and signal powers. Dynamics of phase shifts involves dynamics of the rectangular 1060 nm pulse amplification in the fiber described by Eq. (7.12). The relaxing parts of the phase shifts highlight the increase of the population inversion in the fiber caused by the CW 980 nm radiation at the end of

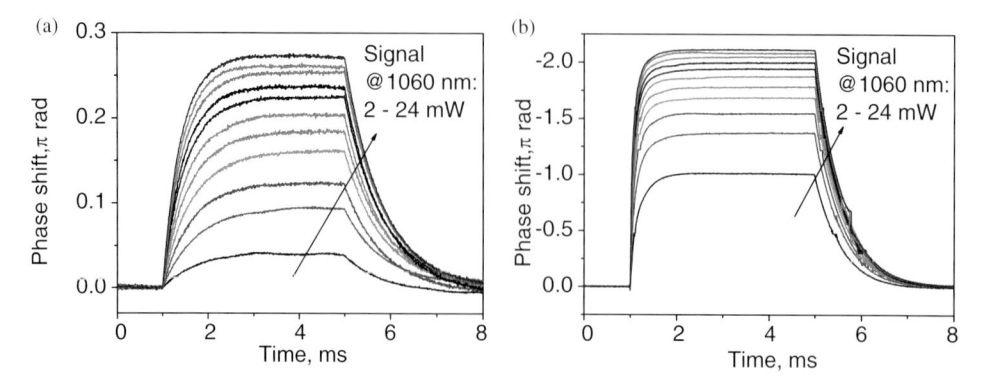

Figure 7.11 Phase shifts induced by 4 ms signal pulses at $\sim$1060 nm alone (a) and in combination with CW 145 mW pumping at 980 nm (b).

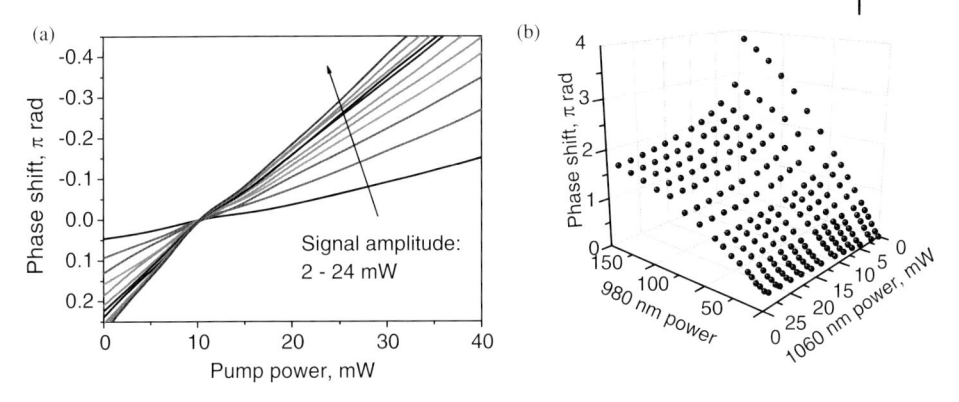

Figure 7.12 (a) Phase shifts induced by 4 ms signal pulses at ~1060 nm as function of the CW pump power and 1060 nm pulse amplitudes. (b) Three-dimensional surface of total steady-state phase shifts induced by combination of 980 and 1060 nm powers.

the signal pulse. Therefore, the relaxing times in this case are pump dependent and in an inverse proportion to CW 980 nm powers.

Figure 7.12 presents the maximal phase shifts induced by the signal pulses at 1060 nm of different amplitudes. The positive index changes observed at low pump powers are associated with predominating absorption of the signal power at $\lambda_S \approx 1060$ nm. Near the amplification threshold (~10 mW of the pump power), the RIC completely vanishes and appears again with negative sign when the amplifier gain grows. The total phase shift induced by two wavelength pumping is shown in Figure 7.12b.

A comparison of the phase shift characteristics with the amplifier gain characteristics shown in Figure 7.13 allowed us to conclude that the steady-state phase shifts $\delta\phi$ induced by 4 ms signal pulses in a qualitative approximation are directly proportional to the change of the population inversion, that is, to the total signal

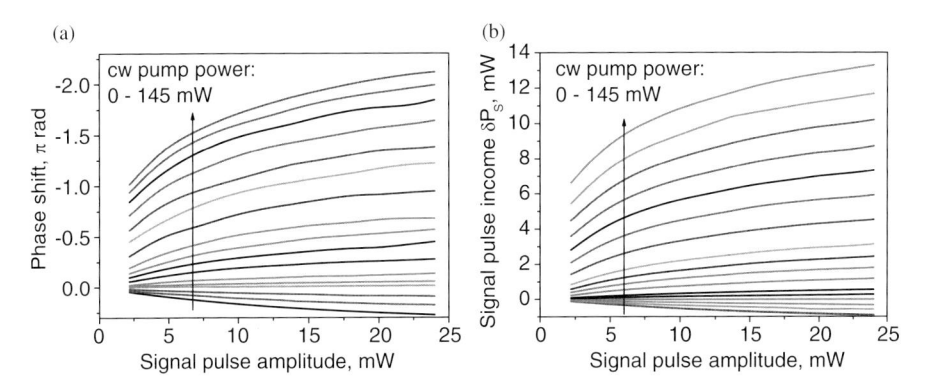

Figure 7.13 The maximal phase shifts induced by 4 ms signal pulses at ~1060 nm (a) and the maximal signal pulse income $\delta P_S = P_{S_{in}} - P_{S_{out}}$ (b) as a function of the signal pulse power at different CW pump power levels.

power income $\delta P_S = P_{S_{out}} - P_{S_{in}}$ during its propagation through the fiber: $\delta\varphi \approx -\tau_{sp}K\delta P_S$, where $P_{S_{in}}$ and $P_{S_{out}}$ are input and output signal amplitudes.

7.3.5
Evaluation of the Polarizability Difference

The factors K measured in Section 7.3.3 allow us to estimate the polarizability difference at 1550 nm [defined as in Eq. (7.12)]. It is found to be the same for all tested fibers. Evaluation of $\bar{\eta}\rho_T(0)$ for a step-index fiber and Gaussian Yb ion distribution with radius equal to $\bar{a}$ ($\bar{\eta} \approx 0.85$) presented in Figure 7.2 gives $\Delta p_{1550} \approx 1.2 \times 10^{-26}$ cm^3. We expect $\sim$20% error in this value due to uncertainty of the doped area size. To measure the polarizability difference dispersion in the spectral range of 1450–1620 nm, fiber B was additionally characterized at different testing wavelengths λ_T. The phase traces shown in Figure 7.14a highlight significant differences in the phase slopes $\delta\varphi/\delta\tau$ referenced to different λ_T. Qualitatively, similar results have been reported earlier for the two-core YDFs [21]. However, in our case, the measured dispersion of the factor $K(\lambda_T)$ is found to coincide with the dependence $\sim \rho_T(0)/\lambda_T$ as it is expressed from the step-index approach (Figure 7.14b). As far as the polarizability difference $\Delta p(\lambda_T)$ is directly proportional to $\sim K^{-1}(\lambda_T)\rho_T(0)/\lambda_T$, the strong wavelength dependencies of $K(\lambda_T)$ and $\sim \rho_T(0)/\lambda_T$ are compensating each other, providing minor polarizability difference dependence on the test wavelength in the measured spectrum range.

The experimentally observed dispersion $\Delta p(\lambda_T)$ has to be compared with the Lorentz line dispersion profiles predicted by Eq. (7.4) for the polarizability difference contributions of near-resonant IR and far-resonance UV transitions of Yb^{3+} ions at ~ 1 and ~ 0.4 μm, respectively. The reconstructed polarizability difference profile matches the UV line wing and has the same value in all investigated spectral ranges.

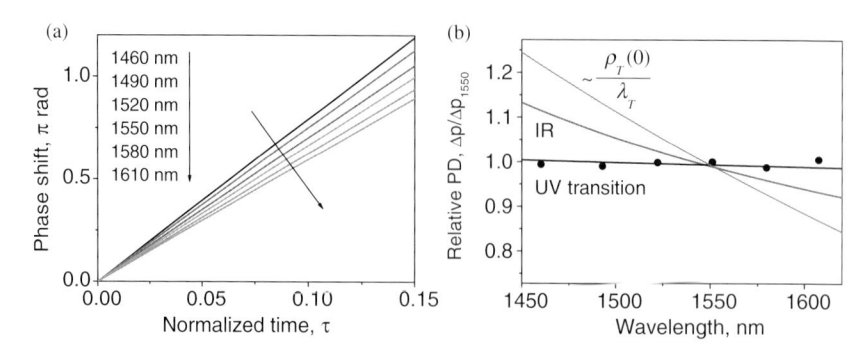

Figure 7.14 (a) Phase shift induced by 145 mW pulses in fiber B at different testing wavelengths. (b) The relative polarizability difference (points) in comparison with wavelength dependencies expressed by the first (resonance) and the second (nonresonance) terms of Eq. (7.4) and the dependence $\sim \rho_T(0)/\lambda_T$.

7.3.6
Comparison of the RIC Effects in Aluminum and Phosphate Silicate Fibers

The RIC effect in commercial aluminum silicate fibers has been quantified in previous sections. Here, we compare the RIC dynamics in aluminum and phosphate silicate YDFs (codoped by $Al_2O_3 + GeO_2$ and P_2O_5, respectively) specified with very different Yb ion lifetimes, ~ 825 and ~ 1460 µs, respectively.

Experimental results are shown in Figure 7.15 for fibers E and F, both fabricated in Fiber Optics Research Center (Russia) with circular cladding profile and specified in Table 7.3. The fibers passed the RIC test in the interferometric setup (Figure 7.6) under similar experimental conditions. The phase shifts shown in Figure 7.15a and b highlight strong RICs induced by 120 mW pulses in both fibers. The different Yb ion lifetimes owned to different fiber samples set different paces to the dynamical processes recorded in the experiment. One can see that the effect in both fibers possesses similar behavior, but with different growing rates, saturating at similar steady-state levels. The decaying parts of the phase traces describe the refractive

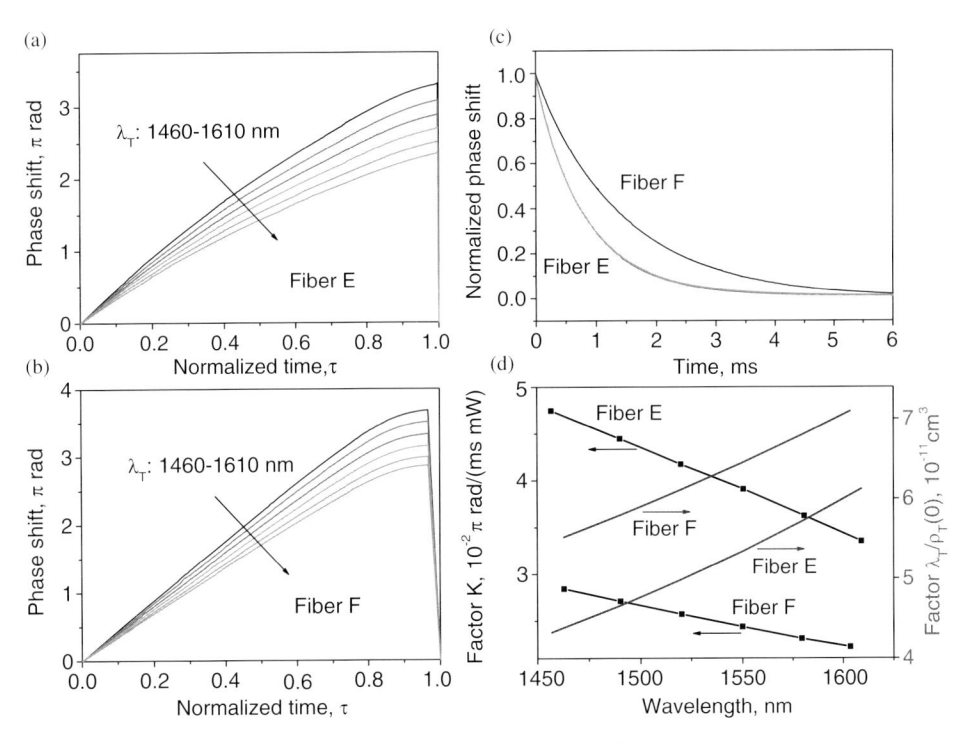

Figure 7.15 Phase shifts induced in aluminum (a) and phosphate (b) Yb-doped fibers by 120 mW pulses at different probe wavelengths λ_T are shown in normalized timescale $\tau = 1 - \exp(-t/\tau_{sp})$. Normalized relaxation parts of all phase shift traces (c), the factor K extracted from the experimental data, and the factor $\lambda_T/\rho_T(0)$ contributing to the polarizability difference value (d).

Table 7.3 Single-mode Yb-doped fibers tested with 980 nm pump pulse excitation.

Parameter	Fiber E	Fiber F
Dopants	Yb + Al + Ge	Yb + P
Peak absorption at 980 nm (α_P) (dB/m)	~ 280	~ 160
Mode-field diameter at λ_S (w) (μm)	~ 4.0	~ 5.0
Cut-off wavelength (λ_c) (nm)	~ 825	~ 950
Lifetime (τ_{sp})	~ 830	~ 1460
Equivalent core radius (a) (μm)	~ 2.0	~ 2.3
Coefficient K at 1550 nm [rad/(ms mW)]	~ 0.035π	~ 0.024π

index relaxation after the pulse excitations are perfectly fitted by the exponential function $\varphi(t) \sim \exp(-t/\tau_{sp})$ with the relaxation constants corresponding to different Yb ion lifetimes τ_{sp} assigned to the samples.

Following the same procedure as described in the previous section, we determine from the slopes $\delta\varphi/\delta t|_{t\to 0}$ the factors $K(\lambda_T) \sim d^{-2}$ for each curve shown in Figure 7.15a and b. Then, we calculate the dependencies $\sim \rho_T(0)/\lambda_T$ (Figure 7.15d) for each fiber. The reproduced polarizability difference values are found to be nearly constants within all probe signal spectrum range: $\sim 1.25 \times 10^{-26}$ cm^3 and $\sim 0.95 \times 10^{-26}$ cm^3 for aluminum and phosphate silicate YDFs, respectively (Figure 7.16). One can see that the polarizability difference values are not so different, despite the Yb ion lifetimes in these fibers that defer almost in two times.

Recently, similar measurements have been performed with gamma-irradiated aluminum silicate YDFs [32]. A gamma-irradiation inducing color center in the silicate glass matrix is found to modify also the RIC response to some extent.

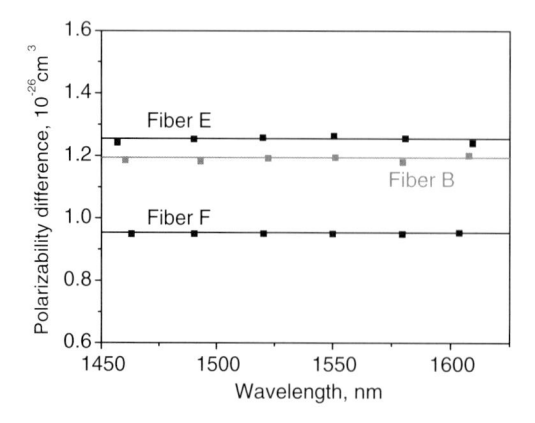

Figure 7.16 The measured polarizability difference values for aluminum silicate (fibers E) and phosphate silicate (fiber F) fibers. The result for fiber B obtained in previous section is shown for comparison.

7.3.7
Conclusion

In conclusion, we have reported a strong RIC at wavelengths of 1460–1620 nm when pumped at the absorption or gain lines of Yb ions. The RIC exhibits a typical excited population dynamics. These facts clearly indicate the electronic mechanism by the polarizability difference of the excited and unexcited Yb ions. The value of the polarizability difference is determined for aluminum silicate and phosphate silicate YDFs. In both cases, the polarizability difference dispersion curve makes evident the nonresonant contribution of the UV transitions to the polarizability difference. In our experiments, the polarizability difference has been directly measured at testing wavelength far from the Yb ion transition resonances. An important parameter related to control of phase shifts in the fiber within its amplification or absorption bands is a ratio between the real and imaginary parts of the dielectric susceptibility $\beta(\lambda)$, as given by the expression (neglecting the thermal part of the susceptibility):

$$\beta(\lambda) = \frac{\Delta \chi_R}{\Delta \chi_{Im}} = \frac{8\pi^2 F_L^2}{n\lambda} \cdot \frac{\Delta p(\lambda)}{\Delta \sigma(\lambda)}. \tag{7.28}$$

It is clear that far from resonances, $\beta(\lambda)$ is rather large. To estimate this value around the resonance wavelengths, both resonance and nonresonance parts of the polarizability difference [Eq. (7.4)] need to be taken into account. In particular, for typical operation wavelengths of Yb-doped amplifier (1030–1100 nm), these parts are comparable and the parameter β could be estimated as $\beta(1064\,\text{nm}) = 9.8$ and $\beta(1080\,\text{nm}) = 24.9$ for phosphate silicate fibers and $\beta(1064\,\text{nm}) = 5.8$ and $\beta(1080\,\text{nm}) = 7.4$ for aluminum silicate fibers. Importantly, it is an order of magnitude higher than the same parameter reported earlier for the bulk laser crystals near gain resonance [33–35]. Therefore, in a YDF amplifier, tuning of the pump power induces phase shifts much more stronger than the changes of the logarithmic gain factor. This fact makes possible a direct pump power phase correction of signals in Yb-fiber amplifiers operating at 1030–1080 nm and has been employed recently for reduction of phase noise in Yb-doped amplifier [36].

7.4
All-Fiber Coherent Combining through RIC Effect in Rare Earth-Doped Fibers

7.4.1
Coherent Combining of Fiber Lasers: Alternative Techniques

Power levels available with CW narrowband fiber sources are limited by nonlinear effects and mainly by stimulated Brillouin scattering (SBS) [37–41]. A way to overcome this limitation is to use coherent combining of single-mode amplifiers, each operating below the SBS threshold. To achieve constructive interference after amplification, that is, to collect power from all channels in one single-mode fiber, the fiber amplifiers must be phase matched together [42].

Different approaches to the constructive phase matching have been discussed: One can take advantage of the self-organization properties of the following: multi-arm cavities [43], multicore fibers delivering supermodes [44], digital holography [45], and SBS phase conjugation [2,3]. A straightforward infallible beam combining can be provided by active phase control of each fiber amplifier by an attached phase modulator [46]. However, piezoelectrical or electro-optical modulators are not perfect solution due to obvious disadvantages of parasitic resonances and integrated bulk and fiber components. In our all-fiber solution (Figure 7.17) [5,47], the rare earth-doped fiber amplifiers operating at λ_S within an YDF transparency band are supplied by the sections of YDFs operating as optically controlled phase modulators. The principle of operation includes RICs induced in YDF by optical pumping at 980 nm alone or in combination with 1060 nm signal discussed in Section 7.3. The method is applicable for Raman, Brillouin, neodymium-, erbium-, thulium-, or holmium-doped fiber amplifiers [48–51].

In the next sections, we verify the validity of this concept demonstrating the coherent combining of two 500 mW EDFAs. The two-level RIC model discussed in Section 7.2 enables a simple algorithm with natural implementation of the RIC effect into an active phase control loop. The abilities of the method to operate against the acoustic phase noise at the rate of $\sim 2.6\pi$ rad/ms (control by one optical signal at 980 nm) and at the rate above $\sim 10\pi$ rad/ms (control by two optical signals at 980 and 1060 nm) are experimentally confirmed.

The experimental setup (Figure 7.18) is not so different from the setup used in the previous part, except the two amplifiers introduced into the interferometer. A master laser diode in combination with a 15 dBm preamplifier delivers single-mode radiation at 1.55 µm with a coherence length of ~ 10 m. The first fiber coupler splits the laser emission into two arms, which are then amplified by two single-mode

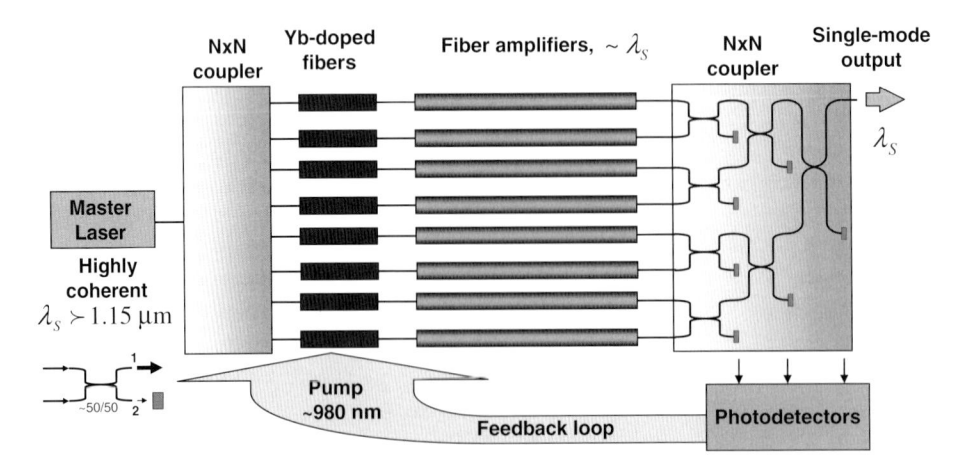

Figure 7.17 Multichannel laser system with coherent beam combining based on RIC in YDF. The type of used optical amplifiers could be adapted for operation at any wavelength λ_s within the Yb-fiber transparency band (1.15–2 µm).

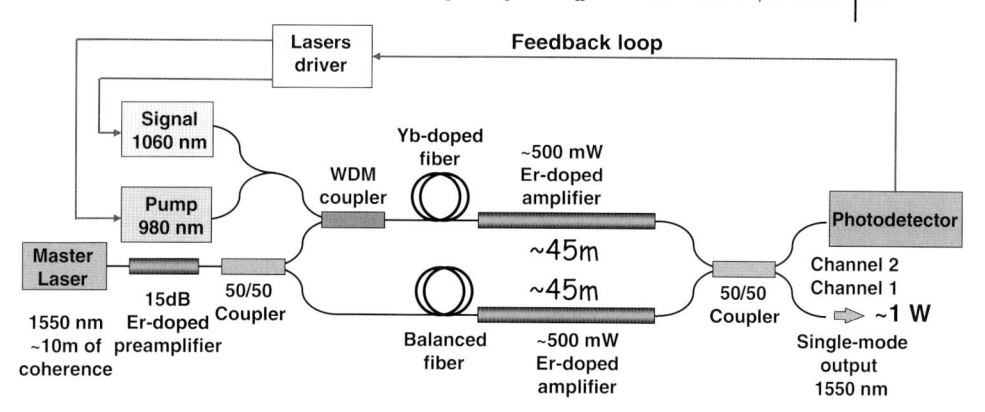

Figure 7.18 Experimental setup used for demonstration of all-fiber coherent combining with one (at 980 nm) and two (980 and 1060 nm) control optical signals.

EDFAs specified with 500 mW output. No spectral broadening of the amplified radiation has been observed at such power level. The EDFAs are supplied by thermoelectric controllers used for low-frequency phase noise elimination.

For fast phase adjustment, one of the arms is supplied with a 2 m length of YDF directly spliced with an amplifier. The YDF has two independent inputs for 1 laser diode operating at 980 and 1060 nm. The use of the highly doped fiber B (see Table 7.2) ensures maximal RIC effect due to total absorption of the pump radiation inside the fiber. Since RIC is directly proportional to the population density of the excited Yb ions, the phase shift induced in the fiber is determined by the laser diode power that could be used to maintain a constructive phase-matched coupling of two intense laser arms in a single-mode fiber (channel 1). The power emitted through channel 2 is used for operation of the feedback loop discussed in the following sections. The relationship between the channel 2 power and phase mismatching is regulated by Eq. (7.26).

7.4.2
Operation Algorithm and Simulated Results

The phase control operation algorithm is based on the steady-state and dynamical characteristics of the electronic RIC discussed in Section 7.3. The steady-state characteristic $\varphi = \phi(P^{980})$ (Figure 7.19a) is evaluated as the phase response to Heaviside step pulses of different amplitude P^{980}. One could tune the phase in the range of up to $\sim 3.75\pi$ rad along the steady-state curve by a simple adjustment of the laser diode power P^{980}. For example, phase switching from $\varphi_1 = \phi(P_1^{980})$ to $\varphi_2 = \phi(P_2^{980})$ is provided by switching the diode power from P_1^{980} to P_2^{980}, but this procedure requires milliseconds to proceed.

Fast dynamical switching is possible within a part of the range covered by the steady-state curve $\varphi = \phi(P^{980})$, for example, within the range of $\sim \pi$ rad marked in

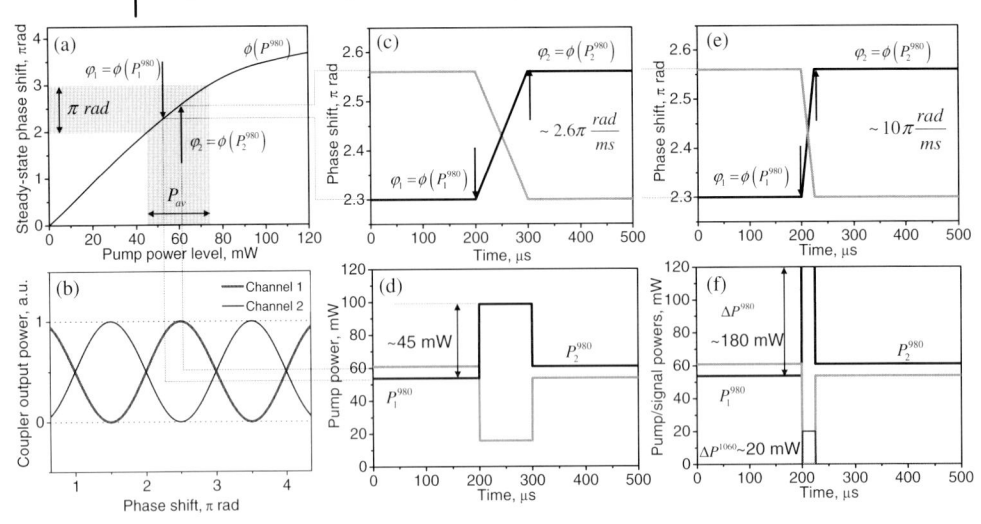

Figure 7.19 Experimental steady-state (a) and dynamical (b) phase characteristics. Phase response (c and e) to a double jump of the 980 nm diode power (d and f): positive (black) and inverted (gray), alone (c and d) and in combination with 1060 nm diode power pulse.

Figure 7.19a that is enough to keep combined laser power near the maximum (Figure 7.19b). Both one wavelength (980 nm) and two wavelength (980 and 1060 nm) control procedures could be implemented for such fast switching.

Let us consider first one signal control procedure employing modulation of 980 nm laser diode alone. In general case of total pump absorption and negligible spontaneous emission, we can extend Eq. (7.27) to all linear parts of the steady-state curve; so the phase response $\Delta\varphi(t)$ to a single positive or negative jump of the diode power ΔP_{980} is expressed as

$$\Delta\varphi(t) = K\tau_{sp}\left[1 - \exp\left(-\frac{t}{\tau_{sp}}\right)\right]\Delta P_{980}, \tag{7.29}$$

where $K \approx 0.056\pi$ rad/(ms mW) at $\lambda_T \approx 1.55$ µm (see fiber B in Table 7.2).

In accordance with Eq. (7.29), a fast positive or negative phase tuning between two steady-state levels (Figure 7.19b and c) from $\varphi_1 = \phi(P_1^{980})$ to $\varphi_2 = \phi(P_2^{980})$ is achievable through two consecutive switchings of the diode power: first, from the level P_1^{980} to a level $P_1^{980} + \Delta P_{980}$ and then to the level P_2^{980}, where ΔP_{980} is the positive or negative power jump available with the laser diode within the used tuning range and $|\Delta P_{980}| \gg |P_2 - P_1|$. The switching time of such dynamical phase change equal to the time τ between two opposite jumps of the diode power, is expressed for small phase steps as $\tau = (\varphi_2 - \varphi_1)/K\Delta P_{980} \ll \tau_{sp}$. The higher the ΔP_{980}, the faster the phase tuning. For the case shown in Figure 7.19, the switching rate of $\sim 2.6\pi$ rad/ms corresponds to the jump of the pump laser with the positive or negative amplitude $\Delta P_{980} \approx \pm 45$ mW.

We can drastically increase the rate of positive phase switching by just employing more powerful laser diode that has a potential for higher positive jumps. Figure 7.19e and f shows the positive switching rate of $\sim 10\pi$ rad/ms achieved with two opposite 980 nm diode power jumps with amplitudes $\Delta P_{980} \approx 180$ mW and $(P_2 - P_1 - \Delta P_{980})$, respectively. Importantly, such tuning does not depend on the P_1^{980} position along the steady-state curve. Therefore, the same rate is provided by similar pulses all over π tuning range.

The situation becomes different for fast negative phase tuning from $\varphi_1 = \phi(P_1^{980})$ to $\varphi_2 = \phi(P_2^{980})$. Formally, such fast tuning is still described by Eq. (7.29). The only difference is that now the diode power jump is negative and its value cannot exceed its absolute power in the current point of the steady-state curve, that is, $|\Delta P_{980}| < P_1$. It means that within the whole tuning range, we are no longer allowed to choose the amplitude $|\Delta P_{980}|$ to be as high as in the case of positive tuning, since now it is limited by the lowest diode power assigned to the range. Higher negative ΔP_{980} is available within smaller tuning ranges. However, the selection of $\sim \pi$ rad tuning range is important to wrap all possible phase shifts. These limitations constrain the maximal negative adjustment rate $\sim 2.6\pi$ rad/ms available with one signal phase control.

In order to increase the negative adjustment rate, we can employ two optical signal control schemes. In this case, the negative jumping of 980 nm laser diode power is accompanied by the second optical rectangular signal at 1060 nm introduced into the YDF synchronously, as shown in Figure 7.20. To describe this process, Eq. (7.29) for pulse duration $\tau \ll \tau_{sp}$ has to be rewritten as

$$\Delta\varphi(t) = K\tau_{sp}\left[1 - \exp\left(-\frac{\tau}{\tau_{sp}}\right)\right]\left(\Delta P_{980} - \frac{\lambda_S}{\lambda_P}\delta P_{1060}\right)$$
$$\approx K\left(\Delta P_{980} - \frac{\lambda_S}{\lambda_P}\delta P_{1060}\right)\tau, \quad (7.30)$$

where $\Delta P_{980} = -P_2$ is the amplitude of negative pump jump from current power P_2 on the steady-state curve to zero, and δP_{1060} is the difference between the output and input powers at 1060 nm in the fiber that could be estimated for short jumps ($\tau \ll \tau_{sp}$) as $\delta P_{1060} \approx (g(P_2^{980}) - 1)\Delta P_{1060}$, where $g(P_2)$ is the total fiber gain at 1060 nm provided by 980 nm power P_2 and ΔP_{1060} is a positive power jump at 1060 nm. Specifically, the phase response to such double pulsing essentially depends on the fiber gain $g(P_2^{980})$ and, therefore, on the position of the initial point ΔP^{980} on the steady-state surface. Therefore, under fixed pulse duration τ, the amplitude of the 1060 nm pulse corresponding to phase switching between two steady-state points have to be adjusted and tabled for each point of steady-state curve.

One can estimate the minimal rate of negative phase tuning for π tuning range shown in Figure 7.19a. The minimal 980 nm power assigned to the range lower boundary is $P_{980} \approx 45$ mW. It determines the value of negative 980 nm power jump at this point $\Delta P_{980} = -P_{980} \approx 45$ mW that caused the negative phase tuning at the rate of $\sim -2.6\pi$ rad/ms. The maximal 1080 nm power jump $\Delta P_{1060} \approx 24$ mW is determined by the maximal power available with the laser diode used in the

7 Refractive Index Changes in Rare Earth-Doped Optical Fibers

(a)

Laser driver → Signal generator → Error signal → PC → Acquisition card (NI)

Pump 980 nm → Power, 980nm → Induced phase shift → Power, 1550 nm → Photodetector

$P_{av} \approx 60\, mW$
$+$
$P_{980\,MEAND}(t)$
$+$
$P_{980\,ERROR}(t)$

$\varphi_{av} \approx \phi(P_{av})$
$+$
$\varphi_{SAW-TOOTH}(t)$
$+$
$\varphi_{ERROR}(t)$

Phase noise
$\varphi_{NOISE}(t)$

(b)

Laser drivers → Signal generators → Error signals → PC → Acquisition card (NI)

Pump 980 nm → Power, 980nm → Power, 1060nm → Induced phase shift → Power, 1550 nm → Photodetector

Signal 1060 nm

$P_{980\,MEAND}(t)$
$+$
$P_{980\,ERROR}(t)$

$P_{1060\,MEAND}(t)$
$+$
$P_{1060\,ERROR}(t)$

$\varphi_{av} \approx \phi(P_{av})$
$\varphi_{SAWTOOTH}(t)$
$+$
$\varphi_{ERROR}(t)$

Phase noise
$\varphi_{NOISE}(t)$

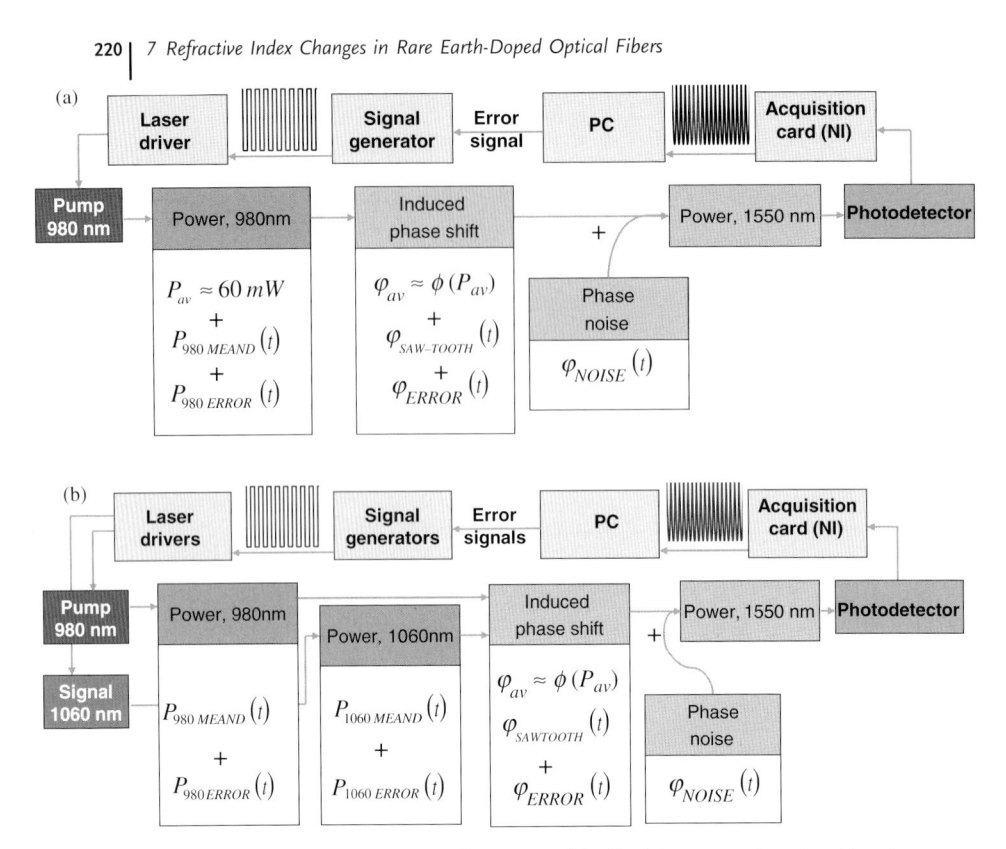

Figure 7.20 The principal scheme of the electrical feedback loop operation. One (a) and two (b) wavelength control configurations.

experiment. At steady-state 980 nm power level of $P_{980} \approx 45$ mW, it causes the phase decrease at the rate of $\sim -7.4\pi$ rad/ms. Therefore, the combined action of the two wavelengths leads to the fast phase tuning at the phase rate of $\sim -10\pi$ rad/ms. At the upper boundary of the π tuning range, at $P_{980} \approx 75$ mW, the amplitude of 980 nm pulse and fiber gain $g(P_{980})$ are both higher. The jumps of the 980 nm power $\Delta P_{980} = -P_{980} \approx 75$ mW and 1060 power $\Delta P_{1060} \approx 15$ mW contribute the total tuning rate of $\sim -10\pi$ rad/ms, providing the rates associated with each pulse of $\sim -3.8\pi$ and $\sim -6.2\pi$ rad/ms, respectively. Within π tuning range, the amplitude of 1060 nm pulse between 14 and 24 mW tabled for each position on the steady-state curve allows keeping the same negative rate of $\sim -10\pi$ rad/ms all over the range. Thus, the use of two wavelength signal control gives a four times increase in tuning rate in comparison to one signal control configuration.

The mission committed to the feedback loop circuit (Figure 7.20) is to support the maximal power level emitted at 1.55 μm through the main system output (channel 1, Figure 7.18). Therefore, the power emitted through the second control output (channel 2) has to be kept as low as possible. This power recorded by the photodetector is used in the feedback circuit.

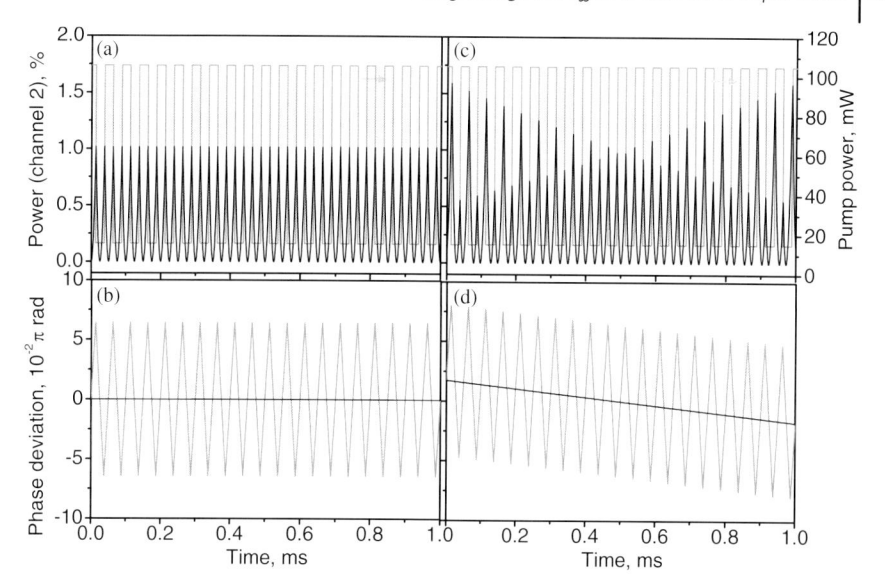

Figure 7.21 Simulated feedback loop operation for cases of compensated (a and b) and uncompensated (c and d) phase noise: (a and c) laser diode (gray) and channel 2 (black) powers; (b and d) reconstructed phase deviation (gray) and reconstructed error signal (black); $P_{av} \approx 60$ mW.

In the case of one 980 nm signal control, the period of data acquisition $\tau = 25$ μs is synchronized with 2.86 MHz acquisition card (National Instruments, NI PCI-6251) analogue output (alone or in combination with a standard pulse generator) that forces the laser driver to emit 50 μs period meander signal with an amplitude $\Delta P_0 = \pm 45$ mW and a controllable DC level P_{av} within the range of 45–75 mW (Figure 7.21a). The modulated diode power induces a fast sawtooth modulation of the phase with a DC level of $\varphi_{av} = \phi(P_{av})$ and the excursion of $\Delta\varphi_0 = K\Delta P_0 \tau \approx 0.06\pi$ rad (Figure 7.21b). Such phase modulation leads to ~100% amplitude modulation of the power in the control output (Figure 7.21c and d), while the modulation of the high-power radiation in the main system output is negligible (~1%). The signal acquired by the photodetector is the result of the superposition between the phase noise and the periodic phase modulation. In the case of right phase matching (when the phase noise is completely compensated), this signal is a perfect sawtooth signal, because the peaks associated with the positive and negative phase changes are of the same amplitude (Figure 7.21a). In contrast, the presence of uncompensated noise causes the signal peaks to spread in two series. Importantly, the phase mismatching φ_{NOISE} is directly proportional to the difference between neighboring peaks (Figure 7.21c). The error signal, produced by a PC from the acquired data, controls the phase φ_{av} through the control of the laser diode current: To compensate the phase mismatching, it produces a smooth correction $\delta P_{av} \rightarrow -P_{ERROR} = -\varphi_{NOISE}/K\tau$ to P_{av} (Figure 7.21a).

For two 980 and 1060 nm signal controls, the principle of operation is nearly the same, but practical realization is more complicated and requires much faster operation of the feedback loop. In this case, the period of data acquisition decreases down to $\tau = 6$ μs. Two ~ 12 μs rectangular signals at 980 and 1060 nm are emitted in antiphase by two laser diodes. The amplitude values P_0^{980} and P_0^{1060} are selected within the range of 180–225 and 14–24 mW for 980 and 1060 nm laser diodes, respectively, in such a way that there combination keeps the phase within full tuning π-range along the steady-state surface $\varphi_{av} = \phi(P_{av}^{980}, P_{av}^{1060})$, as shown in Figure 7.12b. The pulse amplitudes ΔP_0^{980} and ΔP_0^{1060} that are determined functions of the current phase φ_{av} cause a fast sawtooth modulation with the excursion of $\Delta\varphi_0 \approx 0.06\pi$ rad similar to that shown in Figure 7.21, but in approximately four times shorter timescale. The presence of uncompensated noise leads to spreading of signal peaks. The phase mismatching φ_{NOISE} proportional to the difference between neighboring peaks allows elaborating error signals added to ΔP_0^{980} and ΔP_0^{1060}.

7.4.3
Environment Noise in Optical System to be Compensated

The phase control algorithms described in the previous section has been applied to the experimental setup (Figure 7.18) demonstrating reliable operation against the phase noise.

Figure 7.22 presents the operation of the laser configuration without an active feedback. The phase traces have been recorded for a steady-state condition that is achieved after both amplifiers have reached thermal equilibrium (after 2–3 min of their operation with a power of 500 mW). Two classes of traces highlight phase fluctuations, associated with thermal variations (Figure 7.22a) and environmental acoustic vibrations (Figure 7.22b). These two kinds of noises contribute to different time domains and show different scales of the phase excursion that have to be compensated. Thermally induced phase noise is large fluctuations with amplitude up to several $\sim \pi$ rad attained for several seconds. For compensation of these

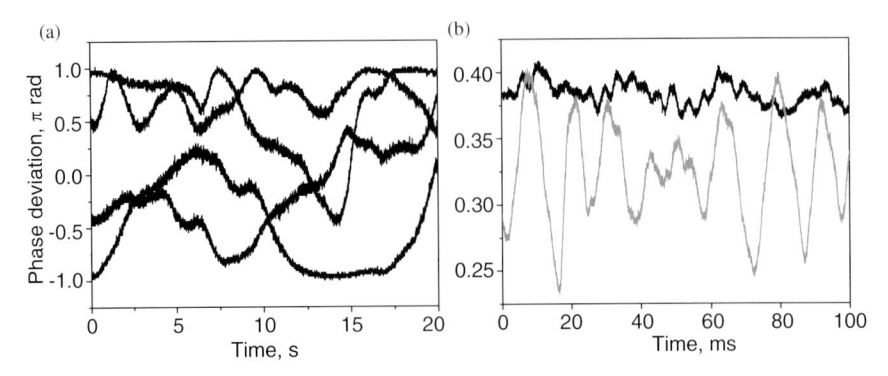

Figure 7.22 Time series of the typical amplifier phase noise. (a) Temperature noise. (b) Acoustic noise: natural (black), and caused by a flick given on the amplifier (gray).

fluctuations, no advanced fast technique is needed: The ordinary thermocontrollers connected with the feedback loop provide perfect suppression of low-frequency thermal noise. In contrast, acoustic phase noise associated mainly with mechanical resonances – noisy equipment, cooling fans, and so on – dominates with excursion rates of $\sim$1 ms and much smaller excursion amplitude, $\sim 0.01\pi$ rad. This should be considered as a minimum requirement for servo loop bandwidth since these measurements were taken in a quiet laboratory setting. Noisier environments would require a commensurate reduction of the feedback loop time, but we have checked that even a flick given on the amplifier causes a phase deviation with the amplitude no larger than $\sim 0.02\pi$ rad at a rate of $\sim 0.02\pi$ rad/ms.

These observations give target parameters for the active phase control to be employed for compensation of the environmental acoustic noise in the system comprising two fiber amplifiers. Note that a typical phase rate value to be compensated is more than 100 times lower than the capability of the proposed one signal control technique as just estimated. However, as the number of parallel amplifiers increases, the phase compensation rates have to be increased proportionally. The algorithm could be applied to the multiamplifier system through time-division multiplexing described in the following section.

7.4.4
Combining of Two Er-Doped Amplifiers through the RIC Control in Yb-Doped Fibers

Coherent combining of two 500 mW Er-doped amplifiers resulting in $\sim$1 W power decoupled through the single-mode fiber output has been successfully demonstrated with the use of 980 nm signal phase control. Typical channel 2 power and phase traces (Figure 7.23a and b) exhibit the features similar to that shown in Figure 7.21c and d. The absolute power variations are about 10–20 mW, that is, $\sim$1–2% of the total power emitted by two amplifiers. One can see how, during the given time series, the initial phase mismatch caused by a flick given on the amplifier compensates due to operation of the feedback loop: The peaks initially separated in two series join together highlighting constructive interference achieved in channel 1. The power characteristics of the combined system (Figure 7.23d) give clear evidence that more than 95% of the radiation generated in two fiber amplifiers is efficiently decoupled through the single-mode output.

The experimental validation of coherent combining concept with phase control by two optical signals has also been performed. The YDF is pumped by double $2\tau \sim 12\,\mu s$ rectangular signals at 980 and 1060 nm emitted in antiphase by two laser diodes. The amplitude values P_0^{980} and P_0^{1060} of the pulses were adjusted around 200 and 20 mW for 980 and 1060 nm laser diodes, respectively. Tuning of these amplitudes changes the phase shift in accordance with the relations discussed above. Figure 7.23c presents the signal recorded by the photodetector from channel 2 at the moment of such tuning. One can see that the signal performance is nearly the same as in the case of one wavelength experiment shown in Figure 7.23b. The signal peaks are spread in two series, so the signal highlights the presence of

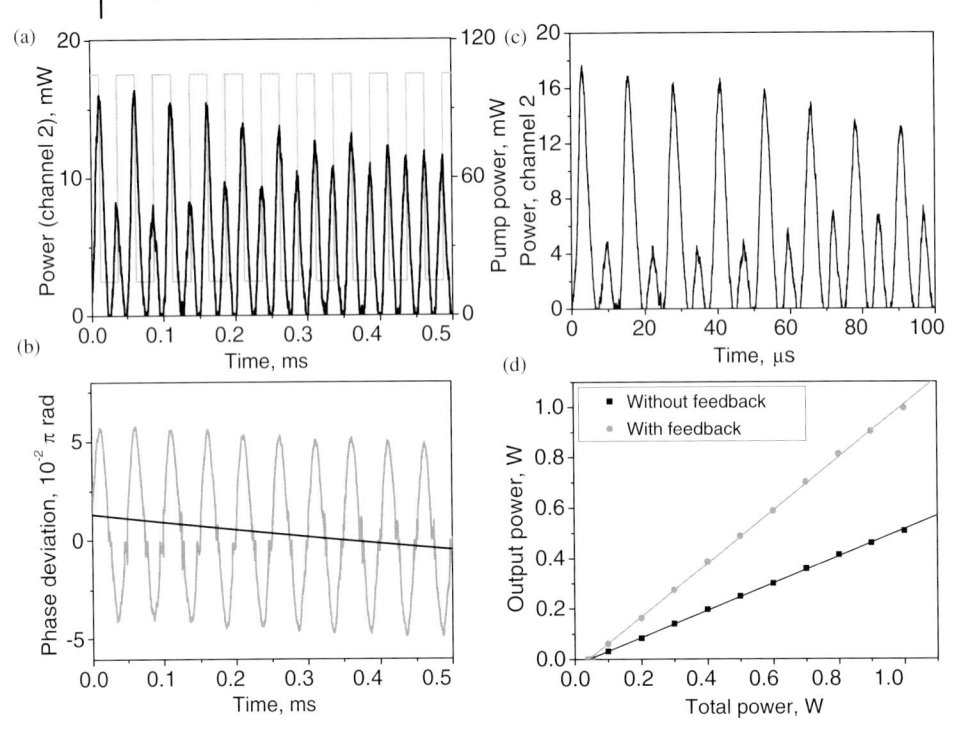

Figure 7.23 Experimental operation of two combined amplifiers against the noise. (a) Photodiode signal (980 nm signal phase control). (b) Reconstructed phase (gray) and generated error signals (black). (c) Photodiode signal (980 and 1060 nm signal phase control). (d) Laser system power characteristics without (gray) and with (black) 980 nm signal phase control.

uncompensated noise directly proportional to the difference between neighboring peaks. However, valid operation of the feedback loop has been prohibited by limitations of the used equipment (the acquisition card and PC used in the experiment are supported operation at the rate of $\tau = 6\,\mu s$).

7.4.5
Extension Algorithm for Combining of N Amplifiers

Although in previous section the validity of the concept has been confirmed by combining of only two 500 mW erbium-doped fiber amplifiers, the technique is extendable for combining of $N = 2^n$ amplifiers. Here, we show a simple algorithm enabling natural implementation of the RIC effect into an active phase control loop. Based on the same technical solution, the algorithm is able to operate against phase noise at the rate of $\sim 2.6\pi$ rad/ms and $\sim 10\pi$ rad/ms shared between n time slots with control by 980 nm laser diode alone and in combination with 1060 nm signal, respectively. Without the loss of generality, only the case of one wavelength control system is considered here.

Let us consider the system combining N amplifiers. It includes $N \times N$ multiplexer built from $(N-1)$ 50/50 optical couplers constituting n levels numbered as $m = 1 \cdots n$ in their pair-to-pair hierarchy. The active phase control has to be implemented to control the phases of beams passing each amplifier and to combine them coherently pair to pair in $(N-1)$ 50/50 couplers (Figure 7.17). Therefore, $N-1$ laser diodes operating at 980 nm are assigned to each identical Yb-doped fibers used as phase shifters. For conformity with the case of two amplifiers ($N=2$) discussed above, we assume that the laser diodes are driven with the time period $\tau = 25$ μs. The feedback loop has to support the maximal power level emitted through each coupler outputs 1, providing minimum power levels detected from the coupler outputs 2 (Figure 7.17). Figure 7.24 shows an example of time-division operation for $8 = 2^3$ amplifiers (three time slots are marked by different tone).

The laser diode attached to kth amplifier emits a controllable DC power P_{av_k} within the range of 45–75 mW and an additional signal $\Delta P_k(t)$ with zero mean spread for $m = 1 \cdots n$ time slots $\Delta P_k(t) = \sum_{m=1 \cdots n} \Delta P_{km}(t)$, where $k = 1 \cdots N$ and $\Delta P_{km}(t)$ is zero or $\pm 45/m$ mW within mth time slot and zero anywhere outside mth time slot. Nonzero modulated diode powers $\Delta P_{km}(t)$ induces positive and negative triangle modulations of the phase with a DC level of $\varphi_{av_k} = \phi(P_{av_k})$ in kth amplifier during mth time slot leading to positive or negative phase excursion $\Delta \varphi_{0_{km}} = K\Delta P_0 \tau \approx 0.06\pi$ rad. Such phase modulation leads to ~100% amplitude modulation of the power in the control coupler outputs during m time slot, while the modulation of the high-power radiation in outputs 1 is negligible. The signals acquired by the

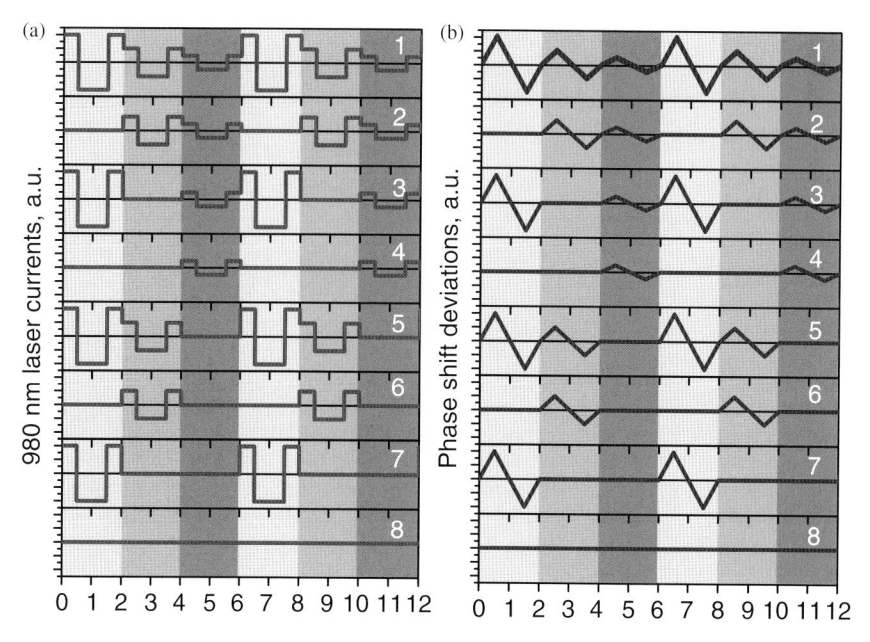

Figure 7.24 Simulated 980 nm laser diode (a) and channel 2 (b) powers for operation of feedback loop in 2^3 amplifier system.

photodetectors of the level m during mth time slot are the result of the superposition between the phase noise and the periodic phase modulation in each amplifier. In the case of right phase matching (when the phase noise is completely compensated), the peaks associated with the positive and negative phase changes belong to the same time slot and have the same amplitude. The presence of uncompensated noise causes the signal peaks belonging to the same time slot spread in two series. The phase mismatching $\varphi_{MIS,p}$ on each coupler is directly proportional to the difference Δ_p between these peaks. The set of these values $\{\Delta_p\}$ is used by PC to reconstruct this mismatching $\{\varphi_{MIS,p}\}$ and generate a set of correction 980 nm powers $\{\delta P_{av,k}\}$ to be added to each amplifier channel.

As a result, time-division multiplexing with n time slots is able to serve beam combining in $(2^n - 1)$ 50/50 couplers, providing the max phase compensation rate $2.6n^{-1}\pi$ rad/ms in each coupler. The algorithm could be naturally extended to two wavelength control systems.

7.4.6
Conclusion

In conclusion, the reported result gives us the basis of the work toward development of the multichannel system shown in Figure 7.17. The method is proved to operate against the noise at a rate of $\sim 2.6\pi$ rad/ms with the control by one optical signal at 980 nm and at the rate above $\sim 10\pi$ rad/ms with control by two optical signals at 980 and 1060 nm. It potentially serves combining of hundreds of amplifiers with noise properties like those shown in Figure 7.22 and opens the potential to produce high-power narrowband radiation through the complete near-infrared employing the YDF phase control in combination with a variety of rare earth-doped and Raman fiber amplifiers, in particular based on LMA fibers. Attractive features of these sources are their compactness, reliability, and all-fiber integrated format.

7.5
Conclusions and Recent Progress

In this chapter, RICs induced in Yb-doped optical fibers by resonance pumping and signal amplification are considered as a side effect of the population inversion mechanism that generally manages operation of the fiber lasers and amplifiers. The fundamentals of the RIC phenomena and details of the effect dynamics are explained in terms of the two-level laser model. The effect is experimentally quantified for commercial aluminum and phosphate silicate fibers demonstrating that the RIC dynamics follows the change of the population of the excited/unexcited ion states with a factor proportional to their polarizability difference. The polarizability difference values for these fibers have been measured.

We propose a simple solution for coherent beam combining of rare earth-doped fiber amplifiers. The RIC effect is employed for an active phase control in all-fiber spliced configuration. A simple algorithm is developed on the basis of the electronic RIC model that allows straightforward implementation of the effect into a feedback loop. Combining of two 500 mW Er-doped amplifiers in a single-mode fiber is demonstrated with optical control by low-power laser diodes. The experimental optical loop is able to operate against the acoustic phase noise within the range of π rad and at a rate of $\sim 2.6\pi$ rad/ms with the control by one optical signal at 980 nm and at the rate above $\sim 10\pi$ rad/ms with control by two optical signals at 980 and 1060 nm. Extension of this approach on N amplifier system is also considered.

Recently, the similar interferometric technique was used for investigation of the thermal and electronic RICs in Er^{3+}/Yb^{3+}-doped fibers [19,52,53] and also in combined Er^{3+}- and Yb^{3+}-doped fibers [54,55]. The electronic RIC value was determined for Yb^{3+}-doped fiber to be about 5×10^{-6} (for doping concentrations of $\sim 3 \times 10^{25}$ ions/m^3) by a spectral interferometer [56]. The contribution of the pump-induced index change to the fiber–amplifier noise and mode instability was discussed [56]. The optical control for the index change (due to pump-controlling population change) was applied for the coherent beam combination of the ytterbium fiber amplifiers [36]. These demonstrations indicate the promising future of the all-optical coherent control for the multichannel fiber amplifier system.

References

1 Richardson, D.J., Nilsson, J., and Clarkson, W.A. (2010) High power fiber lasers: current status and future perspectives. *J. Opt. Soc. Am. B*, **27**, 63–92.

2 Kuzin, E.A., Petrov, M.P., and Fotiadi, A.A. (1994) Phase conjugation by SMBS in optical fibers, in *Optical Phase Conjugation* (eds M. Gower and D. Proch), Springer, pp. 74–96.

3 Ostermeyer, M., Kong, H.J., Kovalev, V.I., Harrison, R.G., Fotiadi, A.A., Mégret, P., Kalal, M., Slezak, O., Yoon, J.W., Shin, J.S., Beak, D.H., Lee, S.K., Lü, Z., Wang, S., Lin, D., Knight, J.C., Kotova, N.E., Sträßer, A., Scheikh-Obeid, A., Riesbeck, T., Meister, S., Eichler, H.J., Wang, Y., He, W., Yoshida, H., Fujita, H., Nakatsuka, M., Hatae, T., Park, H., Lim, C., Omatsu, T., Nawata, K., Shiba, N., Antipov, O.L., Kuznetsov, M.S., and Zakharov, N.G. (2008). Trends in stimulated Brillouin scattering and optical phase conjugation. *Laser Part. Beams*, **26**, 297–362.

4 Fotiadi, A.A., Antipov, O.L., and Mégret, P. (2008) Dynamics of pump-induced refractive index changes in single-mode Yb-doped optical fibers. *Opt. Express*, **16**, 12658–12663.

5 Fotiadi, A.A., Zakharov, N.G., Antipov, O.L., and Mégret, P. (2009) All-fiber coherent combining of Er-doped amplifiers through refractive index control in Yb-doped fibers. *Opt. Lett.*, **34**, 3574–3576.

6 Brown, D. and Hoffman, H. (2001) Thermal, stress, and thermo-optic effects in high average power double-clad silica fiber lasers. *IEEE J. Quantum Electron.*, **37**, 207–217.

7 Davis, M., Digonnet, M., and Pantell, R. (1998) Thermal effects in doped fibers. *J. Lightwave Technol.*, **16**, 1013–1023.

8 Jansen, F., Stutzki, F., Otto, H.-J., Eidam, T., Liem, A., Jauregui, C., Limpert, J., and Tünnermann, A. (2012) Thermally induced waveguide changes in active fibers. *Opt. Express*, **20**, 3997–4008.

9 Digonnet, M.J.F., Sadowski, R.W., Shaw, H.J., and Pantell, R.H. (1997) Resonantly enhanced nonlinearity in doped fibers for low-power all-optical switching: a review. *Opt. Fiber Technol.*, **3**, 44–64.

10 Antipov, O.L., Eremeykin, O.N., Savikin, A.P., Vorob'ev, V.A., Bredikhin, D.V., and Kuznetsov, M.S. (2003) Electronic changes of refractive index in intensively pumped Nd:YAG laser crystals. *IEEE J. Quantum Electron.*, **39**, 910–918.

11 Paschotta, R., Nilsson, J., Tropper, A.C., and Hanna, D.C. (1997) Ytterbium-doped fiber amplifiers. *IEEE J. Sel. Top. Quantum Electron.*, **33**, 1049–1056.

12 Snyder, W. and Love, J.D. (1983) *Optical Waveguide Theory*, Chapman & Hall, London.

13 Jeunhomme, L. (1983) *Single-Mode Fiber Optics*, Marcel Dekker, New York.

14 Bass, M., Van Stryland, E.W., Williams, D.R., and Wolfe, W.L. (1995) *Handbook of Optics*, 2nd edn, McGraw-Hill, New York.

15 Privalko, V.P. (1984) *Handbook of Physical Chemistry of Polymers*, Naukova Dumka, Kiev.

16 Melkumov, M.A., Bufetov, I.A., Kravtsov, K.S., Shubin, A.V., and Dianov, E.M. (2004) *Cross Sections of Absorption and Stimulated Emission of Yb^{3+} Ions in Silica Fibers Doped with P_2O_5 and Al_2O_3*, GPI RAS, Moscow.

17 Carslaw, H.S. and Jaeger, J.C. (1964) *Conduction of Heat in Solids*, Oxford University Press, Moscow.

18 Fotiadi, A.A., Antipov, O.L., Kuznetsov, M.S., Panajotov, K., and Mégret, P. (2009) Rate equation for the nonlinear phase shift in Yb-doped optical fibers under resonant diode-laser pumping. *J. Hologr. Speckle*, **5**, 299–302.

19 Gainov, V.V. and Ryabushkin, O.A. (2011) Effect of optical pumping on the refractive index and temperature in the core of active fibre. *Quantum Electron.*, **41**, 809–814.

20 Dawson, J.W., Messerly, M.J., Beach, R.J., Shverdin, M.Y., Stappaerts, E.A., Sridharan, A.K., Pax, P.H., Heebner, J.E., Siders, C.W., and Barty, C.P.J. (2008) Analysis of the scalability of diffraction-limited fiber lasers and amplifiers to high average power. *Opt. Express*, **16**, 13240–13266.

21 Arkwright, J.W., Elango, P., Atkins, G.R., Whitbread, T., and Digonnet, M.J.F. (1998) Experimental and theoretical analysis of the resonant nonlinearity in ytterbium-doped fiber. *J. Lightwave Technol.*, **16**, 798–806.

22 Wu, B., Chu, P.L., and Arkwright, J.W. (1995) Ytterbium-doped silica slab waveguide with large nonlinearity. *IEEE Photon. Technol. Lett.*, **7**, 1450–1452.

23 Stepanov, S.I., Fotiadi, A.A., and Mégret, P. (2007) Effective recording of dynamic phase gratings in Yb-doped fibers with saturable absorption at 1064 nm. *Opt. Express*, **15**, 8832–8837.

24 Jauregui, C., Eidam, T., Otto, H.-J., Stutzki, F., Jansen, F., Limpert, J., and Tünnermann, A. (2012) Temperature-induced index gratings and their impact on mode instabilities in high-power fiber laser systems. *Opt. Express*, **20**, 440–451.

25 Desurvire, E. (1994) *Erbium-Doped Fiber Amplifiers: Principles and Applications*, John Willey & Sons, Inc., New York.

26 Bochove, E. (2004) Nonlinear refractive index of rare-earth-doped fiber laser. *Opt. Lett.*, **29**, 2414–2416.

27 Barmenkov, Yu.O., and Kir'yanov, A.V., and Andres, M.V. (2004) Resonant and thermal changes of refractive index in a heavily doped erbium fiber pumped at wavelength 980 nm. *Appl. Phys. Lett.*, **85**, 2466–2468.

28 Garsia, H., Johnson, A.M., Oguama, F.A., and Trivedi, S. (2005) Pump-induced nonlinear refractive index change in erbium- and ytterbium-doped fibers: theory and experiment. *Opt. Lett.*, **30**, 1261–1263.

29 Antipov, O.L., Bredikhin, D.V., Eremeykin, O.N., Savikin, A.P., Ivakin, E.V., and Sukhadolau, A.V. (2006) Electronic mechanism of refractive index changes in intensively pumped Yb:YAG laser crystals. *Opt. Lett.*, **31**, 763–765.

30 Margerie, J., Moncorgé, R., and Nagtegaele, P. (2006) Spectroscopic investigation of the refractive index variations in the Nd:YAG laser crystal. *Phys. Rev. B.*, **74**, 235108-10.

31 Messias, D.N., Catunda, T., Myers, J.D., and Myers, M.J. (2007) Nonlinear electronic line shape determination in Yb^{3+}-doped phosphate glass. *Opt. Lett.*, **32**, 665–667.

32 Fotiadi, A.A., Petukhova, I., Mégret, P., Shubin, A.V., Tomashuk, A.L., Novikov, S.G., Borisova, C.V., Zolotovskiy, I.O., Antipov, O.L., Panajotov, K., and Thienpont, H. (2012) Monitoring of gamma-irradiated Yb-doped optical fibers

through pump-induced refractive index changes. *Proc. SPIE*, **8439**, 84390G.

33 Antipov, O.L., Belyaev, S.I., Kuzhelev, A.S., and Chausov, D.V. (1998) Resonant two-wave mixing of optical beams by refractive index and gain gratings in inverted Nd:YAG. *J. Opt. Soc. Am. B*, **15**, 2276–2281.

34 Soulard, R., Moncorgé, R., Zinoviev, A., Petermann, K., Antipov, O., and Brignon, A. (2010) Nonlinear spectroscopic properties of Yb^{3+}-doped sesquioxides Lu_2O_3 and Sc_2O_3. *Opt. Express*, **18**, 11173-80.

35 Soulard, R., Brignon, A., Huignard, J.-P., and Moncorgé, R. (2010) Nondegenerate near-resonant two-wave mixing in diode pumped Nd^{3+} and Yb^{3+} doped crystals in presence of athermal refractive index grating. *J. Opt. Soc. Am. B*, **27**, 2203–2210.

36 Tünnermann, H., Feng, Y., Neumann, J., Kracht, D., and Weßels, P. (2012) All-fiber coherent beam combining with phase stabilization via differential pump power control. *Opt. Lett.*, **37**, 1202–1204.

37 Agrawal, G.P. (2001) *Nonlinear Fiber Optics*, Third edition Academic Press, Boston, Mass.

38 Fotiadi, A.A. and Kiyan, R.V. (1998) Cooperative stimulated Brillouin and Rayleigh backscattering process in optical fiber. *Opt. Lett.*, **23**, 1805–1807.

39 Fotiadi, A.A., Kiyan, R., Deparis, O., Mégret, P., and Blondel, M. (2002) Statistical properties of stimulated Brillouin scattering in singlemode optical fibers above threshold. *Opt. Lett.*, **27**, 83–85.

40 Fotiadi, A.A., Mégret, P., and Blondel, M. (2004) Dynamics of self-Q-switched fiber laser with a Rayleigh-stimulated Brillouin scattering ring mirror. *Opt. Lett.*, **29**, 1078–1080.

41 Fotiadi, A.A. and Mégret, P. (2006) Self-Q-switched Er–Brillouin fiber source with extra-cavity generation of a Raman supercontinuum in a dispersion shifted fiber. *Opt. Lett.*, **31**, 1621–1623.

42 Fan, T.Y. (2005) Laser beam combining for high-power, high-radiance sources. *IEEE J. Sel. Top. Quantum Electron.*, **11**, 567–577.

43 Bruesselbach, H., Jones, D.C., Mangir, M.S., Minden, M., and Rogers, J.L. (2005) Self-organized coherence in fiber laser arrays. *Opt. Lett.*, **30**, 1339–1341.

44 Huo, Y. and Cheo, P.K. (2005) Analysis of transverse mode competition and selection in multicore fiber lasers. *J. Opt. Soc. Am. B*, **22**, 2345–2349.

45 Bellanger, C., Brignon, A., Colineau, J., and Huignard, J.P. (2008) Coherent fiber combining by digital holography. *Opt. Lett.*, **33**, 293–295.

46 Augst, S.J., Fan, T.Y., and Sanchez, A. (2004) Coherent beam combining and phase noise measurements of Yb fiber amplifiers. *Opt. Lett.*, **29**, 474–476.

47 Fotiadi, A.A., Zakharov, N.G., Antipov, O.L., and Mégret, P. (2008) All-fiber coherent combining of Er-doped fiber amplifiers by active resonantly induced refractive index control in Yb-doped fiber. Conference on Lasers and Electro-Optics, San Jose, CA, May 4–9, 2008, Paper CWB2.

48 Bruesselbach, H., Wang, S., Minden, M., Jones, D.C., and Mangir, M. (2005) Power-scalable phase-compensating fiber-array transceiver for laser communications through the atmosphere. *J. Opt. Soc. Am. B*, **22**, 347–353.

49 Goodno, G.D., Book, L.D., and Rothenberg, J.E. (2009) Low-phase-noise, single-frequency, single-mode 608 W thulium fiber amplifier. *Opt. Lett.*, **34**, 1204–1206.

50 Taylor, L., Feng, Y., and Calia, D.B. (2009) High power narrowband 589 nm frequency doubled fibre laser source. *Opt. Express*, **17**, 14687–14693.

51 Fotiadi, A.A., Kuzin, E.A., Petrov, M.P., and Ganichev, A.A. (1989) Amplitude-frequency characteristic of an optical-fiber stimulated Brillouin amplifier with pronounced pump depletion. *Sov. Tech. Phys. Lett.*, **15**, 434–436.

52 Gainov, V.V., Shaidullin, R.I., and Ryabushkin, O.A. (2011) Steady-state heating of active fibres under optical pumping. *Quantum Electron.*, **41**, 637–643.

53 Gainov, V.V. and Ryabushkin, O.A. (2012) Kinetics of index change in core of active fibers doped by Yb^{3+} and Er^{3+} ions under optical pumping. *Opt. Spectrosc.*, **112**, 510–518.

54 Schimpf, D.N., Seise, E., Jauregui-Misas, C., Nodop, D., Limpert, J., and Tünnermann, A. (2011) Refractive index changes due to gain/absorption in Yb-

doped fibers. SPIE Photonics West 2011, Paper 7914-50.

55 Tünnermann, H., Neumann, J., Kracht, D., and Wessels, P. (2011) All-fiber phase actuator based on an erbium-doped fiber amplifier for coherent beam combining at 1064 nm. *Opt. Lett.*, **36**, 448–450.

56 Tünnermann, H., Neumann, J., Kracht, D., and Weßels, P. (2012) Gain dynamics and refractive index changes in fiber amplifiers: a frequency domain approach. *Opt. Express*, **20**, 13551–13559.

8
Coherent Beam Combining of Pulsed Fiber Amplifiers in the Long-Pulse Regime (Nano- to Microseconds)

Laurent Lombard, Julien Le Gouët, Pierre Bourdon, and Guillaume Canat

8.1
Introduction

Long-pulse fiber amplifiers based on a master oscillator (MO) and a pulse shaper are versatile sources, as they allow adjusting or switching parameters such as the oscillator spectral purity, pulse shape, and repetition rate. These sources have many light detection and ranging (LIDAR or LADAR) applications. However, increasing the peak power is complicated by nonlinear effects. Coherent beam combining (CBC) is a way to circumvent this limitation and scale the emitted power.

This chapter investigates the technological issues of active coherent combining of pulsed fiber sources in the long-pulse regime (nano- to microsecond). Various CBC techniques are reviewed and compared in case of pulsed operation. A criterion is proposed to select suitable techniques based on the number of sources and noise bandwidth (BW). Then, pulsed fiber amplifiers main characteristics and limits are discussed. Measurements of the random phase noise and the deterministic in-pulse phase distortion are reported and compared with a theoretical model. In particular, the influence of the in-pulse phase distortion is discussed in both the stimulated Brillouin scattering (SBS) and stimulated Raman scattering (SRS) peak power–limited regimes. Experimental demonstration of CBC of two long-pulse fiber amplifiers with 95% efficiency is also reported.

Over the last decade high-power fiber lasers have undergone a tremendous growth, in terms of both the overall performances and diversity of addressed applications.

This growth has been initiated in the 1980s and 1990s by the development, for optical telecommunications and ultrafast data transfer, of high-transmittance, high-gain, and excellent optical quality rare earth-doped silica fibers for laser signal amplification [1]. During the last few years, manufacturing processes for these laser fibers and related components have undergone great improvement.

After the blowup of the telecommunication industry bubble, alternative domains of application have been investigated to take advantage of the available technologies at eye-safe wavelength, such as LIDAR. Coherent LIDAR systems have since become widespread velocity and distance measurement tools, used for diverse purposes such as wind profiling [2], vibrometry [3,4], vehicles velocimetry [5,6], and high-resolution telemetry [7].

The invention of double-clad fibers was a major milestone in the power increase of fiber sources, unlocking the technological capability to couple extremely high levels of pump power in rare earth-doped optical fibers [8,9]. Nowadays, powers exceeding 20 kW can be delivered by a single continuous wave (CW) ytterbium-doped fiber-amplifying chain [10,11].

Fiber lasers and amplifiers have multiple qualities, setting them apart from the bulk solid-state lasers and suffer less thermal effects than rod, slab, or disk lasers. The main qualities of fibers are their simultaneous high length and flexibility; therefore, the pump absorption and laser gain can be spread over a few meters, while they can be coiled and conveniently packaged in a small volume. Given the high length of doped fibers, the laser gain can be high, as well as the surface for thermal exchange. This allows building very efficient high-power laser sources where heat transfer is easier to manage. This very long active medium is also a waveguide that spatially filters the transverse modes [12]. Thus, fiber sources preserve beam quality more easily than other solid-state lasers. In fiber lasers (i.e., doped fiber inside a cavity), the high length of the active medium also results in spectral narrowing. Finally, in most sources with wavelength in 1–2 μm range, fiber sources can be monolithic, that is, comprise solely fiber components [13]. Such all-fiber systems, once the components are spliced together, are rugged and compact, and very promising for integration on various platforms or use in adverse environments.

However, while rare earth-doped fiber can make excellent CW sources, their advantages can become drawbacks when they are operated in pulsed regime. Indeed, the energy accumulation in this regime induces peak powers that are some orders of magnitude higher than in CW regime. In addition, due to strong confinement of amplified power in the fiber core, the light intensity can become dramatically high, giving rise to nonlinear effects or even leading to damage of the fiber.

Nonlinear effects such as SBS and SRS are strong limitations to the level of peak power emitted by a fiber source, especially for narrow-linewidth amplifiers [14]. As will be detailed in Section 8.4, there are three main options to mitigate these effects: increase the mode area, decrease the interaction length, or lower the nonlinear gain. Several techniques have been developed accordingly, such as large-mode-area (i.e., large core) fibers that maintain single-mode guiding [15], high-order mode guiding [16], high-concentration doping in nonsilica glass fibers [17,18], acoustic wave antiguiding [19], and temperature or strain gradients [20]. To date, some of these techniques pushed the peak power to about 2 kW for diffraction-limited erbium-doped fiber amplifiers (EDFAs) with 100 ns pulse duration [21,22]. For ytterbium doping, the peak powers can reach some tenth of kilowatt [23–25].

One example of typical applications of these state-of-the-art sources is coherent LIDAR measurement, which relies on long-coherence laser pulses. The principle of the measurement is to infer distance information from time of flight and velocity from Doppler shift [26,27]. For this application, high-energy pulses are required to improve the detection range, allowing mapping of atmospheric turbulences like wake vortices or wind shears [28]. These induce hazardous velocity variations for aircrafts approaching airports at low speed and low altitude. In this case, the required velocity resolution is typically 1 m/s (5 km/h). For an eye-safe laser source with wavelength 1.5 µm, this corresponds to a spectral resolution of 1 MHz; and for a Fourier-limited pulse shape, this corresponds to pulse durations of about 1 µs. Typically, pulse energies of 1 mJ at a repetition rate of 5 kHz would enable measurements at more than 10 km for a reasonable speed of 3 horizon scan/min. Narrow linewidth fiber sources have already reached energies of the millijoule class [29,30].

Once nonlinear effects have been reduced through some mitigation technique, if further increase of the peak power is still needed, CBC represents an ultimate solution. The first combining technique that was demonstrated was spectral combining, which is basically wavelength-multiplexed combination [31,32]. Another technique consists in overlapping independent beams from broadband (i.e., low-coherence) lasers in the far field. Both methods demonstrated capability to deliver multikilowatt levels of combined power. However, the resulting broad spectrum makes it incompatible with the previous applications.

Coherent combining may be simply realized by a passive technique. Here, various coherent sources operating at the same wavelength are multiplexed over separated amplifying media and mixed together through a final common optical coupling component [33]. Unfortunately, the damage threshold of this final coupler can be a strong limitation to the level of combined power.

The most versatile approach is coherent combining with active phase control, which requires MOPA laser architecture. The common MO is split into multiple optical paths, each path being amplified by its own amplifier. Then, an active component (most generally an electro-optic phase shifter) is added prior to each amplifier. The phase difference between each amplified path and a reference path is measured, and the proper phase shifts are applied in real time by the electro-optic modulators, (EOMs) in order to phase lock all power amplifier outputs.

This active phase control can be implemented in a configuration where all power amplifier outputs are laid out side by side (tiled aperture). The laser beams are spatially separated in the near field, but they overlap in the far field with an interference pattern. The optimal phase locking of all amplifiers is obtained when the intensity is maximal in the central lobe of the interference pattern. Moreover, if the near-field fill factor is high enough (i.e., the empty space between the adjoining laser beams is minimized), power dissipation in the side lobes of the pattern can be efficiently alleviated.

Let us note that coherent combining in tiled aperture offers further possibilities. Indeed, if the phase piston of each amplifier is separately controlled, coherent adaptive optics, beam steering, or wavefront shaping is also at hand. Target-in-the-loop phase locking (i.e., maximization of the optical intensity on a remote target) can also be achieved. This technique relies on collecting phase difference information from processing the optical signal backscattered by the remote target.

For CW signals, a remarkable power of 105 kW has been obtained in 2009 through active coherent combining of seven Yb:YAG slab amplifiers [34]. With fiber amplifiers, more than 1 kW power has been demonstrated in 2011 [35,36].

The demonstration of coherent combining of pulsed beams has represented a noticeable step, as will be detailed in this chapter. In 2010, an active combining technique (i.e., where the phase of each combined laser source is dynamically controlled) has been successfully tested to perform coherent combining of pulsed fiber amplifiers in long-pulse regime ($\sim$100 ns) [37] and in short-pulse regime ($\sim$100 fs) [38,39].

This chapter investigates the technological issues of active coherent combining of pulsed fiber sources in the long-pulse regime. Section 8.2 benchmarks the various existing techniques and configurations for CBC with active phase control. Limitations on the number of sources and operability of these configurations in the case of pulsed lasers are discussed. Section 8.3 introduces the theory of pulsed amplification in fiber media and describes the impact of gain saturation on the emitted pulse shape. Section 8.4 presents the main nonlinear effects that may limit energy-per-pulse scaling in pulse fiber sources. Section 8.5 focuses on the study of the physical phenomena that can induce phase fluctuations in a pulsed fiber amplifier. Phase noise and in-pulse phase distortion measurements are confronted with theoretical considerations. Experimental demonstration of CBC of two long-pulse (100 ns) fiber amplifiers is finally presented and discussed in Section 8.6. As a conclusion, alternative techniques for energy-per-pulse scaling are proposed in Section 8.7 and avenues for CBC of shorter pulses (down to 1 ns duration) and operation at higher levels of peak power are suggested in Section 8.8.

8.2
Beam Combining Techniques

We discuss in this part the main techniques to achieve active CBC of two or more fiber amplifiers in a master oscillator power fiber amplifier (MOPFA) configuration (CW or pulsed operation). A narrow-linewidth MO is split between several arms, each one including a phase modulator and one or more fiber amplifiers. CBC is performed by stabilizing amplified signal interferences at the output to a constructive or a destructive state. The output beam should have the spatial (e.g., single-mode), spectral (e.g., single-frequency), and polarization (e.g., linear polarization) properties of a single-arm beam, while summing up the powers of all arms.

Depending on how the amplifier outputs are geometrically combined, the various CBC techniques can be divided into two main categories: the tiled aperture configuration and the filled aperture configuration. Similarly, the phase locking techniques can be divided into two categories: direct or indirect. In the direct techniques, all arm phases are measured before compensation, whereas in the indirect technique, only a single intensity detector (power-in-the-bucket) is used before compensation.

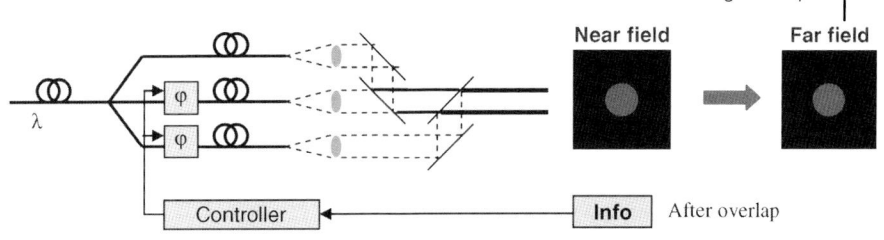

Figure 8.1 Filled aperture configuration.

8.2.1
Filled and Tiled Apertures

Let us assume that the beams to be coherently combined are all single mode, with the same size and divergence. The beams can be overlapped so as to create single-mode beams of the same size and divergence as each individual beams (filled configuration) or placed side by side so as to create a larger beam but with lower divergence (tiled configuration).

In the case of the filled aperture configuration, the beams overlap on a set of beam splitters or on a grating and interfere so that most of the energy is in a single-mode beam (Figure 8.1). The resulting beam is single mode in the near field as well as in the far field. When the phase differences are servo-locked to zero, the power of the output beam is maximized (the other outputs of the interferometer are minimized). Information on individual amplifier phase has to be extracted from the "overlapped" beam.

In the case of the tiled aperture configuration, the output beams are placed side by side. The beams usually overlap in the far field and not in the near field at the amplifiers output. In the example illustrated in Figure 8.2, the far-field pattern is the result of the interference of the various beams and corresponds to the Fourier transform of the near-field pattern. In the case of a square matrix of Gaussian beams,

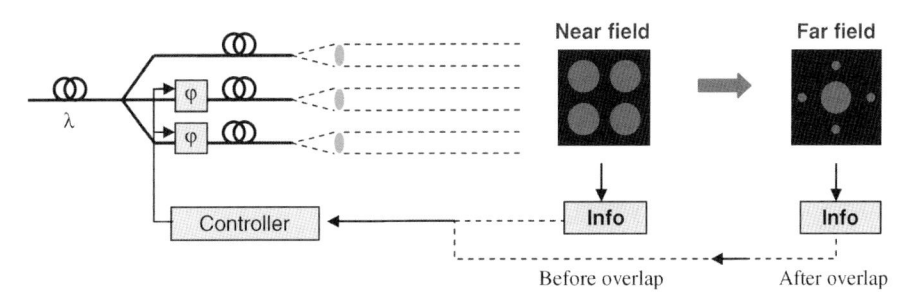

Figure 8.2 Tiled aperture configuration.

the far field is made of a central lobe and side lobes. When the phases are locked, the power in the central lobe is maximized and the power in the side lobes is minimized. The information for the phase of each individual amplifier can be either extracted from the nonoverlapped beams (near field) or from the "overlapped" beam (far field).

The filled configuration is more efficient in terms of output beam quality, but requires that the final overlapping element must bear all the power. On the other hand, the tiled configuration is simpler to implement and easier to align, but yields undesirable side lobes.

8.2.2
Locking Techniques

The goal of all CBC techniques is to maintain to zero the phase difference between the amplifiers, so as to maximize the power of the output beam (filled aperture) or the power of the central lobe power (tiled aperture). Thus, phase locking can be obtained either by monitoring the phase difference between the amplifiers or by maximizing the power of the output beam or central lobe. Two families of CBC techniques can be distinguished:

- Direct measurement of the phase difference between the interfering arms, sometimes with an additional reference arm. These techniques can be referred to as "direct phase locking techniques."
- Phase corrections to optimize a performance criterion, usually the total power-in-a-bucket detector. They can be referred to as "indirect phase locking techniques."

8.2.2.1 Direct Phase Locking Techniques

The direct techniques require spatial separation of the beams, since each arm must be compared with its neighbor or with a reference arm, in order to obtain the set of relative phases $\{\varphi_1, \ldots, \varphi_N\}$ to be zeroed. Figure 8.3 shows typical configurations,

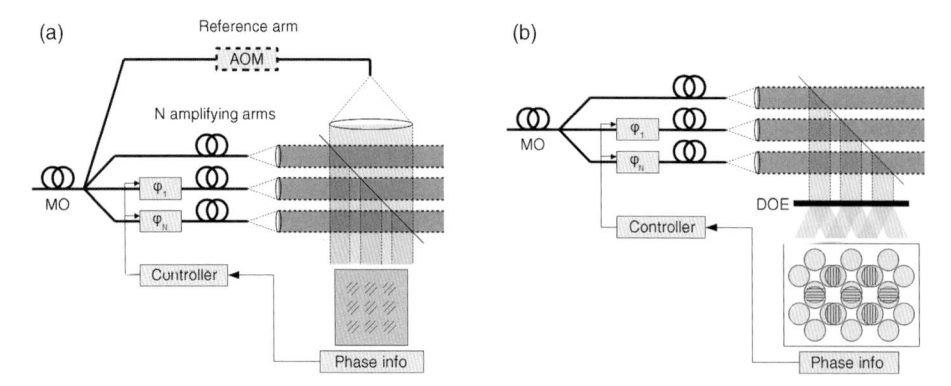

Figure 8.3 Typical direct phase locking technique with a reference arm (a) and without a reference arm (b) DOE: diffractive optical element.

with or without reference arm. Part (a) of the figure has been implemented either with a shifted [40] or nonshifted reference arm [41].

The frequency shift of the reference arm can be obtained with an acousto-optic modulator (AOM). The fringes then move at the AOM frequency (F_{AOM}) and a fast detector associated with phase demodulators is used to determine the oscillating fringes phase (*heterodyne phase measurement*). Other setups use nonshifted reference. In this case, the fringes move at the amplifier-induced phase noise frequency (~0.01–1 kHz range) and can be measured with a camera (*fringe position*). Figure 8.3b can be viewed as a self-referenced version of the left scheme, where fringe patterns in both directions are measured between each beam and its nearest neighbors [*quadri-wave lateral shearing interferometer*]. A matrix of phase gradient in both directions is extracted and inverted to recover the relative phase differences between the various arms. The global piston is not measured in this case and cannot be corrected.

These techniques are well adapted to tiled aperture configuration. The techniques using a camera are also referred to as "collective" techniques as the complexity of the technique is only slightly dependent on the number of fibers.

The Hänsch–Couillaud interferometer is another direct phase locking technique, which takes advantage of polarization to provide simple measurements of the phase difference. However, it requires a polarization analyzer per channel and thus is limited to a low number of amplifiers (*Hänsch–Couillaud interferometer*) [42].

8.2.2.2 Indirect Phase Locking Techniques

In this case, the only available information is usually a power-in-the-bucket intensity that the controller tries to maximize. Figure 8.4 shows two typical implementations of the indirect measurement in filled and tiled aperture pupils. This method is compatible with both configurations, but requires large modulation and detection BWs.

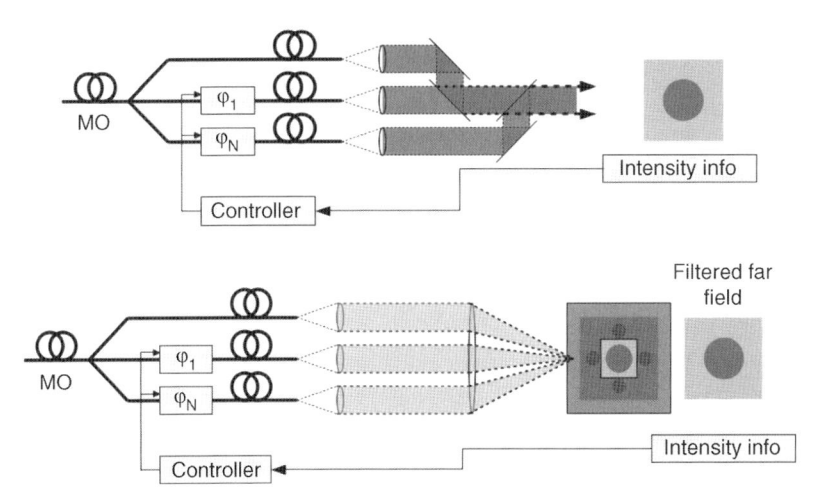

Figure 8.4 Possible configurations for indirect phase measurement. In both cases, the total output power is the information used for phase locking.

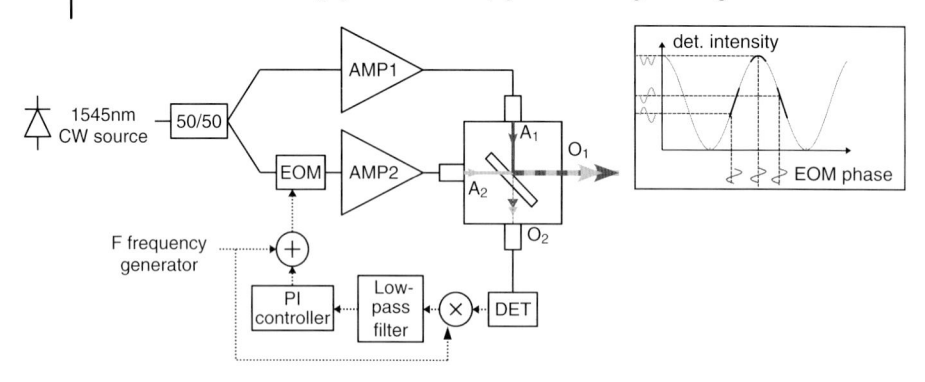

Figure 8.5 LOCSET implementation on a two-arm system.

Two main phase locking methods are commonly used: LOCSET (locking of optical coherence by single-detector electronic-frequency tagging) and SPGD (stochastic parallel gradient descent). In both cases, the useful information must be extracted from a single data: the total power through a hole. The useful information is the $N-1$ relative phases of the $N-1$ arms related to the reference arm.

In the LOCSET technique, the $N-1$ relative phases are measured independently, thanks to a frequency tagging of the various arms. In the simplest case of two arms ($N=2$), Figure 8.5 shows a basic LOCSET implementation: A small phase modulation at frequency F is applied to one of the arms, resulting in a small-intensity modulation in the combined beam. The modulation frequency F is much larger than the maximum phase noise frequency. The modulation amplitude is much smaller than 2π, typically $2\pi/50$. This modulation is detected via a single photodetector. As seen in the inset of Figure 8.5, if the phase difference yields a maximum of intensity, the modulation frequency F of the signal will be converted into $2F$ for the photocurrent. If the phase difference shifts the signal away from the maximum, left or right, the photocurrent will keep the frequency modulation F. Whether the sign of the required correction is positive or negative is determined by the phase difference of 0 or π between the photocurrent and the modulation signal at frequency F.

The output of the low-pass (LP) filter is an error signal proportional to the phase difference $\Delta\varphi$. A simple proportional integrator (PI) controller can then be used to lock $\Delta\varphi$ to 0 and stabilize constructive or destructive interferences on the detector. Frequency multiplexing is used in the case of a large number of amplifiers: Each path is controlled using a unique tagging frequency. Note that time multiplexing has also been proposed for CBC [43].

In the case of N arms, this technique requires 1 detector, $N-1$ frequency generators, multiplexers, and PI controllers. Let us assume that the controllers are digital PI with a cycle time T_{PI} and cycle frequency $f_{PI}=1/T_{PI}$. Digital PI controllers usually present a 63% rise time of $\tau_{rise}=3\times T_{PI}$ and a $-3\,dB$ cutoff frequency of $f_{-3\,dB}\sim f_{PI}/10$. The noise at $f_{-3\,dB}$ will thus be reduced by 3dB with a PI controller with cycle frequency $f_{PI}=10\times f_{-3\,dB}$. If the LP filters are chosen, for example, to cut all frequencies above $f_{PI}/2$, the modulation frequencies $\{F_n\}_{1 \leq n \leq N-1}$ can be spaced by f_{PI}. The modulation frequencies must also be contained within one octave: $F_{N-1} < 2 \times F_1$ to avoid overlap in the intensity spectrum. Thus, for N arms, a choice of modulation frequencies can be $F_n = (N + n - 1) \times f_{PI}$, so that $F_1 = N \times f_{PI}$ and $F_{N-1} = (2 \times N - 2) \times f_{PI}$.

In the SPGD algorithm, the $N - 1$ phase modulators are directly set by a fast controller. One iteration of the algorithm starts with a $N - 1$ phase vector on which a small random perturbation is applied. The new vector yields a new detected intensity. If the intensity is higher, the new vector is kept; otherwise, the perturbation is discarded. The algorithm ultimately converges to the vector maximizing the intensity. The 10–90% rise time is roughly $\tau_{rise} = 10 \times N_{fiber}/f_{cycle}$, leading to a cutoff frequency $f_{SPGD} \sim 10 \times N \times f_{-3 dB}$. Note that some authors [44] show that the rise time could be cut by a factor of 2 to 3 with optimized SPGD algorithms.

8.2.3
Requirements of Various Techniques

This section presents the requirements for a simple setup with N fibers, with phase noise maximum frequency f_{noise}. Actuator BWs are assumed to cover at least the noise frequency up to f_{noise}. In the case of direct detection of individual phase, the controller is assumed to be a PI controller with $f_{PI} \sim 10 \times f_{noise}$. The maximum frequency f_{max} involved in each case is indicated. It will be useful to determine the possible techniques in pulsed operation.

8.2.3.1 Indirect Phase Locking Techniques

Indirect phase locking technique requirements are listed in Table 8.1. The table lists the required number of modulators and their operating frequencies or BW. f_{max} is the maximum frequency involved. The choice of frequencies has been discussed in Section 8.2.2.2.

Some examples are shown in Table 8.2. With a large number of fibers, the required frequency for the modulators scales with the number of fibers, so the

Table 8.1 Indirect phase locking technique requirements.

		Number	Frequencies	Max frequency f_{max}
Frequency tagging (LOCSET)	Modulators	$N-1$	$F_1 = 10 \times N \times f_{noise}$ $F_{N-1} = 10 \times (2 \times N - 2) \times f_{noise}$	$\sim 20 \times N \times f_{noise}$
	Detector	1	$BW > 20 \times N \times f_{noise}$	
	Controllers	$N-1$	$f_{PI} = 10 \times f_{noise}$	
Intensity optimization (SPGD)	Modulators	$N-1$	$BW > 10 \times N \times f_{noise}$	$\sim 10 \times N \times f_{noise}$
	Detector	1	$BW > 10 \times N \times f_{noise}$	
	Controller	1	$f_{SPGD} = 10 \times N \times f_{noise}$	

Table 8.2 Numerical values for Table 8.1 in various configurations.

		$N=3$, $f_{noise}=1$ kHz		$N=20$, $f_{noise}=1$ kHz		$N=10\,000$, $f_{noise}=1$ kHz	
		Nb	f_{max} (kHz)	Nb	f_{max} (kHz)	Nb	f_{max} (kHz)
Frequency tagging (LOCSET)	Modulators	2	60	19	400	10^4	200 000
	Detector	1	60	1	400	1	200 000
	Generators, mixer, and controllers	2	10	19	10	10^4	10
Intensity optimization (SPGD)	Modulators	2	30	19	200	10^4	100 000
	Detector	1	30	1	200	1	100 000
	Controller	1	30	1	200	1	100 000

Cells with technically challenging figures are highlighted in gray. Nb: number.

number of modulators times their frequency scales with N^2. This is the main limitation of these techniques. Both techniques roughly require the same set of modulators, the main difference being that the LOCSET technique uses multiple PI controllers in parallel, while the SPGD technique requires one fast controller. The cells that will lead to technical difficulties are highlighted.

8.2.3.2 Direct Phase Locking Techniques

Direct phase locking technique requirements are listed in Table 8.3. The camera used in the collective methods is referred to as "matrix."

Some examples are shown in Table 8.4. For a very large number of fibers, the collective techniques are attractive, but require high repetition rate cameras.

8.2.4
Case of Pulsed Laser

In the case of pulsed lasers, several issues have to be considered: the short-pulse synchronization, the phase variation inside the pulses, and the direct compatibility of the CW phase control techniques with pulse operation.

A requirement for CBC to work is that the different paths interfere with a good contrast. The phase difference between the combined amplifiers should thus be short compared to the coherence length. Even if no special effort is made for the amplifier length, we can assume that the length difference between the arms is lower than ~1 m in free space. In the case of long narrow-linewidth pulses, the coherence length is usually equal to the physical pulse length. For example, pulse duration of 100 ns corresponds to a pulse length of 30 m, which is large compared to the paths length difference. On the contrary, for short-pulse amplifier (~1 ns and below pulse duration), keeping good interference contrast

Table 8.3 Direct phase locking technique requirements.

Technique	Device	Nb	Required BW or f_{Pl}	Max frequency f_{max}
Heterodyne phase measurement with one frequency-shifted (by F_{AOM}) reference arm ($F_{AOM} \gg 10 \times f_{noise}$)	Modulators	N	$10 \times f_{noise}$	F_{AOM}
	Detectors	N	F_{AOM}	
	Mixers and controllers	N	$10 \times f_{noise}$	
Quadri-wave lateral shearing interferometer with a diffractive element	Modulators	N	$10 \times f_{noise}$	$10 \times f_{noise}$
	Fast camera and image processing controller	1 matrix	$10 \times f_{noise}$	
Fringe position with one reference arm	Modulators	N	$10 \times f_{noise}$	$10 \times f_{noise}$
	Fast camera and image processing controller	1 matrix	$10 \times f_{noise}$	
Hänsch–Couillaud interferometer	Modulators	$N-1$	$10 \times f_{noise}$	$10 \times f_{noise}$
	HC interferometer and detectors	$2 \times (N-1)$	$10 \times f_{noise}$	
	Controllers	$N-1$	$10 \times f_{noise}$	

F_{AOM}: AOM frequency shift, in the case of heterodyne phase measurement. Nb: number.

would require balancing the amplifier paths within centimeter or millimeter range.

Pulse beam combining is also sensitive to any phase fluctuation arising during the pulses (usually shorter than $\sim$10 µs). These very fast phase fluctuations cannot be easily compensated by a controller. Fortunately, if the amplifiers are identical, the in-pulse phase fluctuations are common to all amplifiers and automatically compensate. However, it is useful to quantify these phase fluctuations: Measurements and theoretical considerations are presented in Section 8.5. More generally, Section 8.5 is dedicated to the phase noise to be compensated in fiber amplifiers, both global and specific to long-pulse amplifiers. It should be noted that in the case where the in-pulse phase fluctuations are deterministic, they can be precompensated by fast phase modulators. For example, Palese *et al.* succeeded in applying 15 rad in less than 1 ns with multiple passes through a 40 GHz phase modulator [45].

Assuming that the pulses are synchronized and that the in-pulse phase variations are either common to all amplifiers or precompensated, different techniques can be applied for CBC in the long-pulse regime. Three situations can be identified,

Table 8.4 Numerical values for Table 8.3 in various configurations.

		N = 3, f_{noise} = 1 kHz		N = 20, f_{noise} = 1 kHz		N = 10 000, f_{noise} = 1 kHz	
		Nb	f_{max} (kHz)	Nb	f_{max} (kHz)	Nb	f_{max} (kHz)
Heterodyne phase measurement with one frequency-shifted reference arm (F_{AOM} = 40 MHz)	Modulators	3	10	20	10	10^4	10
	Detectors	3	40 000	20	40 000	10^4	40 000
	Mixers and controllers	3	10	20	10	10^4	10
Hänsch–Couillaud interferometer	Modulators	2	10	20	10	10^4	10
	HC interferometer and detectors	4	10	40	10	2.10^4	10
	Controllers	2	10	20	10	10^4	10
Quadri-wave lateral shearing interferometer	Modulators	3	10	20	10	10^4	10
	Fast camera and image processing controller	1	10	1	10	1	10
Fringe position	Modulators	3	10	20	10	10^4	10
	Fast camera and image processing controller	1	10	1	10	1	10

Cells with technically challenging figures are highlighted in gray. Nb: number.

depending on the value of the pulse repetition frequency (PRF) with respect to the double of the maximum frequency $2 \times f_{max}$ (Shannon):

- CW or high repetition rate (PRF > $40 \times N \times f_{noise}$): Compatibility with all techniques is straightforward, in particular LOCSET and SPGD. A LP filter removing any frequency above $2 \times f_{max}$ will allow considering the emission as CW by the controller. If PRF > $20 \times N \times f_{noise}$, only SPGD is compatible.
- $20 \times f_{noise}$ < PRF < $20 \times N \times f_{noise}$: Compatibility with direct detection techniques is straightforward.
- PRF < $20 \times f_{noise}$: No straightforward compatibility with CW techniques.

Note that the pulse duration has no impact on the choice of the controller technique, except only on the pulse synchronization requirement. In the cases where compatibility with CW techniques is not straightforward, it is sometimes possible to use these techniques anyway. Additional measurements are then necessary, like the phase in-between the pulses. An example will be shown in Section 8.6 where a PRF of 10 kHz will be used with f_{noise} = 3 kHz.

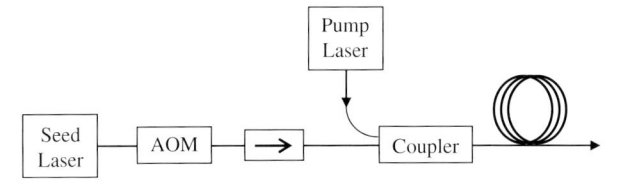

Figure 8.6 Scheme of a generic forward pumped pulsed fiber amplifier. The acousto-optic modulator (AOM) is used to shape the pulse, which can precompensate for pulse distortion (see Section 8.3.4).

8.3 Amplification of Optical Pulse in Active Fiber

A laser amplifier obviously increases the peak power of the incident pulse, but can also affect its temporal shape, due to gain depletion, its spectral distribution, and nonlinear effects. Optical fibers exhibit particular features in this respect, since the population inversion is generally not uniform along the active medium. This part is dedicated to the understanding of the pulse power evolution during the propagation in the fiber. The generic structure of a pulsed fiber amplifier is sketched in Figure 8.6.

Another particular aspect of amplification in optical fibers that will be addressed here is the amplified spontaneous emission (ASE), where the initially isotropic fluorescence is guided in the directions of the fiber and gets amplified, thus depleting the gain [46].

8.3.1 Approximations and Validity Domain of the Calculation

In the most general case, the calculation of the response of an amplifying or absorbing medium to a coherent light field requires the resolution of the Maxwell and optical Bloch equations. We focus here on the case of the incoherent interaction between the optical signal and the resonant medium, which corresponds to the case where the pulse duration τ_p is much longer than the relaxation time T_2 of the atomic polarization. Therefore, the atomic polarization "follows" instantaneously the driving source field [47]. In this case, the evolution of the populations can be described by a set of rate equations that are derived from the energy conservation rule and that only take into account the three contributions of absorption, stimulated emission, and spontaneous emission [48].

In addition, we consider optical pulses that are short compared to the excited-state lifetime T_1. In other terms, the calculation applies to a gain medium that has been already optically pumped and that does not experience spontaneous emission during the pulse propagation. The boundaries of the calculation presented here are thus given by the condition $T_2 \ll \tau_p \ll T_1$. For the example of erbium in aluminosilica fibers, the radiative lifetime is $T_1 \sim 10$ ms and the homogeneous dephasing time is $T_2 \sim 100$ fs at room temperature [49].

Finally, we consider that over the spectral range contained in the pulse, the refractive index of the medium is constant or varies most linearly with the optical frequency. Therefore, group velocity or gain dispersion effects are neglected and cannot be the cause of pulse expansion or compression. In practice, this approximation corresponds to the case of incoming light with narrow spectral width or of short propagation length. Other nonlinear effects such as SBS and SRS or four-wave mixing are neglected in this section, but will be discussed in the next one.

8.3.2
Pulse Propagation in the Resonant Medium

As an optical pulse propagates in the amplifying medium, the peak power increases while the available gain is depleted. This leads to a progressive reduction of the population inversion and lowers the gain for the end of the pulse. The equations describing the variations of the population density $N_2(z,t)$ in the upper level and of the intensity $I(z,t)$ have been extensively described in Refs [50,51] in the case of fiber amplifiers. For pulse duration $\tau_p \ll T_1$ (no feeding or loss mechanism), the equations become

$$\frac{\partial I(z,t)}{\partial t} + c\frac{\partial I(z,t)}{\partial z} = cI(z,t)[\sigma_{em}N_2(z,t) - \sigma_{abs}(N_0 - N_2(z,t))] \tag{8.1}$$

and

$$\frac{\partial N_2(z,t)}{\partial t} = -\frac{I(z,t)}{\hbar\omega}[\sigma_{em}N_2(z,t) - \sigma_{abs}(N_0 - N_2(z,t))], \tag{8.2}$$

where the cross sections σ_{abs} and σ_{em} refer respectively to the absorption and the stimulated emission at the laser frequency ω. The total population N_0 of active ions is considered constant over the amplifier length.

For an arbitrary waveform, the resolution of the coupled Eqs. (8.1) and (8.2) is rather complex, as can be found in Ref. [52]. We now consider the time frame that follows the pulse at the phase velocity c, such that the envelope is always centered at $t = 0$. Here, the signal at position z, in an amplifier with initial population inversion $\Delta_0(z) = N_2(z, -\infty) - N_0(\sigma_{abs}/\sigma_S)$ and input intensity $I_0(t)$, is given by

$$I(z,t) = \frac{I_0(t)}{1 - \left[1 - \exp\left(-\sigma_S \int_0^z \Delta_0(z')dz'\right)\right] \times \exp\left(-\frac{1}{U_{sat}} \int_{-\infty}^t I_0(t')dt'\right)}, \tag{8.3}$$

where $\sigma_S = \sigma_{em} + \sigma_{abs}$, and $U_{sat} = \hbar\omega/\Gamma_S\sigma_S$ is the saturation fluence defined as the energy per unit area that is required to stimulate with a probability of 50% the emission from an atom initially in the excited laser level [53]. The overlap $\Gamma_S = A_{dopant}/A_{mode}$ between the doped region (area A_{dopant}), and the signal mode (area A_{mode}) can be significantly lower than unity in the case of a fiber amplifier, thus increasing the saturation fluence.

Defining the energy of the incoming pulse as $U_{in}(t) = \int_{-\infty}^{t} I_0(t')dt'$ and given the initial gain G_0 that is available over the amplifying medium and the saturation energy U_{sat}, the output pulse shape is given by

$$I(z,t) = \frac{I_0(t)}{1 - \left[1 - G_0^{-1}(z)\right] \times \exp\left\{-\left[U_{in}(t)/U_{sat}\right]\right\}}. \tag{8.4}$$

This simple expression is known as the Franz-Nodvik formula [52]. In the particular case of a fiber amplifier, the population inversion is generally not uniform over the doped fiber. Indeed, it can depend on whether the pumping is co- or contrapropagating. The small-signal gain at the fiber output is then given by $G_0(L) = \exp\left(\sigma_S \int_0^L \Delta_0(z')dz'\right)$. This value and that of U_{sat} may be inferred from measurements, by injecting pulses with a determined shape, for a given pump configuration and using Eq. (8.4) to fit the output pulse shape.

8.3.3
Practical Calculation of the Output Pulse Based on the CW Regime

Using Eq. (8.4), it is possible in principle to calculate the output pulse for a given input pulse shape, knowing the initial gain value G_0. In practice, G_0 depends on the depletion of the gain during the pulses and on the rebuilding of the population inversion in between the pulses, so it may be difficult to determine. However, there are two asymptotic cases where the initial gain G_0 can be determined more easily.

For repetition rates that are low compared to the decay rate γ_{pop} of the population inversion, the population inversion can reach again its initial value between two pulses. Therefore, the gain G_0 between the input and output powers (P_{in} and P_{out}, respectively) is simply the small signal value of the amplifier gain. On the contrary, if the repetition rate is higher than the population inversion decay, the gain cannot "build" fast enough before being depleted by the pulses, so the population inversion reaches a lower steady value. The gain here is identical to that of a CW amplifier whose input power P_{int}^{CW} would be equal to the average power $\langle P_{in}\rangle$ of the input-pulsed signal [54]. The two possible regimes are summarized by the following equations:

$$\text{PRF} \ll \gamma_{pop} : G_0 = \frac{\langle P_{out}\rangle}{\langle P_{in}\rangle}, \quad \text{with} \quad \langle P_{in}\rangle \ll P_{sat} = \frac{U_{sat}A_{dopant}}{T_1}. \tag{8.5}$$

$$\text{PRF} \gg \gamma_{pop} : G_0 = \frac{P_{out}^{CW}}{P_{in}^{CW}}, \quad \text{with} \quad P_{in/out}^{CW} = \frac{1}{T}\int_0^T P_{in/out}(t)dt. \tag{8.6}$$

As for γ_{pop}, its value depends on the saturation of the transition. When the population inversion is low, its lifetime corresponds to the excited-state lifetime T_1. As the flux of resonant photons on the atom becomes higher than the fluorescence decay rate $1/T_1$, the probability of stimulated emission increases, and so does the decay rate of the population inversion. Its value can be expressed as a function of the ratio between input and saturation power and of the initial gain [54]:

$$\gamma_{pop} = \frac{1}{T_1}\left(1 + \frac{P}{P_{sat}} \frac{G_0 - 1}{\ln G_0}\right). \quad (8.7)$$

This value can thus be substantially higher than the spontaneous emission rate. For example, in an EDFA with a 20 µm core diameter fiber, the saturation power $P_{sat} = U_{sat} A_{dopant}/T_1$ is about 60 mW. Considering a typical repetition rate of 20 kHz, an input peak power of 30 W, and a pulse duration of 1 µs, the input average power is 600 mW. For an initial gain $G_0 = 10$, the decay rate of the population inversion is thus increased by a factor of 40 due to the transition saturation.

8.3.4
Pulse Shape Distortion

A straightforward consequence of the gain depletion during the pulse propagation is distortion of the amplified signal compared to the input pulse shape. This distortion appears for high values of the gain or the input power and can be calculated using Eq. (8.4), given the input signal pulse shape, the saturation fluence U_{sat}, and the initial value of the gain G_0. As shown in Figure 8.7, a square pulse undergoes a much more notable change of the shape than a Gaussian pulse. We remind that even though the Gaussian pulse seems to arrive earlier and thus travel faster, this feature only corresponds to the higher amplification of the leading edge compared to the pulse trail.

Just as it is possible to deduce the shape of the output pulse from that of the input pulse, it is also possible to anticipate the input pulse shape that would allow obtaining any arbitrary desired shape $I_{out}(t)$ of the amplified pulse [50]. The required

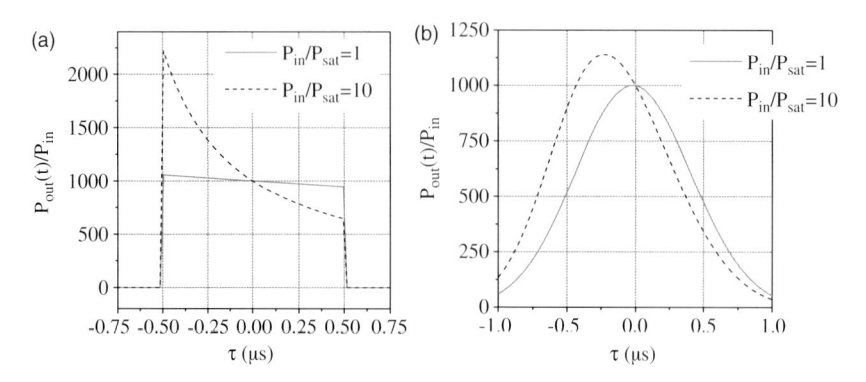

Figure 8.7 Influence of the gain saturation on the pulse shape at the output of the amplifier, in the case of a square (a) or Gaussian (b) input pulse of FWHM duration 1 µs. For both calculations, the initial gain is $G_0 = 1000$ and the input power is equal to (solid line) or 10 times (dotted line) the saturation power P_{sat}.

input pulse-shape is given by [53]

$$I_{in}(t) = \frac{I_{out}(t)}{1 + (G_0 - 1) \cdot \exp\{-[U_{out}(t)/U_{sat}]\}}, \quad (8.8)$$

where $U_{out}(t) = \int_{-\infty}^{t} I_{out}(t')dt'$. As previously mentioned, the values of the saturation energy and the initial gain can be measured by fitting the result of the amplification of a known input pulse shape [55].

8.3.5
Influence of the Amplified Spontaneous Emission

In the absence of nonlinear effects, the energy that can be extracted from the gain medium is ultimately limited by ASE. As the pump power increases and if the pulse period is on the order of magnitude of the radiative lifetime of the population inversion state, the amount of fluorescence increases and starts depleting the population inversion. Therefore, the available gain does not increase anymore, thus reducing the efficiency of the pump power [48].

In this respect, fiber amplifiers present the simultaneous drawbacks of long gain medium where the ASE can build, together with waveguiding of the fluorescence, which increases the probability of gain depletion. The ASE contribution can thus represent one of the limits of the performance of a fiber amplifier, especially at low repetition rate compared to γ_{pop}.

The extractible energy E_{out} can be computed by integration of Eq. (8.4) at the fiber end over the pulse duration. We get the following expression:

$$E_{out} = E_{sat} \ln \left(1 + \left[\exp \frac{E_{in}}{E_{sat}} - 1\right] E_{sat}\right), \quad (8.9)$$

where $E_{in} = U_{in} A_{mode}$ and $E_{sat} = U_{sat} A_{dopant}$. In the limit when $E_{in} \ll E_{sat}$, the expression above then simplifies to [51]

$$E_{out} = E_{in} + E_{sat} \ln G_0. \quad (8.10)$$

This is another way to write that the output energy is the sum of the input energy plus the stored energy above the bleaching inversion level of the fiber [56]. This is an upper bound of the extractable energy.

In a fiber amplifier, the initial gain G_0 has generally a maximum value of about 20–30 dB. Going much beyond this value to store more energy in the medium is very difficult as ASE depletes the population inversion. Any further stored energy, that is, further pump power, will be useless. This limit can thus define the maximum useful value of extractible energy from a fiber. Obviously, the lower the confinement of the signal in the fiber (low numerical aperture (NA)) or the overlap between the dopant and ASE modes, the lower the ASE and the higher the energy extraction. If we neglect E_{in}, we find that the output energy is limited to about five times the saturation energy, as confirmed by Refs [56–58].

Recalling the 20 μm core diameter erbium-doped fiber of the example mentioned above, the saturation energy is about 0.6 mJ, so energy of about 3 mJ should be extractible. However, this estimation does not take into account the nonlinear effects that can take place in the fiber and that generally limit the maximum output peak power in practice.

8.4
Power Limitations in Pulsed Fiber Amplifiers

In the regime of short-pulse amplification, the peak power is ultimately limited by nonlinear effects that are due to the interaction between the confined light and the fiber material. We remind here the origin and consequences of these nonlinearities, focusing on single-frequency fiber amplifiers. We consider a linewidth ranging from a few megahertz to hundreds of megahertz, depending on the application. The corresponding pulse durations lay between 10 ns and 1 μs.

8.4.1
Physical Principle of the Stimulated Brillouin Scattering

As discussed in Section 8.1, in the case of LIDAR sources, the spectral linewidth and pulse duration are directly related to the velocity resolution required. For coherent LIDAR applications, the typical pulse duration lays in the range from 100 ns to 1 μs. In this regime, the main limitation to peak power is SBS effect. Indeed, as will be detailed subsequently, the Brillouin gain is about two orders of magnitude higher than any other nonlinear effects.

The SBS effect is particularly limiting in fibers where high-power optical signals are confined to a small area. As the high electric field propagates in the fiber core, the material experiences a strain (electrostrictive response) that propagates along the fiber axis at the sound velocity. This modulation of the material density ρ modifies the optical refractive index of the medium, forming an index grating that propagates with a wave vector q_B, in the same direction as the optical wave [59]. This grating scatters back a fraction of the input signal (Figure 8.8), thus reducing the output optical power and the total optical efficiency. Moreover, the backscattering of a

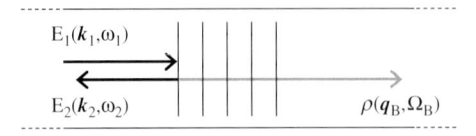

Figure 8.8 Illustration of the phase matching condition between the input and backscattered Stokes optical wave (given by their wave vectors k_1 and k_2 and their angular frequencies ω_1 and ω_2) and the acoustic wave (q_B, Ω_B).

fraction of the signal at the output of the fiber, where it reaches the highest power, can lead to high backward propagating pulses that deteriorate the components at the input of the amplifier.

Since part of the input energy $\hbar\omega_1$ is transferred to the medium to generate the copropagating acoustical phonons, the backscattered light is a Stokes wave of energy $\hbar\omega_2 < \hbar\omega_1$. The value of ω_2 is determined by the phase matching between the input optical field and the acoustic wave, both propagating with velocities that are intrinsic to the fiber medium. As illustrated in Figure 8.8, the wave vectors k_1, k_2, and q_B of the input, Stokes, and acoustic waves, respectively, are related by momentum conservation:

$$k_1 = k_2 + q_B. \tag{8.11}$$

By energy conservation, the angular frequency of the backscattered Stokes field is given by $\omega_2 = \omega_1 - \Omega_B$, where Ω_B is related to the acoustic wave vector by $|q_B| = \Omega_B/v_{ac}$. The sound velocity v_{ac} is related to the material density ρ and the adiabatic bulk modulus K_S through the Newton–Laplace formula [59]:

$$v_{ac} = \sqrt{\frac{K_S}{\rho}}, \tag{8.12}$$

which is an increasing function of the material temperature T (for $T > 100$ K) [60]. The energy and momentum conservation rules then allow writing the Brillouin shift as

$$\Omega_B = \frac{2n_{eff}(v_{ac}/c)\omega_1}{1 + n_{eff}(v_{ac}/c)} \tag{8.13}$$

for a fiber that guides an optical mode with an effective mode index n_{eff} at frequency ω_1. The sound velocity v_{ac} being obviously much smaller than the light speed c, the Brillouin frequency shift can be simplified as

$$\frac{\Omega_B}{2\pi} \approx 2\frac{n_{eff}}{\lambda_1}v_{ac}. \tag{8.14}$$

In the case of a fused silica fiber, where $v_{ac} \approx 6.10^3$ m/s and $n_{eff} \approx 1.45$ for $\lambda_1 = 1.55$ μm, the frequency shift is $\Omega_B/2\pi \sim 11$ GHz and varies by less than 0.5 GHz for the usual values of the fiber parameters.

8.4.2
SBS Gain

The SBS corresponds to a frequency conversion between two optical fields of frequencies ω_1 and ω_2 through the generation of phonons at frequency $\Omega = \omega_1 - \omega_2$. The efficiency of the process is described by a factor g_B that couples the spatial evolution of the optical intensities [59,61]. Due to the viscosity of the material in which the acoustic waves propagate, the material response in the SBS process is not instantaneous. The resulting exponential damping of the phonons implies that the Brillouin gain has a Lorentzian shape:

$$g_B(\Omega) = g_0 \frac{\Gamma_B^2}{4(\Omega_B - \Omega)^2 + \Gamma_B^2}, \tag{8.15}$$

with a FWHM linewidth $\Delta \nu_B = \Gamma_B/2\pi$ that corresponds to the inverse of the phonon lifetime T_B. For bulk silica, the linewidth $\Delta \nu_B \approx 20$ MHz corresponds to a phonon lifetime on the order of 10 ns. The maximum value of the gain is obtained when the forward and backward propagating optical fields at ω_1 and ω_2 are phase matched with the phonons with Brillouin frequency Ω_B. It is given by

$$g_0 = \frac{\pi n_{\text{eff}}^7 p_{12}^2}{c\lambda_1^2 \rho_0 v_{\text{ac}} \Delta \nu_B}, \tag{8.16}$$

where p_{12} is the longitudinal elasto-optic coefficient and ρ_0 is the average material density [61]. The value of g_0 can be directly extracted from differential gain measurements, and is typically $g_0 \approx 2.5 \times 10^{-11}$ m/W for a single-mode fiber [62]. Note that the SBS gain g_0 is inversely proportional to the phonon linewidth $\Delta \nu_B$.

8.4.3
SBS Threshold Input Power

As the Stokes wave propagates back to the fiber input, its amplitude is amplified exponentially (provided the input signal power is not depleted). The SBS threshold power $P_{\text{th}}^{\text{SBS}}$ is defined as the input power for which the backward and forward propagating optical waves have equal powers at the fiber input. Hence, in a passive single-mode fiber, the SBS threshold power has the following expression [61,63]:

$$P_{\text{th}}^{\text{SBS}} = 21 \frac{K A_{\text{eff}}}{g_0 L_{\text{eff}}}, \tag{8.17}$$

where A_{eff} is the effective mode area and L_{eff} is the effective length that characterizes the distance over which the signal is high enough to render the SBS effect efficient. For a fiber amplifier with linear gain g_a, the effective length can be defined as $L_{\text{eff}} = \exp(-g_a L) \cdot \int_0^L \exp(g_a z) dz$ [64]. Finally, the factor K is the polarization dependence factor, ranging from 1 (parallel polarizations for signal and Stokes waves) to 2 (orthogonal polarizations) [65].

The SBS limitation may be illustrated following the example used in Section 8.3, considering an Er:Yb-doped fiber with effective mode area $A_{\text{eff}} \approx 400\,\mu\text{m}^2$, seeded with an average input power of 600 mW. It can be calculated that for a signal absorption of 30 dB/m at 1.55 µm, a fiber length of 7 m allows reaching $G_0 = 10$. In this case, for a copropagating pump power of 20 W at 976 nm, the effective length is $L_{\text{eff}} \approx 5$ m. Therefore, the SBS threshold peak power will be inferior to 300 W, corresponding to pulse energy of 0.3 mJ for the 1 µs pulse duration. However, the

Frantz–Nodvik calculation showed that despite the ASE, the fiber amplifier should be able to yield a maximum energy of 5 mJ. Hence, SBS mitigation techniques must be implemented to make the most of the maximum extractible energy.

8.4.4
SBS Reduction

For fused silica fibers, the SBS gain BW $\Gamma_{B,fiber}$ is larger than the value $\Gamma_{B,bulk}/2\pi =$ 20 MHz of the bulk material, and $\Gamma_{B,fiber}/2\pi$ typically ranges from 50 to 100 MHz [61]. Indeed, the inhomogeneities in a material increases with its length. As the material properties are locally perturbed, the intrinsic sound velocity and thus the local Brillouin frequency are shifted from $\Omega_{B,bulk}$. For a long medium like an optical fiber, where the shift can be larger than $\Gamma_{B,bulk}$, the corollary is a broader Brillouin gain spectrum and, according to Eq. (8.16), a lower maximum value for the global SBS gain, integrated over the fiber length.

The effect of the natural inhomogeneities in a fiber can be accentuated by artificially introducing modulations to the material properties, for example, through temperature [66] or stress modulations [67]. Another way to broaden the BW of the SBS spectrum (and to reduce the gain) is to introduce a longitudinal variation of the core diameter [68,69].

Finally, the core refractive index n_{core} or the NA over the fiber length provides one more degree of freedom on the SBS gain broadening. Indeed, as the NA and the angular acceptance of the fiber increase, the input and Stokes optical waves may form an increasing angle. Therefore, the phase matching condition between the optical wave vectors and the still longitudinal acoustic wave vector q_B occurs for a larger BW, given by [70]

$$\Gamma_{B,fiber} = \sqrt{\Gamma_{B,bulk}^2 + \Omega_B^2 \frac{NA^4}{4n_{core}^4}}. \tag{8.18}$$

The correction starts being significant (i.e., >20%) for NAs higher than 0.25.

8.4.5
Domain of SBS Predominance

Many methods exist to reduce the influence of the SBS effect on fiber amplifiers, based on the expression of the threshold power in Eq. (8.17). Increasing A_{eff}, reducing the SBS gain [71,72] or the effective length [22], thus allows reaching higher peak powers in single-mode fiber amplifiers. However, it is important to note that this effect is only relevant when the input signal spectrum is much narrower than the SBS gain spectrum. Therefore, all optical components see the same phonon wave vector, and the backscattering on the phonon-induced density grating is maximal.

On the contrary, once the linewidth $\Delta\nu_L$ of the input signal becomes of the order of magnitude of the linewidth $\Delta\nu_B$ of the intrinsic SBS gain spectrum, its maximum value g_0 decreases. This effect is well illustrated in the case of an input signal with Lorentzian spectrum profile. The convolution of the two profiles is also Lorentzian, with a width $\Delta\nu'_B = \Delta\nu_B + \Delta\nu_L$, and the SBS gain becomes

$$g'_0 = \frac{\Delta\nu_B}{\Delta\nu_B + \Delta\nu_L} g_0. \tag{8.19}$$

In the particular case of pulsed fiber amplifiers, this expression helps perceiving that the influence of the SBS will vanish for pulses that are short compared to the phonon lifetime. Below a duration that will be calculated subsequently, the main limiting effect on the amplifier efficiency stops being the SBS, becoming instead the SRS.

8.4.6
Physical Principle of the Stimulated Raman Scattering

SRS is a nonlinear process based on the inelastic scattering of the photons on molecules, in a process that forms a propagating grating of nonlinear polarizability. Whether the molecules absorb or yield energy, the scattered photons have an energy that is respectively lower (Stokes) or higher (anti-Stokes) than the input photons [59]. The value of the frequency shift between input and scattered photons is given by the energy of molecular vibrations, which depends on the material composition. In optical fibers, mostly made of fused silica glass, the frequency shift is about 13.2 THz (or 440 cm^{-1}), which corresponds to 0.1 µm for a pump photon at 1.55 µm. For systems at room temperature or below, the predominant Raman conversion is the Stokes transition, so the scattered photons have lower frequency than the input photon and are mainly centered around a wavelength of 1.65 µm. Due to the amorphous nature of fused silica glass, the Raman gain spreads out into a 8 THz wide spectrum, making the nonlinear conversion possible over about 60 nm around 1.65 µm.

The consequence of the SRS effect is similar to that of SBS: The energy transfer from the input signal to the Stokes wave reduces the amplifier efficiency at the required optical frequency. Unlike the SBS effect, for SRS, both forward and backward propagating Stokes waves can be generated. For the forward propagating, the threshold power is defined by the input signal power such that the Stokes wave and the transmitted signal wave have the same power [63] and writes as

$$P_{th}^{SRS} = 16 \frac{KA_{eff}}{g_R L_{eff}}, \tag{8.20}$$

where $g_R \approx 7.10^{-14}$ m/W for bulk silica [61], whereas the other quantities are identical or similar to those defined from Eq. (8.17). The main difference is thus the gain of the Raman interaction, which is more than two orders of magnitude lower than the intrinsic Brillouin gain.

Unlike the Brillouin gain, the value of g_R is not affected by the spectral BW of the signal that generates the polarizability wave. The transition from one limiting effect to the other is close to a pulse duration of about $\tau_p = 10$ ns. Below this duration, the SRS effect is generally predominant. In the particular case of LIDARs, for example, the corresponding resolution in velocity becomes too low compared to the typical interesting signals (wind velocity gradients, wake vortices, etc.). Therefore, the details of the SRS effect are not discussed here.

8.4.7
Maximum Peak Power Achievable

Since the SBS threshold power represents an upper bound for the possible extractible output power, the output peak power $P_{\text{out}}^{\text{max}}$ obtained through the Frantz–Nodvik calculation has to be compared with $P_{\text{th}}^{\text{SBS}}$. This finally tells if the result $P_{\text{out}}^{\text{max}}$ can indeed be obtained in practice.

8.5
Phase Noise and Distortion in Fiber Amplifiers

We now carry out the analysis of the phase at the output of pulse fiber amplifiers. For the purpose of phase combining of various fiber amplifiers, two features must be examined. First, the phase noise, related to optical path variations, usually due to thermal fluctuations, mechanical vibrations, and pump diode intensity noise. The BW of this is typically limited to several 10 kHz. Second, a determinist phase shift induced during the pulse by nonlinear phase and gain distortions specific to pulsed regime. The phase noise during the pulse itself can be neglected because its timescale is usually much larger than the pulse duration.

We first present an experimental analysis of the phase noise in a homemade high-power CW fiber amplifier and in various commercial power amplifiers. Then, we report on the measurements of in-pulse phase distortion. These deterministic variations are induced by gain depletion and Kerr effect, and can thus be compared with theoretical calculations.

8.5.1
Phase Noise Measurement

The considerations on phase noise are carried out on a homemade CW fiber amplifier designed for 100 W operation and on various Watt-level average power commercial fiber amplifiers designed for pulsed operation with 100–200 W peak power level.

Figure 8.9 shows the setup for the phase noise measurement of a 100 W level CW fiber amplifier. The amplifier is a MOPFA operating at 1 μm and consisting of two successive ytterbium-doped fiber amplifiers. The first amplifier has 20 dB gain delivering an output power of up to 3 W, and the second one has 15 dB gain delivering an output power of 100 W. The second-stage fiber is pumped backward at 975 nm. The last 20 mm of the fiber is suspended in the air, while the main part is held between two metal plates. A dichroic mirror reflects the output signal toward an

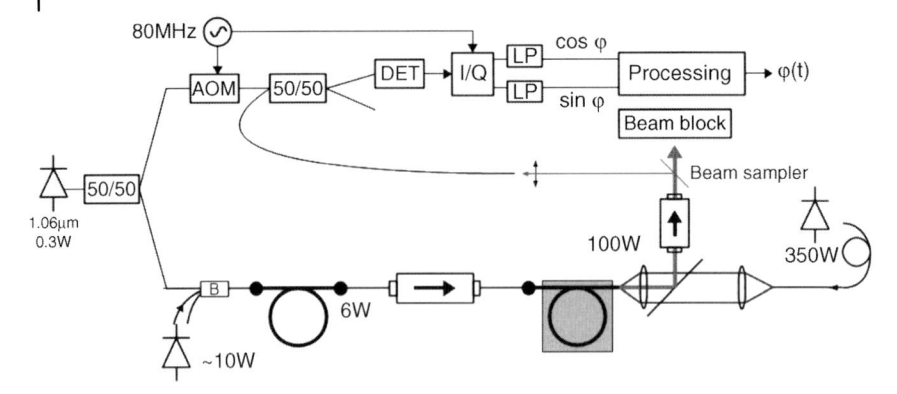

Figure 8.9 100 W fiber amplifier noise measurement.

isolator followed by a beam block. A sampled part of the beam is coupled into a fiber and interferes on a 50/50 coupler with an 80 MHz shifted reference beam. The interference signal is detected (DET) and its phase is recovered through an I/Q (in-phase I and quadrature Q) demodulator, LP filters, and digital processing.

Several phase records are plotted in Figures 8.10 and 8.11 for two timescales. As expected, the phase noise is larger with larger pump power. The four plotted curves correspond to the configurations listed in Table 8.5.

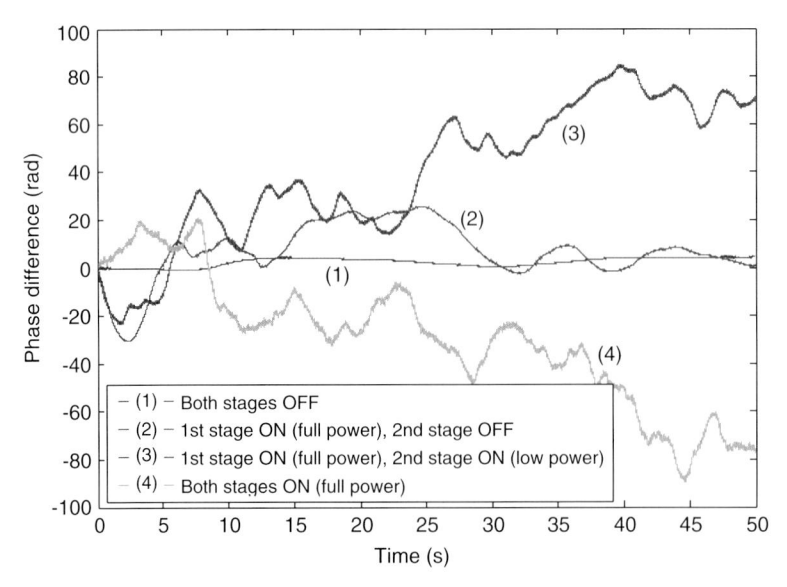

Figure 8.10 Measured phase difference for both stages OFF, first stage ON, or both stages ON. The total record time is 50 s and the sample rate is 5 kHz (both stages OFF) and 40 kHz (first ON and both ON).

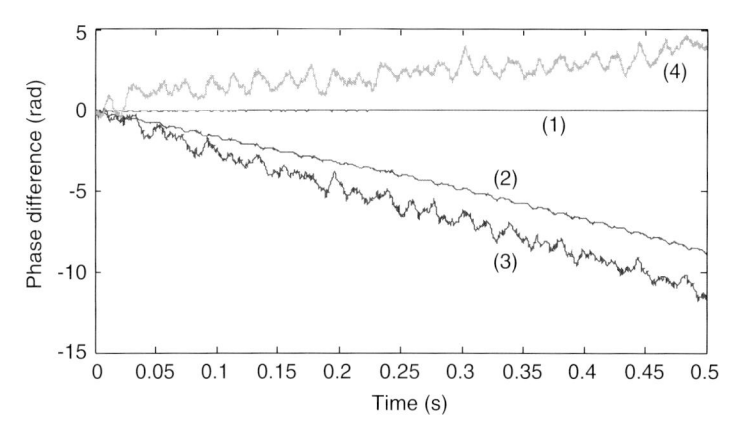

Figure 8.11 Same as previous figure with timescale limited to 500 ms.

The last two cases demonstrate that the main origin of the phase noise is not the level of pump power, but rather the mechanical vibrations due to the commercial pump system. Indeed, the second-stage pump is a 350 W fiber-coupled diode bar, which is packaged in a rack that includes a primary and a secondary water-cooling circuit. The primary water cooling circuit is a recycling water loop circulating between a chiller and the diode. The chiller itself is cooled by a wall-plugged secondary circuit. Although the rack is not directly in contact with the vibration-isolated table, the pump guiding fiber might transmit vibrations, so the damping is not perfect. Some phase noise features are related to the fiber packaging: In this experiment, the first amplifying fiber is loosely coiled on a plate, whereas the second amplifying fiber is sandwiched in a water-cooled metallic plate.

Figure 8.12 shows the power spectral densities of the phase noise. Three regimes can be identified:

- Below 10 Hz – the noise is continuous and mainly due to thermal exchanges between the fiber and its surrounding environment (thermal noise).
- Between 10 Hz and 1 kHz – the noise is spiky and mainly due to acoustic vibrations (fans, water pumps, etc.).
- Above 1 kHz – here can be found electrical noise spikes, for example, noise of the pump diode current, which can be transferred to pump intensity noise and to signal phase noise. This yields low-amplitude, high-frequency phase noise.

Table 8.5 Experimental configurations for phase noise measurement.

Case	Stage 1	Stage 2 cooling	Stage 2 pump
1	OFF	OFF	OFF
2	Full power	OFF	OFF
3	Full power	ON	Low power
4	Full power	ON	Full power

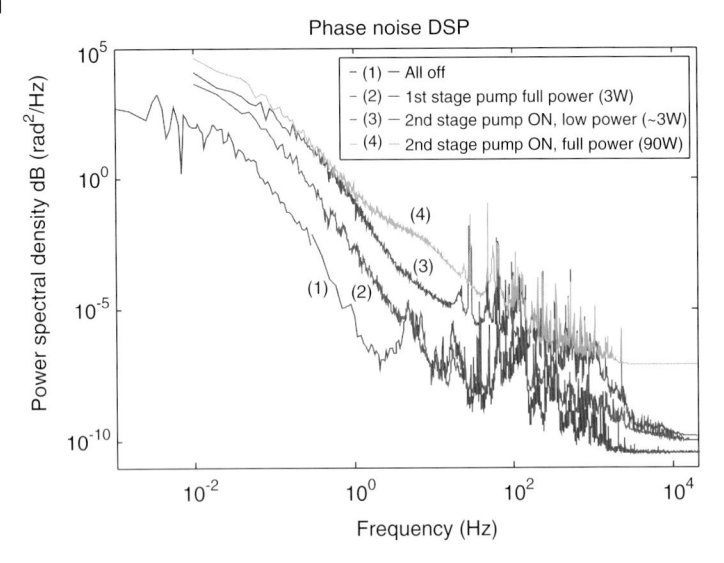

Figure 8.12 PSD of phase noise measured in the three cases.

One can learn from Figure 8.12 that the addition of the first stage mostly affects the low-frequency thermal noise. Almost no vibration is added. On the contrary, switching ON the cooling system of the second-stage pump increases notably the phase noise at all frequencies and in particular at acoustic frequencies. Then increasing the power of stage 2 up to 90 W has only a small effect on the noise spectrum, for frequencies between 0 and 10 Hz. The increase of the noise floor above 1 kHz is due to the photon noise increase on the detector. The second amplifier does not bring additional low-frequency (<1 Hz) noise, because the thermal noise at lower frequencies is damped by the metal mount.

Performance of a beam combining system is usually quantified by the residual phase error $\Delta\varphi_{RMS} = 2\pi/N$. Combining efficiency of 95% roughly corresponds to $\Delta\varphi_{RMS} \sim 2\pi/15$ (also noted $\lambda/15$) and of 99% to $\Delta\varphi_{RMS} \sim 2\pi/30$. $\Delta\varphi_{RMS}$ can be computed from the power spectral density (PSD) (Figure 8.12) by taking the square root of the PSD integral over the frequency.

Figure 8.13 shows the PSD integration from infinity to a given frequency. The value at lowest frequency shows the phase error integrated over all frequencies. They are much larger than a typical target residual phase error of $2\pi/50$ (dashed line). This figure is useful to compute the residual phase error assuming a controller that removes all errors lower than a given frequency and leaves untouched the errors higher than this frequency. According to these measurements, a correction of the frequencies below ~1 Hz is good enough for the case of a one-stage amplifier to obtain the target $2\pi/50$ residual phase error. For a two-stage amplifier, a correction of the frequencies up to ~100 Hz is required.

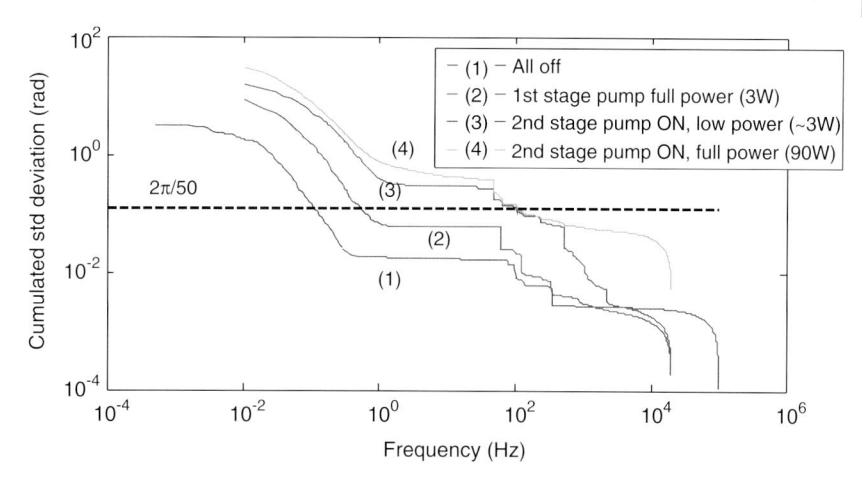

8.5 Phase Noise and Distortion in Fiber Amplifiers | 257

Figure 8.13 Cumulated standard deviation from low to high frequencies. The dashed curve represents the $2\pi/50$ limit.

The same curve measured with three different commercial amplifiers is shown in Figure 8.14. The amplifier that yields the curve (2) delivers pulses with 200 W peak power and is conduction cooled (no fan). Amplifiers (3) and (4) are air cooled with a fan and deliver a lower peak power of 100 W. For these amplifiers, a feedback loop up to ~100 Hz would be preferable, whereas for the amplifier (2) a 1 Hz feedback loop would be good enough to correct the lower noise.

These experiments show that a very careful design of the whole amplifier chain design is required to minimize the phase noise sources. Main phase noise sources are low-frequency thermal noise, medium-frequency vibration noise, and high-frequency electrical noise.

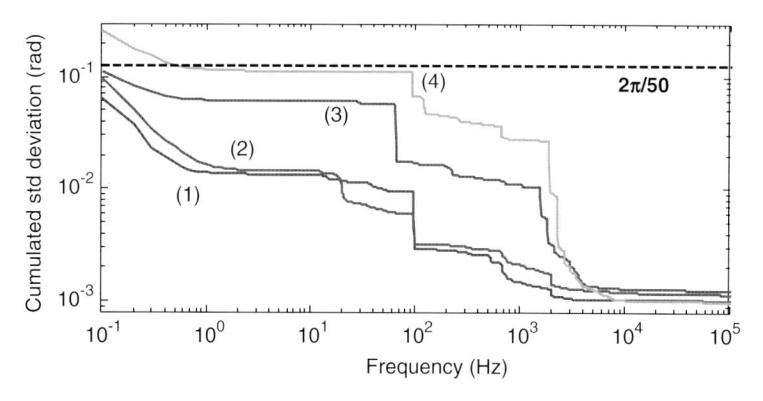

Figure 8.14 Phase noise cumulated standard deviation for (1) no amplifier, (2) a ~200 W peak power amplifier with conduction cooling (no fan), and (3) and (4) two ~100 W peak power pulsed fiber amplifiers. Amplifier (4) has more fans than amplifier (3). The dashed curve shows the $2\pi/50$ limit.

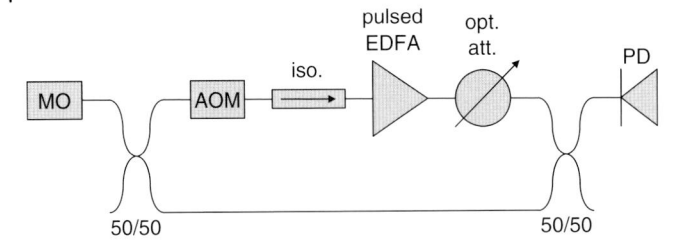

Figure 8.15 Experimental setup for in-pulse phase fluctuation measurements. MO, continuous-wave master oscillator; AOM, acousto-optic modulator for pulse generation; iso: optical isolator; opt. att.: optical attenuator, to balance signal levels on the two optical paths; PD: photodetector measuring the interference signal between the MO cw signal and the amplified pulses.

8.5.2
In-Pulse Phase Shift Measurement

We measured the phase variation inside a pulse for various time and power regimes. The phase during the pulse is expected to be affected by two main effects: the gain variation (through Kramers–Kronig relations) and the pulse intensity variation (through Kerr effect). Section 8.5.2 presents experimental measurements of the phase variation in microsecond pulses and Section 8.5.3 provides theoretical considerations on these phase variations.

The measurement setup is described in Figure 8.15. We use a CW MO at 1.5 μm. The signal is then modulated with an AOM and amplified in an EDFA. The pulsed amplified signal finally interferes with the MO CW reference signal on a photodetector in a classical interferometer design.

The amplified peak power is limited to about 100 W by SBS. The pulse duration can be varied from 300 ns to 10 μs. The input pulse is time shaped with the AOM to maximize the energy extraction with minimal peak power.

Figure 8.16a shows an input pulse with duration 1 μs and repetition rate 3 kHz. The output pulse shape is plotted at the center. As explained in Section 8.3.4, its distortion results from the much higher gain available at the front of the pulse than at its end.

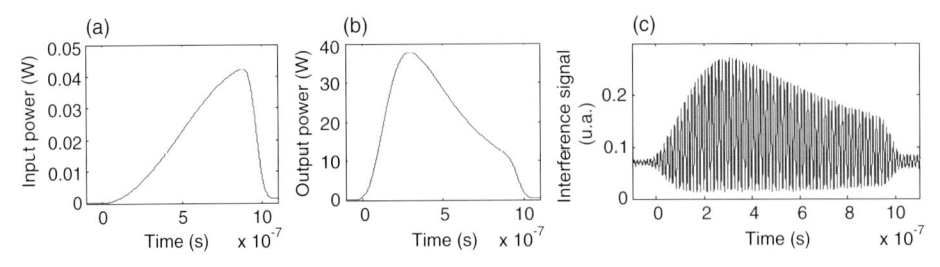

Figure 8.16 Sample experimental signals for 1 μs pulse duration, 3 kHz pulse repetition rate. (a) Input pulse profile. (b) Output amplified pulse. (c) Interference signal.

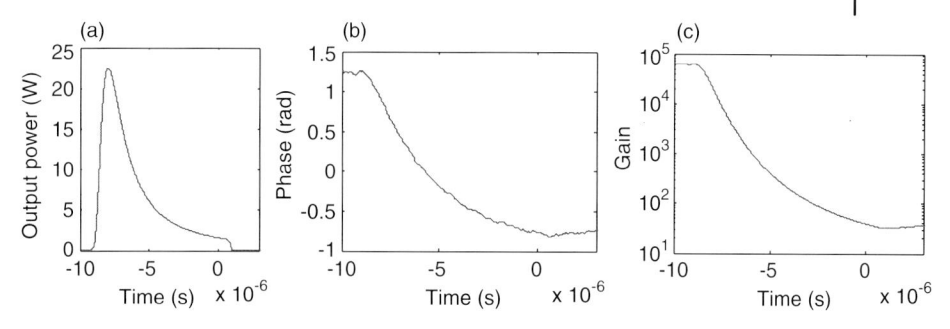

Figure 8.17 (a) Output pulse for 10 μs pulse duration. (b) Computed phase variation of 70 MHz carrier in the interference signal. (c) Net gain computed as the ratio between the input and the output pulses.

The corresponding interference pattern is presented in Figure 8.16c. The intensity of the interference signal on the photodetector can be written as

$$I(t) = E_{01}^2 + E_{02}(t)^2 + 2E_{01}E_{02}(t)\cos(2\pi \cdot F_{AOM} \cdot t + \phi_0 + \delta\phi(t)), \quad (8.21)$$

where E_{01} is the electric field amplitude of the reference beam, while E_{02} refers to the pulsed amplified beam. F_{AOM} is the AOM frequency shift, ϕ_0 is a phase difference offset, and $\delta\phi(t)$ represents the pulsed wave phase fluctuation. The cosine term of this signal, $2E_{01}E_{02}(t)\cos(2\pi \cdot F_{AOM} \cdot t + \phi_0 + \delta\phi(t))$, can be extracted by spectral filtering using Fourier transform.

Along with the 10 μs output pulse shape (Figure 8.17a), Figure 8.17b shows the time evolution of the measured phase variation and the logarithm of the gain (Figure 8.17c). The two curves have very similar shapes, indicating that the phase variation is mainly due to Kramers–Kronig relations (see Section 8.5.3).

Figure 8.18a shows the phase versus gain during the pulse, and Figure 8.18b gives the corresponding output pulse shapes. In case of long pulse, the phase variation scales as the logarithm of the gain. A linear fit of this curve is also plotted, exhibiting a slope of 0.27. This slope is removed from the four phase curves and the result is plotted in Figure 8.18c. The obtained phase variations are very similar to the output pulse intensity shapes.

The linear dependence of phase with the logarithm of the gain can be explained by Kramers–Kronig relations, and the additional phase variation observed for shorter pulses can be explained by Kerr effect. We note that the maximum phase variation measured here is about 2 rad.

8.5.3
In-Pulse Phase Shift Calculation

When a signal propagates into a doped optical fiber, its complex amplitude is affected by the interaction with the gain medium. Simultaneously, large intensities and long interaction lengths can trig nonlinearities such as Raman scattering or Kerr effect even in relatively short fiber lengths [64]. We will first discuss the Kerr-induced

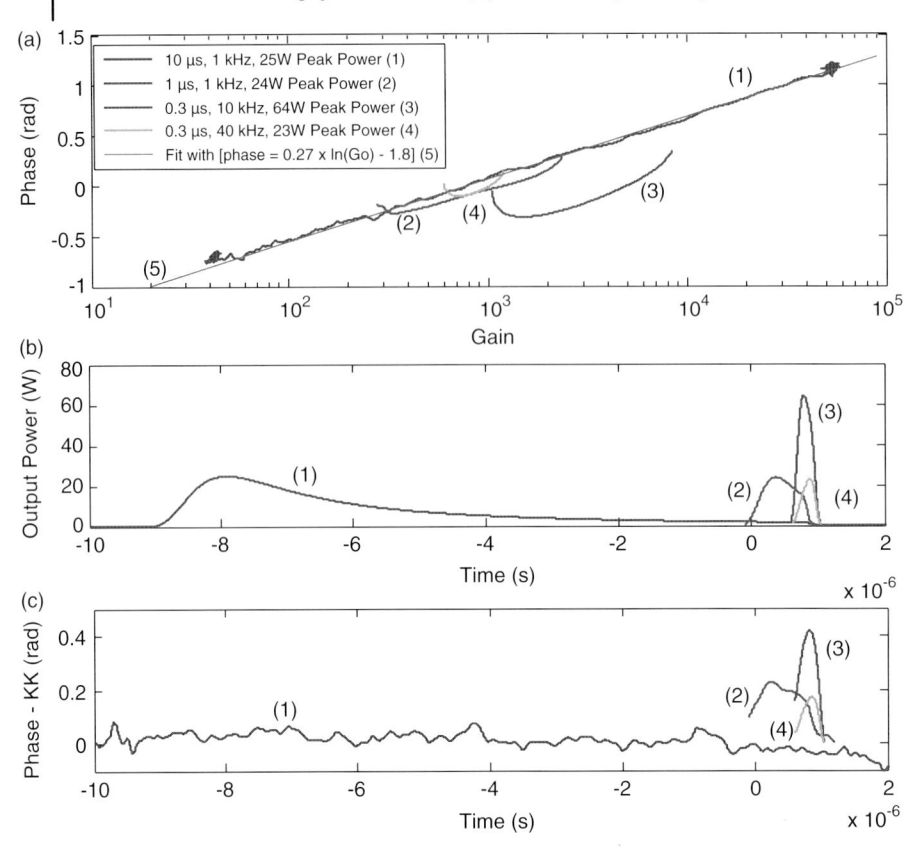

8 Coherent Beam Combining of Pulsed Fiber Amplifiers in the Long-Pulse Regime

Figure 8.18 (a) Pulse phase versus gain for various pulse durations and repetition rates. The 10 μs pulse phase is fitted with [phase $= 0.27 \times \ln(G_0) - 1.8$]. (b) Output pulse shapes for the four cases. (c) Pulse phase fluctuation from part (a) when the fit tendency is removed.

nonlinearity, which is peak power dependent, before examining the Kramers–Kronig contribution showing that it is pulse energy dependent.

8.5.3.1 Kerr-Induced Phase Shift

When a laser pulse travels into an optical fiber, it is affected by variations of the refractive index due to the Kerr optical effect following the relation

$$n(z,t) = n_0 + \Delta n(z,t) = n_0 + n_2 I(z,t), \tag{8.22}$$

where n_0 is the unperturbed refractive index and n_2 is the nonlinear refractive index [61]. By causality, the refractive index variation is null before and after the pulse. It scales as the peak power and reaches its maximum at the pulse maximum. When a pulse propagates into a fiber amplifier with a peak power $P(z,t)$ at position z and time t, it reaches its maximum $P_{\text{out}}(t_1) = P(L,t_1)$ at position $z = L$ and time t_1. The variation of the intensity between initial t_0 time and time t_1 will thus induce a phase shift of the optical wave at pulsation ω. The shift can be related to the peak power $P(z,t)$ and the fundamental mode effective area A_{eff} by

$$\Delta\varphi(\omega,t_1) = \frac{\omega}{c}\int_0^L \Delta n\left(z,t_1-\frac{z}{c}\right)dz = \frac{\omega n_2}{cA_{\text{eff}}}\int_0^L P\left(z,t_1-\frac{z}{c}\right)dz = \frac{\omega n_2}{cA_{\text{eff}}} P_{\text{out}}(t_1) L_{\text{eff}},$$

(8.23)

where the effective length L_{eff} represents the equivalent interaction length for a signal with constant power $P_{\text{out}}(t_1)$. The phase variation should hence follow the pulse shape. This effect is called the self-phase modulation (SPM).

As we focus here on fiber amplifiers for coherent detection LIDARs, the peak power is usually limited by SBS. The threshold is reached for SBS at output peak power P_{max} given by the Smith relation (8.17). Combining Eqs. (8.23) and (8.17), the phase shift generated at peak power P_{max} can be calculated:

$$\Delta\varphi(\omega,t) = 21\frac{\omega n_2}{cg_B} = 42\frac{\pi n_2}{\lambda g_B}.$$

(8.24)

Taking the value of $n_2 = 3 \times 10^{-20}$ m^2/W for pure silica [61], $g_B = 2 \times 10^{-11}$ m/W measured for standard passive fibers at 1.55 µm, relation (8.24) gives $\Delta\varphi_{\text{max}} = 0.1$ rad at wavelength $\lambda = 1.55$ µm. This means that the maximum nonlinear phase shift is limited in this regime to $\Delta\varphi_{\text{max}}$, whatever be the mode area or fiber length.

In erbium–ytterbium, the value of n_2 is generally considered to be larger than in pure silica by a factor of 2 [73], because of the large concentration of codopants. If the Brillouin threshold is not reached in passive pigtails but in the amplifier itself and assuming that g_B is identical in an active fiber, the maximum phase shift thus becomes $\Delta\varphi_{\text{max}} = 0.2$ rad. Although still small, this variation can be measured, as demonstrated in the experimental results above.

As for the coherent combination of pulsed amplifier, we are only concerned with the differential phase shift between several amplifiers. Assuming that the peak power and fiber length may vary by 10% between each amplifier, the differential phase shift will be on the order of $\Delta\varphi_{\text{differential}} = \Delta\varphi_{\text{max}}/10$, which corresponds to $\lambda/300$. In this regime, the SPM-induced phase shift is not expected to be an issue.

For pulse durations shorter than the phonon lifetime (about 15 ns), the SBS threshold rises strongly and will not limit the pulse peak power anymore. In this regime, the peak power can be limited by SRS. Combining the relations (8.20) and (8.23) leads to

$$\Delta\varphi(\omega,t) = 16\frac{\omega n_2}{cg_R} = 32\frac{\pi n_2}{\lambda g_R}.$$

(8.25)

Using the above-mentioned values of $n_2 = 3 \times 10^{-20}$ m^2/W and $g_R = 7 \times 10^{-14}$ m/W for standard passive fibers, Eq. (8.25) yields $\Delta\varphi_{\text{max}} = 28$ rad at wavelength $\lambda = 1.55$ µm. These figures compare well with the results from Palese et al. [45] who measured ~ 7 waves SPM shift in a SRS peak power-limited pulse. Considering the same 10% dispersion of values of the peak powers and fiber lengths between the various amplifiers, the differential phase shift will be on the order of $\Delta\varphi_{\text{differential}} = \Delta\varphi_{\text{max}}/10$, which corresponds to $\lambda/2$. In this regime, the SPM-induced phase shift is expected to be an issue and will require either much tighter control on the dispersion between all amplifiers or intrapulse phase control.

8.5.3.2 Gain-Induced Phase Shift

Along with the Kerr-induced phase shift, another phase shift contribution, proper to doped fiber amplifiers, comes from the ion–photon interaction in the gain medium. Indeed, transitions from the fundamental to the excited state change the polarizability of the fiber doping ions.

For an amplifying or absorbing medium with electric susceptibility $\chi(\omega) = \chi'(\omega) - i\chi''(\omega)$ and linear response to the incident electric field, the causality of the time response leads to the well-known Kramers–Kronig relations between the real and imaginary parts of the frequency response [74].

When the susceptibility is small ($|\chi(\omega)| \ll 1$), the light absorption or amplification little affects the amplitude of the incident light field on a wavelength distance scale. Within this approximation, the refractive index can be related to the real part of χ by

$$n(\omega) \approx n_0 + \frac{\chi'(\omega)}{2n_0}, \tag{8.26}$$

while the intensity absorption can be related to the imaginary part:

$$\alpha(\omega) \approx -\frac{\omega}{c}\frac{\chi''(\omega)}{n_0}. \tag{8.27}$$

Therefore, the real part $\chi'(\omega)$ of the susceptibility describes the optical phase variation, whereas the imaginary part $\chi''(\omega)$ determines the absorption or amplification. $\chi''(\omega)$ can be computed from cross-sections and saturation parameters using fiber amplifier models. $\chi'(\omega)$ can then be computed using Kramers–Kronig relations. Similarly, the Kramers–Kronig relation allows relating the pulse-induced refractive index Δn to the pulse-induced absorption variation $\Delta\alpha$. Using Eqs. (8.26) and (8.27), this relation reads

$$\Delta n(\omega) = \frac{c}{\pi} \text{PV} \int_0^\infty \frac{\Delta\alpha(\omega_1)}{\omega^2 - \omega_1^2} d\omega_1, \tag{8.28}$$

with PV designating the Cauchy principal value.

In the case of a typical fiber amplifier where the signal power is large enough to saturate the transition, the signal gain depends on the pulse energy (Section 8.3.2). The system is thus no longer linear. In this situation, there has been a debate whether the Kramers–Kronig relation can be applied or not. The susceptibility for individual Stark level transitions was computed using the formalism of the density matrix by Desurvire [73]. In this work it is claimed that the Kramers–Kronig relation did not apply in the presence of saturation. Mathematically speaking, the susceptibility expression has one pole both in the upper and lower complex plans, meaning that Kramers–Kronig relation cannot be applied as it relies on Titchmarsh's theorem [75,76].

However, a different point of view has been given in the case of a saturated two-level system and more recently in the case of amplifying fibers [76,77]. According to these analyses, the influence of the pulse-induced saturation on the phase of the amplified pulse is equivalent to the effect on the pulse phase of an amplifier prepared by a stationary perturbation with the same saturating characteristics. This is similar to a linearization procedure. For this equivalent system, Kramers–Kronig relations apply. One could thus rely on Eq. (8.28) to compute the index variation due to inversion population depletion.

Let us now consider a doped fiber amplifier with length L, providing a gain g(z) in a small section dz located at position z along the fiber. We assume that this gain is produced by ions described by an effective two-level system. We name $N_1(z)$ and $N_2(z)$ the populations of the two levels, such that in the quasi-two-level approximation, the total population $N_1(z) + N_2(z) = N_0$ is constant.

The refractive index variation $\Delta n(\omega,z)$ caused by the pulse perturbation is related to the gain variation $\Delta g(\omega,z)$ by the relation

$$\Delta n(\omega, z) \approx \frac{c}{\pi} \text{PV} \int_0^\infty \frac{\Delta \alpha(\omega_1, z)}{\omega^2 - \omega_1^2} d\omega_1 = -\frac{c}{\pi} \text{PV} \int_0^\infty \frac{\Delta g(\omega_1, z)}{\omega^2 - \omega_1^2} d\omega_1. \tag{8.29}$$

The gain of a small fiber section of length dz in the step-index approximation is

$$g(\omega, z) = (\sigma_e(\omega) N_2(z) - \sigma_a(\omega) N_1(z)) \Gamma(\omega), \tag{8.30}$$

where $\Gamma(\omega)$ is the overlap factor between the optical mode and the doped core. Given the quasi-two-level approximation, the gain can be rewritten as a function of the population N_2 in the excited level. Then, one can write the gain variation due to the pulse as a function of the variation of population N_2:

$$\Delta g(\omega, z) = [(\sigma_e(\omega) + \sigma_a(\omega)) \Delta N_2(z)] \Gamma(\omega). \tag{8.31}$$

Figure 8.19 shows the spectrum of the refractive index variation, computed for an Er: Yb-doped phosphosilicate host, for a variation ΔN_2 equal to N_0. The slope is steepest at 1530 and 1545 nm, which correspond to the cross section inflection points. The variation is null at 1535 nm, which corresponds to the cross section maximum.

The phase shift after propagation in the medium is given by the following expression:

$$\Delta \varphi(\omega) = \frac{\omega}{c} \int_0^L \Delta n(\omega, z) dz \tag{8.32}$$

Therefore, using Eqs. (8.29) and (8.31), we obtain the phase shift induced by gain variation during the pulse:

$$\Delta \varphi(\omega, \Omega) = -\frac{\omega}{\pi} \int_0^L \Delta N_2(z) dz \, \text{PV} \int_0^\infty \frac{\sigma_a(\omega_1) + \sigma_e(\omega_1)}{\omega^2 - \omega_1^2} \Gamma(\omega_1) d\omega_1. \tag{8.33}$$

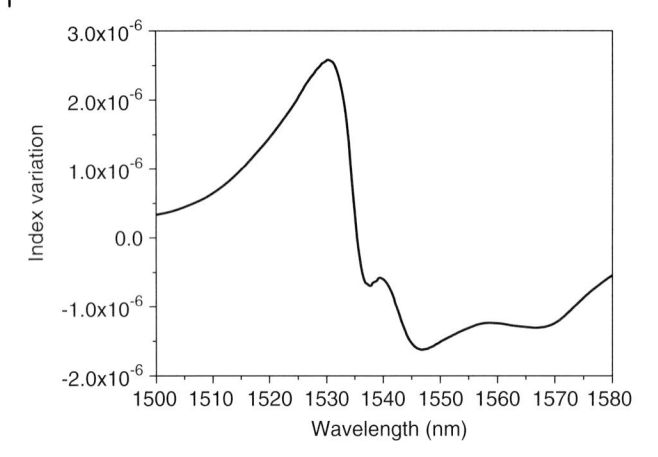

Figure 8.19 Refractive index variation with wavelength induced by polarizability change for erbium–ytterbium in a phosphosilicate host. Absorption at 1532 nm equals 20 dB/m.

In order to obtain an expression for the phase versus gain variation, we can introduce the amplifier initial total gain G_i just before the pulse arrival at time t_i and the final total gain G_f just after the pulse departure t_f at signal wavelength λ_s corresponding to pulsation ω_s:

$$\Delta\varphi(\omega) = -\omega_s \frac{\ln(G_f/G_i)}{\Gamma(\omega_s)[\sigma_a(\omega_s) + \sigma_e(\omega_s)]} \frac{1}{\pi} \text{PV} \int_0^\infty \frac{\sigma_a(\omega_1) + \sigma_e(\omega_1)}{\omega^2 - \omega_1^2} \Gamma(\omega_1) d\omega_1. \tag{8.34}$$

Equation (8.34) allows computing directly the phase variation from the gain variation. The gain is itself a function of the pulse energy.

Eventually, one can conclude that the phase variation should be computed using the Kramers–Kronig transform of the sum of the cross section and that it is proportional to the variation of the logarithm of the total amplifier gain. This means that the phase variation does not depend on the peak power, but rather on the pulse energy.

We used Eq. (8.34) to calculate the phase shift $\Delta\varphi$ generated between the beginning and the end of a pulse, in the case of the pulsed fiber amplifier that has been described in Section 8.5.2. This two-stage amplifier can deliver more than 80 W peak power. We consider rather long pulses to deplete strongly the population inversion and thus obtain a large gain variation $\Delta(\ln G)$. The peak power and the pulse shape for various conditions were computed as a function of the pulse duration between 300 ns and 10 μs for 1 kHz pulse repetition rate, using our fiber amplifier model [51]. We could thus calculate the variations of the logarithmic gain $\ln(G_f/G_i)$ and of the phase using Eq. (8.34). Figure 8.20 shows the excellent agreement between measurements and model. The variation of the phase shift with $\ln(G_f/G_i)$ is close to linear with a slope of 0.26.

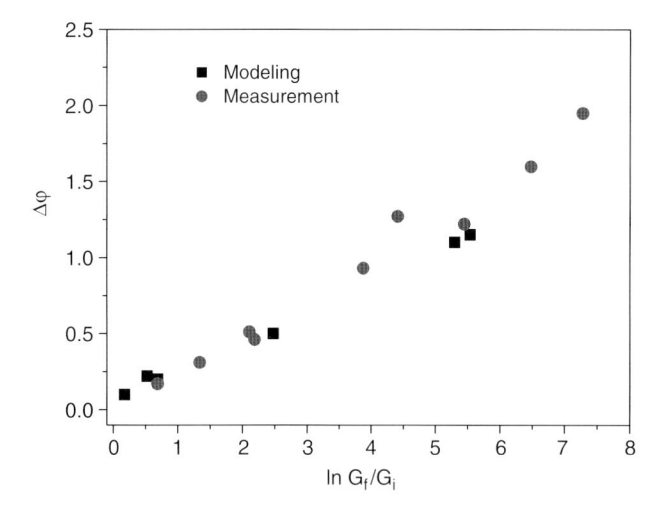

Figure 8.20 Evolution of the phase shift $\Delta\varphi$ with the logarithmic gain variation for various conditions (pulse duration ranging between 300 ns and 10 μs, different pump power, and pulse repetition rate ranging between 1 and 10 kHz). Comparison between measurement results (disks) and modeling results (squares).

In the case of coherent combination of pulsed amplifiers, the phase shift induced by the differential gain depletion between the various amplifiers can affect the relative phase shift between several amplifiers. According to Frantz and Nodvik relation (8.9), the initial and final gains are related to the pulse output energy and fiber saturation energy by

$$G_f = 1 + (G_i - 1)\exp\left(-\frac{E_{out}}{E_{sat}}\right). \tag{8.35}$$

For $G_i \gg 1$ and $G_f \gg 1$, we have

$$\ln\left(\frac{G_f}{G_i}\right) \sim -\frac{E_{out}}{E_{sat}}. \tag{8.36}$$

Assuming that the amplifiers to combine are built with the same fibers, operate at output energies $E_{out} \sim E_{sat}$, and that their output energies differ by about 10%, the differential phase shift will thus be $\Delta\varphi \sim 0.026$ corresponding to $\lambda/200$.

On the other hand, in the case of beam combining of very different amplifiers, or if the phase distortion is a problem for the application, it is possible to introduce a phase distortion precompensation in the pulse. In the broadband femtosecond regime, it is possible to achieve intrapulse chirp using active pulse shaping in the spectral domain. In the narrow-linewidth nanosecond regime, however, pulse phase control can be achieved in the time domain with high-speed electronics and recursive configuration. Indeed, Palese et al. [45] achieved the precompensation of a 15 rad phase distortion on a 1 ns pulse down to residual phase fluctuation of

~1 rad, induced by pulse-to-pulse energy fluctuation. However, this solution is rather complex and at the cost of a signal drop.

To conclude, based on our study of the phase perturbation caused by SPM and inversion depletion during the pulse, we can argue that both effects have negligible impact on the relative phase error in the long-pulse regime compared to the main source of perturbation.

8.6
Experimental Setup and Results of Coherent Beam Combining of Pulsed Amplifiers Using a Signal Leak between the Pulses

In this section, we report the experimental demonstration of CBC of two fiber amplifiers in 100 ns pulse regime using a signal leak between the pulses [37]. As discussed in Section 8.2, the information contained in the pulses is not sufficient for phase correction. Indeed, f_{noise} is on the order of 1 kHz and PRF is 10 kHz: PRF $< 20 \times f_{noise}$. Thus, rather than using the information contained in the pulses, we use a signal leak between the pulses for phase stabilization. Pulses of ~100 W SBS-limited peak power are combined with 95% efficiency, a residual phase error of $\lambda/27$, and no significant beam quality degradation.

Figure 8.21 shows the experimental setup of the CBC of the two pulsed fiber amplifiers. A 1.5 μm CW seed laser with 15 kHz linewidth is preamplified up to 200 mW and modulated with an AOM to produce 70 ns pulses at 10 kHz with Gaussian-like temporal shape and a low-power CW leak. A 50/50 coupler then splits the signal into two arms containing commercial pulse fiber amplifiers. One arm also contains a phase modulator (lithium niobate EOM) that applies a small phase modulation at frequency $F = \sim 1$ MHz. The amplified peak power is limited to ~100 W by SBS arising in the amplifiers [78]. As fiber lengths inside the amplifiers

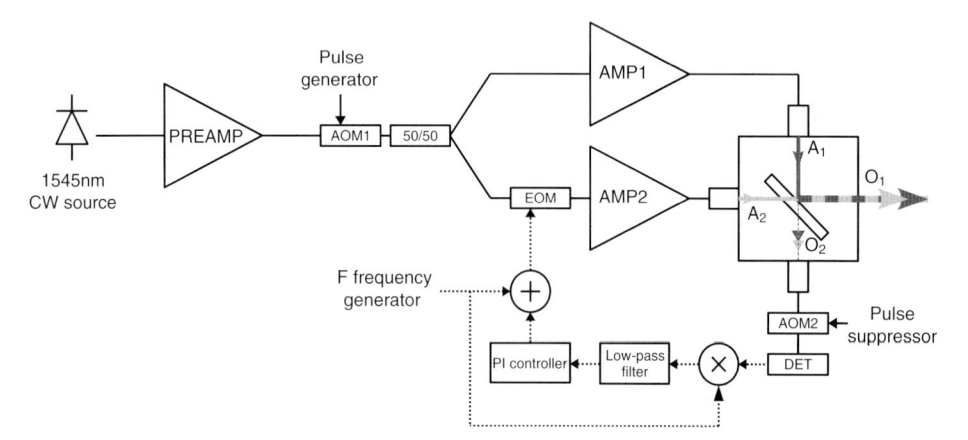

Figure 8.21 Experimental setup of CBC of two pulsed fiber amplifiers. PREAMP, preamplifier; AMP1 and AMP2, amplifiers; A_1 and A_2: amplifier outputs; O_1 and O_2, CBC outputs.

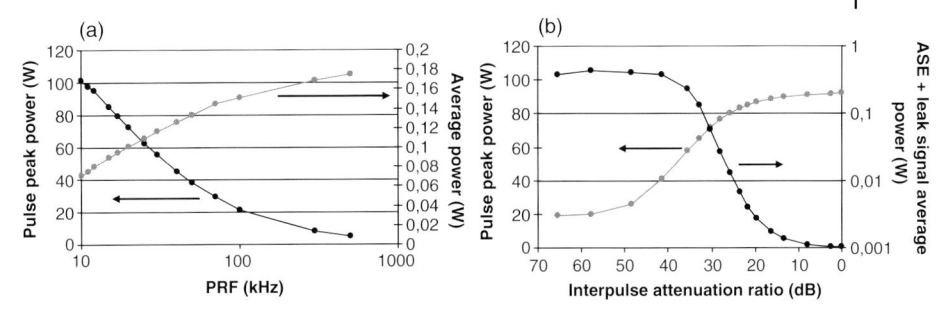

Figure 8.22 Typical performances of the fiber amplifiers (AMP1 and AMP2). (a) Output pulse peak power and total average power versus PRF. (b) Pulse peak power and interpulse average power (ASE + leakage signal) versus interpulse AOM extinction ratio.

are somewhat different, the SBS actual peak power limits are 95 and 123 W, respectively. The signal leak power must be kept as low as possible because the energy extracted by the leak degrades the performances of the pulses and adds noise. A second AOM is added before the detector to suppress the pulses and let only the amplified signal leak reach the detector. The phase stabilization is then performed as in CW regime, using the amplified leak modulated at F.

Figure 8.22 shows the typical performances of the fiber amplifiers for 70 ns pulse signal. The peak and average powers versus PRF are plotted in Figure 8.22a. Below 10 kHz, unwanted SBS effects appear. Figure 8.22b shows the effect of additional interpulse leak on the pulse peak power and on the interpulse signal power (sum of ASE and amplified signal leak) as a function of AOM extinction ratio for a $PRF = 10\,kHz$. An extinction ratio as low as 40 dB does not affect significantly the output peak power. However, an extinction ratio of 60 dB is chosen to keep the interpulse average power as low as possible. The interpulse average power baseline stems from ASE.

The two amplifier outputs (A_1 and A_2) are then collimated with 8 mm aspheric lenses and combined on a bulk beam splitter. Spatial alignment of the two beams is performed by superposing the beams (translation) and removing any fringe pattern (angle). Both beams then overlap along their propagation axes at both outputs O_1 and O_2. Temporal synchronization of the two beams is achieved by balancing the lengths of the two arms so that the amplified pulses reach the beam splitter at the same time (the <2 ns delay time is small compared to 70 ns pulse duration). This setup is equivalent to a Mach–Zehnder interferometer with outputs O_1 and O_2 having the same spatial characteristics as A_1 and A_2. Note that the same result could be obtained with a 50/50 fiber coupler in an all-fiber configuration, but limited to a much lower peak power to mitigate SBS induced by additional fiber length. This configuration is scalable by cascading bulk beam splitters in a similar way as in Ref. [10].

Figure 8.23 shows the interpulse power composed of ASE and signal leak. For comparison, we added the interpulse power in the case of high extinction ratio (no leak case). The integrated average power ratio is 1.5% in the ASE + leak and 98.5% in the pulses.

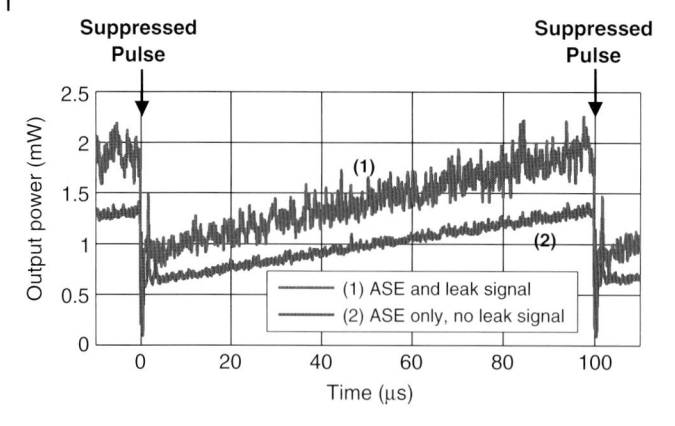

Figure 8.23 Interpulse signal power in A_2 after pulse suppression over a full 100 (s period. (1) ASE and leak signal. (2) ASE only (leak attenuation >70 dB).

CBC is simply achieved by minimizing the leak power at O_2, once the pulse is suppressed, with a frequency tagging controller developed for CW CBC. The signal can be either minimized or maximized. In the first case, the pulses are constructively combined at the output O_1 (and destructively combined at output O_2). The corresponding measured average output powers at O_1 are 146 and 7 mW for constructive and destructive interferences, respectively. The actual reflectivity of the beam splitter for A_1 is 42% (transmission 58%), thus compensating for the slight difference of peak power in the two amplifiers at O_2 output (see Table 8.6). This ensures that very low power can be achieved in O_2 when CBC is effective, thus enabling high efficiency in O_1.

Figure 8.24a shows the pulse profiles in various places. It has a Gaussian shape that is not affected by the amplification. Output peak power is 208 W (O_1 in Figure 8.24), to be compared to 218 W = 95 + 123 W (sum of A_1 and A_2 pulse amplitudes), thus achieving an average power combining efficiency of 95%. The few missing percents stem from nonperfect time and spatial overlap as well as the residual phase error. A 10 kHz triggered real-time acquisition board records ~100 points per pulse. A signal postprocessing then determines the energy of successive pulses. Figure 8.24b shows the pulse energy evolution over 10 s, with and

Table 8.6 Characteristics of the used fiber amplifiers in 70 ns pulse regime.

	Average power (mW)	Peak power (W)	ASE (%)	ASE + leak (%)	Average power @ O_1 (mW)	Average power @ O_2 (mW)
AMP1	67	95	1.0	1.4	28	39
AMP2	86	123	1.1	1.5	50	36

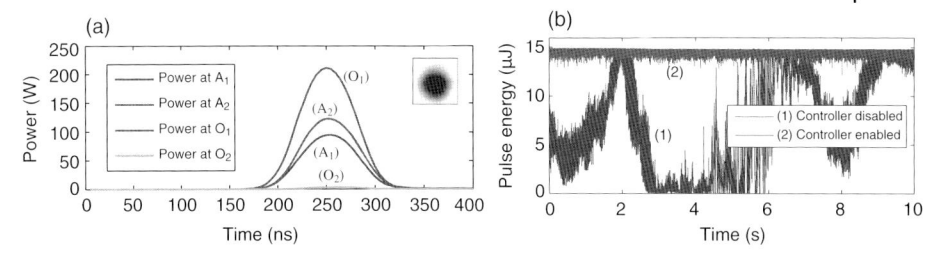

Figure 8.24 (a) Measured pulse profiles before and after combination (constructive and destructive interferences). Output beam profile is shown as an insert. (b) Evolution of combined pulse energy with controller disabled and enabled.

without the controller running. A residual phase error of $\lambda/27$ is evaluated using Eq. (8.37)

$$\Delta\varphi_{RMS} = 2\sqrt{\frac{\Delta V_{RMS}}{V_{MAX}}}, \qquad (8.37)$$

where $V(t)$ is the energy evolution of the pulses, V_{MAX} is its maximum value, and ΔV_{RMS} is its root mean square. We have observed that the combined beam quality is similar to those of the two individual AMP1 and AMP2 single-mode beams. The beam combining does not degrade the beam quality.

This demonstrates the CBC of two SBS-limited fiber amplifiers in the 100 ns pulsed regime. A signal leak between the pulses is used for phase error measurement and correction with the frequency tagging technique. Using fiber amplifiers with peak power limited to 95 and 123 W, respectively, we have obtained an output peak power of 208 W with no significant degradation of the beam quality. A power efficiency of 95% and a residual phase error of $\lambda/27$ have been achieved.

As expected from the results of phase fluctuation measurements during the laser pulse, these phase variation do not impair coherent combining in the case of low-peak-power fiber amplifiers. Coherent combining can be performed very efficiently with classical CW coherent combining techniques and optical and electronic components. This technique can be extended to a larger number of fibers, limited by component requirements (modulators and controllers) and in terms of geometrical combination by the same limits as in CW.

8.7
Alternative Techniques for Pulse Energy Scaling

For the purpose of scaling the energy of the pulses generated from fiber sources, other alternatives to coherent combining are proposed and studied worldwide. All these options aim at raising the threshold of detrimental nonlinear and damaging effects in fiber amplifiers, by either using a single fiber or spreading the power among multiple separate fibers.

In addition to the well-known techniques of SBS mitigation mentioned above (see Section 8.1), beam cleanup and phase conjugation are still other ways to increase the peak power in a single fiber. The principle of beam cleanup is to first generate high-energy pulses from a beam with rather poor spatial quality. Then, a nonlinear process is involved to transfer the high quality of a complementary low-power beam toward the high-peak-power beam [79]. This technique can be implemented through nonlinear SBS [80,81] or SRS [82] coupling. Phase conjugation uses double-pass configuration in a laser amplifier. The amplification induces optical aberrations that deteriorate the spatial quality of the output beam. Thanks to phase conjugation of the amplified laser beam, these aberrations can be theoretically perfectly compensated. Phase conjugation can be obtained by nonlinear wave mixing or SBS in a medium such as a nonlinear crystal or a nonlinear fiber [83].

The main limitation of the techniques based on nonlinear wave mixing arises from the proximity of the nonlinear effect and optical damage thresholds. This results in a limited operating range in terms of peak power. One should also mention adaptative optics, particularly digital holography techniques. This last technique relies on a spatial light modulator to clean up the wavefront of a laser beam, but the BW is limited by the slow feedback loop calculations [84].

Apart from the techniques based on active phase control, passive coherent combining techniques are possible. They rely on the coupling of the power emitted from multiple laser sources into a common device. For example, various laser cavities can be coupled through a common output mirror [85–87]. Another passive coherent combining technique consists in feeding multiple laser amplifiers back with a fraction of their interference signal [88]. These techniques are mainly limited in the number of lasers or amplifiers that can be combined. In all these configurations, a set of laser cavities self-organize to find out and share a common cavity mode. It can be proved both mathematically and numerically that such self-organizing is not possible for more than a few tens of lasers [89,90].

Another way to distribute power between multiple laser media is to use multicore fibers to amplify pulsed laser light. Spreading the power over multiple rare earth-doped cores results in an extended mode area very similar to the extended effective area of LMA fibers.

One possibility consists in dividing the fiber core into small multiple cores close enough to be strongly coupled. This bundle of coupled cores can be designed to guide and amplify one supermode. This results in an increased effective area for the guided mode and an increased level of achievable energy per pulse [29,91].

It is also possible to manufacture multicore fibers with larger and uncoupled cores. In this case, each core acts as an individual amplifying medium in a configuration very similar to that of multiple lasers coupled through a common laser cavity mirror. Beams amplified separately in these fiber cores can be combined using coherent or spectral combining, just as if each core was an individual laser amplifier [92,93].

The major issue of multicore fibers, whatever the core size and the level of core-coupling, is their manufacturing that is a very demanding and complex multistep process. Such fiber amplifiers are most often very expensive components and with potentially high losses.

Last but not least, spectral combining techniques have also demonstrated very good results for pulsed laser beam combination. For many years, they have been the only combining techniques to operate in the pulsed regime. The principle of spectral combining is to perform incoherent combining of multiple beams at different wavelengths. All those beams are eventually overlapped using a diffractive element such as a diffraction grating [32,36], a volume Bragg grating (VBG) [94], or other spectrally selective beam splitter [95].

An appealing feature of spectral combining is that no control is required on the phase of each beam. It can be operated either in CW or pulsed regime. In the pulsed regime, an additional difficulty is the synchronization of the arrival time of the pulses. Main limitations of spectral combining are damage threshold and thermal sensitivity of the diffractive element and cross talk between wavelength channels [94,96]. Furthermore, the broad and discrete spectrum delivered after spectral combining is not suitable for all applications.

8.8
Conclusion

In this chapter, we presented the key issues in the coherent combination of N pulsed fiber MOPFAs. In particular, the PRF must be compared with the highest phase noise frequency to determine the proper technique. For high PRF (PRF $> 40 \times N \times f_{noise}$), all techniques are compatible; for medium PRF ($20 \times f_{noise} <$ PRF $< 20 \times N \times f_{noise}$), only the direct detection techniques are compatible; for low PRF (PRF $< 20 \times f_{noise}$), a signal information is required between the pulses (e.g., signal leak).

For a pulsed fiber amplifier, the output phase fluctuations have two components: the random phase noise and the deterministic in-pulse phase distortion. Measurements of the phase noise and in-pulse phase distortion show that the requirements of coherent combining should be considered as early as the design of each amplifier.

The phase noise is mainly due to background conditions and is identical as in CW regime. The noise is generated by thermal, mechanical, or electrical fluctuations and we show that these contributions should be minimized in the first place.

On the contrary, the deterministic in-pulse phase distortion is specific to the pulsed operation. The peak power of narrow-linewidth fiber pulse amplifiers is usually limited by SBS and SRS. Before that limit, phase distortion occurs for much lower powers, due to gain depletion and SPM. The influence of the distortion can be minimized by using identical amplifiers. The measured in-pulse phase fluctuations are in good agreement with theoretical predictions. The phase distortion induced by gain depletion is not expected to be sensitive to amplifier mismatch, unlike SPM, especially at SRS-limited peak power levels (<10 ns pulse duration).

Eventually, we demonstrate the coherent combination of two narrow-linewidth, SBS-limited pulsed fiber amplifiers at 100 W peak power level. Our LOCSET technique uses a signal leak between the pulses.

Scaling CBC to higher pulse peak power in narrow-linewidth operation is a challenge, particularly in the SRS-limited peak power regime (pulse duration <15 ns). Indeed, in this regime, the use of nonidentical amplifiers will lead to different SPM phase distortions and degrade the beam combining efficiency. Additional fast in-pulse phase precompensation system has been developed and implemented, but is complex and inefficient. The ultimate limit is then the residual random SPM phase fluctuations induced by energy fluctuation.

References

1 Kanamori, H., Yokota, H., Tanaka, G., Watanabe, M., Ishiguro, Y., Yoshida, I., Kakii, T., Itoh, S., Asano, Y., and Tanaka, S. (1986) Transmission characteristics and reliability of pure-silica-core single-mode fibers. *J. Lightwave Technol.*, **LT-4**, 1144–1150.

2 Frehlich, R.G. and Kavaya, M.J. (1991) Coherent laser radar performance for general atmospheric refractive turbulence. *Appl. Opt.*, **30**, 5325–5352.

3 Berni, A.J. (1994) Remote sensing of seismic vibrations by laser Doppler interferometry. *Geophysics*, **59**, 1856–1867.

4 Shang, J., He, Y., Liu, D., Zang, H., and Chen, W. (2009) Laser Doppler vibrometer for real-time speech-signal acquirement. *Chinese Opt. Lett.*, **7**, 732–733.

5 Paul, D.M. and Jackson, D.A. (1971) Rapid velocity sensor using a static confocal Fabry–Perot and a single frequency argon laser. *J. Phys. E*, **4**, 170–177.

6 Barker, L.M. and Hollenbach, R.E. (1972) Laser interferometer for measuring high velocities of any reflecting surface. *J. Appl. Phys.*, **43**, 4669–4675.

7 Pillet, G., Morvan, L., Dolfi, D., and Huignard, J.-P. (2008) Wideband dual-frequency lidar-radar for high-resolution ranging, profilometry, and Doppler measurement. *Proc. SPIE*, **7114**, 71140.

8 Kawakami, S. and Nishida, S. (1974) Characteristics of a doubly clad optical fiber with a low-index inner cladding. *IEEE J. Quantum Electron.*, **10**, 879–887.

9 Po, H., Snitzer, E., Tumminelli, L., Hakimi, F., Chu, N.M., and Haw, T. (1989) Doubly clad high brightness Nd fiber laser pumped by GaAlAs phased array. Optical Fiber Communication Conference, Houstan, Texas, Paper PD7.

10 Shiner, B. (2009) Recent technical and marketing developments in high power fiber lasers. The European Conference on Lasers and Electro-Optics, Munich, Germany, Paper TF1-2.

11 Thieme, J. (2010) Power scaling of fiber lasers. SPIE Photonics West, Paper 7580-19.

12 Gapontsev, V., Gapontsev, D., Platonov, N., Shkurikhin, O., Fomin, V., Mashkin, A., Abramovb, M., and Ferin, S. (2005) 2 kW CW ytterbium fiber laser with record diffraction-limited brightness. Proceedings of CLEO Europe, p. 508.

13 Ehrenreich, T., Leveille, R., Majid, I., Tankala, K., Rines, G.A., and Moulton, P.F. (2010) 1-kW, all-glass Tm:fiber laser. SPIE Photonics West.

14 Jeong, Y., Nilsson, J., Sahu, J.K., Payne, D.N., Horley, R., Hickey, L.M.B., and Turner, P.W. (2007) Power scaling of single-frequency ytterbium-doped fiber master-oscillator power-amplifier sources up to 500 W. *IEEE J. Sel. Top. Quantum Electron.*, **13**, 546–551.

15 Offerhaus, H.L., Broderick, N.G., Richardson, D.J., Sammut, R., Caplen, J., and Dong, L. (1998) High-energy single-transverse-mode Q-switched fiber laser based on a multimode large-mode-area

erbium-doped fiber. *Opt. Lett.*, **23**, 1683–1685.

16 Ramachandran, S., Nicholson, J.W., Ghalmi, S., Yan, M.F., Wisk, P., Monberg, E., and Dimarcello, F.V. (2006) Light propagation with ultralarge modal areas in optical fibers. *Opt. Lett.*, **31**, 1797–1799.

17 Shi, W., Petersen, E.B., Leigh, M., Zong, J., Yao, Z., Chavez-Pirson, A., and Peyghambarian, N. (2009) High SBS-threshold single-mode single-frequency monolithic pulsed fiber laser in the C-band. *Opt. Express*, **17**, 8237–8245.

18 Shi, W., Petersen, E.B., Nguyen, D.T., Yao, Z., Chavez-Pirson, A., Peyghambarian, N., and Yu, J. (2011) 220 µJ monolithic single-frequency Q-switched fiber laser at 2 µm by using highly Tm-doped germanate fibers. *Opt. Lett.*, **36**, 3575–3577.

19 Li, M.-J., Chen, X., Wang, J., Gray, S., Liu, A., Demeritt, J.A., Ruffin, A.B., Crowley, A. M., Walton, D.T., and Zenteno, L.A. (2007) Al/Ge co-doped large mode area fiber with high SBS threshold. *Opt. Express*, **15**, 8290–8299.

20 Kovalev, V.I. and Harrison, R.G. (2006) Suppression of stimulated Brillouin scattering in high-power single-frequency fiber amplifiers. *Opt. Lett.*, **31**, 161.

21 Canat, G., Lombard, L., Bourdon, P., Jolivet, V., Vasseur, O., Jetschke, S., Unger, S., and Kirchhof, J. (2009) Measurement and modeling of Brillouin scattering in a multifilament core fiber. CLEO 2009, Paper JTuB3.

22 Shi, W., Petersen, E.B., Yao, Z., Nguyen, D.T., Zong, J., Stephen, M.A., Chavez-Pirson, A., and Peyghambarian, N. (2010) Kilowatt-level stimulated-Brillouin-scattering-threshold monolithic transform-limited 100ns pulsed fiber laser at 1530 nm. *Opt. Lett.*, **35**, 2418–2420.

23 Limpert, J., Hoffer, S., Liem, A., Zellmer, H., Tünnermann, A., Knoke, S., and Voelckel, H. (2002) 100-W average-power high-energy nanosecond fiber amplifier. *Appl. Phys. B*, **75**, 477–479.

24 Fomin, V., Abramov, M., Ferin, A., Abramov, A., Mochalov, D., Platonov, N., and Gapontsev, V. (2010) 10 kW singlemode fiber laser. Proceedings of 5th International Symposium on High-Power Fiber Lasers and Their Applications.

25 Engin, D., Lu, W., Akbulut, M., McIntosh, B., Verdun, H., and Gupta, S. (2011) 1 kW cw Yb-fiber-amplifier with <0.5 GHz linewidth and near diffraction limited beam-quality, for coherent combining application. *Proc. SPIE*, **7914**, 791407.

26 Biernson, G. and Lucy, R.F. (1963) Requirements of a coherent laser pulse-Doppler radar. *Proc. IEEE*, **51**, 202–213.

27 Goodman, J.W. (1966) Comparative performance of optical-radar detection techniques. *IEEE Trans. Aerosp. Electron. Syst.*, **2**, 526–535.

28 Dolfi-Bouteyre, A., Canat, G., Valla, M., Augere, B., Besson, C., Goular, D., Lombard, L., Cariou, J., Durecu, A., Fleury, D., Bricteux, L., Brousmiche, S., Lugan, S., and Macq, B. (2009). Pulsed 1.5 µm LIDAR for axial aircraft wake vortex detection based on high-brightness large-core fiber amplifier. *IEEE J. Sel. Top. Quantum Electron.*, **15**, 441–450.

29 Canat, G., Jetschke, S., Unger, S., Lombard, L., Bourdon, P., Kirchhof, J., Jolivet, V., Dolfi, A., and Vasseur, O. (2008) Multifilament-core fibers for high energy pulse amplification at 1.5 µm with excellent beam quality. *Opt. Lett.*, **33**, 2701–2703.

30 Akbulut, M., Hwang, J., Kimpel, F., Gupta, S., and Verdun, H. (2011) Pulsed coherent fiber lidar transceiver for aircraft in-flight turbulence and wake-vortex hazard detection. *Proc. SPIE*, **8037**, 80370.

31 Yu, C.X., Augst, S.J., Redmond, S.M., Goldizen, K.C., Murphy, D.V., Sanchez, A., and Fan, T.Y. (2011) Coherent combining of a 4 kW, eight-element fiber amplifier array. *Opt. Lett.*, **36**, 2686–2688.

32 Schmidt, O., Klingebiel, S., Ortac, B., Roser, F., Bruckner, F., Clausnitzer, T., Kley, E.-B., Limpert, J., and Tünnermann, A. (2008) Spectral combining of pulsed fiber lasers: scaling considerations. *Proc. SPIE*, **6873**, 687317.

33 Shirakawa, A., Saitou, T., Sekiguchi, T., and Ueda, K.-I. (2002) Coherent addition of fiber lasers by use of a fiber coupler. *Opt. Express*, **10**, 1167–1172.

34 McNaught, S.J., Komine, H., Weiss, S.B., Simpson, R., Johnson, A.M.F., Machan, J., Asman, C.P., Weber, M., Jones, G.C., Valley, M.M., Jankevics, A., Burchman, D., McClellan, M., Sollee, J., Marmo, J., and

Injeyan, H. (2009) 100 kW coherently combined slab MOPAs. CLEO 2009, Paper CThA1.

35 Flores, A., Shay, T.M., Lu, C.A., Robin, C., Pulford, B., Sanchez, A.D., Hult, D.W., and Rowland, K.B. (2011) Coherent beam combining of fiber amplifiers in a kW regime. CLEO 2009, Paper CFE3.

36 Schmidt, O., Wirth, C., Tsybin, I., Schreiber, T., Eberhardt, R., Limpert, J., and Tünnermann, A. (2009) Average power of 1.1 kW from spectrally combined, fiber-amplified, nanosecond-pulsed sources. *Opt. Lett.*, **34**, 1567–1569.

37 Lombard, L., Azarian, A., Cadoret, K., Bourdon, P., Goular, D., Canat, G., Jolivet, V., Jaouen, Y., and Vasseur, O. (2011) Coherent beam combination of narrow-linewidth 1.5 μm fiber amplifiers in a long-pulse regime. *Opt. Lett.*, **36**, 523–525.

38 Daniault, L., Hanna, M., Lombard, L., Zaouter, Y., Mottay, E., Goular, D., Bourdon, P., Druon, F., and Georges, P. (2011) Coherent beam combining of two femtosecond fiber chirped-pulse amplifiers. *Opt. Lett.*, **36**, 621–623.

39 Seise, E., Klenke, A., Limpert, J., and Tünnermann, A. (2010) Coherent addition of fiber-amplified ultrashort laser pulses. *Opt. Express*, **18**, 27827–27835.

40 Demoustier, S., Brignon, A., Lallier, E., Huignard, J.-P., and Primot, J. (2006) Coherent combining of 1.5 μm Er-Yb doped single mode fiber amplifiers. CLEO 2006, Paper CThAA5.

41 Yu, C.X., Kansky, J.E., Shaw, S.E.J., Murphy, D.V., and Higgs, C. (2006) Coherent beam combining of large number of PM fibres in 2-D fibre array. *Electron. Lett.*, **42**, 1024–1025.

42 Seise, E., Klenke, A., Breitkopf, S., Limpert, J., and Tünnermann, A. (2011) 88 W 0.5 mJ femtosecond laser pulses from two coherently combined fiber amplifiers. *Opt. Lett.*, **36**, 3858–3860.

43 Ma, Y., Zhou, P., Wang, X., Ma, H., Xu, X., Si, L., Liu, Z., and Zhao, Y. (2011) Active phase locking of fiber amplifiers using sine-cosine single-frequency dithering technique. *Appl. Opt.*, **50**, 3330–3336.

44 Redmond, S.M., Kansky, J.E., Creedon, K.J., Missaggia, L.J., Connors, M.K., Turner, G.W., Fan, T.Y., and Sanchez-Rubio, A. (2011) Active coherent combination of >200 semiconductor amplifiers using a SPGD algorithm. CLEO 2011, Paper CTuV1.

45 Palese, S., Cheung, E., Goodno, G., Shih, C.-C., Di Teodoro, F., McComb, T., and Weber, M. (2012) Coherent combining of pulsed fiber amplifiers in the nonlinear chirp regime with intra-pulse phase control. *Opt. Express*, **20**, 7422–7435.

46 Injeyan, H. and Goodno, G. (2011) *High Power Laser Handbook*, McGraw-Hill.

47 Kryukov, P.G. and Letokhov, V.S. (1970) Propagation of a light pulse in a resonantly amplifying (absorbing) medium. *Sov. Phys. Usp.*, **12**, 641–672.

48 Koechner, W. (2006) *Solid-State Laser Engineering*, 6th edn, Springer.

49 da Silva, V.L., Silberberg, Y., Heritage, J.P., Chase, E.W., Saifi, M.A., and Andrejco, M.J. (1991) Femtosecond accumulated photon echo in Er-doped fibers. *Opt. Lett.*, **16**, 1340–1342.

50 Wang, Y. and Po, H. (2003) Dynamic characteristics of double-clad fiber amplifiers for high-power pulse amplification. *J. Lightwave Technol.*, **21**, 2262–2270.

51 Canat, G., Mollier, J.-C., Bouzinac, J.-P., Williams, G.M., Cole, B., Goldberg, L., Jaouën, Y., and Kulcsar, G. (2005) Dynamics of high-power erbium–ytterbium fiber amplifiers. *J. Opt. Soc. Am. B*, **22**, 2308–2318.

52 Frantz, L.M. and Nodvik, J.S. (1963) Theory of pulse propagation in a laser amplifier. *J. Appl. Phys.*, **34**, 2346–2349.

53 Siegman, A.E. (1986) *Lasers*, University Science Books.

54 Canat, G. (2006) Conception et réalisation d'une source impulsionnelle à fibre dopée erbium-ytterbium millijoule de grande brillance spectrale, PhD Thesis. École nationale supérieure de l'aéronautique et de l'espace, Toulouse, France.

55 Schimpf, D.N., Ruchert, C., Nodop, D., Limpert, J., Tünnermann, A., and Salin, F. (2008) Compensation of pulse-distortion in saturated laser amplifiers. *Opt. Express*, **16**, 17637–17646.

56 Renaud, C.C., Offerhaus, H.L., Alvarez-Chavez, J.A., Nilsson, J., Clarkson, W.A., Turner, P.W., and Richardson, D.J. (2001)

Characteristics of Q-switched cladding-pumped ytterbium-doped fiber lasers with different high-energy fiber designs. *IEEE J. Quantum Electron.*, **37**, 199–206.

57 Sintov, Y., Katz, O., Glick, Y., Acco, S., Nafcha, Y., Englander, A., and Lavi, R. (2006) Extractable energy from ytterbium-doped high-energy pulsed fiber amplifiers and lasers. *J. Opt. Soc. Am. B*, **23**, 218–230.

58 Sintov, Y., Glick, Y., Koplowitch, T., and Nafcha, Y. (2008) Extractable energy from erbium–ytterbium co-doped pulsed fiber amplifiers and lasers. *Opt. Commun.*, **281**, 1162–1178.

59 Boyd, R. (2003) *Nonlinear Optics*, 2nd edn, Academic Press.

60 Jagannathan, A. and Orbach, R. (1990) Temperature and frequency dependence of the sound velocity in vitreous silica due to scattering off localized modes. *Phys. Rev. B*, **41**, 3153–3157.

61 Agrawal, G.P. (2007) *Nonlinear Fiber Optics*, 4th edn, Academic Press.

62 Nikles, M., Thevenaz, L., and Robert, P.A. (1997) Brillouin gain spectrum characterization in single-mode optical fibers. *J. Lightwave Technol.*, **15**, 1842–1851.

63 Smith, R.G. (1972) Optical power handling capacity of low loss optical fibers as determined by stimulated Raman and Brillouin scattering. *Appl. Opt.*, **11**, 2489–2494.

64 Jaouën, Y., Canat, G., Grot, S., and Bordais, S. (2006) Power limitation induced by nonlinear effects in pulsed high-power fiber amplifiers. *C.R. Phys.*, **7**, 163–169.

65 van Deventer, M.O. and Boot, A.J. (1994) Polarization properties of stimulated Brillouin scattering in single-mode fibers. *J. Lightwave Technol.*, **12**, 585–590.

66 Imai, Y. and Shimada, N. (1993) Dependence of stimulated Brillouin scattering on temperature distribution in polarization-maintaining fibers. *IEEE Photonics Technol. Lett.*, **5**, 1335–1337.

67 Horiguchi, T., Kurashima, T., and Tateda, M. (1989) Tensile strain dependence of Brillouin frequency shift in silica optical fibers. *IEEE Photon. Technol. Lett.*, **1**, 107–108.

68 Thomas, P.J., Rowell, N.L., van Driel, H.M., and Stegeman, G.I. (1979) Normal acoustic modes and Brillouin scattering in single-mode optical fibers. *Phys. Rev. B*, **19**, 4986–4998.

69 Shiraki, K., Ohashi, M., and Tateda, M. (1995) Suppression of stimulated Brillouin scattering in a fibre by changing the core radius. *Electron. Lett.*, **31**, 668–669.

70 Kovalev, V.I. and Harrison, R.G. (2002) Waveguide-induced inhomogeneous spectral broadening of stimulated Brillouin scattering in optical fiber. *Opt. Lett.*, **27**, 2022–2024.

71 Dragic, D., Liu, C.-H., Papen, G.C., and Galvanauskas, A. (2005) Optical fiber with an acoustic guiding layer for stimulated Brillouin scattering suppression. CLEO/QELS 2005, Paper CThZ3.

72 Li, M., Chen, X., Wang, J., Gray, S., Liu, A., Demeritt, J., Ruffin, A., Crowley, A., Walton, D., and Zenteno, L. (2007) Al/Ge co-doped large mode area fiber with high SBS threshold. *Opt. Express*, **15**, 8290–8299.

73 Desurvire, E. (1990) Study of the complex atomic susceptibility of erbium-doped fiber amplifiers. *J. Lightwave Technol.*, **8**, 517–1527.

74 Hutchings, D.C., Sheik-Bahae, M., Hagan, D.J., and Van Stryland, E.W. (1990) Kramers–Krönig relations in nonlinear optics. *Opt. Quantum Electron.*, **24**, 1–30.

75 Titchmarsh, E. (1986). *Introduction to the Theory of Fourier integrals*, 2nd edn, Oxford University, Clarendon Press.

76 Bisson, J.F. and Kouznetsov, D. (2008) Comments on "Study of the complex atomic susceptibility of erbium-doped fiber amplifiers". *J. Lightwave Technol.*, **26**, 457–459.

77 Bochove, E. (2004) Nonlinear refractive index of a rare-earth-doped fiber laser. *Opt. Lett.*, **29**, 2414–2416.

78 Kulcsar, G., Jaouën, Y., Canat, G., Olmedo, E., and Debarge, G. (2003) Multiple-Stokes stimulated Brillouin scattering generation in pulsed high-power double-cladding Er^{3+}-Yb^{3+}-codoped fiber amplifier. *Photonics Technol. Lett.*, **15**, 801–803.

79 Lombard, L., Brignon, A., Huignard, J.P., Lallier, E., Lucas-Leclin, G., Georges, P., Pauliat, G., and Roosen, G. (2004) Diffraction-limited polarized emission from a multimode ytterbium fiber

amplifier after a nonlinear beam converter. *Opt. Lett.*, **29**, 989–991.

80 Lombard, L., Brignon, A., Huignard, J.P., Lallier, E., and Georges, P. (2006) Beam cleanup in a self-aligned gradient-index Brillouin cavity for high-power multimode fiber amplifiers. *Opt. Lett.*, **31**, 158–160.

81 Steinhausser, B., Brignon, A., Lallier, E., Huignard, J.-P., and Georges, P. (2007) High energy, single-mode, narrow-linewidth fiber laser source using stimulated Brillouin scattering beam cleanup. *Opt. Express*, **15**, 6464–6469.

82 Baek, S.H. and Roh, W.B. (2004) Single-mode Raman fiber laser based on a multimode fiber. *Opt. Lett.*, **29**, 153–155.

83 Harrison, R.G., Kovalev, V.I., Lu, W., and Yu, D. (1999) SBS self-phase conjugation of CW Nd:YAG laser radiation in an optical fibre. *Opt. Commun.*, **163**, 208–211.

84 Paurisse, M., Hanna, M., Druon, F., Georges, P., Bellanger, C., Brignon, A., and Huignard, J-P. (2009) Phase and amplitude control of a multimode LMA fiber beam by use of digital holography. *Opt. Express*, **17**, 13000–13008.

85 Shirakawa, A., Saitou, T., Sekiguchi, T., and Ueda, K.-I. (2002) Coherent addition of fiber lasers by use of a fiber coupler. *Opt. Express*, **10**, 1167–1172.

86 Sabourdy, D., Kermene, V., Desfarges-Berthelemot, A., Lefort, L., Barthelemy, A., Even, P., and Pureur, D. (2003) Efficient coherent combining of widely tunable fiber lasers. *Opt. Express*, **11**, 87–97.

87 Wang, B. and Sanchez, A. (2011) All-fiber passive coherent combining of high power lasers. *Opt. Eng.*, **50**, 111606.

88 Lhermite, J., Desfarges-Berthelemot, A., Kermene, V., and Barthelemy, A. (2007) Passive phase locking of an array of four fiber amplifiers by an all-optical feedback loop. *Opt. Lett.*, **32**, 1842–1844.

89 Bochove, E.J. and Shakir, S.A. (2009) Analysis of a spatial-filtering passive fiber laser beam combining system. *IEEE J. Sel. Top. Quantum Electron.*, **15**, 320–327.

90 Chang, W.-Z., Wu, T.-W., Winful, H.G., and Galvanauskas, A. (2010) Array size scalability of passively coherently phased fiber laser arrays. *Opt. Express*, **18**, 9634–9642.

91 Michaille, L., Bennett, C.R., Taylor, D.M., Shepherd, T.J., Broeng, J., Simonsen, H.R., and Petersson, A. (2005) Phase locking and supermode selection in multicore photonic crystal fiber lasers with a large doped area. *Opt. Lett.*, **30**, 1668–1670.

92 Hartl, I., Marcinkevicius, A., McKay, H.A., Dong, L., and Fermann, M.E. (2009) Coherent beam combination using multi-core leakage-channel fibers. Advanced Solid State Photonics Conference, Paper TuA6.

93 Lhermite, J., Suran, E., Kermene, V., Louradour, F., Desfarges-Berthelemot, A., and Barthélémy, A. (2010) Coherent combining of 49 laser beams from a multiple core optical fiber by a spatial light modulator. *Opt. Express*, **18**, 4783–4789.

94 Sevian, A., Andrusyak, O., Ciapurin, I., Smirnov, V., Venus, G., and Glebov, L. (2008) Efficient power scaling of laser radiation by spectral beam combining. *Opt. Lett.*, **33**, 384–386.

95 Schmidt, O., Wirth, C., Nodop, D., Limpert, J., Schreiber, T., Peschel, T., Eberhardt, R., and Tünnermann, A. (2009) Spectral beam combination of fiber amplified ns-pulses by means of interference filters. *Opt. Express*, **17**, 22974–22982.

96 Drachenberg, D., Divliansky, I., Smirnov, V., Venus, G., and Glebov, L. (2011) High power spectral beam combining of fiber lasers with ultra high spectral density by thermal tuning of volume Bragg gratings. *Proc. SPIE*, **7914**, 79141.

9
Coherent Beam Combining in the Femtosecond Regime

Marc Hanna, Dimitrios N. Papadopoulos, Louis Daniault, Frédéric Druon, Patrick Georges, and Yoann Zaouter

9.1
Introduction

Coherent beam combining of several laser sources is currently considered as a scalable way to increase the achievable power of lasers once single-source limits are reached, while retaining the beam quality of the individual sources. These limits fall into the following categories. For average power-limited sources, the problems are generally related to the thermal behavior of the gain medium under high-power pumping. For peak power-limited sources, the issues are usually connected with optical damage of the materials or detrimental nonlinear effects. Coherent beam combining has been first implemented for CW lasers [1], for which the thermal problems dominate. In this regime, if the spectral properties are not considered, other combining techniques such as spectral combining can be used. Coherent combining has been extended to the pulsed regime shortly after, first for nanosecond pulses (see Chapter 8) [2,3] and then for femtosecond pulses [4,5], allowing the scaling of peak power. These demonstrations have been done with fiber amplifiers, where nonlinear limitations dominate. For ultrashort pulses, coherent combining is the only combining scheme that allows preserving the properties of each combined beam, both for beam quality and spectral/temporal properties.

In this chapter, we will discuss the principles and specifics of coherent combining in the femtosecond regime with the aim of scaling the power of laser systems and describe some experiments that have been performed so far in this field. These experimental results have been mostly obtained in the context of fiber-based amplifier systems, with pulse energies ranging from a few microjoules to a few millijoules. Although there is no demonstration with bulk amplifier systems yet, nothing prevents using coherent combining in this context. The building blocks of a typical system that combines chirped-pulse amplifiers are shown in Figure 9.1. Note that the compressor stage might be taken before the combination step if power handling issues are encountered. In this case, the optical power can be distributed over N compressors.

Coherent Laser Beam Combining, First Edition. Edited by Arnaud Brignon.
© 2013 Wiley-VCH Verlag GmbH & Co. KGaA. Published 2013 by Wiley-VCH Verlag GmbH & Co. KGaA.

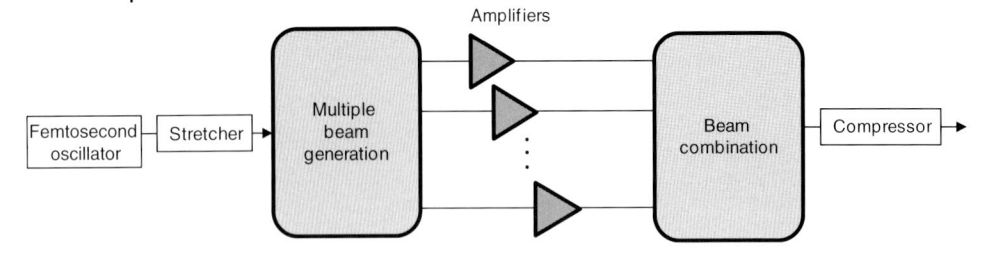

Figure 9.1 Generic femtosecond coherent combining setup.

In the later sections of this chapter, we extend the discussion to the coherent addition of ultrashort optical pulses in different configurations. We first discuss the idea of divided-pulse amplification that makes use of temporal multiplexing to scale down the peak power prior to amplification and temporal coherent recombining. Then, cavity enhancement experiments in the femtosecond regime are presented. In this case, the quantity that is scaled is the intracavity power. Finally, we describe pulse synthesis experiments, where pulses with different spectral contents are coherently combined to generate pulse widths on the order of an optical cycle.

9.2
General Aspects of Coherent Combining over Large Optical Bandwidths

In this section, we lay out some basic considerations in ultrafast optics in order to be able to describe the coherent addition of several ultrashort optical pulses. Then, we examine theoretically how the combining efficiency is related to the pulse parameters and show examples of the effects such as dispersion or self-phase modulation on a combining system performance. Finally, we briefly touch upon the fact that in a real system where the temporal properties of each of the combined beams are not exactly the same, some combining geometries lead to spatiotemporal distortions of the final beam. These considerations prepare the reader for the next section that describes a few actual experiments of coherent beam combining.

9.2.1
Description and Propagation of Femtosecond Pulses

All the spatial issues encountered in CW beam combining are essentially unmodified when considering ultrashort pulses, so we focus here on the temporal/spectral aspects. We will see that spatiotemporal coupling issues might arise in some combining geometries. The femtosecond pulses are described in the time domain using the complex envelope $E(t)$, where the optical carrier frequency has been removed, that can be written as

$$E(t) = \sqrt{I(t)} \exp(-i\varphi(t)), \tag{9.1}$$

where $I(t)$ is the time-dependent optical intensity and $\varphi(t)$ is the temporal phase. This envelope can be Fourier transformed to get an equivalent representation of the pulse in the frequency domain:

$$\tilde{E}(\omega) = \sqrt{S(\omega)}\exp(-i\phi(\omega)), \tag{9.2}$$

where $S(\omega)$ is the power spectrum and $\phi(\omega)$ is the spectral phase. What really distinguishes coherent combining in the femtosecond regime from that in the CW regime is that the spectrum extends over a large bandwidth, so that $\phi(\omega)$ can vary significantly over the optical bandwidth and therefore cannot be considered as a single number. The efficient coherent addition of ultrashort pulses requires a phase matching over the whole optical bandwidth, so that several additional parameters should be controlled. More specifically, it is customary to Taylor expand the spectral phase around the central optical frequency to express it as

$$\phi(\omega) = \phi_0 + \phi_1\omega + \frac{1}{2}\phi_2\omega^2 + \cdots, \tag{9.3}$$

where ϕ_0 is the absolute phase, ϕ_1 is the first-order phase, representing a group delay, and ϕ_2 is the second-order phase, corresponding to a group delay dispersion or linear chirp. As intuitively obvious, combining ultrashort pulses requires that the pulses arrive simultaneously at the combining component so that their group delay ϕ_1 is identical. Moreover, the propagation in a dispersive medium of length z and group velocity dispersion β_2 is easily described in the frequency domain as

$$\tilde{E}_{\text{out}}(\omega) = \tilde{E}_{\text{in}}(\omega)\exp\left(-i\frac{\beta_2}{2}\omega^2 z\right). \tag{9.4}$$

If we consider only the dispersion, we see that the overall accumulated dispersion for each combined beam must be the same, for the spectral phase to match over the whole bandwidth. As pulses get shorter, it might be necessary to include in the analysis dispersion orders greater than 2; however, we will limit our present analysis to the second order of the spectral phase.

We can also take into account the contribution of self-phase modulation (SPM) to the spectral phase of a pulse propagating in a nonlinear medium. If we restrict the discussion to chirped-pulse amplifier (CPA) systems, which is the most used architecture for femtosecond amplifiers [6], SPM can be easily taken into account in the analysis. The CPA architecture consists in introducing a large group delay dispersion to temporally stretch the pulses, thereby reducing the peak power in the amplifier. At the output of the amplifier, the opposite dispersion is introduced to reveal the pulse width and amplified peak power. In such systems, due to the high linear chirp introduced by the stretcher, the temporal profile of the pulse propagating inside the amplifier is a scaled version of the spectral shape. In this case, for moderate SPM effects, the nonlinear contribution to the spectral phase can be expressed as proportional to the spectral shape [7]. With a spectrum shape normalized to 1, the proportionality coefficient is directly the B-integral experienced by the stretched pulse:

$$B = \int_0^L \gamma P_{peak}(z) dz, \quad (9.5)$$

where γ is the nonlinear coefficient given by $\gamma = (n_2 \omega_0)/cA_{eff}$ and $P_{peak}(z)$ is the peak power of the pulse at location z in the fiber. In this expression, n_2 is the nonlinear index of the medium, ω_0 is the optical angular frequency, c is the speed of light in vacuum, and A_{eff} is the effective area of the beam propagating in the medium. Considering that the spectral shape remains unchanged upon propagation, the nonlinear contribution to the spectral phase corresponds to the B-integral multiplied by the spectral shape function Stilde(omega), normalized to unity:

$$\tilde{E}_{out}(\omega) = \tilde{E}_{in}(\omega) \exp\left(-iB\hat{S}(\omega)\right). \quad (9.6)$$

This contribution to the spectral phase, in a real system, can be different for each combined beam, and might therefore lead to a spectral phase mismatch, potentially reducing the combining efficiency. Depending on the specific spectral shape, the contribution of the nonlinear phase to various polynomial orders of the spectral phase can vary, and higher order phase terms can become important in highly nonlinear systems.

9.2.2
Coherent Combining over a Large Bandwidth

The various coherent combining geometries that have been used in CW systems such as tiled and filled apertures can be directly transposed for femtosecond sources. We leave aside the spatial aspects for the moment to focus on the spectral/temporal specificities of beam combining for short pulses. Assuming a perfect spatial combination of the beams, the field at the output of the combination system is just the coherent sum of the electric fields emitted by each branch of the system. Assuming N combined arms and an equal power in each arm, the combined average power can be written as

$$P_{combined} = \frac{1}{N} \int \left| \sum_n \tilde{E}_n(\omega) \right|^2 d\omega, \quad (9.7)$$

and the combining efficiency is the ratio of the combined output power to the total power at the output of the system:

$$\eta = \frac{P_{combined}}{\sum_n \int |\tilde{E}_n(\omega)|^2 d\omega}. \quad (9.8)$$

These expressions show the importance of the relative spectral phase between the electric fields to be combined. The combining efficiency is an essential parameter.

Experimentally, both spatial and temporal mismatches impact the combining efficiency, and typical combining efficiencies for experiments in the femtosecond regime so far are around 90%. Efficiencies higher than 95% have been experimentally reached in well-controlled completely linear systems [4], where spatial aspects are limiting. When nonlinearity of the amplifiers increase, the spectral phase mismatch is more difficult to control and starts decreasing the experimentally accessible efficiency. In the case of extremely nonlinear systems where the stretcher is removed entirely to allow spectral broadening [8], the approximation made above to evaluate the SPM-induced spectral phase is no longer valid. Coherent combining systems can still be used, and a passive system in this regime has been demonstrated [9].

The combining efficiency η is an average power quantity, and might not be the most relevant number to quantify the performances of the combining, in particular when scaling the peak power is the main objective. In this case it is better to use a temporal quantity, defined as the ratio of combined peak power to the sum of peak powers of the individual beams to be combined. For instance, when there is residual random group velocity dispersion among the arms to be combined, it is easy to see that the impact of such a second-order phase distribution is not the same on the peak and average combined power. The peak power efficiency is defined as the ratio between the combined peak power and the sum of the incoming peak powers:

$$\eta_{\text{peak}} = \frac{\max\left((1/N)\left|\sum_n E_n(t)\right|^2\right)}{\sum_n \max\left(|E_n(t)|^2\right)}. \tag{9.9}$$

This quantity is more difficult to measure precisely in practice, necessitating a complete characterization of the ultrafast electric field method such as frequency-resolved optical gating (FROG) [10] or spectral phase interferometry for direct electric field reconstruction (SPIDER) [11].

9.2.3
Influence of Spectral Phase Mismatch on the Combining Efficiency

To illustrate quantitatively the impact of spectral phase mismatch on the combining efficiencies, results of numerical simulations of the coherent combining process in the femtosecond regime are presented here. For spatial aspects, analytical estimates are available in the literature to assess the impact of random tilts and offsets on the combining efficiency [12]. For the temporal aspects, an analytical estimate is available in the case of the combination of two elements with deterministic values of the mismatch in dispersion, group delay, and B-integral [13]. Here, we consider a CPA system that combines 10 beams with no spatial mismatches and look at the influence of random Gaussian distributions of the group delay, group velocity dispersion, and B-integral on the combining efficiencies, in terms of both average

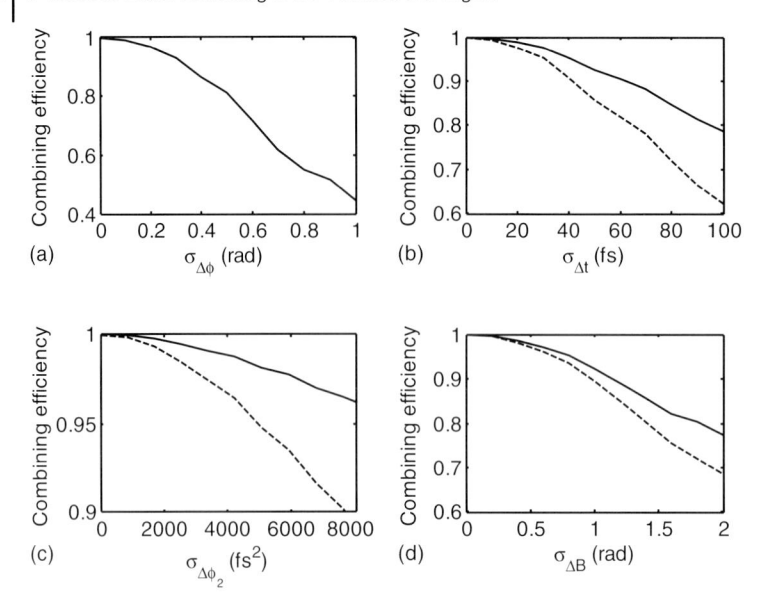

Figure 9.2 Impact of temporal defects on the combining efficiencies (solid line: average power, dashed line: peak power). (a) Absolute phase. (b) Group delay. (c) Group delay dispersion. (d) B-integral. All calculations were made with a pulse width of 200 fs.

and peak powers, using Monte Carlo simulations and the expressions of paragraph 9.2.1. The initial considered pulse width is 200 fs, typical of high-power Yb systems. The results are plotted in Figure 9.2.

The graph in Figure 9.2a shows the influence of a nonperfect active stabilization system, resulting in a random residual zeroth-order phase with standard deviation given by $\sigma_{\Delta\varphi}$. As intuitively expected, the impact is exactly the same as in the CW regime. For all the other graphs in Figure 9.2b–d, we consider that the absolute phase is perfectly corrected by an active stabilization system, so that there is no zeroth-order phase mismatch. These results show that the group delay must be controlled to within 20% of the pulse width to limit the impact on the efficiency to less than 10%. Translating into length units, this corresponds to a system where the combined paths are not different from each other by more than about 10 μm for the considered pulse width of 200 fs. The impact of mismatched group velocity dispersion is not critical for this pulse width: A dispersion of 4000 fs^2 decreases the average power combination efficiency by only 1.5% and corresponds to the propagation in 20 cm of glass. The sensitivity to mismatched B-integral is shown in Figure 9.2d. Depending on the overall nonlinearity level at which the system is operated, this sensitivity can be more detrimental. In particular, since the B-integral depends on the power, fluctuations of power translate into B-integral changes and might lead to significant degradation of the combining efficiency. For low nonlinearity systems with $B < \pi$, however, this effect is negligible, and it has been confirmed experimentally.

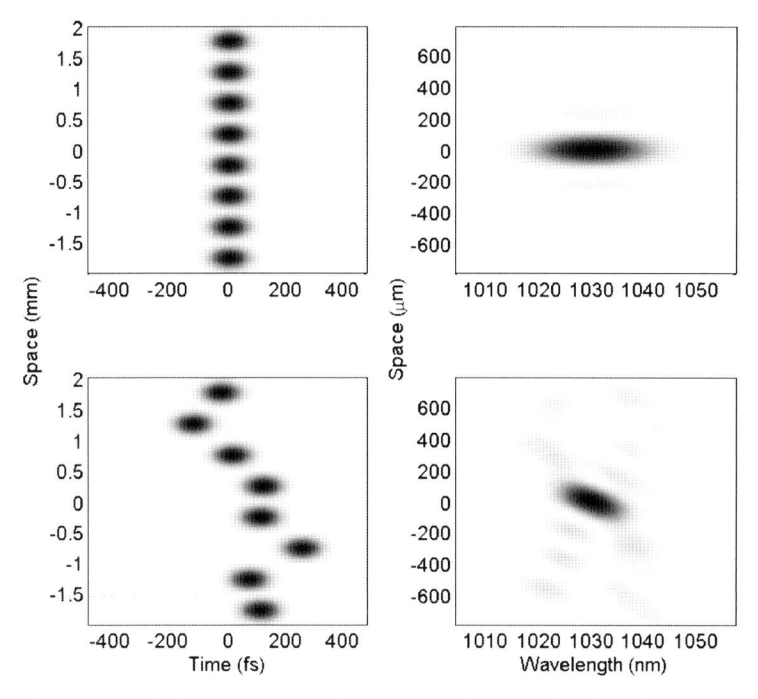

Figure 9.3 *Left column*: Intensity in the space–time domain in the tiling plane (*top*: no group delay fluctuations, *bottom*: presence of group delay fluctuations). *Right column*: Resulting space–wavelength intensity distribution at the focus of a 60 cm lens.

9.2.4
Space–Time Effects

Although we have left aside the spatial aspects up to now, they are of course of utmost importance to design an efficient femtosecond combining system. Let us consider a system in which the combination is achieved in tiled aperture architecture. If all the temporal/spectral aspects are perfectly controlled, the loss in combination efficiency related to spatial tilts and offsets is the same as for a CW system. However, when temporal imperfections are present, the fact that each beam is located at a specific position in the tiled beam leads to space–time coupling, even in the absence of spatial imperfections.

As an example, Figure 9.3 shows a system in which eight beams composed of 100 fs pulses with a 250 μm width are tiled, spaced 500 μm apart from each other. This beam is focused using a 60 cm focal length lens, and the effect of a large group delay discrepancy, with a standard deviation of 100 fs, is examined at the focus. The top row is the simulation corresponding to a perfect system, for reference. It is well known that a beam exhibiting pulse front tilt leads to spatial chirp at the focus of a lens [14]. Here, the nature of the space–time coupling is more intricate, but induces a similar effect, with a wavelength–space intensity distribution that is coupled, resulting in a spatial chirp of 0.2 nm/μm at the focus. Depending on the target application, these space–time coupling effects can be detrimental, and are increasingly important when the pulse width gets shorter, in the sub-100 fs regime.

9.3
Coherent Combining with Identical Spectra: Power/Energy Scaling

In this section, experimental implementations of coherent beam combining systems in the femtosecond regime are described and are separated into two categories. The first one corresponds to active systems, in which there is a feedback mechanism on the relative zeroth-order phase between the beams. The second one corresponds to systems in which the combined beams travel through the same overall optical path, but go through the elements in a different order, resulting in a passive automatic canceling of the path fluctuations. For both types of systems, experimental demonstrations have been mostly done with only two beams, although a first demonstration with four actively stabilized beams has been recently reported by the University of Michigan [15].

9.3.1
Active Techniques

9.3.1.1 Experimental Implementations
The first coherent combining demonstrations in femtosecond regime have used active techniques [4,5]. In the experiments performed so far, an active stabilization of the zeroth-order phase is only performed, while the other parameters such as group delay and group delay dispersion are adjusted in a static manner. It was shown, however, that such a restricted feedback is appropriate for the pulse widths considered, ranging from 200 fs to 1 ps. Two experimental techniques for the combination/feedback loop have been demonstrated.

The first technique [4] was demonstrated by the Friedrich Schiller University in Jena (Germany), and uses a polarizing beam splitter (PBS) as a combining element. The two beams to be combined are set to have orthogonal polarization state. At the output of the PBS, all the optical power is combined in a single beam regardless of the relative phase between them. However, the combined part of this beam corresponds to the part of the beam that possesses a linear polarization state oriented at 45° from the PBS axes. The feedback involves a measurement of the polarization state of the output beam and a Hänsch–Couillaud system [16] to generate the error signal. The feedback is implemented on a piezo-mounted mirror located in one of the amplifying arms. This technique has led to the generation of 550 fs pulses with energy of 3 mJ [17]. This experimental demonstration is depicted in Figure 9.4.

The system includes an oscillator, two stretchers, three amplification stages, and two acousto-optic modulators that allow operating the system at a repetition rate of 10 kHz with sufficient average power to seed the final amplification stages. An active

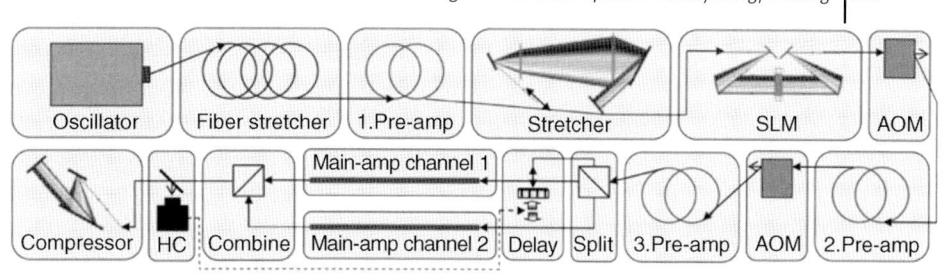

Figure 9.4 Experimental setup of the active system based on the polarization combining technique. SLM, spatial light modulator; AOM, acousto-optic modulator; HC, Hänsch–Couillaud detector. After Ref. [17].

spectral phase control is implemented through the use of a spatial light modulator inserted in a zero-dispersion line. After the last preamplifier, the seed signal is divided into two arms that include two similar amplifiers. One of the arms includes a delay line that is set manually to match the group delays of both arms coarsely. The delay line also includes a piezo-mounted mirror used as a feedback element for the active optical phase control. The active phase stabilization scheme bandwidth is 1 kHz. The combined amplifiers are made of two 80 cm long rod-type fibers exhibiting a mode field diameter of 75 µm, pumped by 915 nm laser diodes. The choice of wavelength is made to maximize the population inversion in the amplifiers, thereby increasing the saturation energy. After amplification and recombination of both arms, the output signal is sent to a compressor with 80% efficiency.

Obtained experimental results are shown in Figure 9.5. The spectra of both channels and the recombined one are nearly identical, and all exhibit a steepening on the long wavelength side. This effect is observed because the extracted energy is high, leading to significant saturation of the gain. In CPA architecture, this leads to a modified spectral shape because wavelengths arriving first are more amplified.

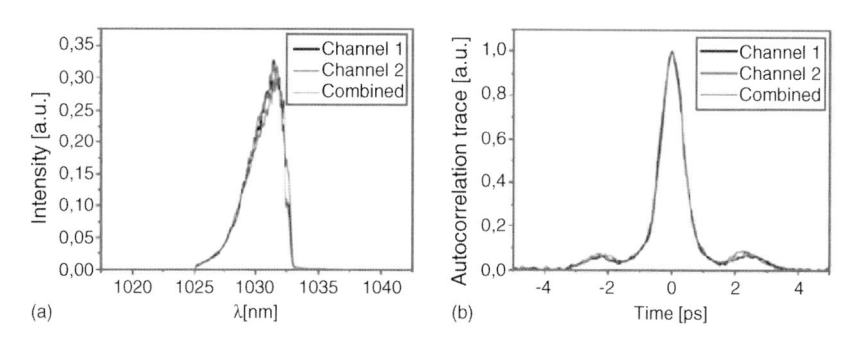

Figure 9.5 (a) Spectra of both channels and recombined beam at 2.1 mJ output energy (15 kHz repetition rate). (b) Autocorrelations of both channels and recombined beams at 3 mJ output energy (10 kHz repetition rate). After Ref. [17].

The autocorrelations obtained at maximum output energy (Figure 9.5b) for both channels and the recombined beam are also very similar and show a FWHM of 800 fs. This output energy corresponds to a B-integral of 9 rad, a rather high value that causes the creation of a significant pedestal. The combining efficiencies for this experiment range from 89% at 2.1 mJ to 84% at 3 mJ compressed output energy. Note that this value of energy is the highest ever reported for a fiber system in the femtosecond regime, meaning that coherent combining is already an enabling technology. Another remarkable result from this work is that, using careful housing and thermal management of the system, the relative phase fluctuations to be compensated remain within one wavelength (2.3 rad peak to peak over 10 min). This shows that even for much shorter pulses, group delay compensation might not be needed. However, the polarization combining used in this work implies that no more than two beams can be combined in one step, so the cascading of combining systems is required to scale the number of combined beams.

The experimental setup corresponding to the second implementation [5] was developed in our research group at the Laboratoire Charles Fabry, Institut d'Optique (France) and is described in Figure 9.6. The overall setup is similar to the first implementation presented above, with two amplifiers being combined in a Mach–Zehnder interferometer geometry. The main difference lies in the active phase control system, which is based on the frequency tagging, that is, LOCSET, method demonstrated in the context of CW combining systems (see Chapter 2) [18]. As a consequence, the recombining element is a 50/50 beam splitter. The frequency tagging technique consists in adding a small-voltage modulation at 250 kHz to the control signal that drives the phase modulator, yielding a small modulation (0.1 rad)

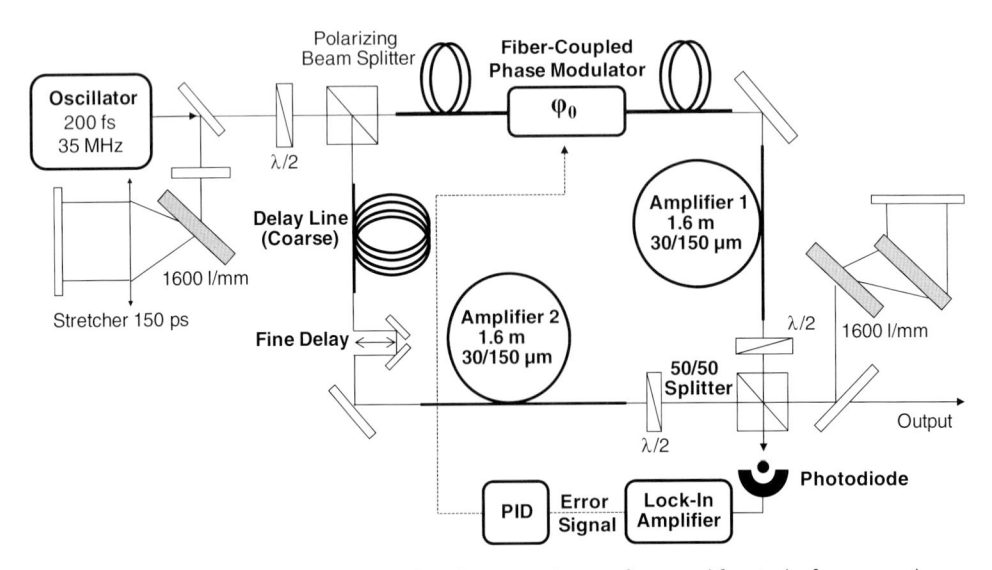

Figure 9.6 Experimental setup for coherent combining of two amplifiers in the femtosecond regime.

of the relative phase between both arms. At the output of the photodiode, a lock-in amplifier allows the detection of the small induced amplitude modulation. The resulting error signal is directly proportional to the optical phase difference between the arms, and is fed back to the phase-controlling element through a digital controller. The overall feedback loop bandwidth in this case is 10 kHz.

The experiment consists in a femtosecond Yb:KYW oscillator delivering 200 fs pulses at 1030 nm and 35 MHz repetition rate is followed by an acousto-optic modulator and a stretcher that broadens the pulses to 150 ps. A half-wave plate with a PBS is used to divide the seed into two arms with adjustable input powers. The first arm consists of a fiber-coupled $LiNbO_3$-integrated phase modulator that will enable the feedback mechanism on the phase. It is followed by a 1.6 m Yb-doped 30/150 μm large mode area (LMA) fiber amplifier. The second arm contains a 2.4 m single-mode fiber, a free space delay line, and another 1.6 m of the same LMA fiber. The single-mode fiber is used to match the group velocity dispersion of both arms and also acts as a coarse delay matching element. The outputs of the fiber amplifiers are collimated and overlapped on a 50/50 beam splitter, which is the combining element in this system. The constructive output of the beam splitter goes to the compressor. The rejected output is followed by a photodiode that detects an error signal to be minimized.

Analysis of the error signal shows an RMS residual phase noise of $\lambda/40$ in locked operation, assuming that the intensity noise is negligible. No readjustment of the delay line is necessary to keep the system locked over hours, indicating that the group delay drift is small compared to the pulse duration. Most of the phase noise content is below 1 kHz in the laboratory environment of this experiment. The system is first operated in linear regime at 35 MHz repetition rate. A combined output power of 12.5 W is measured before compression, corresponding to 92% combining efficiency. To achieve this efficiency, the spatial overlap of the beams at the output of the beam splitter must be very carefully optimized. Any difference between the wavefronts, such as defocus and tilt, or in the beam profiles, such as ellipticity and beam position, results in a drastically reduced combining efficiency. Both channel beams and the combined beam exhibits an M^2 factor of less than 1.15. The temporal/spectral characterization of the combined beam is presented in Section 9.3.1.2, along with a more detailed measurement of combining effects in the spectral domain.

9.3.1.2 Measurement of Spectral Phase Mismatch

We now describe a way to identify and quantify the various contributions to the nonperfect combining efficiency. This is done through the measurement of the relative spectral phase between both arms of the combining system. This technique allows both a precise balance of the SPM effects in a nonlinear amplifying setup and the identification of the spectral and spatial contributions to the non-perfect combining efficiency [19].

In the setup shown in Figure 9.6, by changing the reference phase of the lock-in amplifier by π, it is possible to switch the 50/50 beam splitter outputs, so that both the combined and rejected beams can be characterized. At the output, the spectra of

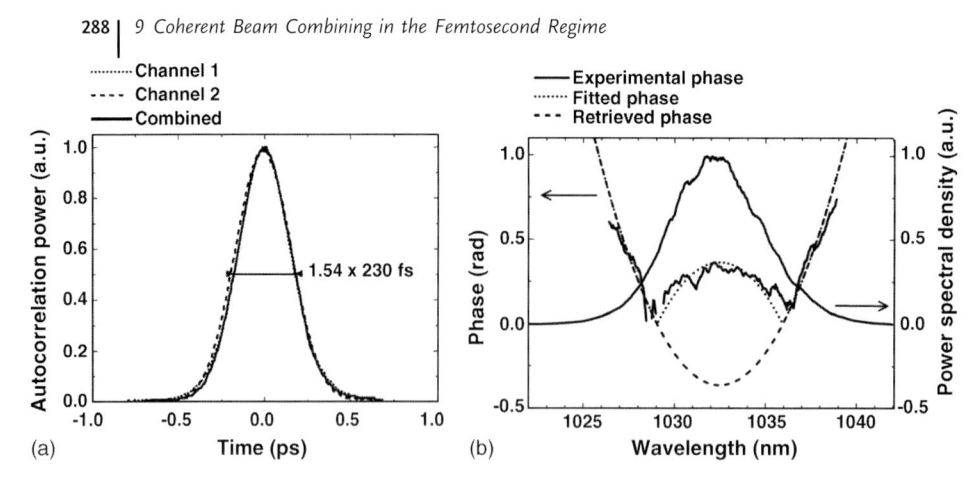

Figure 9.7 (a) Autocorrelations of channels 1 and 2 and combined beam in linear regime. (b) Output spectrum and differential phase in the linear regime. After Ref. [19].

channels 1 and 2 along with the combined and the rejected beams are measured. The spectral intensity at the output $I(\omega)$ is given by the standard interference expression for each frequency component:

$$I(\omega) = \frac{1}{2}[I_1(\omega) + I_2(\omega)] + \sqrt{I_1(\omega)I_2(\omega)}\cos[\Delta\phi(\omega)], \quad (9.10)$$

allowing the retrieval of the magnitude of the relative spectral phase $\Delta\phi(\omega)$ between the incident beams with spectral intensities $I_1(\omega)$ and $I_2(\omega)$. As opposed to single-pulse complete characterization techniques such as frequency-resolved optical gating, this method is differential and includes zeroth- and first-order contributions of the relative spectral phase. It is therefore well suited to measuring precisely the spectral phase mismatch.

At 35 MHz repetition rate, that is, in the linear regime, the autocorrelation of the recombined pulse is shown in Figure 9.7a, along with the autocorrelations of the separate channels, which are very similar. The inferred recombined pulse width is 230 fs assuming a sech-square shape.

The absolute value of the relative phase is retrieved using the method described above and is shown in Figure 9.7b along with the combined spectrum. Since the amplification regime is linear, the curve is fitted with the absolute value of a parabola $\phi(\omega) = \phi_0 + \phi_1\omega + \phi_2\omega^2$ that takes into account the zero-order phase, delay and group velocity dispersion discrepancies, respectively. We observe mostly a mismatch in the second-order spectral phase, due to mirrors with uncontrolled dispersion. To control the validity of these measurements, the recombined spectrum can be reconstructed using the experimental spectra of the two channels and the fitted spectral phase. The result is in excellent agreement with the experiment and corresponds to a combining efficiency of 97%. This implies that 3% of the overall combining losses are due to spectral effects, while the remaining losses are related to polarization and spatial issues such as wavefront and beam profile discrepancies and are estimated here to 5%.

To investigate the nonlinear regime, the repetition rate is reduced to 1 MHz by the acousto-optic pulse picker. In this case, each channel experiences *a priori* the same SPM effects, which should not affect the combining efficiency. In practice, the accumulated nonlinear phase can be different because of the different injection or pumps powers, leading to a significant drop of combining efficiency. The spectral phase measurement provides a way to evaluate and compensate for this effect. To take into account the nonlinear contribution to the differential spectral phase, the fit expression is modified as

$$\phi(\omega) = \phi_0 + \phi_1\omega + \frac{1}{2}\phi_2\omega^2 + \Delta B I_N(\omega), \tag{9.11}$$

where ΔB is the B-integral discrepancy and $I_N(\omega)$ is the normalized fit of one of the incoming spectra. Observation of this spectral phase allows adjusting the level of nonlinearity by tuning the injection and pump powers [20], along with the delay, in order to minimize the rejected power. In this situation, the amplifiers do not deliver the same power and the related efficiency loss is 0.5%. The obtained combining efficiency is 91% with 9.2 W output power, that is, only 1% less than in linear regime. The autocorrelations of each channel and the combined beam are shown in Figure 9.8a. The incoming and combined pulses share the same shape, distorted by the high B-integral of 6 rad without any active correction of spectral phase, and also the same compression point. The autocorrelation FWHM is 410 fs, yielding pulse duration of 320 fs. The combined spectra are plotted with the relative spectral phase in Figure 9.8b. The efficiency calculated from the spectral data is 96%, showing that the spatial overlap-related losses still remain around 5%. Balancing the nonlinearities is gainful even in the case where it results in an unbalanced output

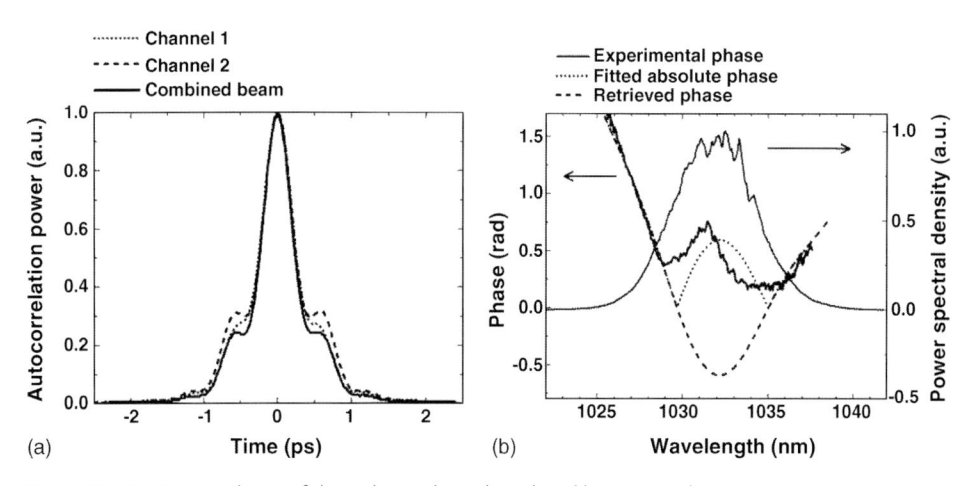

Figure 9.8 (a) Autocorrelation of channels 1 and 2 and combined beam in nonlinear regime. (b) Spectrum of the combined beam and relative spectral phase between both channels. After Ref. [19].

power, because the efficiency is much more impacted by phase mismatch than by intensity mismatch.

9.3.2
Passive Coherent Combining Techniques: Path-Sharing Network

9.3.2.1 Principle

With the term passive coherent combining, we intend to describe a specific coherent beam combination configuration where phase stabilization is an inherent characteristic of the system. Therefore, the combination of N independent laser amplifiers is accomplished with no use of phase detection or active feedback mechanism. Although passive coherent combining inside laser cavities has been extensively studied (see Chapters 10–13) [21,22], passive techniques to combine amplifiers that are seeded by a common oscillator have only been suggested very recently. This idea is theoretically directly applicable to any kind of amplification system in any kind of operation mode ranging from single-frequency CW systems to ultrafast femtosecond amplifiers. In the latter case, as will be shown subsequently, the individual beams are automatically locked in phase over all orders.

The principle of operation of such a combining configuration is quite simple. The general idea is based on the creation of an amplification network where an arbitrary large number of independent but ideally identical laser amplifiers are organized in such a way that an input beam is progressively split into halves that are forced to travel exactly the same overall distance through the amplification network, but through unique and entirely different pathways. Each one of the input pulse replicas after completing a journey through the network having been amplified to the same level and accumulating the same phase is eventually coherently recombined to a single output. The simplest implementation of such a network is naturally the combination of only two beams. Although one can imagine different realizations, the simplest among them however is the Sagnac interferometer type. Figure 9.9a gives a schematic of such a system.

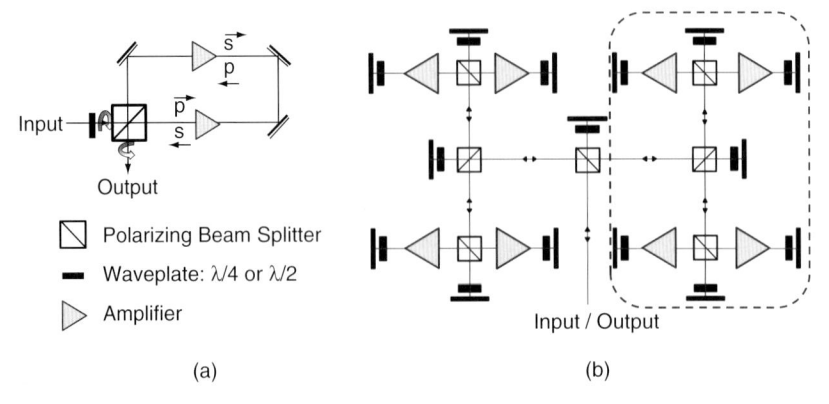

Figure 9.9 The passive coherent combination concept for two (a) and N (b) multipass amplifiers.

As splitting element for the input pulses, we use in this example a PBS in combination with a waveplate ($\lambda/4$ or $\lambda/2$). The two polarization states (s and p) of the input pulse are forced to follow counter-propagating directions in the same amplification loop. Each one of the replicas is amplified in both amplifiers and reaches the same energy going back to the PBS. Since both replicas are amplified to the same level and travel over the same optical path (accumulating the same phase), here they constructively interfere toward the second output of the PBS in an orthogonal polarization state to the input pulse. In this realization the amplification network is simply a common loop for the two amplifiers, while the splitting and the recombination is based on the polarization state of the initial beam and the two amplified replicas. It is clear that phase matching in this configuration is automatically achieved over all phase orders, an especially interesting property for very large bandwidth ultrafast pulse amplification. Although the amplifiers are assumed to be polarization independent in this example, the principle is easily extended to seed the Sagnac in both directions with the same polarization state. In this case, the use of an optical isolator allows final extraction of the recombined beam, as described hereafter in the experiments.

The generalization of the idea to an arbitrary number of amplifiers is possible. An example of $N = 8$ amplifiers network is shown in Figure 9.9b (the intermediate step of the combination of four amplifiers is indicated by the dashed line frame). Here, an elementary cell consisting of two amplifiers, similar to the one already described, is interleaved, forming a rather complex network. It is straightforward to see that again the input pulse progressively enters this network and successively split into replicas that in overall travel the same distance and go through all amplifiers. The amplification of each replica takes place in a highly ordered manner that guarantees their identical amplification until the final output of the network. This means that when one replica is amplified by the nth amplifier, this amplifier has already amplified $n - 1$ other replicas and that its energy before and after amplification is equal to the energy of all other replicas before and after their respective nth amplification.

It is important to note that apart from the completely passive character of the coherent combination of the independent amplifiers and the absence of phase detection or feedback stabilization electronics, there is an additional fundamental difference between this implementation and the previously described active configurations. In fact, sharing the same amplification network, each amplifier now becomes a multipass amplifier of N passes, with N being the number of combined amplifiers. This specificity has not only the positive aspects but also the negative aspects, especially regarding the scaling of the combination to a large number of amplifiers. It is rather evident that this implementation would be better applied in low-gain amplification systems where an increase in the number of passes is required to improve the energy extraction efficiency of each individual amplifier. This way not only the output of N amplifiers is coherently added but also the operation performance of each amplifier could be improved. On the other hand, it is also clear that an increase in the number of combined amplifiers is accompanied by an increase in the overall beam path and therefore an increase in the system's complexity. Furthermore, if a single amplifier induces a high level of nonlinear effects, this accumulation is not desirable. Finally, if for the combination of N amplifiers we assume N splitting/combining elements with L the losses per element, we can approximate the total anticipated energy evolution with the number of amplifiers to follow a rise proportional to $N \times (1 - L)^N$. In such a case, the energy of the combined system ceases to increase when the number of amplifiers is greater than $N_{max} \geq (1 - L)/L$. At $L = 5\%$, this corresponds to $N = 19$, while a more conservative estimation of the combination efficiency at $L = 10\%$ restricts the applicability of the passive amplification network to $N = 8$ independent amplifiers.

9.3.2.2 Experimental Demonstrations

So far only two experiments have been realized confirming the idea of passive coherent combination of two independent femtosecond ytterbium-doped fiber amplifiers. In a first proof-of-principle experiment, based on standard LMA fibers, essentially the same type as in Figure 9.6, the remarkable robustness of the combination of the two amplifiers reaching up to 96% combination efficiency has been demonstrated for 250 fs multimicrojoule level pulse up to 10 W average power [9]. These experiments have also provided the first proof of the applicability of the passive combination architecture under the presence of relatively high nonlinearities due to SPM and corresponding B-integrals on the order of 15 rad.

In a more recent experiment [23], we used the passive combining architecture and rod-type fiber amplifiers to generate high temporal quality 300 fs pulses with 650 µJ of energy per pulse at a repetition rate of 92 kHz. This corresponds to an average power of 60 W and a peak power in excess of 2 GW. Furthermore, at 2 MHz of repetition rate, combining average powers of 135 W before and 105 W after compression have been measured. These results have clearly demonstrated the potential of the technique in a high-energy, high-power fiber system.

Figure 9.10 shows the experimental setup used. To maintain a good temporal pulse quality simultaneously with a high energy level, a moderately nonlinear fiber CPA is implemented in which the impact of nonlinearities is partially compensated

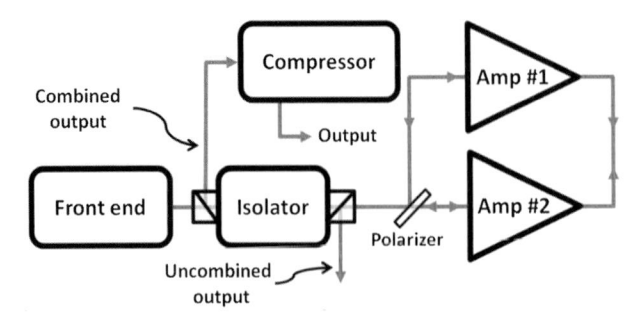

Figure 9.10 Experimental setup of high-energy passive coherent combination Yb:doped fiber CPA amplification system. After Ref. [23].

by the dispersion mismatch of the stretcher and compressor units. The two amplifiers to be combined are built within the Sagnac interferometer that is seeded, through an optical isolator, by a front end. This front end is composed of a passively mode-locked ultrafast oscillator, a pulse picker, a pulse stretcher, and a single-mode fiber preamplifier. The repetition rate of the oscillator of ~25 MHz can be down-counted by the pulse picker from 92 kHz to 25 MHz. The stretching ratio is designed to preserve a pulse duration of ~600 ps after the power amplifiers when operated at maximum gain. An optical isolator is placed after the front end to prevent any optical feedbacks coming from the power amplifiers. It is also used to separate the combined and the uncombined outputs of our passively coherently combined amplification scheme. Then, the beam is split by a polarizer and each secondary beam seeds a power amplifier made of ytterbium-doped rod-type photonic crystal fibers. The mode field diameter of the fibers is 85 μm, the pump cladding diameter is 285 μm, and the length is 1 m that ensures sufficient pump absorption and optical efficiency.

The outputs of each amplifier are connected to each other to close the Sagnac interferometer. Both beams are then recombined on the polarizer, with linear polarizations orthogonal to each other. If both beams are in-phase, they form a combined output beam propagating back to the isolator with a linear polarization state at an angle of 45°, similar to the experiment in Ref. [17]. This beam is therefore transmitted back through the rotator and ejected by the polarizer located at the input of the isolator. The uncombined part of the beam is removed by the polarizer located at the output of the isolator. Measurement of the average power at both outputs allows computing the combination efficiency defined as the ratio between the combined power and the total power.

The path length of one round-trip in the Sagnac interferometer is 5.5 m, making this setup immune to all phase noises with frequency contents below 27 MHz. This large bandwidth essentially removes all environmental sources of noise such as acoustic and thermal fluctuations of the relative phase. To highlight the robustness of this architecture with respect to phase noise, we measure the pulse-to-pulse stability of the combined beam at maximum energy to be 2% RMS, just slightly higher than the 1.1% RMS measured at the front end output. This very small decrease of the pulse-to-pulse stability for a nonlinear FCPA, built on an uncovered breadboard, clearly indicates that phase noise has very little impact, if any, on the global coherent combination stability. This is of particular importance if lasers using passive coherent combination architecture are to be used in harsh environments or if the pulse-to-pulse stability is of particular importance for the targeted scientific or industrial applications. Finally, the coherently combined beam is sent through a 79% transmission compressor.

Figure 9.11 shows the amplifier characteristics at 92 kHz repetition rate with the combined and uncombined average powers measured at the output ports of the isolator. At the maximum total pump power of 160 W, we reach 75 W of passively coherently combined power, that is, 815 μJ, with a slope efficiency close to 60%. Despite the B-integral value of 7 rad per round-trip, the combining efficiency is over 90%.

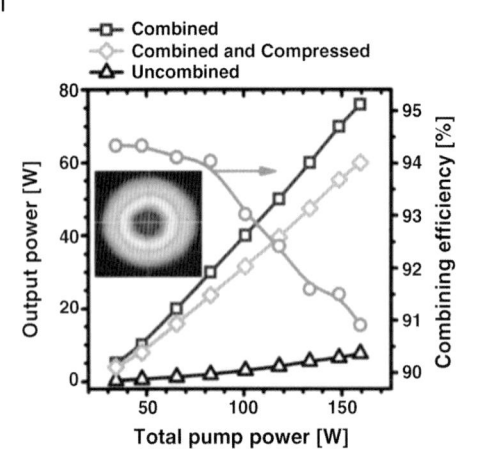

Figure 9.11 *Left scale*: Output powers of the amplifier as a function of pump power at a repetition rate of 92 kHz. *Right scale*: Combining efficiency as a function of pump power. *Inset*: Beam profile taken at maximum energy at the output of the system. After Ref. [23].

The compressibility of the amplified pulses at maximum energy is also investigated. Figure 9.12 shows the autocorrelation traces at 650 µJ compressed energy (60 W average power) acquired with an intensity autocorrelator and independently measured with a second-harmonic generation frequency-resolved optical gating apparatus (FROG), with a FWHM of 400 fs. Figure 9.11 also shows the retrieved temporal profile of the pulse from the FROG measurement, with duration of 300 fs, corresponding to a time–bandwidth product (TBP) of 0.75, and peak power in excess

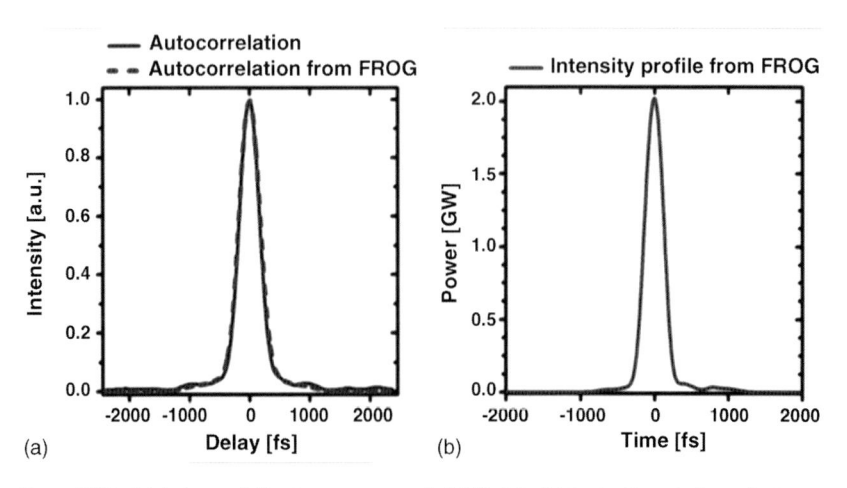

Figure 9.12 (a) Autocorrelation traces measured at 650 µJ (solid line) with an independent autocorrelator (dashed) from a FROG measurement. (b) Retrieved temporal profile of the pulse from the FROG measurement. After Ref. [23].

of 2 GW. The spectral bandwidth measured at maximum compressed energy is 8.5 nm. The beam quality after compression shows no degradation with increasing combined average power with constant values of $M_x^2 = 1.25$ and $M_y^2 = 1.15$.

Application of the passive combination with more than two amplifiers could be a next step toward further increase of the output energy of ultrafast amplification systems. However, in the case of fiber amplifiers, the interest is rather limited since, as it has been already noted, it would result in the increase of the number of passes per fiber amplifier and therefore the accumulated nonlinearities. In other systems however such as bulk media-based multipass amplifiers, especially when limited by low gain and pure energy extraction capacity, such a perspective seems very appealing.

9.4
Other Coherent Combining Concepts

We now describe in less detail coherent combining ideas that have been applied in the femtosecond regime, but are different from standard beam combining, where N similar beams are used to provide a single powerful output beam. First, one can separate temporally several replicas of the pulse to be amplified and recombine them at the output. The second idea deals with coherent superposition of pulses inside an optical resonator. The third concept consists in combining several femtosecond pulses with different spectra to generate a single, possibly shorter and more powerful pulse at the output.

9.4.1
Temporal Multiplexing: Divided Pulse Amplification

The divided pulse amplification (DPA) architecture was proposed in 2007 at Cornell University (USA) [24] in the context of high-power picosecond pulse sources. Indeed, the CPA idea is difficult to apply for pulses longer than a few picoseconds, because practical devices introducing enough dispersion to stretch such narrow-band pulses do not exist. To circumvent this problem, DPA consists in splitting the input pulses in the time domain into several replicas prior to amplification instead of stretching them, as shown in Figure 9.13. This process reduces the peak power by a factor equal to the number of replicas. After amplification, a coherent recombination of the replicas into a single output pulse is necessary. DPA can therefore be seen as a time-multiplexing architecture or as a coherent combining system in the time domain. It is therefore not strictly speaking beam combining, since there is a single spatial beam, but can nevertheless be used to avoid nonlinearity or damage issues.

The splitting and recombining processes can be based on the use of highly birefringent crystals that exhibit a large group velocity difference between their optic axes. By cascading a sequence of N such crystals (e.g., YVO_4) periodically rotated at 45°, with lengths appropriately chosen with respect to the pulse width and to the total

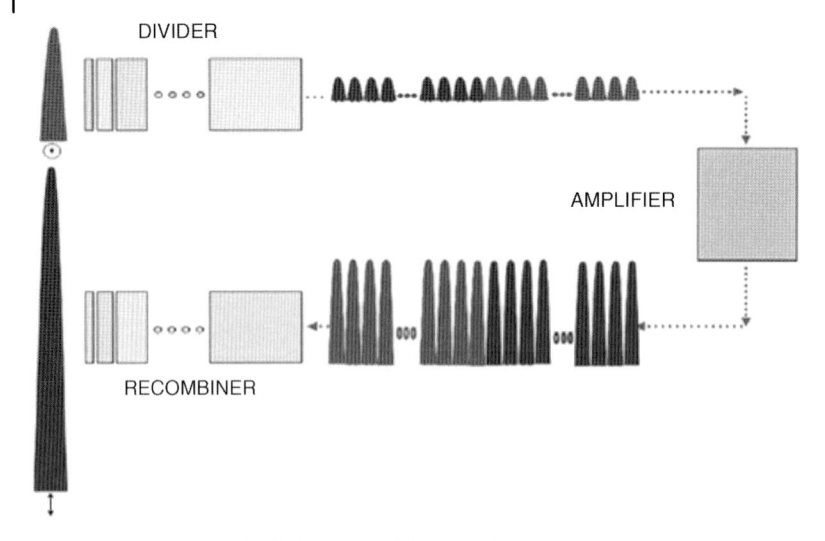

Figure 9.13 Principle of divided-pulse amplification. After Ref. [24].

number of pulses, a series of 2^N pulses can easily be generated. Importantly, the same arrangement can be used to recombine all the pulses in a single one at the output of the amplifier. Although it is not a requirement, experimental demonstrations so far have been made with a unique sequence of crystals to divide and combine the pulses in a counterpropagating geometry.

In a recent experiment [25], 32 replicas of the input pulses are generated to amplify the 2.2 ps pulses to 1 MW in a LMA Yb-doped fiber amplifier, the highest peak power ever generated in this pulse duration range. This scheme can be advantageously used in the context of ultrashort parabolic fiber amplifiers [26], with typical pulsewidths at the output of the amplifier of few ps, while the pulses can typically be recompressed to sub-100 fs durations [27]. The DPA scheme can also be used in conjunction with CPA [28], but in practice the delay between replicas must be larger than the stretched pulse width, which becomes quickly impractical to implement using birefringent crystals. Freespace delay lines are used in this case.

9.4.2
Passive Enhancement Cavities

The idea of using an optical cavity to enhance the power emitted by a laser source was coined long ago, but has been applied to ultrafast lasers only during last decade. A schematic of the principle of cavity enhancement is shown in Figure 9.14. The laser beam is fed to an optical cavity, and a strong intracavity optical signal can build up as a result of the coherent addition between the external field and the intracavity field. In single-frequency regime, the only requirement to obtain resonant enhancement is that the optical cavity length corresponds to a multiple of the wavelength. In

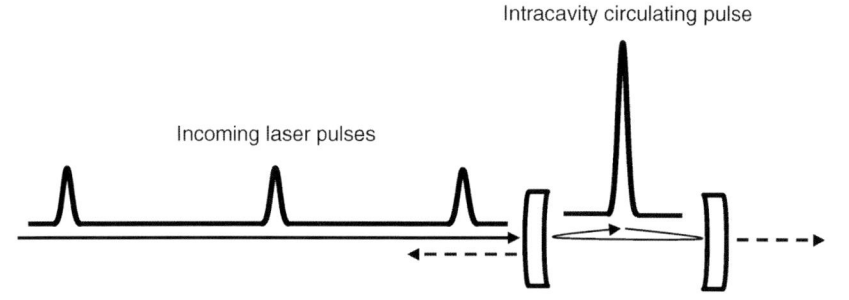

Figure 9.14 Passive enhancement cavity principle.

the femtosecond regime, the frequency comb corresponding to the input excitation spectrum must coincide with the cavity-defined frequency comb. This is equivalent to saying that the phase matching condition must be fulfilled over a large bandwidth.

In addition to the zeroth-order phase condition, the enhancement cavity must be excited with a pulsed beam in a period corresponding to a multiple of the cavity round-trip time. In the frequency domain, this ensures that the frequency steps between two teeth of the combs are matched. This additional requirement corresponds to the first-order phase, or group delay, matching condition.

The optical bandwidth over which the frequency combs of the feeding laser and the cavity overlap is essentially determined by the dispersion in the cavity that results in a nonuniformity of the cavity frequency comb. This effect reduces the intracavity power enhancement that can be expected, and increases the pulse width of the circulating pulse. In a CW regime, and when the impedance matching condition is satisfied, the enhancement is given by F/π, where F is the finesse of the cavity. A description of the effect of cavity dispersion on the actual circulating pulse is given in Ref. [29].

Two types of experiments have been reported that make use of the cavity enhancement idea. The first type consists in amplifying the optical pulses by stacking them onto each other for a certain number of round-trip times, and then ejecting them out of the cavity using an electro- or acousto-optic cavity dumper [30]. The output beam has a reduced repetition rate, but features higher pulse energy, and the cavity has acted as an energy concentrator in time.

The second type of experiments consists in locating the final interaction, where high-intensity pulses are needed and used, inside the enhancement cavity. If this interaction has low efficiency, it contributes little to the losses of the cavity and still allows efficient pulse buildup. Another condition to use this architecture is that the interaction should leave the optical beam intact, so that spatial mode matching between the cavity and the input beam is still possible. Possible applications where this geometry can be successfully used include high-harmonic generation [31] and Compton or Thomson scattering.

Cavity enhancement factors of several hundreds were demonstrated in temporal regimes ranging from sub-100 fs pulses to picosecond pulses [32,33], allowing the optical peak power in the cavity to be scaled to hundreds of megawatt directly at the oscillator repetition rate of several tens of megahertz. In several experiments, optical intensities of several 10^{14} W/cm^2 at the intracavity focus were obtained. In terms of average intracavity power, the performances reached so far lie in the 1–100 kW range. Further power scaling is limited by thermo-optic effects at the cavity mirrors, optical damage, and dispersion control.

9.4.3
Coherent Combining with Disjoint Spectra: Ultrafast Pulse Synthesis

The possibility of combining ultrafast pulses with different spectral content to synthesize much shorter pulses was suggested in 1990 [34]. This idea necessitates that the combined spectrum exhibit optical coherence across the entire spectral content and that the relative phase between the initial pulses be controlled. Hence, pulse synthesis requires both coherent combining (phase control between the pulses to be combined) and spectral combining (an optical element that allows the spatial combination of different central wavelengths). Because mode-locked lasers directly emitting few-cycle pulses are routinely used in laboratories, pulse synthesis has been mostly used to reach the single or suboptical cycle regime, with little considerations about the power scaling aspect. However, this technique might be used in the future to achieve both ultrashort and energetic pulses, a difficult task to achieve with standard optical amplification techniques because of the gain narrowing phenomenon.

One of the pioneer experiments in this area consisted in phase locking two independent Ti:sapphire oscillators with spectra centered at 760 and 810 nm [35]. In this case, the oscillators must be stabilized and synchronized in both carrier envelope phase and repetition rate. In the time domain, this ensures that the relative optical phase and delay between the pulses remain constant from pulse to pulse. In the frequency domain, it corresponds to the generation of an optical comb defined by a single offset frequency and repetition frequency. This first experiment demonstrated the coherent addition of different spectral contents and the shortening of the combined pulse compared to the parent pulses.

Since then, a number of experiments have been carried out in the same direction [36], using various sources such as optical parametric oscillators or amplifiers. Recently, an experiment was carried out using a single femtosecond erbium-doped fiber oscillator used as a seed for two separate supercontinuum generation stages [37]. These stages are optimized, filtered, and compressed independently to generate two femtosecond pulses with spectral content centered at 1.12 and 1.77 μm, each spanning several hundreds of nanometers. Since the oscillator is common in this case, only the relative phase fluctuations between the arms must be controlled. In the experiment, this parameter was not submitted to active feedback, but the setup is stable enough to allow characterization of a synthesized pulse in free-running operation. This results in the generation of a 4.3 fs pulse, corresponding to a single-cycle pulse at the central wavelength.

Another recent experiment [38] achieves pulse synthesis from a Ti:sapphire oscillator that is split into two separate optical parametric chirped pulse amplifiers (OPCPA) operating at 900 nm and 2.2 µm, respectively. Since the oscillator is shared, the repetition frequency is automatically matched. The offset frequencies (or equivalently the carrier envelope phase) are stabilized in both OPCPA stages. In addition, a balanced cross-correlator measures the relative delay between the pulses, which acts as an error signal to adjust a piezo-mounted mirror in one of the branches. This feedback results in the stabilization of the relative delay to within 250 as. Overall, the stabilization mechanisms allow the synthesis of coherent spectral content spanning 1.8 octaves, and the individual phases of both arms can be adjusted to control the shape of the synthesized electric field. This experiment results in the generation of energetic pulses of 15 µJ, potentially exhibiting a high-field transient of duration equal to 0.8 optical cycles at the centroid wavelength.

The characterization of such broadband coherent radiation is often difficult, and even more so if the carrier envelope phase, an important quantity for few-cycle pulses, is to be measured. Another difficulty is that the energy contained in the pulse can be low enough to make difficult the nonlinear processes that are often necessary to carry out the optical characterization methods. The approach pursued in Refs [37,38] is to separately fully characterize the arms that are combined using a standard technique such as frequency-resolved optical gating or spectral interferometry. After this characterization, the remaining quantities that must be measured to fully characterize the synthesized pulse are the relative phase and delay. If the carrier envelope phase value is to be measured, the phase of each arm and the delay are necessary. In all cases, the full characterization of such short transients of optical radiation remains a challenge.

9.5
Conclusion

The possibility to scale the average and peak powers of ultrafast laser sources beyond a single-amplifier technical limitation using coherent beam combining is now established. The rather low sensitivity of combining efficiency with respect to temporal parameters such as group delay dispersion for pulses in the 100 fs range implies that combining a large number of amplifiers seems realistic. New tools are being developed in this direction such as the active stabilization of the group delay [39], in addition to the optical phase. The additional possibility to combine power scaling and pulse synthesis [40], with a control over the intensity and phase of several spectral bands, opens up fascinating perspectives. Spatiotemporal analysis and characterization will become increasingly important as the combined pulses get shorter, down to the few-optical cycle regime.

References

1 Fan, T.Y. (2005) Laser beam combining for high-power, high-radiance sources. *IEEE J. Sel. Top. Quantum Electron.*, **11**, 567–577.
2 Cheung, E.C., Weber, M., and Rice, R.R. (2008) Phase locking of a pulsed fiber amplifier. Advanced Solid-State Photonics, Nara, Japan, January 2007, Paper WA2.
3 Lombard, L., Azarian, A., Cadoret, K., Bourdon, P., Goular, D., Canat, G., Jolivet, V., Jaouën, Y., and Vasseur, O. (2011) Coherent beam combination of narrow-linewidth 1.5 μm fiber amplifiers in a long-pulse regime. *Opt. Lett.*, **36**, 523–525.
4 Seise, E., Klenke, A., Limpert, J., and Tünnermann, A. (2010) Coherent addition of fiber-amplified ultrashort laser pulses. *Opt. Express*, **18**, 27827–27835.
5 Daniault, L., Hanna, M., Lombard, L., Zaouter, Y., Mottay, E., Goular, D., Bourdon, P., Druon, F., and Georges, P. (2011) Coherent beam combining of two femtosecond fiber chirped-pulse amplifiers. *Opt. Lett.*, **36**, 621–623.
6 Strickland, D. and Mourou, G. (1985) Compression of amplified chirped optical pulses. *Opt. Commun.*, **56**, 219–221.
7 Galvanauskas, A. (2002) Ultrashort-pulse fiber amplifiers, in *Ultrafast Lasers: Technology and Applications* (eds M.E. Fermann, A. Galvanauskas, and G. Sucha), CRC Press, p. 209.
8 Zaouter, Y., Papadopoulos, D.N., Hanna, M., Boullet, J., Huang, L., Aguergaray, C., Druon, F., Mottay, E., Georges, P., and Cormier, E. (2008) Stretcher-free high energy nonlinear amplification of femtosecond pulses in rod-type fibers. *Opt. Lett.*, **33**, 107–109.
9 Daniault, L., Hanna, M., Papadopoulos, D.N., Zaouter, Y., Mottay, E., Druon, F., and Georges, P. (2011) Passive coherent beam combining of two femtosecond fiber chirped-pulse amplifiers. *Opt. Lett.*, **36**, 4023–4025.
10 Kane, D. and Trebino, R. (1993) Characterization of arbitrary femtosecond pulses using frequency-resolved optical gating. *IEEE J. Quantum Electron.*, **29**, 571–579.
11 Iaconis, C. and Walmsley, I.A. (1998) Spectral phase interferometry for direct electric-field reconstruction of ultrashort optical pulses. *Opt. Lett.*, **23**, 792–794.
12 Goodno, G.D., Shih, C.-C., and Rothenberg, J.E. (2010) Perturbative analysis of coherent combining efficiency with mismatched lasers. *Opt. Express*, **18**, 25403–25414.
13 Klenke, A., Seise, E., Limpert, J., and Tünnermann, A. (2011) Basic considerations on coherent combining of ultrashort laser pulses. *Opt. Express*, **19**, 25379–25387.
14 Akturk, S., Gu, X., Gabolde, P., and Trebino, R. (2005) The general theory of first-order spatio-temporal distortions of Gaussian pulses and beams. *Opt. Express*, **13**, 8642–8661.
15 Siiman, L.A., Chang, W.-Z., Zhou, T., and Galvanauskas, A. (2012) Coherent femtosecond pulse combining of multiple parallel chirped pulse fiber amplifiers. *Opt. Express*, **20**, 18097–18116.
16 Hänsch, T.W. and Couillaud, B. (1980) Laser frequency stabilization by polarization spectroscopy of a reflecting reference cavity. *Opt. Commun.*, **35**, 441–444.
17 Klenke, A., Seise, E., Demmler, S., Rothhardt, J., Breitkopf, S., Limpert, J., and Tünnermann, A. (2011) Coherently-combined two channel femtosecond fiber CPA system producing 3mJ pulse energy. *Opt. Express*, **19**, 24280–24285.
18 Shay, T., Benham, V., Baker, J.T., Sanchez, A.D., Pilkington, D., and Lu, C.A. (2007) Self-synchronous and self-referenced coherent beam combination for large optical arrays. *IEEE J. Sel. Top. Quantum Electron.*, **13**, 480–486.
19 Daniault, L., Hanna, M., Lombard, L., Zaouter, Y., Mottay, E., Goular, D., Bourdon, P., Druon, F., and Georges, P. (2012) Impact of spectral phase mismatch on femtosecond coherent beam combining systems. *Opt. Lett.*, **37**, 650–652.
20 Jiang, S., Hanna, M., Druon, F., and Georges, P. (2010) Impact of self-phase modulation on coherently combined fiber chirped-pulse amplifiers. *Opt. Lett.*, **35**, 1293–1295.

21 Sabourdy, D., Kermene, V., Desgarges-Berthelemont, A., Lefort, L., Barthélémy, A., Mahodaux, C., and Pureru, D. (2002) Power scaling of fibre lasers with all-fibre interferometric cavity. *Electron. Lett.*, **38**, 692–693.

22 Ishaaya, A.A., Davidson, N., and Friesem, A.A. (2009) Passive laser beam combining with intracavity interferometric combiners. *IEEE Sel. Top. Quantum Electron.*, **15**, 301–311.

23 Zaouter, Y., Daniault, L., Hanna, M., Papadopoulos, D., Morin, F., Hönninger, C., Druon, F., Mottay, E., and Georges, P. (2012) Passive coherent combination of two ultrafast rod type fiber chirped pulse amplifiers. *Opt. Lett.*, **37**, 1460–1462.

24 Zhou, S., Wise, F.W., and Ouzounov, D.G. (2007) Divided-pulse amplification of ultrashort pulses. *Opt. Lett.*, **32**, 871–873.

25 Kong, L.J., Zhao, L.M., Lefrancois, S., Ouzounov, D.G., Yang, C.X., and Wise, F.W. (2012) Generation of megawatt peak power picosecond pulses from a divided-pulse fiber amplifier. *Opt. Lett.*, **37**, 253–255.

26 Papadopoulos, D.N., Zaouter, Y., Hanna, M., Druon, F., Mottay, E., Cormier, E., and Georges, P. (2007) Generation of 63fs 4.1 MW peak power pulses from a parabolic fiber amplifier operated beyond the gain bandwidth limit. *Opt. Lett.*, **32**, 2520–2522.

27 Daniault, L., Hanna, M., Papadopoulos, D. N., Zaouter, Y., Mottay, E., Druon, F., and Georges, P. (2012) High peak-power stretcher-free femtosecond fiber amplifier using passive spatio-temporal coherent combining. *Opt. Express*, **20**, 21627–21634.

28 Zaouter, Y., Guichard, F., Daniault, L., Hanna, M., Morin, F., Hönninger, C., Mottay, E., Druon, F., and Georges, P. (2013) Femtosecond fiber chirped- and divided-pulse amplification system. *Opt. Lett.*, **38**, 106–108.

29 Petersen, J. and Luiten, A. (2003) Short pulses in optical resonators. *Opt. Express*, **11**, 2975–2981.

30 Potma, E.O., Evans, C., Xie, X.S., Jones, R.J., and Ye, J. (2003) Picosecond-pulse amplification with an external passive optical cavity. *Opt. Lett.*, **28**, 1835–1837.

31 Jones, R.J., Moll, K.D., Thorpe, M.J., and Ye, J. (2005) Phase-coherent frequency combs in the vacuum ultraviolet via high-harmonic generation inside a femtosecond enhancement cavity. *Phys. Rev. Lett.*, **94**, 193201.

32 Pupeza, I., Eidam, T., Rauschenberger, J., Bernhardt, B., Ozawa, A., Fill, E., Apolonski, A., Udem, T., Limpert, J., Alahmed, Z.A., Azzeer, A.M., Tünnermann, A., Hänsch, T.W., and Krausz, F. (2010) Power scaling of a high-repetition-rate enhancement cavity. *Opt. Lett.*, **35**, 2052–2054.

33 Hartl, I., Schibli, T.R., Marcinkevicius, A., Yost, D.C., Hudson, D.D., Fermann, M.E., and Ye, J. (2007) Cavity-enhanced similariton Yb-fiber laser frequency comb: 3×10^{14} W/cm^2 peak intensity at 136 MHz. *Opt. Lett.*, **32**, 2870–2872.

34 Hänsch, T.W. (1990) A proposed sub-femtosecond pulse synthesizer using separate phase-locked laser oscillators. *Opt. Commun.*, **80**, 71–75.

35 Shelton, R.K., Ma, L.-S., Kapteyn, H.C., Murnane, M.M., Hall, J.L., and Ye, J. (2001) Phase-coherent optical pulse synthesis from separate femtosecond lasers. *Science*, **293**, 1286–1289.

36 Sun, J. and Reid, D.T. (2009) Coherent ultrafast pulse synthesis between an optical parametric oscillator and a laser. *Opt. Lett.*, **34**, 854–856.

37 Krausst, G., Lohss, S., Hanke, T., Sell, A., Eggert, S., Huber, R., and Leitenstorfer, A. (2010) Synthesis of a single cycle of light with compact erbium-doped fibre technology. *Nat. Photonics*, **4**, 33–36.

38 Huang, S.-W., Cirmi, G., Moses, J., Hong, K.-H., Bhardwaj, S., Birge, J.R., Chen, L.-J., Li, E., Eggleton, B., Cerullo, G., and Kärtner, F.X. (2011) High-energy pulse synthesis with sub-cycle waveform control for strong-field physics. *Nat. Photonics*, **5**, 475–479.

39 Benjamin Weiss, S., Weber, M.E., and Goodno, G.D. (2012) Group delay locking of coherently combined broadband lasers. *Opt. Lett.*, **37**, 455–457.

40 Chang, W.-Z., Zhou, T., Siiman, L. A., and Galvanauskas, A. (2013) Femtosecond pulse spectral synthesis in coherently-spectrally combined multi-channel fiber chirped pulse amplifiers. *Opt. Express*, **21**, 3897–3910.

Part Two
Passive and Self-Organized Phase Locking

10
Modal Theory of Coupled Resonators for External Cavity Beam Combining

Mercedeh Khajavikhan and James R. Leger

10.1
Introduction

Laser radiance, defined as the power per unit area per unit solid angle, directly relates to two laser characteristics: output power and beam quality. In many applications, radiance is the quantity of practical interest since it is directly proportional to the ability to concentrate power at a large distance from a collimating optic of fixed diameter or at the focal plane of a lens with a fixed numerical aperture. Unfortunately, the radiance of virtually all lasers is limited by practical considerations of aberration, nonlinearities, thermal management, and optical damage. Coherent beam combining offers a potential route to high radiance that circumvents many of these practical limitations [1,2].

Coherent beam combining methods can be categorized into three general classes: (i) master oscillator–power amplifier architectures [3,4], (ii) beam coupling in nonlinear media [5], and (iii) coupled laser cavities [6,7]. This chapter is devoted to the analysis of several architectures belonging to the third category. We concentrate entirely on a linear analysis of these coupled laser cavities, and do not consider intensity-specific or gain-specific effects such as optical nonlinearities and gain pulling. This simplification allows us to employ a modal approach to describe various resonator characteristics and results in a deep understanding of the resonator characteristics, including the number of allowed modes, the modal discrimination, the oscillation frequency of the allowed modes, and the sensitivity to path length changes in the beam combining optics. We should be aware, however, that nonlinearities can play a significant role in a real high-power laser system and must be included in a more complete model.

We begin this chapter with a short review of coherent beam combining requirements. We then describe the mathematical framework of our modal analysis and discuss the role of wavelength agility in compensating for path length errors. The majority of the chapter is then devoted to exploring the modal behavior of two classes of coupled cavities: those that increase radiance by increasing the power per unit area (superposition techniques) and those that increase radiance by increasing the

Coherent Laser Beam Combining, First Edition. Edited by Arnaud Brignon.
© 2013 Wiley-VCH Verlag GmbH & Co. KGaA. Published 2013 by Wiley-VCH Verlag GmbH & Co. KGaA.

power per unit solid angle (parallel coupling techniques). Along with the theoretical development, we provide experimental data that illustrate the predicted performance of a few representative architectures.

10.2
Coherent Beam Combining Requirements

To appreciate the coherent beam combining problem, it is instructive to assess the physical limits of beam combining dictated by the radiance theorem. A classical interpretation of the radiance theorem implies that it is impossible to increase the power per unit area per unit solid angle of a spatially incoherent light field with a passive optical system. However, for laser systems, it is more appropriate to describe the theorem in terms of optical modes, where a mode is a deterministic solution of the electric field in the wave equation that satisfies the appropriate boundary conditions. It is generally represented by a complex function that describes the magnitude of the field at all points and the fixed phase relationships between all points in the mode. In this context, the radiance theorem states that, given N uncorrelated optical modes, the power per mode cannot be increased by any optical system that is in thermal equilibrium. Since the phases of the fields from mutually incoherent lasers vary in time with respect to one another, the light from an ensemble of these beams must be described as containing multiple spatial modes, one from each laser. Using this picture, we conclude radiance can be increased only by reducing the number of modes in the system by establishing some degree of mutual coherence between the original lasers. The radiance can be maximized by forcing all the individual laser modes into a single coherent state by establishing a non-time-varying phase relationship between each of the laser sources. Note that this condition does not require the lasers to be monochromatic.

There are several nonclassical ways to channel optical power from individual lasers into a reduced number of modes. One is to use the multiple lasers as a pump to invert a secondary gain medium. This architecture is used, for example, in a diode-pumped solid-state laser to increase the radiance of the semiconductor diode laser array. Of course, all the problems and limitations of a single-gain element laser are still present. Alternatively, the individual gain media can be placed in a common cavity with coupling between the laser elements to create a coupled oscillator. The advantage of this architecture is that the power is now distributed between multiple gain media. In general, the light field from the coupled lasers consists of a linear combination of "supermodes" corresponding to the specific normal modes of the coupled array. In our formulation, we assume that each individual laser is oscillating in a single spatial mode (such as a Gaussian distribution), and the term "supermode" is used to describe a coherent collection of the individual laser modes. In this approach, coherence is established by providing supermode-dependent cavity losses. If only one supermode is allowed to exceed the lasing threshold, the phases from all the lasers that participate in the supermode are locked together in a coherent state. From this point of view, coupling between lasers is necessary simply to produce a supermode that contains an ensemble of many lasers. Of course, allowing additional supermodes to lase degrades the coherence; in the limit where all supermodes lase, the source can be considered to be totally mutually incoherent and no increase in radiance is possible. Thus, it is clear that the ability to discriminate between supermodes is an important attribute of the various cavity architectures. In addition, fundamental supermode loss and shape, as well as supermode discrimination are strongly influenced by path length errors across the laser array. Our modal theory can be used to explore these effects.

The above coherence requirement ensures that the power per mode increases as more lasing elements are added to the array. However, it does not necessarily ensure that the power per unit area per unit solid angle (the conventional definition of radiance) is optimum in any sense. To achieve this, the supermode must be converted into the desired form by manipulating its intensity and phase. The actual optimal shape is dependent on the particular application. To maximize on-axis irradiance in the far field or at the focal plane of a lens, the field emanating from a finite-size aperture (containing the combined laser beams) must be coupled into a far-field delta function in an optimal manner. Because of the Fourier relationship between the near and far fields, the optimum near-field distribution is one that approximates a plane wave over the extent of the optical aperture. Thus, the supermode should be converted into a shape that *uniformly* illuminates the optical exit aperture in both amplitude and phase [1]. On the other hand, if one desires a mode with a low M^2, the beam should be converted into a Gaussian distribution [8]. There are many techniques to perform these conversions in a quasi-lossless manner by manipulating the phase of the beam in one or more locations [9]. This step is essential to achieve a useful output and true high radiance. Here too, path length errors in the combining process can lead to laser outputs that, although perhaps coherent, contain distorted wavefronts and result in low-radiance beams.

10.3
General Mathematical Framework of Passive Laser Resonators

As already mentioned, our analysis is limited to a linear description of the optical cavity, and thus can employ linear systems modeling techniques. We start by assuming that the coupled cavity consists of a combination of N individual gain elements, where each gain element is designed to support a single spatial mode (often a Gaussian or quasi-Gaussian distribution). Thus, the ith supermode of the coupled cavity can be expressed as an N-dimensional vector $\vec{v}_i$, where the complex vector components describe the amplitude and phase of each individual gain element. The main goal of the analysis is to solve for the cavity supermodes and their associated losses subject to various operating conditions.

The cavity supermodes are spatiotemporal electric field distributions that, after traveling one round-trip inside the cavity, completely reproduce themselves except for a possible change in overall magnitude. To find these supermodes, we start by calculating the matrix $\boldsymbol{M}_{rt}$ that describes the effect of propagating a vector description of the light distribution through one cavity round-trip. M_{rt} consists of the product of matrices corresponding to the various optical elements, propagation effects, and phase errors encountered in the cavity round-trip. The supermodes are then found by solving the eigenequation:

$$M_{rt}\, \vec{v}_i = \lambda_i\, \vec{v}_i, \tag{10.1}$$

where $\vec{v}_i$ is the ith eigenvector of the round-trip propagation matrix M_{rt} and λ_i is its corresponding complex eigenvalue.[1] The physical meaning of $|\lambda_i|$ is apparent from Eq. (10.1) and it can be interpreted as the amplitude attenuation factor of the supermode $\vec{v}_i$ after executing a single round-trip in the cavity. Thus, $1 - |\lambda_i|^2$ describes the power loss to the ith supermode after one round-trip. The phase component of the eigenvalue represents a phase shift of the supermode upon one round-trip for the assumed wavelength, meaning that lasing must occur at a different wavelength such that this phase component becomes an integer multiple of 2π. We consider the effect of this eigenvalue phase in Section 10.3.2.

Since the vectors $\vec{v}_i$ are orthonormal and complete, any light field distribution in the cavity can be expressed as a linear superposition of these vectors. Consequently, as the cavity light field builds up from noise when the laser gains are increased, the first distribution to achieve the lasing threshold will be the one with the lowest round-trip loss or the largest eigenvalue $|\lambda_1|$. The coherent lasing state will be described by this fundamental supermode $\vec{v}_1$, and the power extracted from each individual gain element will be proportional to the squared magnitude of the associated components in $\vec{v}_1$. Fundamental supermode vectors that have approximately equal coefficients will extract the power most efficiently from the entire array, whereas those with nulls at particular gain elements will not allow those elements to contribute to the coherent state. In most practical cases, $\vec{v}_1$ is a reasonably accurate description of the fundamental spatial supermode even when gain nonlinearities are considered. However, the losses associated with a particular supermode may be significantly affected by nonlinearities such as gain broadening and spatial and spectral hole burning. These effects may favor one set of supermodes and suppress the rest. A more complete analysis must take these nonlinearities into account.

10.3.1
Coherent Beam Combining by a Simple Beam Splitter

As a simple example of the analytical technique, we start by modeling coherent beam combining by a beam splitter or 3 dB coupler (Figure 10.1). The beam splitter and the fiber-coupled cavities are optically equivalent, and the choice between them is dictated by practical considerations. The basic operation can be understood by applying the time reversal property. If we illuminate the beam splitter in Figure 10.1a from the right with a single beam, the element produces a transmitted beam (traveling to the left) and a reflected beam (traveling down). These two beams are mutually phase-coherent

---

[1] In this chapter, the symbol λ_i with a subscript is reserved for eigenvalues. The symbol λ with no subscript will be used in subsequent sections to indicate optical wavelength.

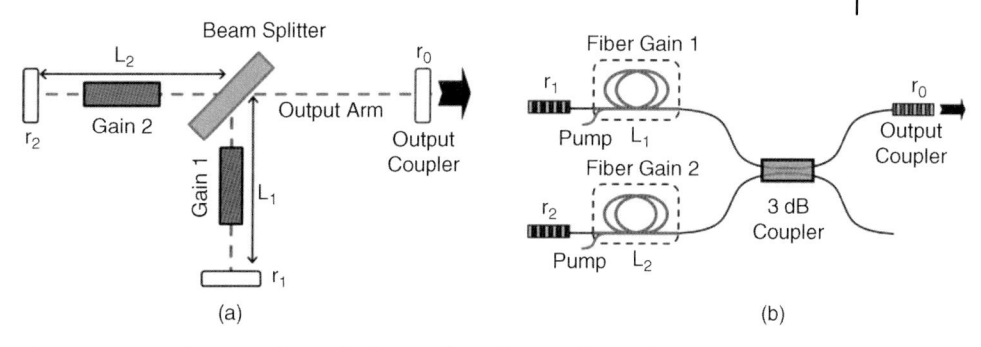

Figure 10.1 (a) Coherent combining by a beam splitter. (b) Equivalent architecture utilizing a 3 dB fiber coupler.

(since they originate from the same source) and have a unique phase relationship between each other. Therefore, time reversal suggests that these two mutually coherent beams can be combined into one by the beam splitter. This is indeed a general property of a beam splitter, where we note that the exact phase relationship between the two beams dictates how much of the light comes out of each port. Although this simple explanation justifies the basic operation of the beam splitter resonator, a more detailed description of the resonator characteristics requires a deeper understanding of its modal structure. This is provided by the eigenvalue equation describing light propagation inside the resonator.

The eigenequation of the beam splitter cavity is obtained by finding a matrix that describes light propagation through one cavity round-trip. To construct this round-trip matrix, we assume the beams begin their journey directly after reflection from the mirrors in the gain arms. They first acquire specific phase shifts proportional to the optical path lengths as they travel along the gain arms (represented by matrix Φ), and are redirected into the vertical and right-hand ports of the beam splitter in accordance with the beam splitter scattering matrix S. Only the beams from the right-hand port are reflected back (by virtue of the output mirror) and are subsequently redirected by the beam splitter into the left-hand and lower resonator arms according to the transpose scattering matrix S^T. The beams travel once more through the gain arms Φ to finally reach the points of departure after being reflected back by the mirrors. The round-trip matrix of the cavity M_{rt} is a product of all these matrices:

$$M_{rt} = \Phi S^T R S \Phi, \qquad (10.2)$$

where

$$\Phi = \begin{pmatrix} e^{j(\phi_1/2)} & 0 \\ 0 & e^{j(\phi_2/2)} \end{pmatrix}, \qquad (10.3)$$

$$R = \begin{pmatrix} r_0 & 0 \\ 0 & 0 \end{pmatrix}, \qquad (10.4)$$

and $\phi_i = 2kL_i$, where k is the wave number of the light, L_i is the length of the ith gain arm, and r_0 is the amplitude reflection coefficient of the output mirror. The scattering matrix for a lossless and reciprocal beam splitter S is given by [10]

$$S = \begin{pmatrix} r & jt \\ jt & r \end{pmatrix}, \tag{10.5}$$

where $r^2 + t^2 = 1$. In this scattering matrix, r is the amplitude reflection coefficient and t is the amplitude transmission coefficient of the beam splitter. While the exact values of r and t depend on a particular beam splitter under study, fundamental properties govern the relationships between the elements of the matrix. For example, reciprocity requires a *symmetric* scattering matrix and a lossless beam splitter requires a *unitary* matrix [11].

Assuming a beam splitter with a 50:50 power splitting ratio, $r = t = 1/\sqrt{2}$ and the round-trip matrix M_{rt} can be simplified to

$$M_{\text{rt}} = \begin{pmatrix} \frac{1}{2} r_0 e^{j\phi_1} & \frac{j}{2} r_0 e^{j[(\phi_1+\phi_2)/2]} \\ \frac{j}{2} r_0 e^{j[(\phi_1+\phi_2)/2]} & -\frac{1}{2} r_0 e^{j\phi_2} \end{pmatrix}. \tag{10.6}$$

By noting the linear dependence between the two columns, it is easy to recognize that M_{rt} is unity rank, meaning that it has only one nonzero eigenvalue. Physically, this corresponds to a resonator that supports only one supermode. By solving the characteristic equation of the matrix, we obtain the single eigenvector $\vec{v}_1$ and its corresponding eigenvalue λ_1 of

$$\vec{v}_1 = \begin{pmatrix} \frac{1}{\sqrt{2}} \\ \frac{j}{\sqrt{2}} e^{j\Delta\phi/2} \end{pmatrix} \tag{10.7}$$

and

$$\lambda_1 = -jr_0 e^{j[(\phi_1+\phi_2)/2]} \sin\left(\frac{\Delta\phi}{2}\right), \tag{10.8}$$

where $\Delta\phi = \phi_1 - \phi_2$. Equation (10.7) shows that the intensities of the fields in the two gain arms are equal, maximizing the power extracted from the two gain media. Equation (10.8) can be used to predict the cavity loss as a function of the phase error $\Delta\phi$ (resulting from a path length difference $L_1 - L_2 = \Delta\phi/2k$). In particular, the power loss due to this phase error (where we set $r_0 = 1$) is given by $1 - |\lambda_1|^2 = \cos^2(\Delta\phi/2)$. Clearly, when the phase difference between the two arms is π, the loss goes to zero, whereas when the phase difference is zero, the loss becomes unity as the power exits the other cavity port and is not returned to the resonator.

10.3.2
Effect of Wavelength Diversity [12]

As is clear from the previous discussion, the relative phases of the gain media must be properly established to achieve high-efficiency coherent beam combining in an external cavity. Unfortunately, many gain media change their physical length and index as a function of temperature, pump current, and other effects so that the correct phase state cannot always be guaranteed. It is sometimes necessary to include active feedback systems in the cavity to ensure that the proper phase state is maintained. Active phase adjustment technologies are reviewed in Part One of this book. However, passively coupled laser arrays can sometimes adjust their phase by simply changing their operating wavelength [13,14]. This section explores the method of phase adjustment by this "wavelength diversity."

In describing the wavelength diversity effect, it is convenient to express the wavelength in terms of the wave number k (equal to 2π/wavelength). In particular, if the two gain arms are of very different lengths, a relatively small change in wave number can give rise to a large relative phase change at the beam splitter. In this way, some amount of path length phase error compensation can be obtained by selecting the proper lasing wave number. We can use the modal picture to quantify this behavior by observing the cavity eigenvalue as a function of path length difference and wave number simultaneously. Although this is a general phenomenon, we illustrate its basic characteristics by considering the beam splitter resonator of Section 10.3.1.

To achieve resonance, a circulating cavity mode must satisfy the self-consistency condition, namely, that the field undergoes a total phase shift of $2\pi m$, where m is an integer. As a result, the laser selects wave numbers that generate real and positive eigenvalues (the so-called longitudinal modes of the laser). Fluctuations in temperature, the linewidth enhancement factor, and hole burning mechanisms can introduce phase changes inside the cavity that temporarily or permanently modify the wave numbers at which the self-consistency condition is satisfied. Applying the self-consistency condition to Eq. (10.8), we notice very distinct behaviors in the limits of small and large path length differences between the gain arms ($\Delta L = L_2 - L_1$). To distinguish between these two regimes, we divide the overall path length difference between the gain arms into two parts: (i) a fixed path length difference ΔL_0 that is chosen by the designer, and (ii) a small unintentional variation of the path length δL that takes on values on the order of a few wavelengths and might have been caused either by imperfections in the fabrication or by random environmental changes. Furthermore, we assume the gain bandwidth of the laser medium supports lasing at frequencies with wave numbers k_i ranging over a spectrum of $2\Delta k_{max}$ such that $k_0 - \Delta k_{max} < k_i < k_0 + \Delta k_{max}$ and k_0 is the center of the gain bandwidth. For a typical neodymium-doped fiber laser, Δk_{max} is on the order of $1400 \, \text{cm}^{-1}$ for a lasing wavelength of around 1.06 μm. For a Nd:YAG crystal lasing at the same 1.06 μm spectral line, Δk_{max} is approximately $16 \, \text{cm}^{-1}$.

If we assume lasing operation at a specific wave number k_i, Eq. (10.8) can be written as

$$\lambda_1 = -jr_0 e^{j(k_0 - \Delta k_i)(2L_1 + \Delta L_0 + \delta L)} \sin\{(k_0 - \Delta k_i)(\Delta L_0 + \delta L)\}, \tag{10.9}$$

where $\Delta k_i = k_0 - k_i$ and the allowed values of k_i must be selected to satisfy the consistency condition $\text{Arg}\{\lambda_1\} = 2\pi m$, where m is an integer. Note that in this equation, $k_0 \Delta L_0 = \phi_0$ is a constant phase shift and $\Delta k_i \delta L$ is small in comparison to $k_0 \delta L$ and will be ignored.

The limits of $\Delta k_{\max} \Delta L_0$ distinguish the two regimes of operations. For $\Delta k_{\max} \Delta L_0 \ll \pi$, the magnitude of the eigenvalue is

$$|\lambda_1| \approx r_0 |\sin(k_0 \delta L + \phi_0)|, \quad \Delta k_{\max} \Delta L_0 \ll \pi. \tag{10.10}$$

For a neodymium-doped fiber laser, this regime is defined by $\Delta L_0 \ll 22\,\mu m$, and for a Nd:YAG crystal, $\Delta L_0 \ll 2\,mm$. Obviously, in this regime, the eigenvalue magnitude shows no wavelength dependence (i.e., the loss for all the allowed longitudinal modes is equal). Since under these conditions spatial modes entirely characterize the cavity's modal behavior, we call this regime of operation "spatial mode dominated."

On the other hand, when $\Delta k_{\max} \Delta L_0 \gg \pi$, the magnitude of the eigenvalue becomes

$$|\lambda_1| \approx r_0 |\sin(k_0 \delta L - \Delta k_i \Delta L_0 + \phi_0)|, \quad \Delta k_{\max} \Delta L_0 \gg \pi. \tag{10.11}$$

At a particular path length error δL, changing longitudinal modes (and therefore changing the lasing wave number Δk_i) can have a profound influence on the magnitude of the eigenvalue. In this regime, the rectified sinusoidal behavior of Eq. (10.11) as a function of wavenumber Δk_i has a sufficiently small period ($\pi/\Delta L_0$) that one longitudinal mode is very likely to have low loss (eigenvalue close to unity) for one path length, while another mode will have low loss when the path length is increased by δL. The laser then simply switches longitudinal modes to maximize the eigenvalue and minimize the phase error loss. A change in δL then only produces a shift of the sinusoidal function in Eq. (10.11). In Figure 10.2, we show the eigenvalue magnitudes of the allowed longitudinal modes for two path length errors (δL's). We refer to this limit ($\Delta k_{\max} \Delta L_0 \gg \pi$) as the "longitudinal mode dominated regime" because the behavior of the cavity is ultimately determined by the longitudinal modes. Of course, this analysis assumes that the separation between longitudinal modes $2\pi/(L_1 + L_2)$ is sufficiently small to allow many modes to lase within the gain bandwidth $2\Delta k_{\max}$. This is very often the case in fiber laser systems, where the length of each fiber (L_1 and L_2) is typically on the order of meters.

To check the accuracy of this analysis and study the effect of path length errors in the transition region between the spatial and longitudinal mode dominated regimes ($\Delta k_{\max} \Delta L_0 \sim \pi$), an experiment was conducted using two coupled polarization modes in an Nd:YAG laser to demonstrate the effect of changes in ΔL_0 in Eq. (10.11) [12]. Figure 10.3 shows the output power as a function of the path length phase difference for the nominal path length difference (ΔL_0) of 0.0 mm

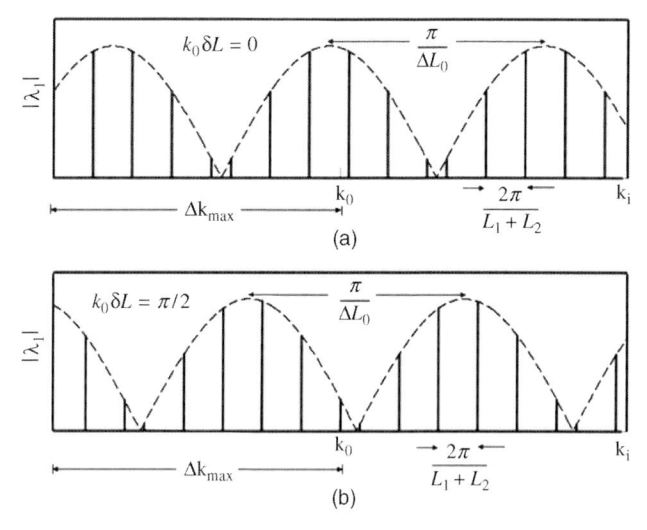

Figure 10.2 Allowed eigenvalues in the longitudinal mode-dominated regime. The solid vertical lines indicate longitudinal modes. (a) Path length error δL produces a phase error of $k_0 \delta L = 0$. (b) Path length error δL produces a phase error of $k_0 \delta L = \pi/2$ [12].

(dotted curve), 0.4 mm (dashed curve), and 1.0 mm (solid curve). It is clear from this figure that as the nominal path length difference between the gain arms (ΔL_0) increases, the output power becomes less and less sensitive to random path length variations as the system transitions from the spatial mode-dominated regime to the longitudinal mode-dominated regime. Recall the condition for the longitudinal

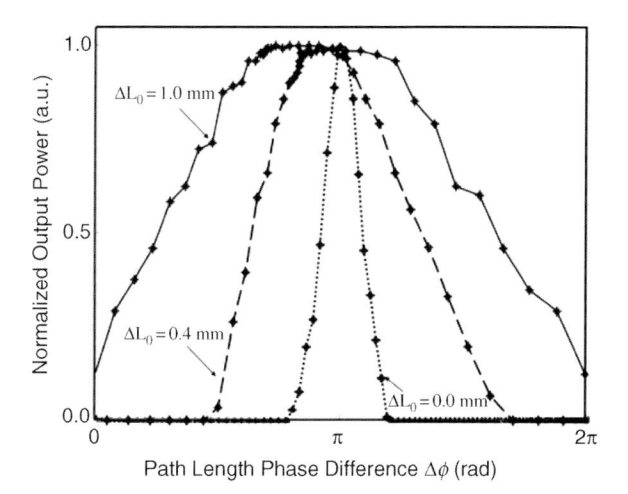

Figure 10.3 Measured total output power from coherently combining two lasers as a function of path length error. The three curves correspond to the transition from the spatial mode-dominated regime ($\Delta L_0 = 0.0$ mm) to the longitudinal mode-dominated regime ($\Delta L_0 = 1.0$ mm) [12].

mode regime is $\Delta k_{max}\Delta L_0 \gg \pi$. Assuming a gain bandwidth of 0.6 nm for the 1064 nm Nd:YAG laser line (corresponding to a Δk_{max} of 16 cm^{-1}), the value for $\Delta k_{max}\Delta L_0$ is given by 0.6 and 1.6 rad for the path length differences of 0.4 and 1.0 mm, respectively. Hence, this coupled system is not quite in the full longitudinal mode-dominated regime, and the output power is still somewhat influenced by path length errors. Nevertheless, the influence is significantly reduced for $\Delta L_0 = 1$ mm, indicating that longitudinal mode diversity is able to partially correct for the path length errors.

Wavelength diversity has been demonstrated in many common cavity structures and is the reason behind the initial success of passive coherent beam combining [13–15]. Its success depends on the gain bandwidth of active elements, the total number of longitudinal modes that coincide with this gain bandwidth, and the number of gain elements. When only two gain elements are involved and the product of gain bandwidth (Δk) and path length difference (ΔL_0) is large, it is easy to find common longitudinal modes to correct for the random path length errors. As the number of gain elements increases, the rate of success in finding common longitudinal modes that corrects all the path length errors reduces. A more thorough understanding of the modal characteristics of external cavities will allow us to better engineer these cavities for a particular laser system. The remainder of this chapter is devoted to the study of different external laser cavities and their modal structure.

10.4
Coupled Cavity Architectures Based on Beam Superposition

The radiance of a laser beam (i.e., power per unit area per unit solid angle) can be increased by either increasing the power per unit area (while keeping the power per solid angle fixed) or increasing the power per unit solid angle (while keeping the power per unit area fixed). We have organized the various beam combining architectures in the remainder of this chapter to coincide with these two routes to increased radiance. The first architecture allows the multiple laser beams to overlap in a common area, thereby increasing the total power per unit area. We call this type of beam combining *beam superposition* and describe it in the current section. The second type of beam combining couples multiple beams in a parallel fashion (maintaining the same power per unit area), but reduces the effective solid angle divergence by establishing coherence across the entire laser array. We call this type of beam combining *parallel coupling*, and describe several examples of this architecture in Sections 10.5 and 10.6.

10.4.1
Generalized Michelson Resonators

The beam splitter cavity described in Section 10.3.1 is the simplest example of a general class of superposition architectures known as Michelson resonators. A more

10.4 Coupled Cavity Architectures Based on Beam Superposition

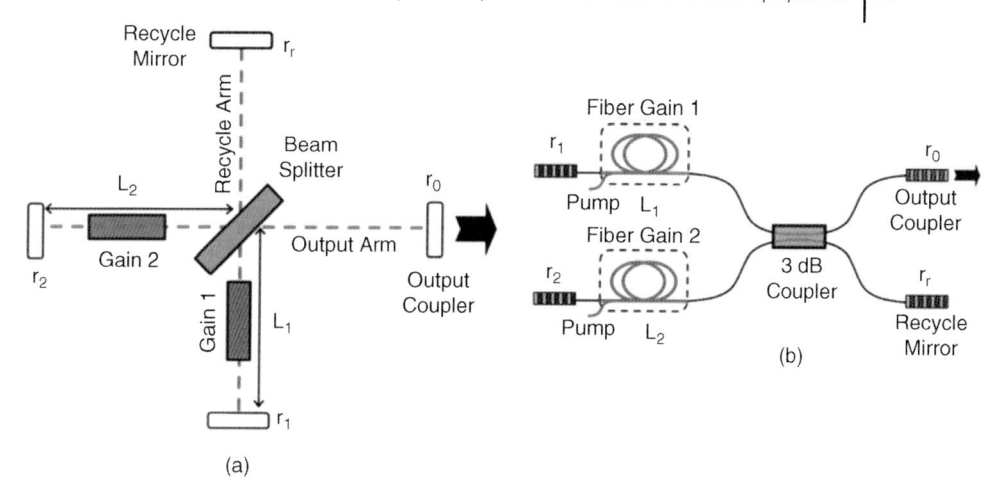

Figure 10.4 Extended Michelson resonator. (a) Beam splitter realization employing a recycle mirror. (b) Equivalent fiber architecture employing a recycle Bragg reflector.

general example of a Michelson resonator is shown in Figure 10.4. Comparing this cavity with the one in Figure 10.1, we see that an additional mirror (called a recycle mirror) has been placed in the previously open port. Since all four ports of the splitting/combining element are contributing to the formation of the cavity, we call this cavity an extended Michelson resonator.

The eigenequation of the extended Michelson cavity is obtained as in Section 10.3.1, where Eq. (10.4) is modified to represent a mirror with amplitude reflection coefficient r_o in the output arm and r_r in the recycle arm:

$$R = \begin{pmatrix} r_o & 0 \\ 0 & r_r \end{pmatrix}. \tag{10.12}$$

Again, assuming a 50 : 50 power beam splitter with $r = t = 1/\sqrt{2}$, the round-trip matrix M_{rt} can be simplified to

$$M_{rt} = \begin{pmatrix} \frac{1}{2}(r_o - r_r)e^{j\phi_1} & \frac{j}{2}(r_o + r_r)e^{j[(\phi_1+\phi_2)/2]} \\ \frac{j}{2}(r_o + r_r)e^{j[(\phi_1+\phi_2)/2]} & \frac{1}{2}(r_r - r_o)e^{j\phi_2} \end{pmatrix}. \tag{10.13}$$

In writing this matrix, we have assumed that the optical path lengths of the output and the recycling arms are precisely equal. The more general case where these two path lengths are allowed to vary is addressed in Ref. [16]. The eigenvectors and eigenvalues of this modified resonator now become

$$\vec{v}_1 = \begin{pmatrix} v_{11} \\ v_{21} \end{pmatrix}, \quad \vec{v}_2 = \begin{pmatrix} v_{12} \\ v_{22} \end{pmatrix}, \quad \text{where} \quad v_{11} = v_{12} = \frac{1}{\sqrt{2}}, \tag{10.14}$$

$$\left.\begin{array}{l}v_{21}\\v_{22}\end{array}\right\} = \frac{j}{\sqrt{2}(r_0+r_r)}\left[(r_0-r_r)e^{-j(\Delta\phi/2)}+j(r_0-r_r)\sin\left(\frac{\Delta\phi}{2}\right)\right.$$

$$\left.\pm j\sqrt{(r_0-r_r)^2\sin^2\left(\frac{\Delta\phi}{2}\right)+4r_0r_r}\right], \quad (10.15)$$

and

$$\left.\begin{array}{l}\lambda_1\\\lambda_2\end{array}\right\} = \frac{-j}{2}e^{j[(\phi_1+\phi_2)/2]}\left\{(r_0-r_r)\sin\left(\frac{\Delta\phi}{2}\right)\pm\sqrt{(r_0-r_r)^2\sin^2\left(\frac{\Delta\phi}{2}\right)+4r_0r_r}\right\}. \quad (10.16)$$

The phase error loss $(1-|\lambda_i|^2)$ is plotted in Figure 10.5. First we notice that, unlike the simple beam splitter resonator, the extended Michelson has two distinct supermodes. By increasing the value of the recycling mirror, it is apparent in Figure 10.5a that the fundamental (highest eigenvalue) supermode loss due to the path length phase error $\Delta\phi = 2k(L_2-L_1)$ can be reduced, making the system more tolerant to random path length changes. However, this improvement comes at the expense of introducing a competing supermode, where losses of the two supermodes are compared in Figure 10.5b. Second, since the eigenvectors are normalized, $v_{11}=v_{12}=1/\sqrt{2}$ implies that $|v_{21}|=|v_{22}|=1/\sqrt{2}$ as well. This implies that, as in the simple beam splitter resonator of Section 10.3.1, the intensities in both gain arms (arms L_1 and L_2 of Figure 10.4) are equal, independent of both the path length phase error between the gain arms and the choice of the reflectivity of the recycle mirror. This intensity distribution is important since, if the fundamental supermode can access most of the gain from both arms, it can clamp the gain and prevent the higher order supermode from oscillating. By restricting the resonator to a single supermode, coherence is established across the array.

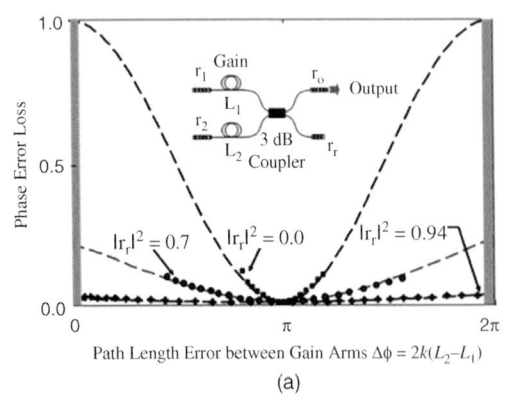

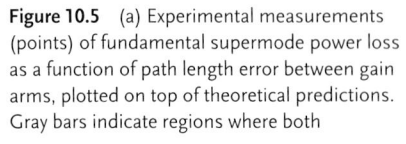

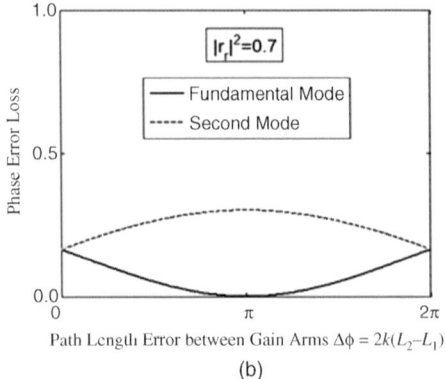

Figure 10.5 (a) Experimental measurements (points) of fundamental supermode power loss as a function of path length error between gain arms, plotted on top of theoretical predictions. Gray bars indicate regions where both supermodes were observed simultaneously. (b) Comparison between theoretical fundamental and second supermode loss as a function of path length error for a recycle mirror reflectivity of $|r_r|^2 = 0.7$ [12].

The operating characteristics of both the simple beam splitter resonator and the extended Michelson resonator have been measured experimentally using a polarization multiplexed Nd:YAG laser resonator. Accurate measurements of the amplitudes and phases of both supermodes as well as the modal power loss $(1 - |\lambda_i|^2)$ have been made as a function of path length phase error. Details can be found in Ref. [12]. The experiment was specifically designed to operate in the spatial mode-dominated regime ($k\Delta L \ll 2\pi$ as defined in Section 10.3.2) to remove the effects of wavelength tuning for path length phase error compensation. The two amplitude components of the fundamental supermode were shown to be identical (within experimental error) for the full range of path length phase errors and recycling mirrors tested. In addition, the phase between the two components was shown to follow the expected relationship shown in Eqs. (10.14) and (10.15). The modal power loss due to phase errors (phase error loss) was measured as a function of path length error $\Delta\phi = 2k(L_2 - L_1)$ and recycling mirror reflection coefficient (r_r) and is shown in Figure 10.5a. The measured phase error loss of the fundamental supermode followed the predicted loss value given by $1 - |\lambda_1|^2$, where λ_1 is given by Eq. (10.16). Note that single supermode performance was observed over most of the path length error range. However, when the path length phase error was chosen close to an integer multiple of 2π, the resonator coherence was lost. This is indicated in Figure 10.5a as a narrow gray band around zero and 2π. The reason for this is apparent from Figure 10.5b, where the theoretical phase error loss to both the fundamental mode and second supermode is plotted from Eq. (10.16). Clearly, for path length phase errors that take on integer values of 2π, the two supermodes become degenerate in loss and the cavity cannot discriminate between them. Consequently, both supermodes lase simultaneously and the coherence is destroyed.

The Michelson cavity can be extended to combine more than two gain elements by a number of means. It is instructive to examine two straightforward extensions shown in Figure 10.6 for combining beams from four gain elements. The first architecture, shown in Figure 10.6a, is a binary tree where each node of the tree is replaced by a beam splitter. The second architecture in Figure 10.6b consists of a linear chain of beam splitters. The detailed lasing characteristics of each cavity can

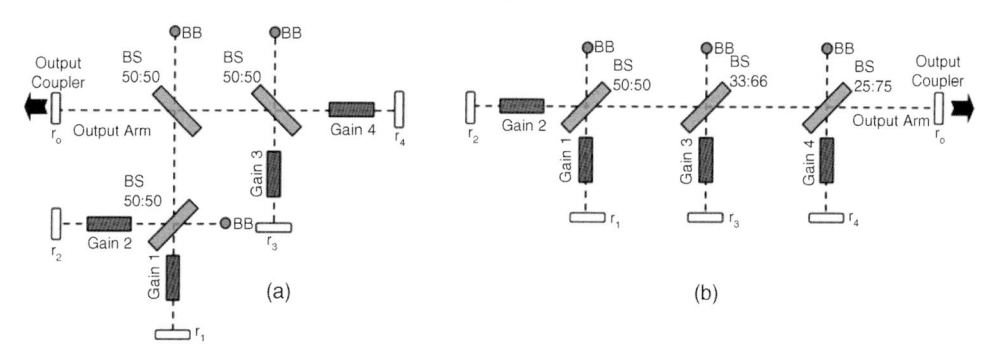

Figure 10.6 (a) Binary tree configuration for combining four gain elements. All beam splitters have intensity splitting ratios of 50 : 50. (b) Linear chain configuration to combine four gain elements. The intensity splitting ratios of the beam splitters vary along the chain. BB, beam block; BS, beam splitter.

be explored by calculating the eigenvectors and eigenvalues of the round-trip propagation matrix M_{rt}, formed by properly multiplying the scattering and propagation matrices of the individual elements of the cavity. It is easy to show that both these architectures support only a single supermode. This supermode can be designed to have equal light intensity in each gain arm by adjusting the splitting ratios of the beam splitters. The binary tree architecture requires all splitting ratios to be 50:50, whereas the beam splitting ratios of the linear chain are a function of position. The path length error loss associated with each configuration as a function of the various path length errors can also be calculated in a straightforward manner. Proper adjustment of the path lengths results in no power loss in the arms with beam blocks. Note that to combine N gain elements, both configurations require the same number of mirrors $(N+1)$, beam splitters $(N-1)$, and beam blocks $(N-1)$.

10.4.2
Grating Resonators

In the previous section, we used $(N-1)$ simple beam splitters or 3 dB couplers to combine N beams by arranging the elements in a binary tree or series configuration. Unfortunately, the beam splitter configuration is quite cumbersome as it requires aligning multiple optical elements. The equivalent 3 dB coupler version using fiber optics does not suffer from these alignment issues. However, the final combined beam must pass through a single fiber, where it is subject to many of the same nonlinearities and power limitations of a single fiber laser. An alternative method that avoids both these issues uses a multiorder diffraction grating designed to efficiently split a single laser beam into N roughly equal output beams. There are several optical elements that can achieve this. One such element is a Dammann grating [17], shown in Figure 10.7. Early Dammann grating designs consisted of

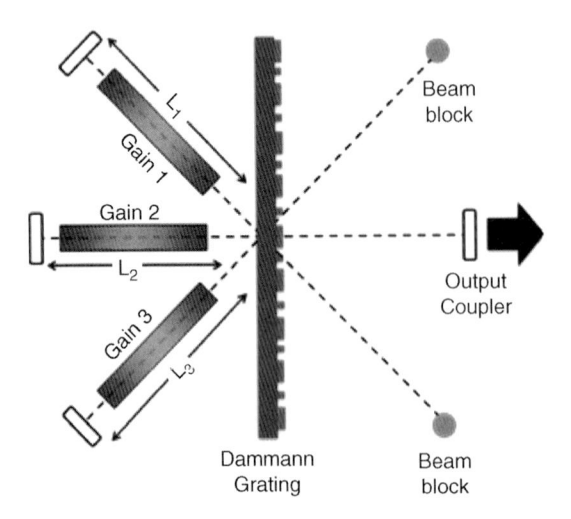

Figure 10.7 Dammann grating cavity.

binary phase steps in a periodic pattern, where each period contained a carefully chosen internal structure. Since the pattern was periodic, the diffraction necessarily consisted of discrete diffraction orders. The internal structure could then be chosen to equalize the intensity of N diffraction orders while suppressing all others. The advantage of this type of Dammann grating is its ease of fabrication. However, some small amount of light is generally contained in unwanted diffraction orders, reducing the overall efficiency of the grating. With the advent of continuous surface relief fabrication methods, more advanced Dammann gratings can now be designed with a continuous (but periodic) phase pattern to improve efficiency [18,19].

Figure 10.7 shows a Dammann grating used in a common cavity to form a Dammann resonator [20]. In this configuration, the Dammann grating acts as a multiarm beam splitter. Similar to the simple beam splitter case, a single light beam traveling from right to left is split into N mutually coherent beams with a fixed phase relationship between each beam. Unlike the beam splitter case, however, a small amount of light is directed to unwanted diffraction orders (not shown in Figure 10.7), resulting in a grating efficiency η (defined as the ratio between the sum of the power in the N desired orders and the total incident power) that is less than unity. If light from only the N desired beams is redirected back onto the grating from the left-hand side (and the proper phase relationship is maintained), reciprocity can be used to show that the grating combines the beams into a single on-axis beam with the same efficiency [21]. Thus, a Dammann resonator contains N low-radiance arms on the left-hand side and one high-radiance arm on the right-hand side.

The scattering matrix for an ideal ($\eta = 1$) Dammann grating can be derived with the help of the properties of lossless, reciprocal, and time-reversible systems. In particular, the matrix must be both reciprocal ($s_{ij} = s_{ji}$) and unitary ($S S^\dagger = I$), where the dagger denotes the Hermitian transpose of the matrix. As an example, a six-port ideal Dammann grating (three input ports and three output ports) is given by

$$S = \begin{pmatrix} \frac{1}{\sqrt{3}} & \frac{1}{\sqrt{3}} e^{j(2\pi/3)} & \frac{1}{\sqrt{3}} e^{j(2\pi/3)} \\ \frac{1}{\sqrt{3}} e^{j(2\pi/3)} & \frac{1}{\sqrt{3}} & \frac{1}{\sqrt{3}} e^{j(2\pi/3)} \\ \frac{1}{\sqrt{3}} e^{j(2\pi/3)} & \frac{1}{\sqrt{3}} e^{j(2\pi/3)} & \frac{1}{\sqrt{3}} \end{pmatrix}. \quad (10.17)$$

The round-trip propagation matrix M_{rt} of this ideal grating cavity is then given by

$$M_{rt} = \frac{1}{3} r_0 \begin{pmatrix} e^{j(4\pi/3)} e^{j\phi_1} & e^{j(2\pi/3)} e^{j[(\phi_1+\phi_2)/2]} & e^{j(4\pi/3)} e^{j[(\phi_1+\phi_3)/2]} \\ e^{j(2\pi/3)} e^{j[(\phi_1+\phi_2)/2]} & e^{j\phi_2} & e^{j(2\pi/3)} e^{j[(\phi_2+\phi_3)/2]} \\ e^{j(4\pi/3)} e^{j[(\phi_1+\phi_3)/2]} & e^{j(2\pi/3)} e^{j[(\phi_2+\phi_3)/2]} & e^{j(4\pi/3)} e^{j\phi_3} \end{pmatrix},$$

$$(10.18)$$

where the definitions of ϕ_i are as before. An inspection of this matrix reveals that every column is identical up to a complex multiplier, indicating that the matrix is of unity rank. Thus, as in the case of the simple beam splitter cavity, the matrix contains only one eigenvector with a nonzero eigenvalue and the system can lase in a single spatial supermode only. The single eigenvector $\vec{v}_1$ and eigenvalue λ_1 associated with this supermode are given by

$$\vec{v}_1 = \begin{pmatrix} \frac{1}{\sqrt{3}} e^{j[(\phi_1 - \phi_2)/2]} e^{j(2\pi/3)} \\ \frac{1}{\sqrt{3}} \\ \frac{1}{\sqrt{3}} e^{j[(\phi_3 - \phi_2)/2]} e^{j(2\pi/3)} \end{pmatrix} \tag{10.19}$$

and

$$\lambda_1 = \frac{r_0}{3} e^{j(4\pi/3)} e^{j\phi_2} \left\{ e^{j(\Delta\phi_1)} + e^{j(2\pi/3)} + e^{j(\Delta\phi_3)} \right\}, \quad \Delta\phi_i = \phi_i - \phi_2. \tag{10.20}$$

Scattering matrices of real Dammann gratings depend on the actual grating design. However, the real Dammann grating cavity still maintains its single-mode characteristics. The cavity loss $(1 - |\lambda_1|^2)$ obtained from Eq. (10.20) is plotted in Figure 10.8 as a function of path length error. The contour plot in Figure 10.8a shows that there is only one phase state between elements that minimizes the loss (ideally to zero), whereas there are two states that maximize the loss (to unity). Figure 10.8b shows the effect of varying the phase in arm 3 when arm 1 and arm 2 are both set to $\pi/3$ rad, whereas Figure 10.8c shows this effect when arm 1 is set to π and arm 2 is set to $\pi/3$ rad. These performance characteristics as a function of path length errors can be generalized to any superposition element, such as an N-beam Dammann grating.

When the phase conditions are not met, the lost light exits the cavity through the secondary ports shown on the right-hand side in Figure 10.7 (containing beam blocks) as well as higher order diffraction angles (not shown). We note that, as in the extended Michelson cavity, it is possible to partially overcome the phase error loss due to random path length variations by adding recycling mirrors to these ports to recover some of the power. In general, the cavity will now contain N supermodes and we must rely on the gain clamping properties of the active media to ensure single spatial supermode operation.

The Dammann grating method can easily be extended to two dimensions, allowing beam combining from two-dimensional arrays. Both rectangular and hexagonal symmetries are possible. By using more general diffractive optics techniques [19,22], it is also possible to design more complex diffractive structures that produce arrays of spots at more general angles.

Volume Bragg gratings (volume holograms) offer an alternative method of producing a multiport beam splitter. By using the volume diffraction effect, it is theoretically possible to couple light from one diffraction order to another with 100% efficiency [23]. Multiport devices can be made exposing a suitable holographic material to a single reference wave and multiple object waves spaced at predetermined angles (with all beams present simultaneously and mutually coherent). Alternatively, the grating can be made by a series of multiple exposures, each with a common reference wave angle and a different object wave angle. One of the principal advantages of a multiplexed volume Bragg grating is that the angles

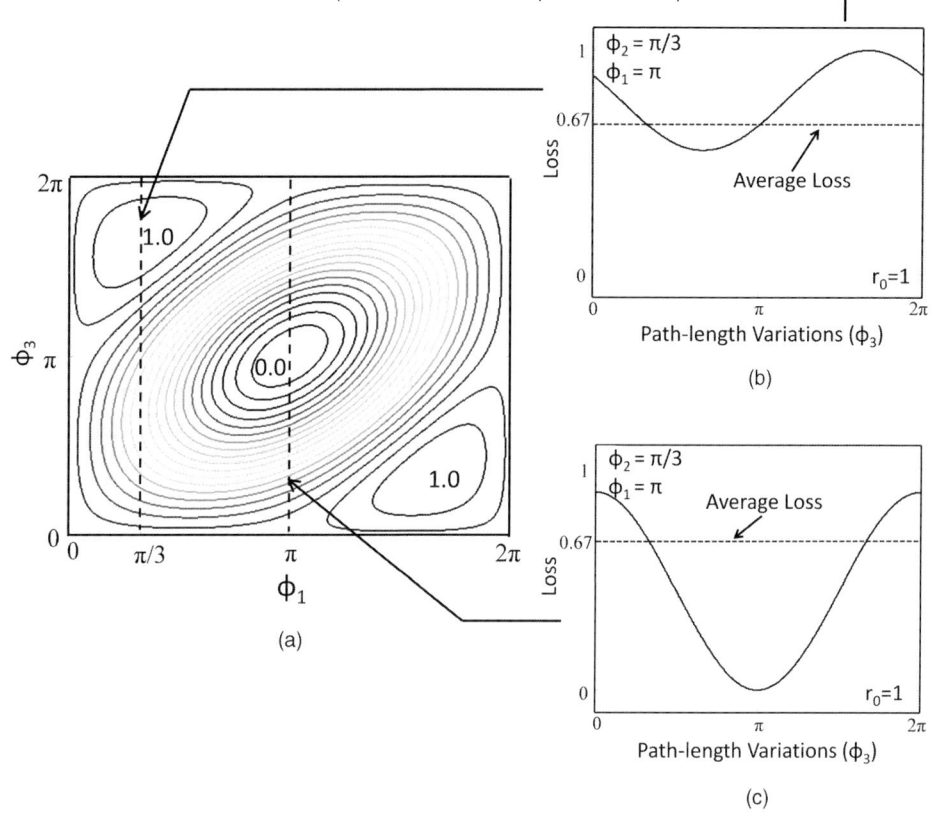

Figure 10.8 Effect of phase errors on the modal loss of an ideal six-port Dammann grating. (a) Modal loss caused by phase errors in input arm 1 (ϕ_1) and input arm 3 (ϕ_3) for $\phi_2 = \pi/3$. (b) Modal loss as a function of phase error in input arm 3, where both ϕ_1 and ϕ_2 have a value of $\pi/3$ rad. (c) Modal loss as a function of phase error in input arm 3, where $\phi_1 = \pi$ and $\phi_2 = \pi/3$ rad [16].

between the various output beams can be chosen by the designer, and can be quite general. The idealized scattering matrix that describes this element is, however, identical to the Dammann grating matrix, and the modal performance in a resonator is described by the equations derived previously in this section.

10.5
Parallel Coupled Cavities Based on Space-Invariant Optical Architectures

In this section, we discuss a general class of coupled cavities where coupling from one gain element to the next occurs by local or quasi-local interaction. Figure 10.9 shows three examples of parallel coupled cavities. In Figure 10.9a, evanescent coupling between fiber waveguides causes light to couple from one guide to its neighbor (and possibly beyond). The spacing between guides and the properties of

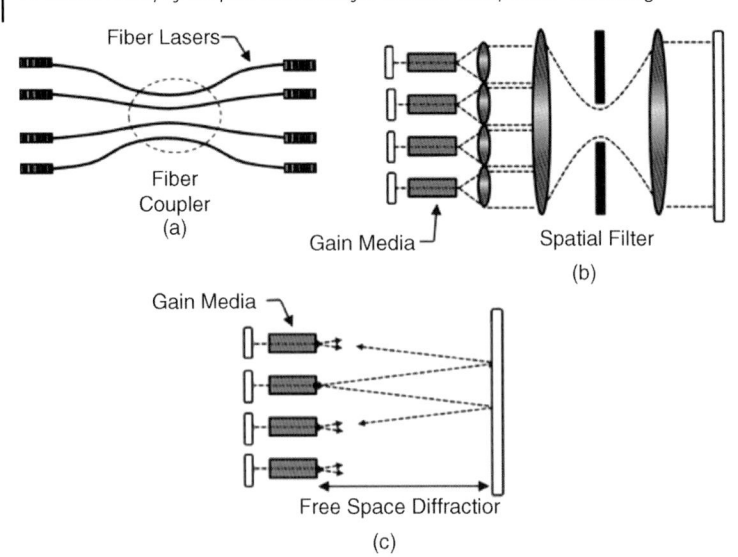

Figure 10.9 Examples of space-invariant coupled laser arrays. (a) Evanescent coupling between waveguides. (b) Coupling by spatial filtering. (c) Coupling through free space diffraction.

the guiding structures govern the strength of the coupling, and coupled mode theory [24,25] or numerical simulation tools employing the finite difference time domain (FDTD) method or finite difference method (FDM) must be used to properly describe the exchange of energy between guides [26]. Figure 10.9b shows a spatially filtered imaging system. The light from the output apertures of the laser gain media is imaged onto itself using an afocal imaging system and a single end mirror. The presence of the spatial filter in the back focal plane of the afocal system causes the image to blur out across the array, coupling light from one aperture to its neighbors [27–32]. The final configuration in Figure 10.9c shows lasing channels coupled by free space diffraction. In this cavity, the degree of coupling depends on the length of the free space section, the size of the lasing apertures, and their spacing [33].

Each of the systems in Figure 10.9 can be designed to be space invariant by choosing all elements and element spacing to be identical. Space invariance implies that the coupling from any gain element to any other gain element is not a function of absolute element locations, but only of the relative distance between the elements. For example, the coupling between the first and third gain elements in Figure 10.9c is the same as the coupling between the second and fourth elements, and so on.

The amplitude and phase of the light field coupled from the ith to the jth gain elements in any coupled resonator are described by the coefficients m_{ij} of the system matrix M. For the space-invariant resonators of Figure 10.9, the coefficients m_{ij} of the system matrix M are only functions of relative distance between elements $(i-j)$. This gives rise to a system matrix M that has identical values along all its diagonals. Such a matrix is called Toeplitz and it has special properties that we can exploit in analysis. Section 10.5.1 takes advantage of these properties to analyze a space-invariant cavity in the weak coupling limit.

In making the space-invariant assertion above, we have tacitly assumed that there are no path length phase errors across the array. Such phase errors destroy the space invariance property, since the additional phase error is now a function of absolute location. The overall round-trip propagation matrix M_{rt} is no longer Toeplitz, and a different mathematical approach must be used to arrive at an analytical solution. In Section 10.5.2, we analyze the resonator architecture shown in Figure 10.9b where path length phase errors are introduced. The analysis is restricted to two coupled lasers to arrive at analytical results. Finally, in Section 10.5.3, we study the properties of the resonator shown in Figure 10.9c in the strong coupling limit. For this analysis, we employ the theory of Talbot imaging to calculate the resonator properties at specific propagation distances.

We note that changing the spacing of the waveguides in Figure 10.9a or the size and spacing of the laser apertures in Figure 10.9b and c in a nonuniform manner allows one to modify the coefficients of the M matrix in a general way (subject to the constraints dictated by the particular coupling architecture). The coupling strengths then become a function of absolute location and the system response becomes space variant. Since the matrix M largely determines the supermode shapes and attenuation factors, these space variant modifications can be used to engineer supermode shape and attenuation [6,34]. However, analytical methods generally do not exist to calculate the modal performance in these more general cases, and the designs are usually performed by numerical techniques.

10.5.1
Space-Invariant Parallel Coupled Resonators with Weakly Coupled Cavities

In general, there is no analytical solution for decomposing a Toeplitz matrix into its characteristic eigenvectors and eigenvalues. Thus, a complete analysis of the space-invariant coupled cavity is not possible. However, approximate solutions can be derived for weak coupling (i.e., small values of m_{ij} that are off the principal diagonal). In particular, if the coupling is so weak that only nearest-neighbor terms need to be considered, the matrix M takes on a tridiagonal form:

$$M = \begin{pmatrix} 1 & \alpha & 0 & 0 & \cdots \\ \alpha & 1 & \alpha & 0 & \\ 0 & \alpha & 1 & \alpha & 0 \\ \vdots & & & \ddots & \\ & & 0 & \alpha & 1 & \alpha \\ & & 0 & 0 & \alpha & 1 \end{pmatrix}, \qquad (10.21)$$

where α is the complex amplitude of the field that is coupled from one laser to its nearest neighbor. We also assume that there are no path length errors or non-uniformities across the array of the gain elements, so the round-trip propagation matrix M_{rt} is equal to the system matrix M. As before, we are interested in solutions to the equation

$$M_{rt} \vec{v}_i = \lambda_i \vec{v}_i, \qquad (10.22)$$

where the vectors $\vec{v}_i$ correspond to the eigenvectors of the matrix M_{rt} with λ_i their corresponding eigenvalues. The solution to Eq. (10.22) for a tridiagonal Toeplitz matrix is shown to be given by [35]

$$\vec{v}_i[k] = \frac{\sqrt{2}}{\sqrt{N+1}} \sin\left(\frac{\pi}{N+1} ik\right) \qquad (10.23)$$

and

$$\lambda_i = 1 + 2a \cos\left(\frac{\pi i}{N+1}\right). \qquad (10.24)$$

It is apparent from Eq. (10.23) that all supermodes have a sinusoidal amplitude profile, with the first supermode having the shape of a half sine wave. This nonuniformity implies that the gain cannot be completely extracted from all the elements in the array. In addition, we can see from Eq. (10.24) that larger real values of α (and hence stronger and longer range coupling) result in larger modal discrimination. Note that the Toeplitz matrix in Eq. (10.21) has not been normalized and so the resulting eigenvalues can take on values greater than unity in Eq. (10.24).

The above analysis can be applied to any space-invariant configuration (examples of which are shown in Figure 10.9a–c) as long as the coupling is sufficiently small. In particular, it has been used extensively to analyze the supermodes of evanescently coupled semiconductor laser arrays (similar to Figure 10.9a) [6].

Figure 10.10 shows the supermodes calculated from the three eigenvectors of a $N = 3$ parallel coupled cavity with relatively weak coupling ($\alpha = 0.2$) and no path length phase errors. Figure 10.10 also shows the effect of the coupling phase on the eigenvalue. In general, when α is real and positive, the $i = 1$ mode corresponding to the in-phase supermode has the largest eigenvalue (i.e., it is the dominant mode) and subsequent mode numbers $i = 2, 3, 4, \ldots$ have monotonically lower values. Conversely, when α is negative, the $i = N$ mode corresponding to the out-of-phase supermode has the largest eigenvalue, and subsequent mode numbers $i = N-1, N-2, N-3, \ldots$ have monotonically lower values. When α is completely imaginary, the eigenmodes of the in-phase

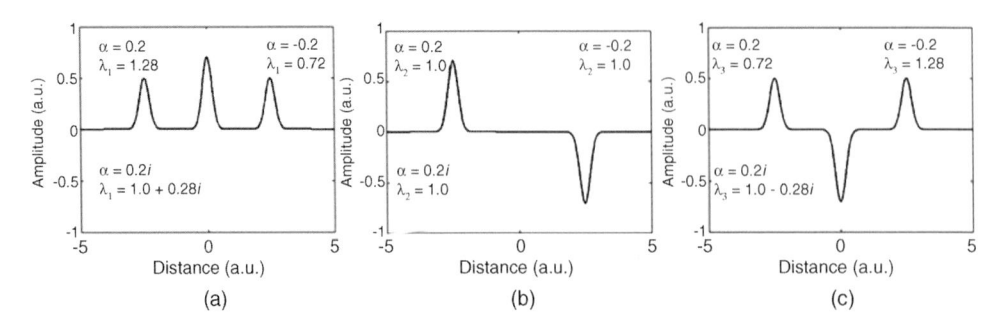

Figure 10.10 Eigenmodes of a space-invariant cavity containing three gain elements in the weak coupling regime. (a) The in-phase supermode. (b) An intermediate supermode. (c) The out-of-phase supermode.

and out-of-phase modes become degenerate in magnitude, as do the eigenmodes of the ith and $(N+1-i)$th modes, for all values of i.

10.5.2
Spatially Filtered Resonators and the Effect of Path Length Phase Errors [36]

We now concentrate on the spatial filtering configuration shown in Figure 10.9b. This system can be understood at the conceptual level by considering the interference pattern that occurs on the left-hand side of the spatial filter (Figure 10.9b). If the lasers contain no mutual coherence, the pattern at the spatial filter is a superposition of the intensity generated by each single-gain aperture. However, if the lasers are all mutually coherent, have a fill factor of unity (meaning the gain media spacing is identical to the individual output aperture widths), and have identical phases, the wavefront appears to come from a much larger single aperture, producing a correspondingly smaller spot at the spatial filter. The filter passes more of the light and the overall round-trip loss is lower for this coherent state. Thus, the first supermode to reach lasing threshold will contain all lasers coherently locked in phase with each other.

The modal theory permits a more thorough understanding of spatial filtering. In particular, the eigenvectors and eigenvalues of the round-trip matrix determine the supermode shape and modal loss under a variety of conditions. The spatial filtering architecture is inherently space invariant (assuming identical laser apertures and spacing) because the laser array and spatial filter are placed in the Fourier planes of the first lens (i.e., the front and back focal planes). Changes in the spatial filter size and shape affect all elements equally, preserving the Toeplitz characteristic of the system matrix M. For weak coupling and no path length errors, the results from the previous section can thus be applied. Path length errors, however, destroy the space invariance of the round-trip matrix M_{rt}, and the results of Section 10.5.1 are no longer applicable. Therefore, to study the effects of these path length errors, we consider a simple two-laser case. We note that the assumption of weak coupling is no longer needed in this analysis, and the coupling strength can be quite general.

A more detailed layout of the spatially filtered common cavity laser system we wish to analyze is shown in Figure 10.11. The cavity consists of two waveguide-based gain media (e.g., rare earth-doped single-mode fibers) separated by a distance 2Δ, two independent end mirrors, and a common output mirror. Spatial filtering is performed by placing an aperture in the back focal plane of an afocal imaging system. We model the aperture as having a Gaussian transmittance for mathematical simplicity. The phase difference between the two channels is introduced by the displacement between the two end mirrors $\Delta z = \lambda(\phi_1 - \phi_2)/(4\pi)$, where λ is the wavelength of the laser light.

The initial field in plane 1 emanating from the two optical waveguides is modeled by two normalized Gaussian distributions, each with beam waist ω:

$$\tilde{u}_1(x) = \sqrt{\frac{1}{\omega}}\sqrt{\frac{2}{\pi}}\exp\left[-(x-\Delta)^2/\omega^2\right] \tag{10.25}$$

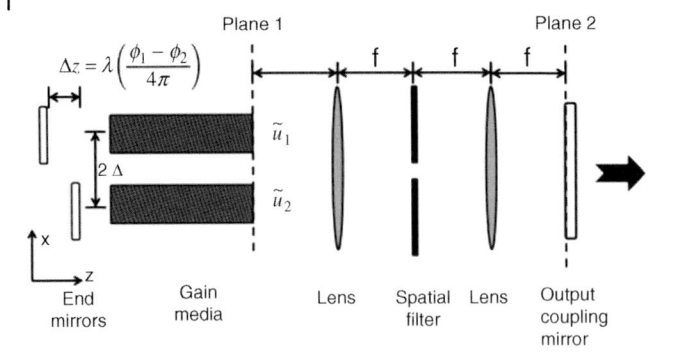

10 Modal Theory of Coupled Resonators for External Cavity Beam Combining

Figure 10.11 Phase locking two gain media by spatial filtering.

and

$$\tilde{u}_2(x) = \sqrt{\frac{1}{\omega}}\sqrt{\frac{2}{\pi}}\exp\left[-(x+\Delta)^2/\omega^2\right]. \tag{10.26}$$

The amplitude transmittance of the Gaussian spatial filter is given by

$$t(x) = \exp\left(\frac{x^2}{a^2}\right), \tag{10.27}$$

where a describes the size of the filter.

The round-trip $ABCD$ matrix of the system (from plane 1 to plane 2 and back to plane 1) is [32]

$$\begin{pmatrix} A & B \\ C & D \end{pmatrix} = \begin{pmatrix} 0 & f \\ -\frac{1}{f} & 0 \end{pmatrix}\begin{pmatrix} 1 & 0 \\ -\frac{j\lambda}{\pi a^2} & 1 \end{pmatrix}\begin{pmatrix} 0 & f \\ -\frac{1}{f} & 0 \end{pmatrix}\begin{pmatrix} 0 & f \\ -\frac{1}{f} & 0 \end{pmatrix}\begin{pmatrix} 1 & 0 \\ -\frac{j\lambda}{\pi a^2} & 1 \end{pmatrix}\begin{pmatrix} 0 & f \\ -\frac{1}{f} & 0 \end{pmatrix}$$

$$= \begin{pmatrix} 1 & j\frac{2f^2\lambda}{\pi a^2} \\ 0 & 1 \end{pmatrix}.$$

$$\tag{10.28}$$

The coefficients of this matrix can be used to calculate the diffraction pattern $\tilde{v}(x)$ of the initial light distribution $\tilde{u}(x)$ from a particular waveguide through the equation:

$$\tilde{v}(x) = \int_{-\infty}^{\infty} \tilde{K}(x,\xi)\tilde{u}(\xi)d\xi, \tag{10.29}$$

where

$$\tilde{K}(x,\xi) = \sqrt{\frac{j}{B\lambda}}\exp\left[\frac{-j\pi}{B\lambda}(Ax^2 - 2x\xi + D\xi^2)\right]. \tag{10.30}$$

In our case, the four $ABCD$ matrix elements are given by $A = 1$, $B = j(2f^2\lambda/\pi a^2)$, $C = 0$, and $D = 1$.

The light that couples back into a particular waveguide is given by an overlap integral between the diffraction pattern after one round-trip $\tilde{v}(x)$ and the initial waveguide mode $\tilde{u}(x)$. We calculate the general coupling coefficient c_{ij} corresponding to the field coupled into the *ith* waveguide from an initial field emanating from the *jth* waveguide. The coefficients of the coupling matrix are given by

$$c_{ij} = \int_{-\infty}^{\infty} \tilde{u}_i^*(x)\tilde{v}_j(x)dx. \tag{10.31}$$

The entire round-trip propagation matrix $\mathbf{M}_{rt}$ must include the additional path length phase errors represented by physical shifts of the end mirrors, where the phase errors are expressed as ϕ_1 and ϕ_2:

$$\mathbf{M}_{rt} = \begin{bmatrix} c_{11}e^{j\phi_1} & c_{12}e^{j\phi_1} \\ c_{21}e^{j\phi_2} & c_{22}e^{j\phi_2} \end{bmatrix}. \tag{10.32}$$

Note that $\mathbf{M}_{rt}$ is no longer Toeplitz in general due to the path length errors.

By solving for the eigenvalues and eigenvectors of the round-trip propagation matrix in Eq. (10.32), the modal properties can be obtained. The two eigenvalues (related to the round-trip supermode loss L by $L = 1 - |\lambda_i|^2$), are given by

$$\lambda_{1,2} = \frac{\exp(j\bar{\phi})}{\sqrt{1+\delta/2}} \left[\cos\left(\frac{\phi}{2}\right) \pm \sqrt{\exp\left[-\frac{8\Delta^2}{(2+\delta)\omega^2}\right] - \sin^2\left(\frac{\phi}{2}\right)} \right], \tag{10.33}$$

where

$$\bar{\phi} = \frac{\phi_1 + \phi_2}{2}, \quad \phi = \phi_1 - \phi_2, \quad \text{and} \quad \delta = \frac{2f^2\lambda^2}{\pi^2\omega^2 a^2}.$$

Note that λ_1 and λ_2 are eigenvalues corresponding to the positive and negative values of the square root in Eq. (10.33), whereas λ is the wavelength. The corresponding normalized eigenvectors are

$$v_1 = \left\{ \frac{R_+}{\sqrt{R_+^2+1}}, \frac{1}{\sqrt{R_+^2+1}} \right\},$$

$$v_2 = \left\{ \frac{R_-}{\sqrt{R_-^2+1}}, \frac{1}{\sqrt{R_-^2+1}} \right\}, \tag{10.34}$$

where

$$R_{+,-} = \left[-j\sin\left(\frac{\phi}{2}\right) \pm \sqrt{\exp\left[-\frac{8\Delta^2}{(2+\delta)\omega^2}\right] - \sin^2\left(\frac{\phi}{2}\right)} \right]$$

$$\cdot \exp\left(\frac{j\phi}{2}\right) \exp\left[\frac{4\Delta^2}{(2+\delta)\omega^2}\right]. \tag{10.35}$$

The two sets of eigenvalues and eigenvectors correspond to the two supermodes supported by this cavity. When there is no phase error [$\phi = 0$ in Eq. (10.35)], the two eigenvectors in Eq. (10.34) represent the traditional symmetric and antisymmetric supermodes seen in many coupled oscillator systems.

Of particular interest is the condition that if

$$\sin^2\left(\frac{\phi}{2}\right) \geq \exp\left[-\frac{8\Delta^2}{(2+\delta)\omega^2}\right], \tag{10.36}$$

then Eq. (10.33) predicts that $|\lambda_1| = |\lambda_2|$. Since the round-trip supermode loss is given by $L = 1 - |\lambda_i|^2$, Eq. (10.36) describes a phase error region where the two supermodes have the same power loss per round-trip. Thus, simple gain clamping will not be able to suppress multimode oscillations in this region. However, the phase of the eigenvalues and hence the lasing frequency changes as a function of the path length error, meaning the two supermodes in this region can be distinguished by frequency. When the path length phase error is smaller than that indicated by Eq. (10.36), the two eigenvalues take on real and different values, allowing the cavity to discriminate between the supermodes by gain.

An experimental test bed similar to the one described in Section 10.4.1 was assembled to measure the modal properties of a cavity coupled by spatial filtering. The amplitude and phase of a specific supermode state were measured interferometrically and conformed to the values predicted by Eqs. (10.34) and (10.35). In particular, the in-phase and out-of-phase modes were clearly seen when the path length error was zero. Quantitative measurements were also made of the supermode loss as a function of path length error when the spatial filter was adjusted to promote the in-phase mode. The resulting data are plotted in Figure 10.12 along with a plot of the theoretical supermode loss $L = 1 - |\lambda_i|^2$ from Eq. (10.33). One obvious feature of the curve is the critical point predicted by the analytical theory in Eq. (10.36). This point separates the small and large path length error regions. In the small path

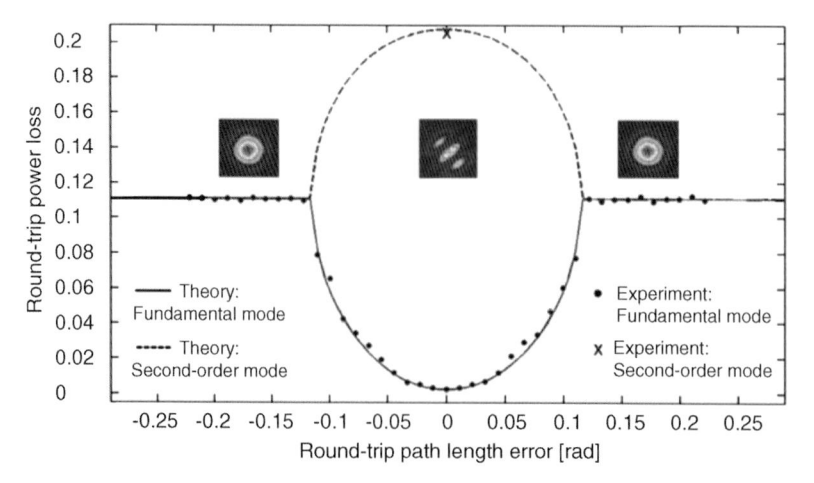

Figure 10.12 Phase error loss versus path length error of a spatially filtered two-beam cavity [36].

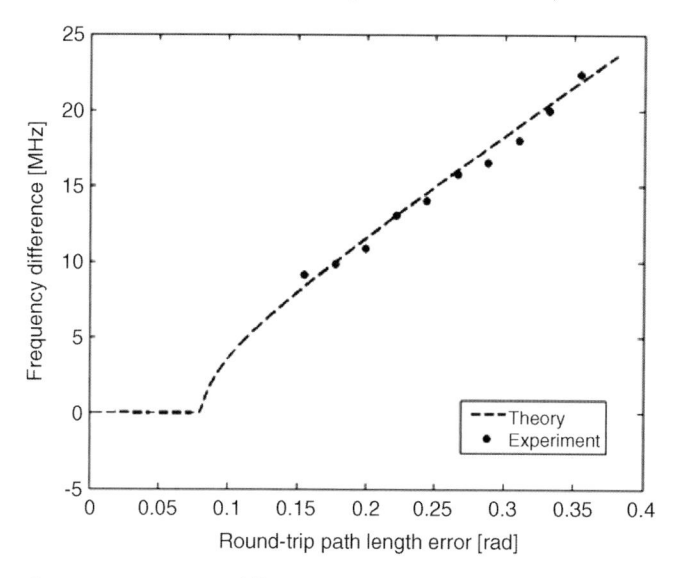

Figure 10.13 Frequency difference between the two supermodes of a spatially filtered two-beam cavity [36].

length error region, the two supermodes have dissimilar phase error loss, and only the less lossy supermode can lase. In the large path length error region, both supermodes have the same loss and have an equal chance of lasing. The far-field patterns are shown as insets in the figure, where the small path length error region is shown to lase in an in-phase supermode and the large path length region is shown as an incoherent state. The experimentally measured loss to the second supermode is also plotted at a path length error of zero.

As predicted by Eq. (10.33), the two supermodes in the large phase error region (i.e., when Eq. (10.36) is satisfied) have the same power loss per round-trip. However, their frequencies begin to separate from each other as the path length phase error increases. Since in this region the cavity supports two supermodes of slightly different frequencies, we have used the beating between the two supermodes to measure the frequency difference. Figure 10.13 shows the measured beat frequencies as a function of round-trip path length error, together with the predicted frequency difference from the phase of Eq. (10.33).

10.5.3
Talbot Resonators

Figure 10.9c shows a laser array that is coupled by free space diffraction. Because diffraction is inherently space invariant, the analysis of Section 10.5.1 is appropriate (assuming identical, equally spaced gain elements) if the coupling between elements is weak and there are no path length phase errors. In this section, we consider the opposite limits, where the coupling due to free space propagation extends over many gain elements and path length errors are present.

Our analysis is based on the theory of Talbot self-imaging [37,38]. If a paraxial coherent light field in a homogeneous medium is periodic in both transverse x- and y-directions (and the periods satisfy a particular relationship with each other), it is easy to show that the field must also be periodic in the third or z-direction. Thus, the periodic field at a particular plane of propagation will exactly repeat itself at multiples of a specific propagation distance, called the Talbot distance, simply through Fresnel diffraction. This can be understood by noting that a *periodic* field propagating in a homogeneous medium can always be decomposed into a set of plane waves traveling at *discrete* angles (a grating being the simplest example). Thus, the propagation distance that results in a phase delay between the on-axis and first off-axis plane waves of 2π will shift higher order plane waves by integer multiples of 2π, resulting in the exact same phase relation between the plane waves in the new plane as the initial plane and hence the same light distribution. Talbot planes can be found for both rectangular and hexagonal periodic arrays. In the following, we restrict ourselves to the one-dimensional Talbot cavity for simplicity.

The Talbot distance for a one-dimensional array is given by $Z_T = 2(2\Delta)^2/\lambda$, where 2Δ is the period of the array and λ is the wavelength of light. If one observes the field at *one-half* of a Talbot distance, the resulting field appears to form another image. However, this fractional Talbot field is shifted by exactly one-half period. At smaller integer fractions of the Talbot length, repeated images can be observed [39]. For example, at one quarter of the Talbot length, both the properly registered and the shifted images occur superimposed on top of each other (with a phase shift between them). At one-eighth of the Talbot length, four copies of the original field are produced with equal spacing between each copy (each with its own phase shift) [40].

The self-replicating property of the Talbot array has been utilized to design a common cavity architecture known as a Talbot cavity [41,42], shown in Figure 10.14. When an array of gain apertures produces a mutually coherent field with identical phase across all apertures, free space propagation through a round-trip distance of an integer multiple of Talbot lengths will produce a field that exactly matches the original field (forming a Talbot self-image), allowing the light to efficiently couple back into the gain apertures. An incoherent ensemble, on the other hand, will have

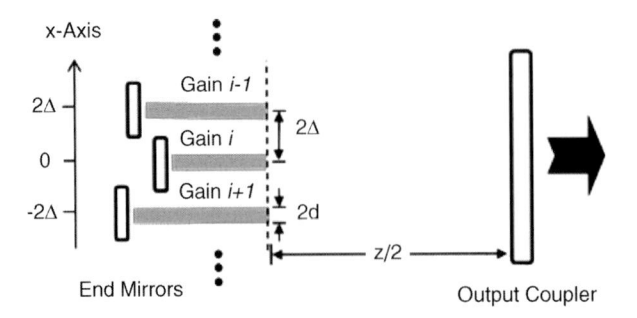

Figure 10.14 Schematic of a one-dimensional Talbot cavity. The rectangular gain sections of varying lengths represent waveguides from fiber or semiconductor lasers that have differing optical path lengths.

no such self-imaging property and the light will be spread across the entire array, reducing the coupling efficiency and increasing the laser threshold. Thus, the first supermode to lase is the coherent state that involves all the lasers in the array. Since smaller apertures give rise to wider diffraction angles, decreasing aperture size increases the amount and range of coupling (for a fixed aperture spacing). From the Talbot imaging point of view, we can hypothesize that a reduction in aperture size also gives rise to better modal discrimination, since the loss to the incoherent state (which spreads out over the entire array) increases as the percentage of laser aperture area decreases. In contrast, the coherent state (lowest-order supermode) couples efficiently to the gain media by Talbot imaging, with the primary loss coming from imperfect Talbot imaging at the edges of the array [40]. We will quantify some of these ideas in the subsequent analysis.

A second Talbot condition is worth noting. Instead of a periodic array of identically phased apertures, if the array contains a phase shift of π between each aperture, the Talbot conditions change such that a *true* Talbot image occurs at a distance of both $Z_T/2$ and Z_T [1,43]. Thus, it is clear that the cavity in Figure 10.14 will be degenerate to the in-phase and out-of-phase modes when the round-trip propagation distance is $z = Z_T$. However, this same cavity will only have one low-loss supermode (the out-of-phase mode) when this propagation distance is $z = Z_T/2$.

A deeper insight into the operating characteristics of the Talbot cavity can be obtained by applying the modal analysis of the previous sections. General numerical solutions are straightforward and will be explored at the end of this section. However, to develop an analytical solution, several assumptions have been made: (i) The individual gain elements in Figure 10.14 are waveguides that support a single Gaussian mode. (ii) The gain elements have no intrinsic coupling (i.e., the coupling between the gain elements is exclusively provided through the Fresnel propagation in the unguided part of the cavity). (iii) The end mirrors and the output coupler mirror are sufficiently wide to present no diffraction loss.

To find the round-trip propagation matrix of the cavity, we start by assuming that each of the gain elements contains a single Gaussian mode $u(x)$ given by

$$u(x) = e^{-\pi(x/d)^2}. \tag{10.37}$$

The distribution of the laser array light field in the aperture plane on the right-hand side of the gain media (along the dotted line in Figure 10.14) is then given by

$$p(x) = \sum_{n=1}^{N} u(x - (N - 2n + 1)\Delta), \tag{10.38}$$

where we consider N gain elements separated by 2Δ. After propagating a round-trip distance z in the cavity (from the aperture plane of the gain elements to the mirror on the right-hand side and back again), the field distribution generated by the mth gain element is given by

$$q(x - (N - 2m + 1)\Delta) = \frac{e^{jkz}}{j\lambda z} \int u(x - (N - 2m + 1)\Delta) e^{(j\pi/\lambda z)(x-x')^2} dx', \tag{10.39}$$

where we have assumed paraxial wave propagation. To determine the elements of the system matrix $M(m,n)$, we calculate the normalized overlap integral between the round-trip field generated by the mth gain element in Eq. (10.39) with the nth waveguide mode of Eq. (10.38). This is given by

$$M(m,n) = \frac{\int u(x-(N-2n+1)\Delta)q^*(x-(N-2m+1)\Delta)dx}{\sqrt{\int |u(x-(N-2n+1)\Delta)|^2 dx}\sqrt{\int |q(x-(N-2m+1)\Delta)|^2 dx}}. \qquad (10.40)$$

To model optical path length differences between waveguides in the gain region, we include phase errors for each waveguide ϕ_m. The elements of the resulting round-trip propagation matrix are

$$M_{rt}(m,n) = \sqrt{j2}\frac{de^{-jkz}}{\sqrt{\lambda z + j2d^2}}e^{-2\pi\frac{(2\Delta)^2 d^2 (m-n)^2}{4d^4+(\lambda z)^2}}e^{+j\pi\frac{\lambda z(2\Delta)^2(m-n)^2}{4d^4+(\lambda z)^2}}e^{j(\phi_m+\phi_n)}. \qquad (10.41)$$

As described in the previous sections, the modal response of the cold cavity can be fully determined by an eigen decomposition of the round-trip propagation matrix M_{rt}. For the Talbot cavity, this analysis helps to understand the impact of the fill factor (d/Δ), the finite number of array elements N, the cavity length z (which can be arbitrary, in general, for a common cavity that relies on free space propagation for coupling between the gain elements), and the phase errors ϕ_m. Here, we study two special cases where the round-trip cavity propagation distances are one-half a Talbot length and a full Talbot length, both in the limit of low fill factor ($d \ll \Delta$) and a finite number of elements (N). In addition, our initial assumption of a paraxial field sets a lower limit on d for the technique to produce a high-quality Talbot image [40].

If $Nd/\Delta \ll 1$, the full Talbot cavity ($z = z_T = (2(2\Delta)^2)/\lambda$) round-trip propagation matrix simplifies to

$$M_{rt}(m,n) = \sqrt{j}\frac{de^{-jkz}}{2\Delta}\begin{cases} e^{j(\phi_m+\phi_n)} & |m-n| : \text{even} \\ je^{j(\phi_m+\phi_n)} & |m-n| : \text{odd} \end{cases}. \qquad (10.42)$$

It is easy to see that only two rows of the matrix are linearly independent. This means that the matrix is of rank two, and the cavity supports only two supermodes with nonzero eigenvalues. To calculate the eigenvalues and the corresponding eigenvectors, we further assume that the path length errors are zero. This matrix then becomes

$$M_{rt} = \sqrt{j}\frac{de^{-jkz}}{2\Delta}\begin{bmatrix} 1 & j & 1 & \cdots \\ j & 1 & j & \\ 1 & j & 1 & \\ \vdots & & & \ddots \end{bmatrix}. \qquad (10.43)$$

When the dimension N of the matrix in Eq. (10.43) is even, the matrix is seen to be circulant (i.e., each successive row is a shifted version of the previous row, where elements that shift out from the right reenter (circulate) on the left). Since it is well known that the eigenvectors of circulant matrices are given by the basis functions of the discrete Fourier transform [44], one can calculate the eigenvalues of M_{rt} by simply taking the discrete Fourier transform of one of its rows. The two basis functions of the discrete Fourier transform that yield nonzero eigenvalues are the constant basis function [1 1 1 1 ...] and the highest frequency basis function [1 −1 1 −1 ...]. The two eigenvalues can then be calculated by the inner products of these two basis functions with a single row of the matrix M_{rt} in Eq. (10.43):

$$\lambda_{1,2} \cong \frac{1}{2}\sqrt{j}\frac{de^{-jkz}}{2\Delta}N(1\pm j), \qquad (10.44)$$

where λ_1 corresponds to the positive sign (constant basis function), λ_2 corresponds to the negative sign (oscillating basis function), and N is the total number of array elements. The corresponding eigenvectors are simply normalized versions of the two basis functions:

$$\vec{v}_1 = \frac{1}{\sqrt{N}}\begin{pmatrix} \vdots \\ 1 \\ 1 \\ 1 \\ 1 \\ \vdots \end{pmatrix}, \quad \vec{v}_2 = \frac{1}{\sqrt{N}}\begin{pmatrix} \vdots \\ 1 \\ -1 \\ 1 \\ -1 \\ \vdots \end{pmatrix}. \qquad (10.45)$$

The eigenmode solution confirms that the two supermodes of the full Talbot cavity are the in-phase mode (where all sources have the same phase) and out-of-phase modes (where sources oscillate between a phase of 0 and π). The magnitudes of the two eigenvalues in Eq. (10.44) are identical, implying that without additional spatial filtering, discrimination between the two supermodes is impossible and full coherence cannot be obtained. It is possible, however, to provide additional spatial filtering (e.g., at the half-Talbot plane) to break this degeneracy [43]. A similar procedure can be employed to solve for the eigenvalues and eigenvectors of Eq. (10.43) when N is odd. In this case, the two eigenvalues and eigenvectors are given by

$$\lambda_{1,2} = \frac{1}{2}\sqrt{j}\frac{de^{-jkz}}{2\Delta}\left(N\pm j\sqrt{N^2-2}\right) \qquad (10.46)$$

and

$$\vec{v}_1 = \frac{1}{\sqrt{N-1}}\begin{pmatrix} \vdots \\ \frac{-j+\sqrt{N^2-2}}{N+1} \\ 1 \\ \frac{-j+\sqrt{N^2-2}}{N+1} \\ \vdots \end{pmatrix}, \quad \vec{v}_2 = \frac{1}{\sqrt{N-1}}\begin{pmatrix} \vdots \\ \frac{-j-\sqrt{N^2-2}}{N+1} \\ 1 \\ \frac{-j-\sqrt{N^2-2}}{N+1} \\ \vdots \end{pmatrix}.$$

$$(10.47)$$

The half-Talbot cavity $\left(z = z_T/2 = (2\Delta)^2/\lambda\right)$ in the same limit of $Nd/\Delta \ll 1$ results in a round-trip propagation matrix $\mathbf{M}_{rt}$ of

$$M_{rt}(m,n) = \sqrt{2j}\frac{d}{2\Delta}e^{-jkz}\begin{cases} e^{j(\phi_m+\phi_n)} & |m-n| : \text{even} \\ -e^{j(\phi_m+\phi_n)} & |m-n| : \text{odd} \end{cases}. \tag{10.48}$$

This matrix is easily seen to be unity rank, since all rows are linear combinations of each other. Consequently, the matrix contains only one supermode with a nonzero eigenvalue. Since in general the trace of a square matrix is equal to the sum of its eigenvalues, an expression for the single nonzero eigenvalue in our case is simply given by the trace of $\mathbf{M}_{rt}$ in Eq. (10.48), or

$$\lambda_1 = \sqrt{2j}\frac{d}{2\Delta}e^{-jkz}\sum_{i=1}^{N} e^{j2\phi_i}. \tag{10.49}$$

In addition, the single eigenvector corresponding to the single nonzero eigenvalue can simply be extracted as a normalized vector from any one of the (linearly dependent) rows of $\mathbf{M}_{rt}$, or

$$\vec{v}_1 = \frac{1}{\sqrt{N}}\begin{pmatrix} \vdots \\ -e^{j\phi_{m-1}} \\ e^{j\phi_m} \\ -e^{j\phi_{m+1}} \\ \vdots \end{pmatrix}, \tag{10.50}$$

where the constant phase term has been dropped.

Several observations can be drawn from these last two equations. First, the eigenvalue in Eq. (10.49) is a function of the fill factor d/Δ, indicating that the round-trip amplitude is attenuated by this factor. This loss can be partially offset by using arrays containing many lasers, as the eigenvalue is directly proportional to N (assuming there are no path length phase errors). These two effects can be easily understood by the theory of Talbot imaging. Although Talbot theory predicts proper imaging independent of the fill factor (within the paraxial limitation), there is a tacit assumption that the array is infinitely long, since true periodic functions must be infinite in length. The edge effects that result from the finite array are responsible for the attenuation in Eq. (10.49) and clearly become more substantial as the fill factor is reduced. Increasing N reduces the influence of these edge effect losses on the overall supermode because the edge loss becomes proportionately smaller as N increases. A more complete discussion on edge losses in a Talbot array can be found in Ref. [40].

A second observation is that the half-Talbot cavity maintains its single-mode performance even in the presence of path length phase errors, where Eq. (10.50) shows the field of the only possible oscillating supermode. However, phase errors can significantly attenuate this mode after one round-trip, as seen in the random-walk amplitude accumulation of Eq. (10.49). Thus, although this type of cavity will oscillate in a coherent state under any phase error conditions, low phase errors are

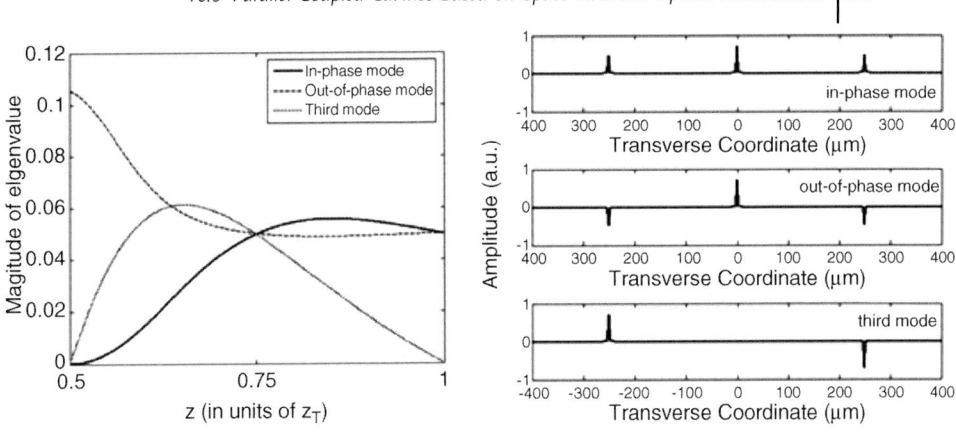

Figure 10.15 (a) Eigenvalue magnitudes of a three-laser Talbot cavity with cavity lengths ranging from $z_T/2 < z < z_T$. (b) Amplitude distribution of the three supermodes when $z = z_T$.

required for efficient coherent beam combining. Of course, the wavelength diversity described in Section 10.3.2 may partially compensate for this phase error loss.

Figure 10.15 shows a numerical solution for three gain channels ($N = 3$) with a fill factor $d/\Lambda = 0.05$. The corresponding three eigenvalues are plotted in part (a) as a function of round-trip propagation length ranging from a half-Talbot distance to a full-Talbot distance, where we have assumed that all phase values $\phi_i = 0$. The half-Talbot cavity is seen to oscillate primarily in a single supermode, whereas the full Talbot cavity supports both the out-of-phase and in-phase supermodes simultaneously. Note that intermediate propagation distances allow other modes to oscillate (supermode number 3 in this case).

The assumption that $Nd/\Lambda \ll 1$ is quite severe. From Eqs. (10.44), (10.46,) and (10.49), it is clear that the eigenvalue $\lambda_1 \ll 1$ under this condition, indicating that the cavity has high loss to the fundamental supermode. When more practical values are chosen for the fill factor and array size, the above analysis is no longer accurate and numerical modeling is required. Nevertheless, there are some characteristics that remain unchanged. First, the fundamental (highest eigenvalue) supermode of the half-Talbot cavity is an out-of-phase mode as predicted above. The eigenvalue of the in-phase mode, although not zero, is the lowest of all the supermodes. Second, the fundamental supermodes of the full Talbot cavity remain the in-phase and out-of-phase modes, although these modes are no longer exactly degenerate, with one supermode having a marginally larger eigenvalue than the other [45]. There are also some differences seen with practical Talbot cavities. Most strikingly, cavities with larger fill factors generally contain N supermodes, and the distribution of power within a single supermode is no longer uniform. As noted in Section 10.5.1, when the nearest-neighbor coupling becomes completely imaginary (as is the case for the full Talbot cavity), the ith and $(N+1-i)$th supermodes contain quasi-degenerate eigenvalues for all N values of i. Finally, the discrimination between adjacent supermodes is generally enhanced as the fill factor d/Δ is reduced, while a larger number of lasers N reduce this discrimination [46].

10.6
Parallel Coupled Resonators Based on Space-Variant Optical Architectures: the Self-Fourier Cavity

The parallel coupled cavities shown in Figure 10.9 are all based on physical phenomena (evanescent coupling, spatial filtering, and free space diffraction) that are inherently space invariant. We now turn to an inherently space-variant architecture that is based on the Fourier transform: the self-Fourier cavity. To see that a system described by a Fourier transformation is space variant, it is only necessary to consider the effect of shifting the system input. The Fourier shift theorem states that a spatial shift of the input function does not shift the output at all, but rather multiplies the output by a linear phase (where the slope of the phase is proportional to the input shift). Clearly, this new output function is not a shifted version of the original output, and therefore the system is space variant.

Self-Fourier resonators are designed to produce a two-dimensional spatial Fourier transform of the light exiting a laser array after one round-trip in the resonator. Self-Fourier functions, which maintain the same functional form when Fourier transformed, serve as the fundamental supermodes of such a resonator. The well-known examples of self-Fourier functions include the Dirac delta comb function and the Gaussian function. It can be shown that an infinite number of distinct families of these functions exist [47].

Figure 10.16 shows a simple self-Fourier cavity. A lens of focal length $2f$ is placed in contact with an output mirror at the end of the cavity. The light enters the lens, is

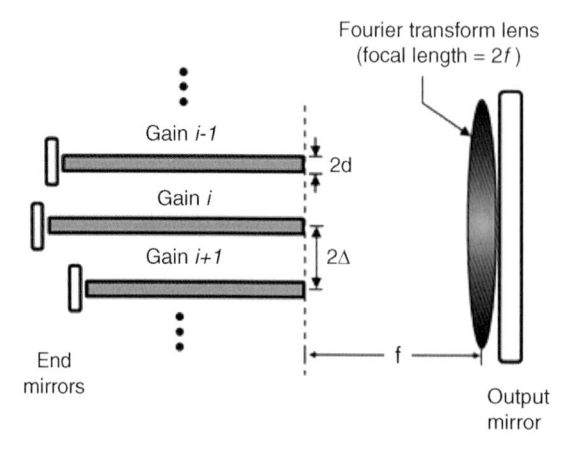

Figure 10.16 Self-Fourier cavity. Light exiting the gain apertures is located in the front focal plane of the double-passed lens, whereas the return light entering the laser apertures is in the back focal plane.

10.6 Parallel Coupled Resonators Based on Space-Variant Optical Architectures: the Self-Fourier Cavity

reflected by the mirror, and passes through the lens a second time, giving the lens–mirror combination an effective double-pass focal length of f. The distance between the gain apertures and the lens–mirror combination is also equal to f. Thus, the gain apertures appear to be in the front and back focal planes of a single equivalent lens with focal length f. The mathematical relationship between the field $u(x,y)$ exiting the gain apertures and the field $q(x',y')$ entering these apertures after reflecting off the lens–mirror pair is given by [48]

$$q(x',y') = \frac{1}{j\lambda f} \int\int u(x,y) \exp\left(-j2\pi \frac{xx'+yy'}{\lambda f}\right) dx dy, \qquad (10.51)$$

where λ is the wavelength of light. Equation (10.51) is recognized as a spatial Fourier transform if we make the correspondence $\xi = x'/\lambda f$ and $\eta = y'/\lambda f$, where ξ and η are the Fourier variables. The self-Fourier scaling rule (where a self-Fourier function has the same spatial scale before and after transformation) can easily be determined by substituting a comb function, $\text{comb}[(x/2\Delta),(y/2\Delta)]$, into Eq. (10.51). The comb function, consisting of an infinite array of delta functions located at integer multiples of 2Δ in the x- and y-directions, has a Fourier transform given by $\mathfrak{F}\{\text{comb}[(x/2\Delta),(y/2\Delta)]\} = (2\Delta)^2 \text{comb}(2\Delta\xi, 2\Delta\eta)$, where ξ and η are the independent variables in the Fourier domain, and the symbol $\mathfrak{F}$ indicates a two-dimensional spatial Fourier transform [48]. This comb function becomes self-Fourier if the spatial distribution of delta functions after Fourier transformation is equal to the spatial distribution of the original comb function. Thus, $1/2\Delta = 2\Delta/\lambda f$, or

$$2\Delta = \sqrt{\lambda f}. \qquad (10.52)$$

Rather than the array of Dirac delta functions above, an array of single spatial mode waveguides is more accurately modeled as an array of equally spaced Gaussian distributions, one from each single spatial mode gain channel. It can be shown [49,50] that if this array of Gaussians is sufficiently large, a self-Fourier function can be established by multiplying the array by a Gaussian envelope. Unlike the previous parallel coupled cavities described in Section 10.5 in which the strongest coupling occurs between nearest-neighbor gain regions, the self-Fourier cavity couples all apertures most strongly to the central gain region. This means that the cavity is inherently space variant, resulting in a non-Toeplitz round-trip propagation matrix M_{rt}. In addition, proper design can provide coupling across the entire array.

It is easy to draw the conclusion that a light field that is truly self-Fourier will serve as a supermode of this system. However, important questions remain unanswered, such as whether this mode is the only supermode of the cavity, how much power loss this mode experiences in one round-trip in the cavity, and the effect of random path length variations of the gain elements on the modal response of the cavity. Our goal in this section is to answer these questions by providing a systematic modal analysis of the cavity.

To find the round-trip propagation matrix of this cavity M_{rt}, we first establish a relationship between the initial field at the right-hand side of the gain elements $p(x)$ and the field at the same location after one round-trip in the cavity $q(x)$. In this analysis, we assume the gain elements support single-mode Gaussian beams that propagate with no diffraction (e.g., those contained in a waveguide). A one-dimensional analysis is employed for simplicity. The field distribution of N gain elements oscillating with equal strength is given by

$$p(x) = \sum_{m=1}^{N} u(x - (N - 2m + 1)\Delta), \quad (10.53)$$

where $u(x) = e^{-\pi[(x/d)]^2}$ is a Gaussian function with the beam waist of $d/\sqrt{\pi}$ and 2Δ is the distance between the neighboring gain elements.

The output apertures of the gain array are placed in the front focal plane of a lens of equivalent double-pass focal length f. Reflection from the flat mirror directly following this lens generates the spatial Fourier transform of the light field back at the gain apertures. The return field $q(x)$ from a single gain element located at $x = (N - 2m + 1)\Delta$ is given by

$$q(x - (N - 2m + 1)\Delta) = \frac{d}{j\lambda f} e^{-\pi[(dx/\lambda f)]^2} e^{-j2\pi[((N-2m+1)\Delta x)/\lambda f]}, \quad (10.54)$$

where λ is the wavelength of light. Note that the Gaussian term in Eq. (10.54) is not a function of the initial location of this single gain element (making the system space variant). However, its location does change the phase tilt seen in the last term of the equation. This tilt allows only part of the light to couple back into the nth gain element. The amount of coupling is given by a normalized mode overlap integral:

$$g(x) = \frac{\int u(x - (N - 2n + 1)\Delta) q^*(x - (N - 2m + 1)\Delta) dx}{\sqrt{\int |u(x - (N - 2n + 1)\Delta)|^2 dx} \sqrt{\int |q(x - (N - 2m + 1)\Delta)|^2 dx}}. \quad (10.55)$$

The round-trip propagation matrix must also account for phase terms that are proportional to the path length error in each gain arm. Assuming a complete set of phase errors ϕ_m, the round-trip propagation matrix M_{rt} can be expressed as

$$M_{rt}(m, n) = \frac{d\sqrt{2\lambda f}}{\sqrt{d^4 + (\lambda f)^2}} e^{-\pi\Delta^2 d^2[((N-2m+1)^2+(N-2n+1)^2)/(d^4+(\lambda f)^2)]} \quad (10.56)$$

$$\times e^{-j2\pi[(\Delta^2 \lambda f(N-2m+1)(N-2n+1))/(d^4+(\lambda f)^2)]} e^{j(\phi_m+\phi_n)}.$$

We start by assuming an odd number of laser gain channels N so that the central gain channel is located at $x = 0$. We will consider the case for an even number of channels at the end of this derivation. If we apply the self-Fourier scaling rule of Eq. (10.52) and assume $\pi d \ll \Delta$ (i.e., a small fill factor), the second exponential term takes on values in its exponent that are approximately integer multiples of 2π (for odd N). Consequently, the exponential is always unity for all values of m and n. The individual rows of the resultant matrix can then be seen to be linearly dependent, implying that the rank of the round-trip propagation matrix M_{rt} is unity. Thus, under this condition, the cavity supports only one supermode with a nonzero eigenvalue. The eigenvector associated with this nonzero eigenvalue $\vec{v}_1$ can be computed by extracting the first column of the matrix $M_{rt}(i,1)$ and normalizing the result. The N individual components v_i of the eigenvector $\vec{v}_1$ are then given by

$$v_i = \sqrt{\frac{d}{\sqrt{2}\Delta}} e^{j\phi_i} e^{-\pi d^2[((N-2i+1)^2)/16\Delta^2]}. \tag{10.57}$$

We note that the shape of the supermode is a Gaussian function, as expected.

Because the matrix M_{rt} is unity rank, the nonzero eigenvalue is also easy to compute. As before, we use the result that the trace of a square matrix (given by adding all the values along the major diagonal) is also equal to the sum of its eigenvalues. It follows that since there is only one nonzero eigenvalue in a rank one matrix, this single eigenvalue will be equal to the trace of M_{rt}, or

$$\lambda_1 = \sum_{m=1}^{N} M_{rt}(m,m) = \frac{d}{\sqrt{2}\Delta} \sum_{i=(-N+1)/2}^{(N-1)/2} e^{j2\phi_i} e^{-\pi(d^2/2\Delta^2)i^2}. \tag{10.58}$$

In the absence of gain media path length errors (i.e., $\phi_i = 0$ for all values of i), the eigenvalue reaches a maximum that approaches unity for sufficiently large N.

It should be noted that as the number of gain elements increases, the laser array fill factor d/Δ must necessarily decrease. This can be seen by noting the expression for the Gaussian envelope in Eq. (10.56). We see that the width of the return Gaussian distribution $m\Delta$ is equal to $\lambda f/d$. Since we require $\lambda f = (2\Delta)^2$, this implies that $m\Delta = 4\Delta^2/d$. To ensure that this light couples across the entire array of length $N\Delta$, we have the following requirement for the array fill factor (d/Δ):

$$\frac{d}{\Delta} = \frac{4}{N}. \tag{10.59}$$

Thus, large arrays must have low fill factors and subsequent beam conditioning optics is usually required.

Finally, we note that our assumption that the number of gain elements N was odd is not required in general. When N is even, the phase factor in Eq. (10.56) contains an oscillating sign. The cavity characteristics are the same, but the components v_i of the fundamental eigenvector $\vec{v}_1$ are now given by

$$v_i = \sqrt{\frac{d}{\sqrt{2}\Delta}} e^{j\phi_i} e^{-\pi d^2[(N-2i+1)^2/16\Delta^2]} (-1)^i. \tag{10.60}$$

This out-of-phase mode shifts the Fourier transform of the Gaussian array by one-half period, as required by an even number of gain elements.

The supermode of the self-Fourier cavity with $N=7$, $d=47.8\,\mu m$, and $2\Delta=249\,\mu m$ is depicted in Figure 10.17. The resulting eigenvalue in this case is $\lambda_1 = 0.98$.

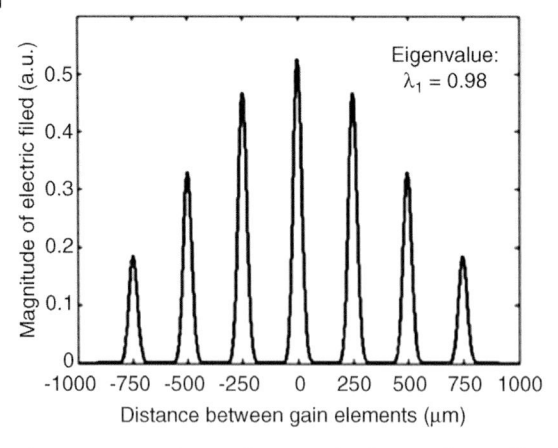

Figure 10.17 Fundamental supermode of a seven-laser self-Fourier cavity ($N=7$).

Table 10.1 Comparison of coherent beam combining architectures.

	Number of supermodes	Fundamental supermode shape	Fundamental supermode loss	Effect of path length errors	Supermode degeneracy
Simple beam splitter	1	Uniform	None	Eq. (10.8)	None
Generalized Michelson[a]	2	Uniform	None	Eq. (10.16)	At a single-phase error
Dammann/volume grating	1	Uniform	Approximately none	Eq. (10.20)	None
Evanescent coupling	N	Sinusoidal	None	—	None
Spatial filtering	N	Sinusoidal	Spatial filter absorption	Eq. (10.33)[b]	None
Full Talbot ($Nd/\Delta \ll 1$)	~2	Sinusoidal	Edge loss	—	Yes
Half Talbot ($Nd/\Delta \ll 1$)	~1	Sinusoidal	Edge loss	Eq. (10.49)	None
Half Talbot ($Nd/\Delta \gtrsim 1$)	N	Sinusoidal	Edge loss	—	None
Self-Fourier[c]	~1	Gaussian	Approximately none	Eq. (10.58)	None

a) Generalized Michelson contains two lasers only.
b) Equation valid only for two lasers.
c) Small fill factor assumption.

10.7
Conclusion

The characteristics and merits of several different external cavity resonators for coherent beam combining have been systematically studied using modal methods. A general observation is that coupling extending across larger numbers of lasers in the array tends to produce higher modal discrimination. We have been particularly interested in the performance of these resonator architectures as a function of path length errors in the various gain channels.

We must emphasize once again that our analysis has been completely restricted to linear operation, and that much more complex behavior is expected (and has been observed) when the gain media contain nonlinear terms. In addition, the analysis of the cavities has been carried out for a single fixed wavelength. We showed in Section 10.3.2 that allowing the wavelength to change can reduce path length phase errors and greatly improve the cavity performance. Our purpose of analyzing the cavities in this way was to allow a comparison between various architectures, and is not meant to suggest that this is the actual performance one could expect in a real system. Nevertheless, this systematic study has revealed several fundamental characteristics of the external cavities. The basic results of these findings are summarized in Table 10.1.

Acknowledgments

The authors would like to thank graduate students Brad Tiffany and Chenhao Wan for their contributions to this chapter. Much of the work reported here has been funded by the Air Force Office of Scientific Research and Cymer, Inc. Parts of references 12, 16, and 36 were reproduced in this chapter by permission of the IEEE.

References

1 Leger, J.R. (1993) External methods of phase locking and coherent beam addition, in *Surface Emitting Diode Lasers and Arrays* (eds G. Evans and J. Hammer), Academic Press, New York.

2 Fan, T.Y. (2005) Laser beam combining for high-power, high-radiance sources. *IEEE J. Sel. Top. Quantum Electron.*, **11**, 567–577.

3 Andrews, J.R. (1986) Traveling-wave amplifier made from a laser diode array. *Appl. Phys. Lett.*, **48**, 1331–1333.

4 Cheung, E.C., Ho, J.G., Goodno, G.D., Rice, R.R., Rothenberg, J., Thielen, P., Weber, M., and Wickham, M. (2008) Diffractive-optics-based beam combination of a phase-locked fiber laser array. *Opt. Lett.*, **33**, 354–356.

5 Segev, M., Weiss, S., and Fischer, B. (1987) Coupling of diode laser arrays with passive phase conjugate mirrors. *Appl. Phys. Lett.*, **50**, 1397–1399.

6 Botez, D. (1994) Monolithic phase-locked semiconductor laser arrays, in *Diode Laser Arrays* (eds D. Botez and D. Scifres), Cambridge University Press, Cambridge, UK.

7 Leger, J.R. (1994) Micro-optical components applied to incoherent and coherent laser arrays, in *Diode Laser Arrays* (eds D. Botez and D. Scifres), Cambridge University Press, Cambridge, UK.

8 Siegman, A.E. (1998) How to (maybe) measure laser beam quality, in *Diode Pumped Solid State Lasers (DPSS) Lasers: Applications and Issues* (ed. M.W. Dowley), Optical Society of America, Washington, DC, pp. 184–199.

9 Leger, J.R. (1996) Laser beam shaping, in *Microoptics* (ed. H.P. Herzig), Taylor & Francis, London.

10 Siegman, A.E. (1986) *Lasers*, University Science Books, Sausalito, CA.

11 Haus, H. (1984) *Waves and Fields in Optoelectronics*, Prentice-Hall, New Jersey.

12 Khajavikhan, M., John, K., and Leger, J.R. (2010) Experimental demonstration of reduced path length sensitivity in coherent

13. Shirakawa, A., Saitou, T., Sekiguchi, T., and Ueda, K. (2002) Coherent addition of fiber lasers by use of a fiber coupler. *Opt. Express*, **10**, 1167–1172.
14. Sabourdy, D., Kermene, V., Defarges-Berthelemot, A., Lefort, L., and Berthelemy, A. (2002) Efficient coherent combining of widely tunable fiber lasers. *Opt. Express*, **11**, 87–97.
15. Rediker, R.H., Rauschenbach, K.A., and Schloss, R.P. (1991) Operation of a coherent ensemble of five diode lasers in an external cavity. *IEEE J. Quantum Electron.*, **27**, 1582–1593.
16. Khajavikhan, M. and Leger, J. (2009) Modal analysis of path length sensitivity in superposition architectures for coherent laser beam combining. *IEEE J. Sel. Top. Quantum Electron.*, **15**, 281–290.
17. Dammann, H. and Klotz, E. (1977) Coherent optical generation and inspection of two-dimensional periodic structures. *Opt. Acta*, **24**, 505–515.
18. Mait, J. (1996) Fourier array generators, in *Microoptics* (ed. H.P. Herzig), Taylor & Francis, London.
19. Sinzinger, S. and Jahns, J. (2003) *Microoptics*, Wiley-VCH Verlag GmbH, Weinheim, Germany.
20. Leger, J.R., Swanson, G.J., and Veldkamp, W.B. (1987) Coherent laser addition using binary phase gratings. *Appl. Opt.*, **26**, 4391–4399.
21. Veldkamp, W.B., Leger, J.R., and Swanson, G.J. (1986) Coherent summation of laser beams using binary phase gratings. *Opt. Lett.*, **11**, 303–305.
22. Gerchberg, R.W. and Saxton, W.O. (1972) A practical algorithm for the determination of the phase from image and diffraction plane pictures. *Optik*, **35**, 237–246.
23. Kogelnik, H. (1969) Coupled-wave theory for thick hologram gratings. *Bell Syst. Tech. J.*, **48**, 2909.
24. Kapon, E., Katz, J., and Yariv, A. (1984) Supermode analysis of phase-locked arrays of semiconductor lasers. *Opt. Lett.*, **9**, 125–127.
25. Butler, J.K., Ackley, D.E., and Botez, D. (1984) Coupled-mode analysis of phase-locked injection laser arrays. *Appl. Phys. Lett.*, **44**, 293–295.
26. Cooper, M.L. and Mookherjea, S. (2009) Numerically-assisted coupled-mode theory for silicon waveguide couplers and arrayed waveguides. *Opt. Express*, **17**, 1583–1599.
27. Philipp-Rutz, E.M. (1975) Single laser beam of spatial coherence from an array of GaAs lasers: free-running mode. *J. Appl. Phys.*, **46**, 4551–4556.
28. Leger, J.R. (1989) Lateral mode control of an AlGaAs laser array in a Talbot cavity. *Appl. Phys. Lett.*, **55**, 334–336.
29. Diadiuk, V., Liau, Z.L., Walpole, J.N., Caunt, J.W., and Williamson, R.C. (1989) External-cavity coherent operation of InGaAsP buried-heterostructure laser array. *Appl. Phys. Lett.*, **55**, 2161–2163.
30. Golubentsev, A.A., Kachurin, O.R., Lebedev, F.V., and Napartovich, A.P. (1990) Use of a spatial filter for phase locking of a laser array. *Sov. J. Quantum Electron.*, **20**, 934–938.
31. Bochove, E.J. and Shakir, S.A. (2009) Analysis of a spatial-filtering passive fiber laser beam combining system. *IEEE J. Sel. Top. Quantum Electron.*, **15**, 320–327.
32. Wan, C., Tiffany, B., and Leger, J.R. (2011) Analysis of path length sensitivity in coherent beam combining by spatial filtering. *IEEE J. Quantum Electron.*, **47**, 770–776.
33. Mehuys, D., Mitsunaga, K., Eng, L., Marshall, W.K., and Yariv, A. (1988) Supermode control in diffraction-coupled semiconductor laser arrays. *Appl. Phys. Lett.*, **53** (13), 1165–1167.
34. Striefer, W., Osinski, M., Scifres, D.R., Welch, D.F., and Cross, P.S. (1986) Phase-array lasers with a uniform, stable supermode. *Appl. Phys. Lett.*, **49**, 1496–1498.
35. Kouachi, S. (2006) Eigenvalues and eigenvectors of tridiagonal matrices. *Electron. J. Linear Algebra*, **15**, 115–133.
36. Wan, C. and Leger, J.R. (2012) Experimental measurements of path length sensitivity in coherent beam combining by spatial filtering. *IEEE J. Quantum Electron.*, **48**, 1045–1051.
37. Talbot, H.F. (1836) Facts relating to optical science No. IV. *Philos. Mag.*, **9**, 401–407.

38 Rayleigh, J.W.S. (1881) On copying diffraction-gratings, and on some phenomenon connected therewith. *Philos. Mag.*, **11**, 196–205.

39 Winthrop, J.T. and Worthington, C.R. (1965) Theory of Fresnel images: I. Plane periodic objects in monochromatic light. *J. Opt. Soc. Am.*, **55**, 373–381.

40 Leger, J.R. and Swanson, G.J. (1990) Efficient array illuminator using binary optics phase plates at fractional Talbot planes. *Opt. Lett.*, **15**, 288–290.

41 Golubentsev, A.A., Likhanskii, V.V., and Napartovich, A.P. (1987) Theory of phase locking of an array of lasers. *Sov. Phys. JETP*, **66**, 676–682.

42 Leger, J.R., Scott, M.L., and Veldkamp, W.B. (1988) Coherent addition of AlGaAs lasers using microlenses and diffractive coupling. *Appl. Phys. Lett.*, **52**, 1771–1773.

43 Leger, J.R. and Griswold, M.P. (1990) Binary-optics miniature Talbot cavities for laser beam addition. *Appl. Phys. Lett.*, **56**, 4–6.

44 Gray, R.M. (2000) *Toeplitz and Circulant Matrices*, Stanford University ISL.

45 Leger, J.R., Mowry, G., and Chen, D. (1994) Modal analysis of a Talbot cavity. *Appl. Phys. Lett.*, **64**, 2937–2939.

46 Mehuys, D., Striefer, W., Waarts, R.G., and Welch, D.F. (1991) Modal analysis of linear Talbot-cavity semiconductor lasers. *Opt. Lett.*, **16**, 823–825.

47 Horikis, T.P. and McCallum, M.S. (2006) Self-Fourier functions and self-Fourier operators. *J. Opt. Soc. Am. A*, **23**, 829–834.

48 Goodman, J.W. (2005) *Introduction to Fourier Optics*, 3rd edn, Roberts and Company Publishers, Greenwood Village, CO.

49 Corcoran, C.J. and Pasch, K. (2005) Modal analysis of a self-Fourier laser cavity. *J. Opt. Soc. Am. A*, **7**, L1–L7.

50 Corcoran, C.J. and Durville, F. (2005) Experimental demonstration of a phase-locked laser array using a self-Fourier cavity. *Appl. Phys. Lett.*, **86**, 201118-1-3.

11
Self-Organized Fiber Beam Combining

Vincent Kermène, Agnès Desfarges-Berthelemot, and Alain Barthélémy

11.1
Introduction

In the past few years, fiber laser has become a kilowatt-level bright source of continuous-wave (CW) light radiation (10 kW single-mode fiber laser is commercially available www.ipgphotonics.com). However, several limitations appear at higher power level due to the high-field confinement in the fiber core, such as fiber end damage, optical nonlinearities (self-focusing, stimulated scatterings, etc.), and thermal issues. Once the limit power per fiber is reached, beam combining appears as a last option. It consists of the coherent addition of the outputs of multiple laser sources either in a single powerful output beam or on a given remote target in the far field with respect to the laser array exit. Among the various techniques explored for achieving coherent beam combining, one of the simplest techniques developed so far is passive combining. Here, passive combining means self-organized and is opposed to active combining involving electro-optic servo loop control. This chapter focuses specifically on passive coherent combining of fiber lasers. A typical system is based on a set of fiber amplifiers put in parallel, which are associated with a single composite cavity. Phase-dependent coupling among the different arms is exploited to produce phase-dependent losses with a minimum for in-phase operation. Laser emission corresponds to the fields experiencing the highest gain–loss difference so that coherent combining occurs spontaneously. The frequency spectrum brings the degree of freedom required by the self-organization process. The emitted frequencies adjust continuously to preserve in-phase amplified beams, leading to stable coherent summation despite environmental perturbations.

The basic principles of such a generic phase locking process are described through the typical example of a Michelson-type laser in Section 11.2. In Section 11.3, the properties of passively cophased fiber lasers are analyzed in terms of spectral behavior, stability, and dynamics. Most of the investigated architectures exhibit simple and straightforward spatial coherence. With laser beam arrays delivered by sources phase locked through mutual injection, the situation gets

Coherent Laser Beam Combining, First Edition. Edited by Arnaud Brignon.
© 2013 Wiley-VCH Verlag GmbH & Co. KGaA. Published 2013 by Wiley-VCH Verlag GmbH & Co. KGaA.

more complex, as discussed in Section 11.3.6. In Section 11.4, scaling of the number of parallel amplifying arms is discussed with respect to the evolution of the combining efficiency. In Section 11.5, the behavior of passively phased lasers used in pulsed regime is reported. It is shown that phase-locking building time is short enough to suit the fast dynamics of a Q-switch operating regime. Finally, specific laser geometry is described where spatial and spectral phase locking work together. It is based on a couple of doped fiber lasers and includes a saturable absorber mirror. Experiments have demonstrated the realization of a coherently combined source of ultrashort pulses.

It is worth mentioning that this chapter does not aim at providing a complete overview of passive fiber laser coherent combining. Instead, it introduces few basic ideas to help readers develop a passively combined laser and reports some of our experimental results on the topic.

11.2
Principles of Passively Combined Fiber Lasers

11.2.1
Different Configurations

The various arrangements (geometry) used to passively combine fiber lasers can be cast into three main categories: lasers with multiple separate fiber amplifiers delivering a single output beam (sometimes called filled aperture systems) [1–5], lasers with a set of parallel fiber amplifiers delivering a coherent beam array (sometimes called tiled aperture) [6–13], and multicore fiber lasers [14–19]. Some typical examples are schematically shown in Figure 11.1. Most of the time people try to preserve the benefit of guided waves in the whole laser cavity. Fiber Bragg gratings (FBGs) were used as cavity mirrors, polarization control was made by fiber loops, and adding and splitting of intracavity fields were based on fiber couplers (FCs). Multiple port couplers as well as cascades of standard 50/50 couplers were found in practical implementations. However, some other coupling techniques are based on diffraction by a diffractive element [1] or just by free space propagation [13]. In addition, at high power levels, fiber couplers are often avoided to prevent damages and are replaced by standard beam splitters. Polarization management issues are also relaxed in free space by comparison with fibers. These are some of the reasons why in a large number of investigated architectures, one finds a combination of waveguides and free space optics. Even with multicore fiber laser, a free space part is often required to provide proper feedback and coupling among the fields of different cores [14]. The case of multicore fiber laser is very specific. The various arms of the already described cavity are replaced by the different supermodes of the multicore fiber with their various propagation constants. The role of the cavity here is to favor oscillation of the sole fundamental supermode that features the highest brightness. Filtering can rest, for example, on self-imaging through multimode interferences [19].

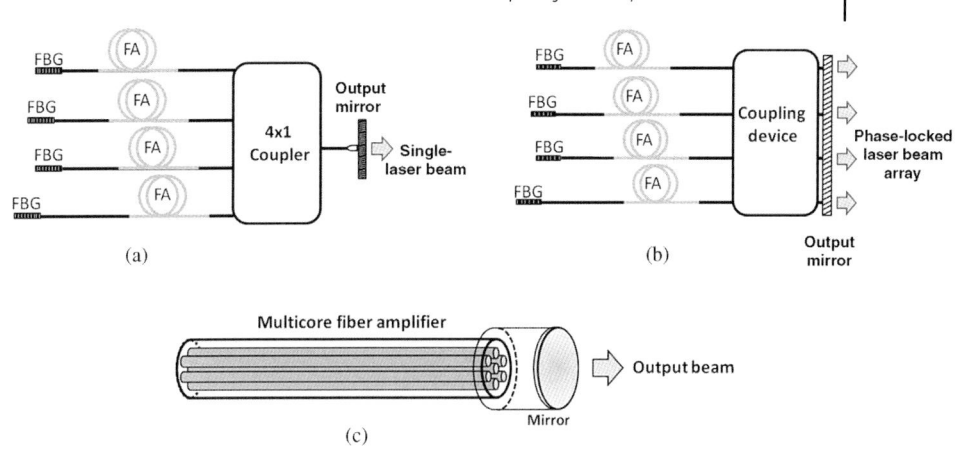

Figure 11.1 Schematics of few generic examples of fiber laser cavities for passive coherent combining. (a) Filled aperture architecture and single output beam. (b) Tiled aperture architecture and multiple output beams. (c) Multicore fiber. FBG, fiber Bragg grating mirror; FA, fiber amplifiers. Output mirror stands for laser output coupler with partial reflection.

Most of the initial works performed on fiber laser combining belong to the category with a single common output. In view of power scaling, the configuration delivering an array of output beams appeared later and more appropriate because it relaxes the thermal constraints and damages risks. To improve the beam brightness in case of a poor filling factor in the tiled beam array, one can use appropriate phase plates [20]. Although the simplified cavity schemes, given in Figure 11.1, represent linear cavity, similar operating principles have been demonstrated in ring configurations [8,19].

11.2.2
Principles

For coherent addition of lasers, the primary requirement is that all laser branches oscillate on the same set of frequencies and that their phase relationships are locked. In the context of passive laser combining, a compound cavity built around an interferometric arrangement satisfies these requirements. For example, with a Michelson interferometer including an amplifier in each arm (Figure 11.1a, configuration reduced to two branches), one obtains a type of Fox–Smith resonator. The intracavity beam splitter (or coupler) permits waves to share a common path and to exchange part of their field. So, the laser oscillates on the set of frequencies that are common to the two subcavities' frequency combs through a kind of Vernier effect (Figure 11.2).

The period of the spectral modulation $c/2\Delta L$ (c is the speed of light) is inversely proportional to the subcavities' length difference ($\Delta L = |L_1 - L_2|$), which was used initially to filter out a single longitudinal mode in gas lasers [21]. For a resonator with two amplifying arms only, the filtering is connected with the fact that radiation coming

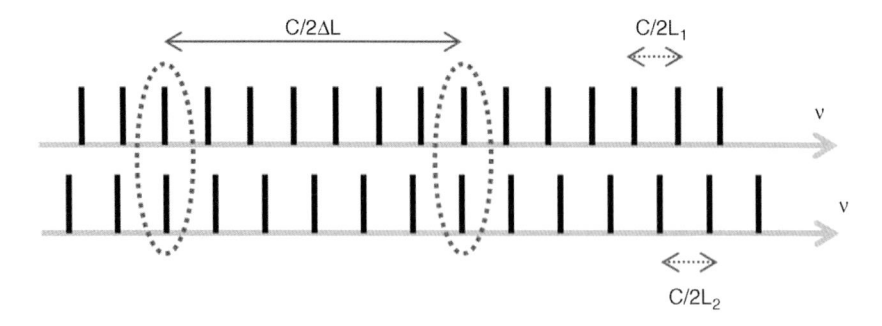

Figure 11.2 Selection of longitudinal modes (Vernier effect) common to two cavities of respective length L_1 and L_2 (encircled by dotted line; $\Delta L = |L_1 - L_2|$).

from each amplifier must interfere at the exit of the splitter so as to coherently combine with the common port closed by a partially reflecting mirror. The laser selects the longitudinal modes with the lowest cavity losses, which are being derived from cold cavity computations already described. The approach can be extended to the cavity for combining a larger number of lasers, and hence with a larger number of subcavities. A signature of phase locking can be observed in the radio frequency (RF) spectrum of the output laser signal after detection by a fast photodiode.

As an example, Figure 11.3 compares the RF spectra of 12 individual lasers under both independent operation (upper part) and phase-locked operation (bottom part).

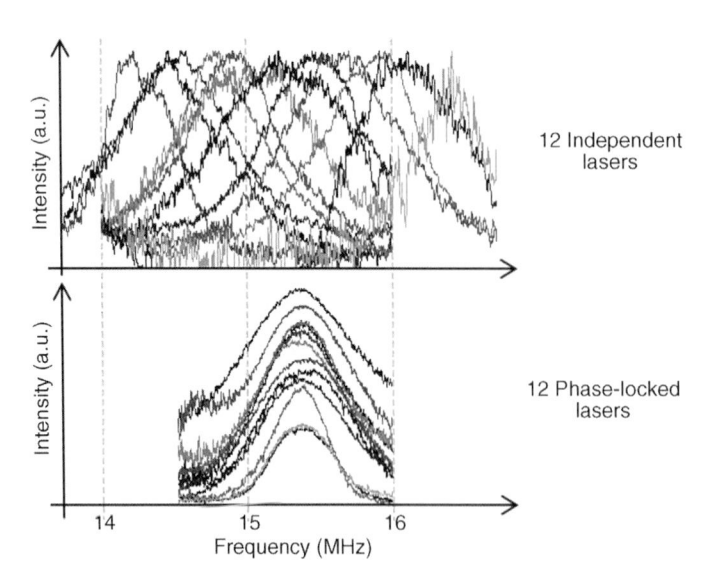

Figure 11.3 RF spectra in an array of 12 beams from a composite resonator made around a set of 12 parallel fiber amplifiers (zoom on the fifth harmonics). *Upper part*: Individual operation of the laser arms showing their different cavity lengths. *Bottom part*: Passively phase-locked operation indicating oscillation on a common set of frequencies.

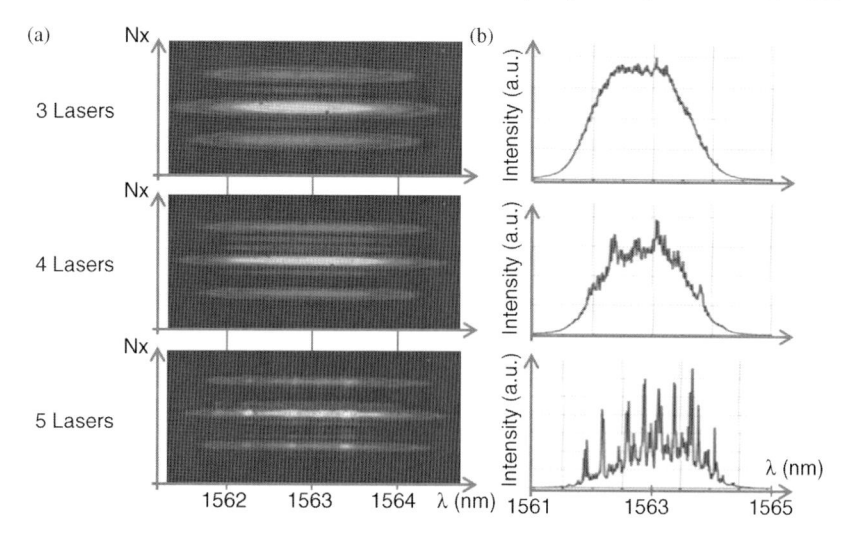

Figure 11.4 Frequency-resolved experimental far-field patterns (a) for arrays of 3 (top), 4 (middle), and 5 (bottom) phase-locked erbium fiber lasers [$N_x = \sin(\theta)/\lambda$, θ denotes the angular domain]. Corresponding spectral intensity profiles are shown in part (b). As the number of lasers in the array increases, the laser spectrum becomes more complex with widely separated spikes and the far-field lobes become sharper.

The situation where each laser works independently gives a different RF peak for each laser element. The frequency of the RF peak being inversely proportional to the cavity round-trip time, it provides a simple measurement of the different subcavity lengths (here ranging from 75 to 86 m). Once the composite cavity is adjusted to couple the individual lasers and make them operate coherently, all emitters deliver radiations carrying the same RF components, indicating that they share the same set of resonance frequencies. It also applies to other kinds of configurations (Figure 11.1b and c). The occurrence of resonances shared by all laser arms, namely, mode congruencies, becomes sparser as the resonator gets more complex. The evolution is almost exponential with respect to the number of laser ways. As an example, Figure 11.4 reports some experimental data recorded with erbium-doped fiber laser array. Figure 11.4a shows images recorded with an imaging spectrograph. They display on the vertical axis the plane wave spectrum ($N_x = \sin(\theta)/\lambda$, where θ denotes the angular domain and λ the wavelength) of the exit beam array and on the horizontal axis the laser spectrum. In Figure 11.4b, a cross section of the same laser spectra has been plotted for a better reading. The number N of lasers in the array varies from 3 (on top) to 5 (at the bottom). For $N = 3$ and to a lesser extent when $N = 4$, the common modes are so closely spaced that they are not resolved by the spectrometer. The envelope of the laser bandwidth only was measured in such cases.

Wavelength filtering is, however, clearly visible on the strongly and sharply modulated spectrum, corresponding to $N = 5$. At the same time, the increase in the array size shrank the main far-field peaks, indicating that phase locking was effective. At some point, with a great number of amplifiers, it is even impossible to find a frequency in common with the entire set of elementary cavities and this raises the losses of the laser. The consequence is also a drop in the efficiency of the power summation (combining efficiency), which will be discussed later. A specific difficulty raised by the coherent combining of fiber lasers comes from the fact that it is almost impossible to get fiber amplifiers identical in optical path length within a fraction of wavelength. Rare earth-doped fiber amplifiers are usually between few meters and few tens of meters in length. They include passive elements such as couplers for the pump power and isolators that are spliced to the doped fiber. In addition to the difficulty of cutting two fiber pieces of exactly equal length, the problem mostly originates in the fact that optical fibers are sensitive to thermal and mechanical effects. The phase of an optical field amplified in a doped fiber therefore varies strongly, depending on the pump power, acoustic noise, environmental vibrations, and amplifier gain. Consequently, a composite laser cavity including many fiber amplifiers is characterized by many subcavities of different lengths, which may vary over time by several wavelengths in a random way. The next section explains how coherent combining in self-organized systems can remain efficient under these changes in length.

11.3
Phase Coupling Characteristics

11.3.1
Power Stability

In a nonprotected environment, the passive coherent combining techniques make use of the spectrum self-adjustment to maintain in real time in-phase emission. A small change in length (even lower than the wavelength) of laser arms shifts the longitudinal modes of the elementary lasers. Then, new spectral lines replace the previous ones in the laser spectrum. They are in connection with the modal congruencies of the new set of longitudinal modes yielding the highest combining efficiency. The best practical configuration is obtained when several resonant modes are kept in the laser gain bandwidth, whatever be the environmental conditions. In the case of small path length differences between arms, the gap between common modes can be close to or even greater than the gain bandwidth. The detrimental result is that under unwanted optical length changes, the resonant frequencies may overlap the gain bandwidth or may fall out, leading to strong power fluctuations. To illustrate these two laser behaviors, let us choose the simplest configuration by considering the Mach–Zehnder fiber laser [4]. The sets of common modes are periodically spaced by $\Delta\lambda = \lambda_0^2/\Delta L$, where λ_0 is the average wavelength and ΔL is the path length difference between the two legs of the interferometer. If $\Delta\lambda$ is greater than the laser gain bandwidth $\Delta\lambda_G$, the power is expected to be unstable, as it can occur for an interferometric device in a nonprotected environment. On the contrary, if $\Delta\lambda$ is significantly smaller than $\Delta\lambda_G$, the average number of longitudinal modes in the laser bandwidth weakly varies under environmental perturbations. Nevertheless,

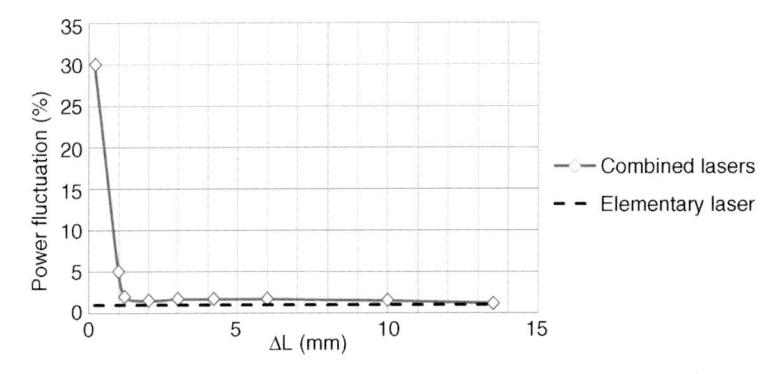

Figure 11.5 Power fluctuations of a coherently combined laser versus path length difference between the two amplifying arms (Er-doped fiber) of the cavity. *Dashed line:* power fluctuation of a single laser.

the combined power is stable, despite the lack of servo control. These simple rules can serve to design an efficient and stable laser combining, including several fiber amplifiers in a compound cavity. Figure 11.5 gives an experimental validation of this simple approach. Power fluctuations of a pair of erbium-doped fiber lasers combined in a Mach–Zehnder cavity are plotted versus ΔL. The laser bandwidth and central wavelength are fixed here by a Bragg grating serving as an end mirror ($\lambda_0 = 1549$ nm, $\Delta \lambda_G = 2$ nm). The fluctuations were integrated in 1 s with a fast InGaAs photodiode of 1 GHz bandwidth. The curve has to be compared with the dashed line obtained with a standard individual erbium fiber laser. The double amplifier laser undergoes strong power fluctuations when ΔL is below 1 mm. This value gives a spectral modulation $\Delta \lambda$ of 2.4 nm, which is about $\Delta \lambda_G$ in agreement with the previous rules. On the contrary, when ΔL is large ($\Delta \lambda \ll \Delta \lambda_G$), a stable coherent power summation is observed without electronic feedback loop, demonstrating that the passive technique can be effective.

This applies only for a small number of lasers. For others, as discussed in Section 11.4, combining efficiency as well as power stability deteriorates.

11.3.2
Cophasing Building Dynamics

An important feature of beam combining techniques is their cophasing dynamics. It is related to the time needed to recover in-phase emission after a perturbation coming from the environment. For active methods, it is relatively easy to estimate the building time of coherence, thanks to the time response of the different components and devices in the electro-optic feedback loop such as detection device, phase modulators, and error signal measurement device. An error signal with a step-like profile is applied to the servo loop and the time required for a new steady-state operation is measured. For passive methods, it is possible to derive techniques analogous to those used in electronic devices. The laser can be disturbed by a fast

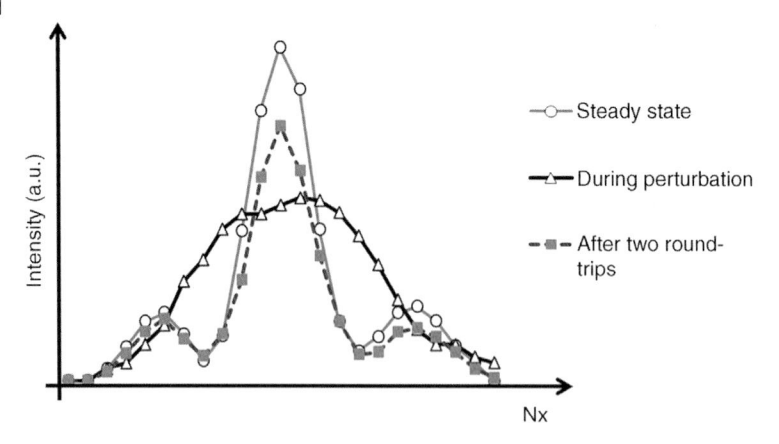

Figure 11.6 Evolution of the two-element fiber laser in far field before, during, and after phase perturbation ($N_x = \sin(\theta)/\lambda$, θ denotes the angular domain).

phase change to characterize its dynamic behavior in the presence of environmental perturbations. Nonlinear cross-phase modulation is one possible way to make a sudden phase shift on one arm of the laser. A short excitation pulse of duration smaller than the cavity photon lifetime and at a wavelength different from that of laser is coupled into one fibered arm in which the laser field travels. During the propagation of the high-intensity excitation beam, the refractive index is changed by Kerr effects and the laser field undergoes a phase shift on duration close to that of the pulse width. We carried out such an experiment in a ring cavity made of two parallel ytterbium-doped fibers [22]. The excitation pulse came from a microchip Nd:YAG laser delivering pulses of 1 ns duration and about 100 W peak power at a low repetition rate of 7.2 kHz. The interaction of this powerful pulse with the laser field along 2 m long fiber induced a phase shift up to 5π. A time-resolved analysis permitted observing of how the combined beam in the far field returned to the in-phase state after the pulse perturbation. A fast photodiode scanned the far field, where the beams coming from both the disturbed and the unperturbed arms overlapped. The curve with circular dots in Figure 11.6 was recorded in the initial steady state and served as a reference. The high-contrast fringes with maximum intensity on the optical axis indicated the in-phase emission. Then, the microchip was turned on and, for different positions of the photodiode in the far field, the intensity was recorded according to time during several cavity round-trips of the laser field. The processing of these data gave the evolution of the far-field pattern during and after the ultrashort phase unbalances.

One has to note that during perturbation, fringes completely disappeared. The probe pulse has a Gaussian temporal profile that makes the phase perturbation time dependent. In other words, the induced phase shift changes along the pulse duration. The temporal resolution of the photodiode is not sufficient to measure these changes. Then the detector integrates the far-field evolution for phase shifts ranging from 0 to 5π, which gives the uniform profile of the curve with circular dots.

The important result is that only a few number of round-trips are necessary to recover the in-phase situation after a strong phase deviation. Indeed, after two round-trips only, the far-field pattern was almost identical to that of the initial state. This experimental study highlights the robustness and fast action of passive coherent combining techniques to compensate for environmental perturbations. The very short response time we measured is in good agreement with the simulations made by Wu et al. [23]. They showed by use of a model, which couples rate equations and the Schrödinger equation for the laser fields, that phase locking between arms of the laser array is established very early and lead to a collective gain dynamic. The very fast response of the passive architectures associated with collective gain dynamics is promising for phase locking of a laser array, delivering nanosecond pulses in Q-switched regime. As we will discuss later, a few results of passive phase locking of fiber lasers in Q-switched regime showed up to now long pulses of about 0.5–10 μs. With a proper choice of laser parameters (amplifying fiber length, pumping level, and response time of Q-switch element), one can expect that passively phase-locked laser array delivers Q-switched nanosecond pulse.

11.3.3
Frequency Tunability

A high-power tunable laser is of interest for various applications such as biomedical sensing and optical communications. Unlike bulk solid-state lasers using crystalline hosts, most of the fiber lasers offer a broad emission bandwidth. This feature associated with the capability of fiber lasers to deliver high power with a good beam quality made them attractive to break into the market of tunable lasers. Furthermore, it appeared interesting to investigate the capability of passively combined lasers to be tuned on their gain bandwidth. The spectral filtering that occurs when lasers are phase locked depends on the different optical lengths of the multiarm lasers. For example, in the Mach–Zehnder configuration depicted in Figure 11.7 (as it is explained in Section 11.3.1), the coherent combining leads to a modulated spectrum with a periodicity of $\Delta\lambda = \lambda_0^2/\Delta L$, which can be orders of magnitude smaller than

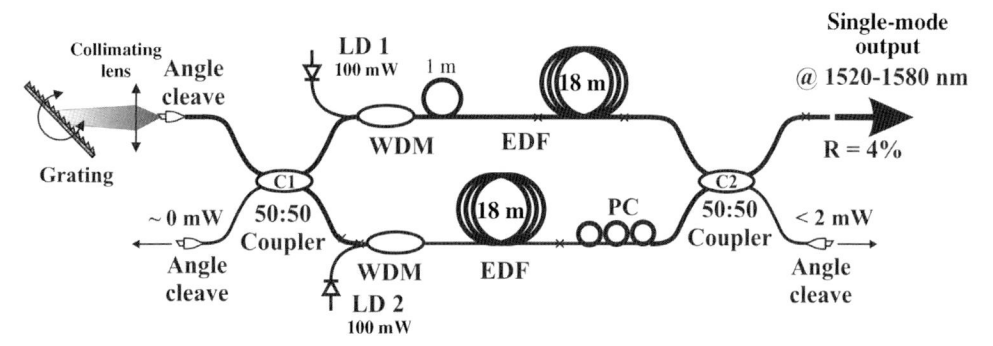

Figure 11.7 Tunable Mach–Zehnder fiber laser. (LD 1, LD 2: 980 nm pump laser diodes; EDF, erbium-doped fiber; WDM, wavelength-division multiplexer; PC: polarization controller).

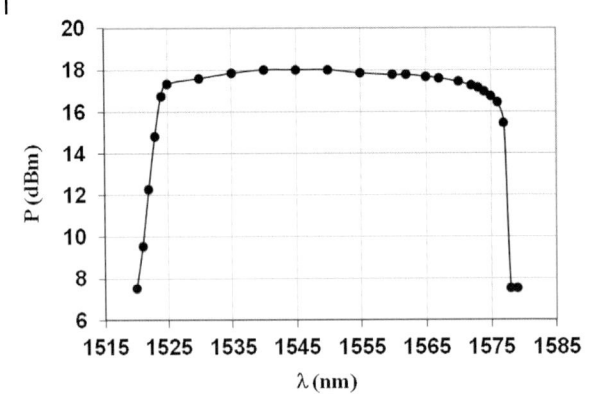

Figure 11.8 Output power versus laser wavelength delivered by the laser of Figure 11.7.

the laser gain bandwidth. So, the tunability of passively combined erbium-doped fiber lasers was demonstrated [24,25]. The cavity with parallel amplifying arms has two common mirrors, placed on two ports of the 50:50 couplers, C1 and C2. In Figure 11.7, the common rear mirror is a diffraction grating in Littrow configuration, whereas the output coupler is made by cleaving one output port of the coupler C2. Wavelength tuning was achieved by rotating the grating. Then the emitted power was measured for several orientations of the grating (Figure 11.8). The results show that the Mach–Zehnder erbium-doped fiber laser is tunable on a wide band of about 60 nm, from 1520 to 1580 nm. The total leakages at the angle-cleaved ports of the couplers account for 3% of the global output power. All these results prove that passive coherent combining is usable with tunable lasers. In other words, it can be an alternative way to perform a high-power laser source that is tunable on a broad spectral band.

Tunability is however restricted to a low number of combined lasers; otherwise, the scarcity of modes makes the system unstable and tuning discontinuous.

11.3.4
Effect of Laser Gain Mismatched on Combining Efficiency

Another interesting feature of coherent combining techniques is their low sensitivity to power disparity. Indeed, such disparities in emitted powers can occur because of nonidentical pumping levels or insertion loss or polarization states between interfering beams. To illustrate this point, Figure 11.9 considers the simplest case of the coherent combining of two beams with different power ratios R. It shows that combining efficiency is not significantly disturbed by a moderate change in power ratio between the two elementary lasers. For example, if one of the two lasers delivers twice the power of the other, combining efficiency is reduced only by 3%. This behavior is due to the interferometric process that adds laser fields rather than intensities.

Extension to more general situations and to a larger number of lasers can be found in Ref. [26].

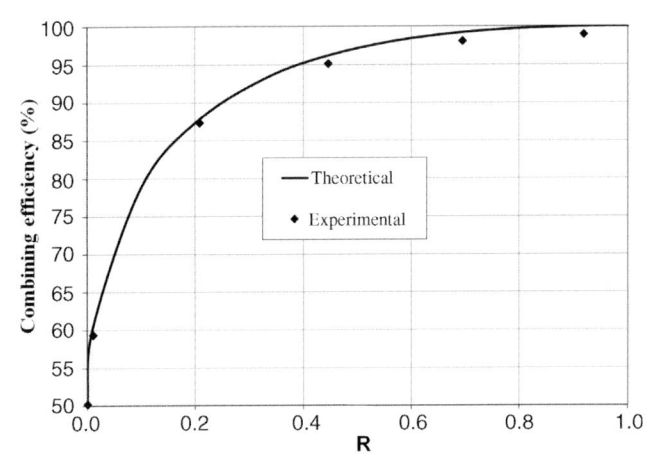

Figure 11.9 Combining efficiency of two lasers versus their power ratio R.

11.3.5
Pointing Agility

In the case of tiled aperture architecture (Figure 11.1b), the emission is made of multiple parallel output beams. Combining occurs outside the laser cavity, in the far field. The coherent summation of beam set results in a structured pattern (vertical axis in Figure 11.4a). This combined beam is made up of main lobes (array lobes) and side lobes inside an envelope, depending upon the numerical aperture of each elementary output beams. The number of main lobes in the far field is inversely proportional to the filling rate (ratio between the size of the output beams and their spacing) in the near field. At 30% of filling rate, the combining beam exhibits essentially one main lobe. The number of side lobes depends on the number of output beams and their distribution in the near field. In the case of multiple amplifiers (Figure 11.1b), the coupling device can be a spatial filter performed by a single-mode fiber. In a ring configuration [8], this fiber feeds back the common signal to the multiple amplifiers. The input face of this fiber is located in the far-field plan of the multiple output beams, into the cavity. Phase locking occurs when the mode of this fiber matches the main lobe of the combined beam. Considering N output beams of radius ω_0 at e^{-2} in intensity, distributed along an axis (1D arrangement) of period Λ, the divergence of the main lobe in the far field is $\Theta_{ML} = \lambda/(N \times \Lambda)$. It must be fitted to one of the feedback fiber modes $\theta_{FF} = \pm\lambda/(\pi \times \omega_{FF})$, where ω_{FF} is the radius of the feedback fiber mode. The position of the main lobe of the combined beam is centered on the core of the single-mode feedback fiber to minimize losses into the laser cavity. These intra-cavity losses are minimal when the fiber core is located at the maximum of the combined beam envelope (Figure 11.10). In this case, all output beams are in-phase. Although it is possible to shift the feedback fiber input along its plane, the main lobe of the combined beam follows the fiber core and shifts inside the

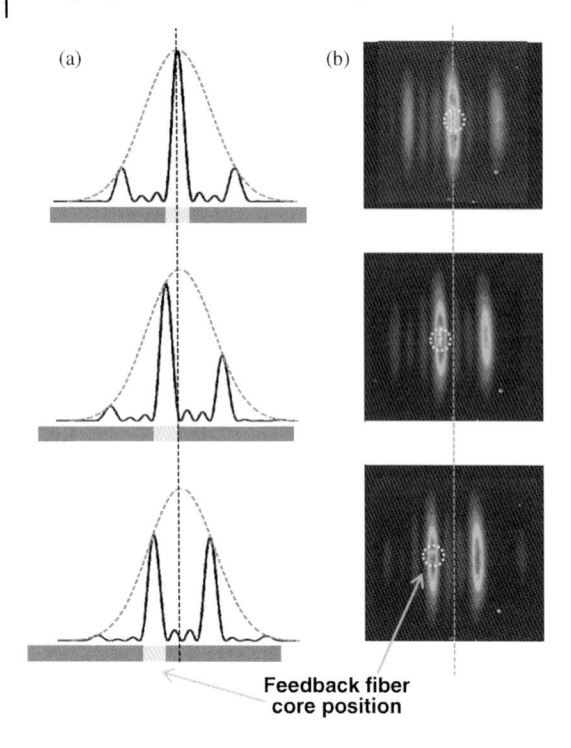

Figure 11.10 (a) Numerical far-field profiles of four phase-locked beams at three different positions of the feedback fiber core (indicated by the gray block). (b) Experimental measurements of far-field patterns for the same positions of the feedback fiber core (indicated by white dotted circles).

envelope. The main output beam pointing can be easily tuned inside its envelope, which is delimited by the numerical aperture $\theta_{\omega_0} = \pm\lambda/(\pi \cdot \omega_0)$ of the beams in the near field.

It has to be considered that periodically (each $\Theta_{shift} = \lambda/\Lambda$) the feedback fiber shifting generates a far-field profile identical to the one when all the output beams are in-phase (main lobe of the far field is centered) but with higher intracavity losses. Thus, laser-pointing agility θ_{tune} is really efficient in an angular domain limited by $\pm\lambda/(2 \times \Lambda)$. We find the well-known behavior of antenna array in the radar area.

11.3.6
Coherence Properties of Multiple Beams Phase Locked by Mutual Injection Process

In tiled aperture architectures (Figure 11.1b), the coupling process is performed by a spatial and spectral filtering or by mutual injection. In the first case, the beams overlap in a common part of the compound cavity, which is the filtering plane. Then, beams are obviously cophased in this common particular plane, providing

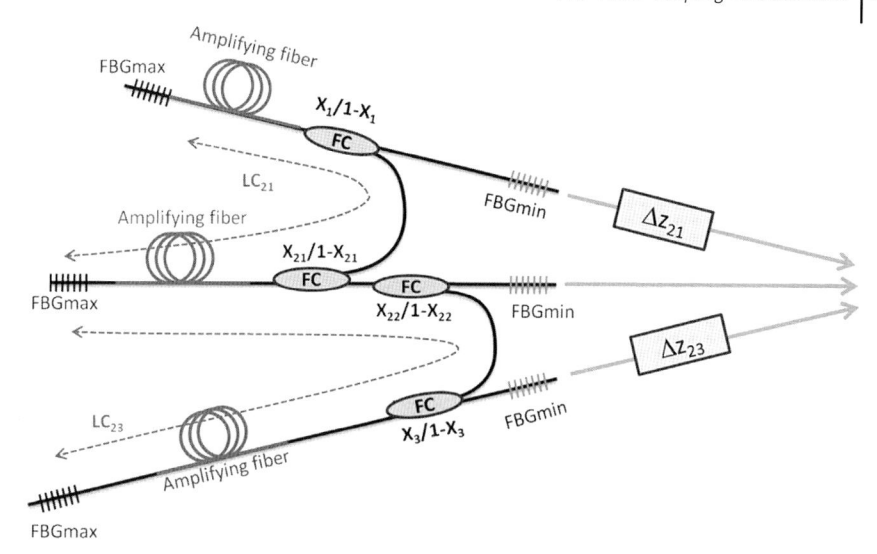

Figure 11.11 Scheme of a three-amplifier laser in the mutual injection configuration. FBG, fiber Bragg grating; FC, fiber coupler; $X_i/1 - X_i$, cross-coupling coefficients; LC_{ij}, subcavity lengths; Δz_{ij}, delay line outside the laser cavity.

synchronous multiple output emission. In the case of mutual injection, the coherent summation occurs on the couplers used to phase lock neighboring lasers. As they are distributed in the compound cavity, the coherence properties induced by this configuration are complex and not straightforward. In the following is described how to anticipate the extra-cavity areas where coherent beam superposition can be observed [27].

Such a passive combining technique is based on optical field sharing between neighboring lasers, using fiber couplers that connect the elementary lasers. An example of fiber laser made up of three amplifiers phase locked by mutual injection is shown in Figure 11.11. Numerous studies have reported some promising results with such a kind of laser, but they also mention some instability of the far-field pattern (changing contrast and shifting fringes) [11]. In fact, the place outside the cavity, where all the beams interfere efficiently, depends on the geometric characteristics of the laser, and the in-phase plane is not equidistant to all the laser outputs as is most often agreed.

Considering the laser configuration depicted in Figure 11.11, made up of three cavities of lengths L_i ($i \in [1,2,3]$), we also notice two other subcavities of lengths LC_{21} and LC_{23}. The coherence features of this complex coupled system can be described by an impulse response approach. An initial pulse of time duration $\delta \tau$, as short as the inverse of the coherence length $L_{co} = \lambda^2/\Delta\lambda_G$ ($\Delta\lambda_G$ laser gain bandwidth) of the laser considered, is launched in one of the cavity arms. This pulse propagates all along the subcavities before leaving the laser from its three output ports in the form of pulse trains $U_i(t)$, with complex field at the output i. We consider a symmetric cavity where the splitting ratio of the fiber couplers becomes $X_1 = X_3$ and

$X_{21} = X_{22} = X_2$ (see Figure 11.11). The impulse response is then given by the following coupled equations:

$$U_1(t) = U_{1(2L_1)}G^2X_1^2 + U_{2(L_1+LC_{21}+L_2)}G^4X_2^2(1-X_2)X_1(1-X_1)$$
$$+ U_{3(L_1+LC_{21}+LC_{23}+L_3)}G^6X_1^2(1-X_1)^2X_2(1-X_2)^2,$$
$$U_2(t) = U_{1(2L_2)}G^2X_2^2 + U_{1(L_1+LC_{21}+L_2)}G^4X_1(1-X_1)X_2^2(1-X_2)$$
$$+ U_{3(L_2+LC_{23}+L_3)}G^4X_1(1-X_1)X_2^3(1-X_2),$$
$$U_3(t) = U_{1(2L_3)}G^2X_1^2 + U_{2(L_2+LC_{23}+L_3)}G^4X_1(1-X_1)X_2^3(1-X_2)$$
$$+ U_{1(L_1+LC_{21}+LC_{23}+L_3)}G^6X_1^2(1-X_1)^2X_2(1-X_2)^2,$$

(11.1)

where $U_{i(L)}$ denotes the complex field U_i after a propagation of distance L and G is a constant gain applied every time the field passes through one of the amplifying media.

The coherence properties of the mutual injection configuration are analyzed by calculating the correlation product between all the pairs of output pulse trains $U_i(t)$, considering a delay τ_i at each laser output ($\Delta\tau_{ij} = \tau_i - \tau_j$ delay between the outputs i and j). One can deduce the intensity pattern I_c of the combined beams as a function of the different time delays $\Delta\tau_{ij}$:

$$I_c(x, y, \Delta\tau_{ij}) = \left\langle \left| \sum_{i=1}^{N} U_i(x, y, t + \tau_i) \right|^2 \right\rangle,$$

(11.2)

where x,y are the spatial coordinates, t represents the time, and $<>$ denotes the time averaging.

The contrast of the interference pattern I_c is calculated as a function of the distance Δz_{ij} that is connected to the delay time $\Delta\tau_{ij}$ between $U_i(t)$ and $U_j(t)$ by $\Delta\tau_{ij} = \Delta z_{ij}/c$ (c is the speed of light). The resulting 3D pattern is shown in Figure 11.12. Several coherence peaks appear, depending on the subcavity lengths. They are on either side of the position $\Delta z_{21} = \Delta z_{23} = 0$, where all the beams should be combined at an equidistant point of the three laser outputs. The case for which beams have equal propagation lengths from output to combining plane does not lead to stable and contrasted interference patterns.

We recognize some specific delays in Figure 11.12 that rely periodically on the coherence peaks, such as $\Delta L_{21} = LC_{21} - L_1 - L_2$ and $\Delta L_{23} = LC_{23} - L_2 - L_3$. Some of these peaks are high enough to consider that coherent combining can be efficient for these settings. Experimental results were obtained with clad-pumped ytterbium-doped fibers as amplifying media. The coupling ratios X_1 and X_2 were 70 and 80%, respectively. Laser subcavity lengths measured $L_1 = 10.2$, $L_2 = 12$, $L_3 = 9.5$, $L_{21} - 22.3$, and $L_{23} = 21.5$ m. The three laser outputs were distributed on the vertices of a triangle to form a two-dimensional structured pattern in the combined plane where all the beams overlap. Two delay lines adjusted the path length difference between the output beams. In general case, the interference between the three output beams was not contrasted, fringes also shifted in time. But varying the delays, it was possible to observe different situations of interference contrasts and stability. Position, where the pattern was

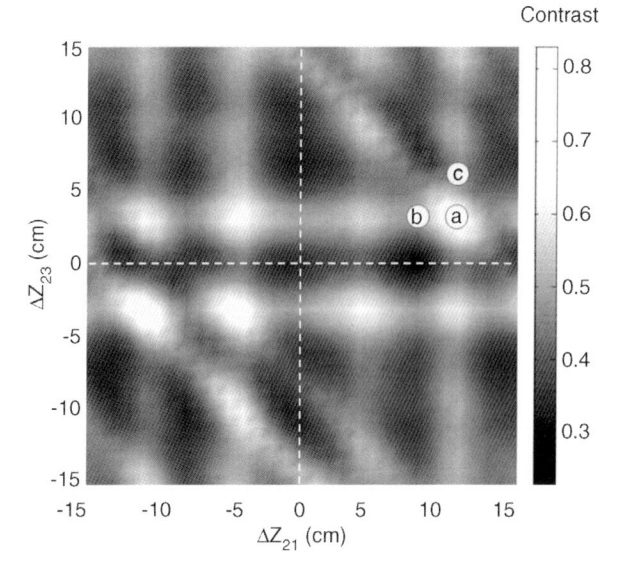

Figure 11.12 Coherence map of three-amplifier laser coupled by mutual injection calculated as a function of the delay line lengths between output beams 1 and 2 (horizontal axis) and between output beams 2 and 3 (vertical axis). Points a–c mark the position of the experimental measurements shown in Figure 11.13.

contrasted, agreed with the ones deduced from calculation (Figure 11.12). Figure 11.13 shows examples of experimental interference patterns corresponding to three particular set of delays Δz_{ij} analogous to the points a–c in Figure 11.12.

Thus, multiple laser emissions can be combined efficiently by the mutual injection process. Nevertheless, this technique requires external delay lines to recover mutual coherence of the output beams. Peaks of coherence can also be enlarged by reducing the laser bandwidth.

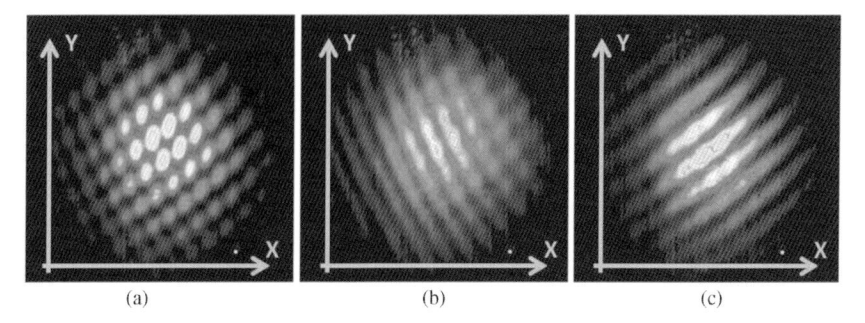

Figure 11.13 Experimental three-wave interference patterns for different adjustments of the delay lines. They are in perfect agreement with numerical expectations of Figure 11.12: (a) at the peak of coherence $\Delta z_{21} = +11$ cm, $\Delta z_{23} = +4$ cm, (b) $\Delta z_{21} = +11$ cm, $\Delta z_{23} = +6$ cm, and (c) $\Delta z_{21} = +9$ cm, $\Delta z_{23} = +4$ cm.

11.4
Upscaling the Number of Coupled Lasers

11.4.1
Phasing Efficiency Evolution

Upscaling the output power of the coupled laser cavity can be performed by increasing the power available from each elementary amplifier. A change in phasing efficiency is not expected from that evolution, provided all cavity elements are designed to withstand the increase in power density and the associated thermal effects. Fiber Kerr nonlinearity, however, may alter the spectral features of different resonator arms because four wave mixing introduces coupling among the longitudinal modes. A broadening of the resonances results from that coupling, as it was observed in practice as well as in simulations. These may be considered as the preliminary stages of optical turbulence effects identified in ultralong fiber lasers [28]. A controversy remains on the positive or negative impact of Kerr nonlinearity on phase locking and combining efficiency of the laser [23,29].

Most of the time, power scaling relies on an increase in the number of combined lasers. The compound laser operates on the frequencies that are common to all the sets of axial resonance associated with the different subcavities. An ideal situation would correspond to a laser with all parallel arms exactly identical in length so that the longitudinal modes are identical for all subcavities as well as for the compound resonator. In this case, increasing the number of elementary sources would not introduce penalty on the combining efficiency. This is unlikely to occur with fiber amplifiers. Therefore, it becomes obvious that with the increase in the number of lasers to be coupled in a common cavity, it will become more difficult to find common frequencies. The filtering due to the multiplication of Vernier effect would result in scarcity of resonance. A drop off in laser efficiency starts when no actual common mode can be found in the emission bandwidth. At the same time, the laser combined power becomes unstable. Simulations have shown that the relative power fluctuations evolve according to N^3, where N denotes the amplifier array size [30].

11.4.2
Main Influencing Parameters

Few analytical approaches have tried to predict the reduction in coherent summation, but none is able to describe completely the actual evolution. Some estimates apply to a few number of lasers where phasing efficiency remains high [31,32], while others apply to a large number of lasers where combining is weakly efficient [33]. Based on simple reasoning using the smallest common multiplier, the probability of finding a common longitudinal mode among N different lasers in a given bandwidth can be shown to be bounded by $\exp(-N)$. More rigorously, based on the approach used for speckle analysis, Siegman [31] established some evolution of the probability of loss in a multiarm interferometer from which we can derive the following expression for the combining efficiency of N lasers:

$$\eta = 1 - 4\pi\{M\Gamma[1 + (N-1)/2]\}^{2/(N-1)}/N^{N/(N-1)}, \tag{11.3}$$

where M stands for the number of possible frequencies for laser oscillation in a given bandwidth Δv_G. M can be approximated by $\Delta v_G \cdot L/c$ for a set of fibers with average round-trip length L. M is the only relevant parameter in all published studies. In fact, since with fiber lasers the root-mean-square deviation in length σ_L is most of the times well below the average length L, it often appears more appropriate to take into account the fact that close to a common resonance, several modes are almost cophased. As pointed out by Rothenberg [33], the frequency filtering performed by the multiarm interferometer has resonance peaks with a bandwidth given by c/σ_L, which is usually larger than the free spectral range c/L in the fiber laser. Therefore, it is in most cases more correct to replace L by σ_L in Eq. (11.3).

However, as already discussed in Section 11.3.1, considering robustness, one conclusion deduced from the various theoretical works is that for obtaining highest phasing efficiency, the best way is to design a laser with the largest $\Delta v_G \cdot L$ product (or $\Delta v_G \cdot \sigma_L$, owing to the situation).

Numerical simulations of the laser operation confirm the above-mentioned conclusions. It is straightforward to compute the spectral response of a multiarm resonator for a given set of parameters and then derive the expected combining efficiency with a simple laser model. The plot in Figure 11.14 indeed indicates that the efficiency of the coherent summation decreases continuously for about eight lasers. Experiments we have carried out with up to 12 fiber lasers as well as those reported with a different configuration by Chang et al. [30] with up to 16 lasers have

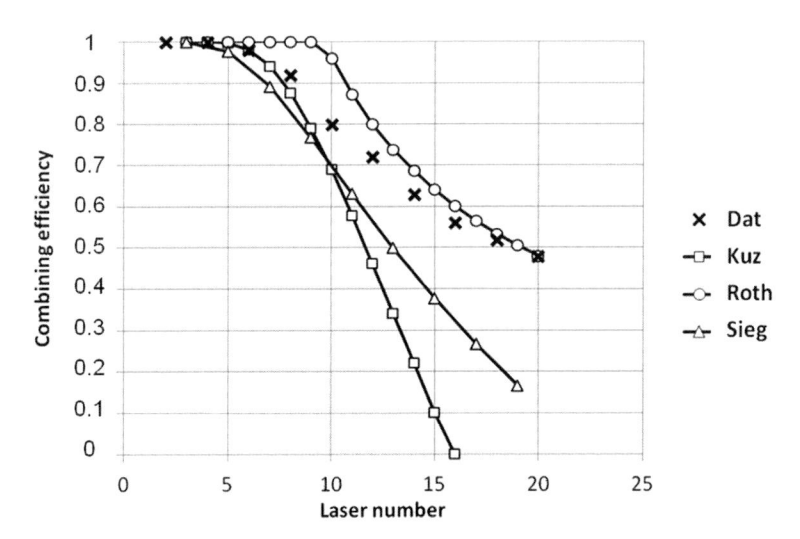

Figure 11.14 Coherent combining efficiency versus the number of lasers N. Data from simulation (dat) compared to various models from Kuznetsov (Kuz) [32] and Rothenberg (Roth) [33] and derived from Siegman's (Sieg) paper [31].

demonstrated a similar behavior. Enhancement of the performances can be obtained through a broadening of the laser bandwidth as well as through a choice of fibered paths with a larger length difference. However, the gain obtained by an increase in the parameter $\Delta v_G \cdot L$ is no longer significant when the number of lasers is high. It seems that an extended coupling among the different amplifying channels produced by the coupling device at each cavity round-trip offers some advantage in comparison to just a N–1 coupling as in a generalized Michelson interferometer [34]. Overcoming the limitations due to the scarcity of mutual resonance seems possible through the involvement of nonlinear effects (phase conjugation mirrors, coupled nonlinear cavity, etc.), which remains to be demonstrated with fiber lasers.

11.5
Passive Combining in Pulsed Regime

Coherent combining techniques have been widely investigated in the past decade in the CW regime. In pulsed regime, new geometries of optical fibers have been proposed to reduce high peak power in the fiber core. However, multiple constraints in the fiber-amplifying design limit their average and peak output powers in the conventional MOPA scheme (serial amplification). Recently, coherent combining techniques were explored in pulsed regimes [35–38]. In this case, the main difficulty is to compensate for the phase differences at the output of the multiple parallel amplifying arms in real time all along the pulse duration. Thus, the active technique needs to adapt the electronic feedback loop bandwidth to the pulse duration, feedback loop including phase analysis, phase recovery, and phase compensation [36]. In some cases, gain–phase connection described by Kramers–Kronig relationship can be very strong, leading to a significant phase evolution on the duration of the amplified pulse. This kind of fast phase distortion due to strong gain depletion is difficult to compensate by servo loop control. In this section, we report some characteristics of passive coherent combining techniques applied to different pulsed operating regimes: from Q-switched to mode locked regimes.

11.5.1
Q-Switched Regime

The principle of passive coherent combining process is not modified by the laser-operating regime. We mention that passive combining is based on single-laser cavity made up of several parallel amplifying arms that are phase locked by some common intracavity filtering. Such kind of laser architecture can easily include a modulator to generate a pulse train. In spite of multiple amplifying arms, it is possible to use a single modulator, preferably inserted on the common path of the filled configurations in Figure 11.1a. In addition, in some ring version of the compound cavity, the modulator does not have to withstand the power coming from all the amplified beams just before they exit the laser. The output coupling, which can be very high in fiber laser, decreases dramatically the power in the cavity and thus passes through the modulator.

11.5 Passive Combining in Pulsed Regime | 363

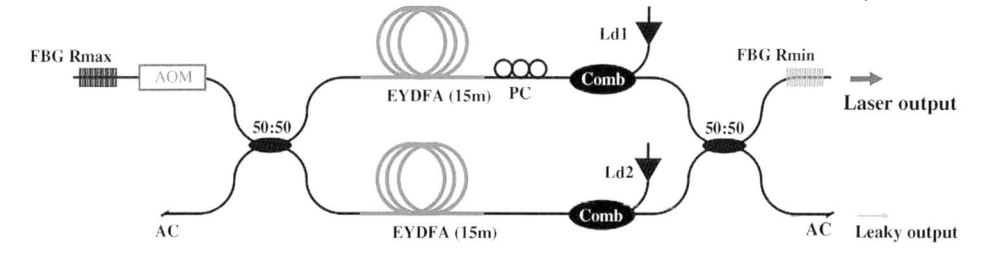

Figure 11.15 Mach–Zehnder interferometer laser made up of double-clad fiber amplifier. EYDF, erbium/ytterbium-doped fiber; comb pump–signal combiner; AOM, acousto-optic modulator; 50/50 balanced fiber coupler; LD, laser diode; PC, polarization controller; AC, angle-cleaved fiber; FBG, fiber Bragg grating.

Let us consider an experimental setup of the linear version of the Mach–Zehnder interferometer fiber laser (Figure 11.15). The laser oscillates between a pair of fiber Bragg gratings, the amplification being distributed between the two parallel codoped fibers (EYDF). The laser bandwidth is centered at 1550 nm. The acousto-optic modulator (AOM) is located in the back common arm of the laser, close to the highly reflective rear mirror (FBG Rmax). The acousto-optic device modulates the cavity losses giving rise to a pulsed operating regime at a frequency rate of 10–65 kHz.

Pulse duration is about three times larger than the cavity round-trip time, leading to a modulated pulse profile. Figure 11.16 b shows a typical pulse profile from the main output of the laser at a frequency rate of 50 kHz. It has to be compared with that of Figure 11.16a obtained from the same laser with only one amplifying fiber operating at the same frequency rate.

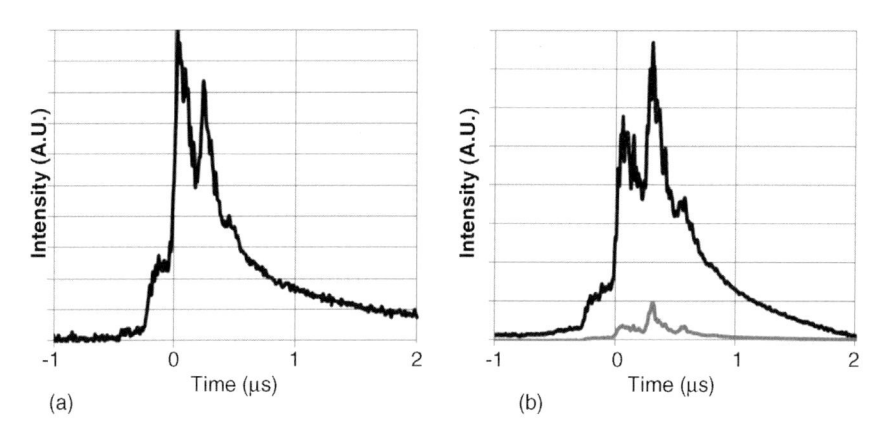

Figure 11.16 Pulse profiles at 50 kHz measured at the output of (a) a conventional laser with only one amplifying fiber and (b) the Mach–Zehnder laser. *Black curve*: main output; *gray curve*: leakage output.

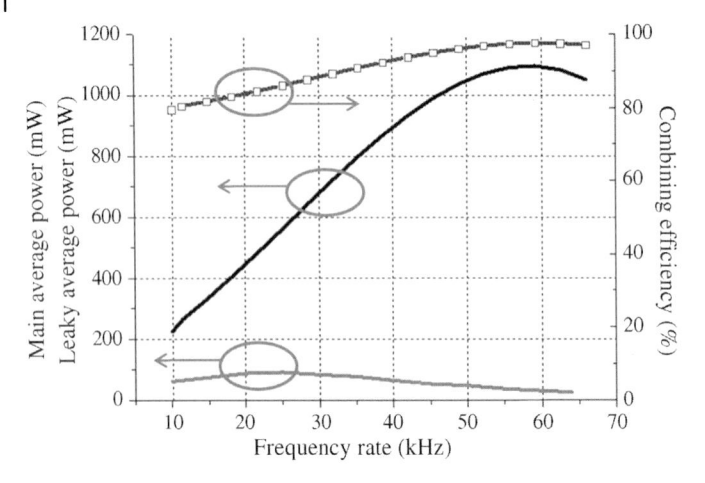

Figure 11.17 Average powers from the main (black curve) and leaky outputs of the Mach–Zehnder laser (gray curve), and combining efficiency (square symbol).

The cavity length is long due to the double-clad fiber pumping scheme. So, the cavity round-trip time is close to 250 ns, producing pulses of long duration. As shown in Figure 11.16, the modulation visible on pulse profile is due to the cavity round- trip and not due to the interferometer configuration of the laser. As the effective length of the interferometer cavity is the average length of the two amplifying subcavities (Figure 11.3), the pulse profile from both lasers are very similar, exhibiting identical period of modulation (250 ns) and duration (360 ns for the single-amplifier laser, 400 ns for the two-element laser at FWHM).

The power from the leakage output of the interferometer laser is very low compared to the one from the main output. At 50 kHz, the energy from the leakage output is less than 1 µJ, while the energy at the main output reaches 21 µJ with more than 95% of coherent combining efficiency.

Figure 11.17 shows the evolution of the coherent combining efficiency as a function of the pulse frequency rate. Maximum at high repetition rate when the pulse duration is long (550 ns at 65 kHz), it decreases slowly to 80% at 10 kHz when the pulse width drops to 280 ns. The modulation contrast of the pulse envelope increases because of high inversion population and the transient regimes become prevalent. Combining efficiency slightly decreases, but keeps a high value.

These results show that coherence between fields from both amplifying arms of the Mach–Zehnder builds essentially faster than the round-trip time of the cavity. The phase locking process is very fast, but it may slightly evolve on the duration of the pulse when some imbalance introduced by the gain occurs between the laser arms. It happens only at a low repetition rate, when the high population inversion leads simultaneously to significant phase change and short pulse duration. However, it was shown experimentally that Q-switching and passive coherent combining are compatible, in particular at high repetition rates, leading to pulses of 130–400 ns duration.

11.5.2
Mode-Locked Regime

If the previous study on Q-switched regime demonstrates that building time of the coherent combining process is about a cavity round-trip time, is it short enough to be compatible with mode locking operation? As previously described, interferometer laser generates a spectral filtering and delivers a modulated spectrum depending on its subcavity differential lengths. Consequently, the spectral modulation tailors the temporal profile of the laser emission. Whatever CW or pulsed regimes are considered, the interferometer laser emission generates structured temporal emission depending on path length difference between amplifying arms. Let us consider the case of two amplifiers such as Michelson or Mach–Zehnder interferometer laser of two subcavity lengths L_1 and L_2 (through amplifiers 1 and 2, respectively), leading to path length difference $\Delta L = |L_2 - L_1|$. The spectral response $h(\nu)$ of the interferometer cavity is given by

$$h(\nu) = g_{\Delta \nu}(\nu) \cdot [\text{III}_{\Delta \nu_c}(\nu) \otimes f_{\delta \nu}(\nu) \cdot \text{III}_{\Delta \nu_L}(\nu)], \tag{11.4}$$

where $\otimes$ denotes the convolution operation, $g_{\Delta \nu}(\nu)$ is the function that describes the spectral envelope of the emission of $\Delta \nu$ bandwidth, $f_{\delta \nu}(\nu)$ fits one spectral modulation of $\delta \nu$ bandwidth fixed by the interferometer architecture of the cavity, $\Delta \nu_c = c/\Delta L$ is the modulation period due to the interferometer configuration of the cavity, and $\Delta \nu_L = 2c/(L_1 + L_2)$ denotes the free spectral range of the interferometer laser. The frequency combs $\text{III}_i(\nu)$ is given by

$$\text{III}_i(\nu) = \sum_{k \in \mathbb{N}} \delta(\nu - k \cdot i), \tag{11.5}$$

where $I = [\Delta \nu_c, \Delta \nu_L]$ and δ is the Dirac function.

All these characteristics are reported in Figure 11.18.

In the time domain, the corresponding function $H(t)$ is as follows:

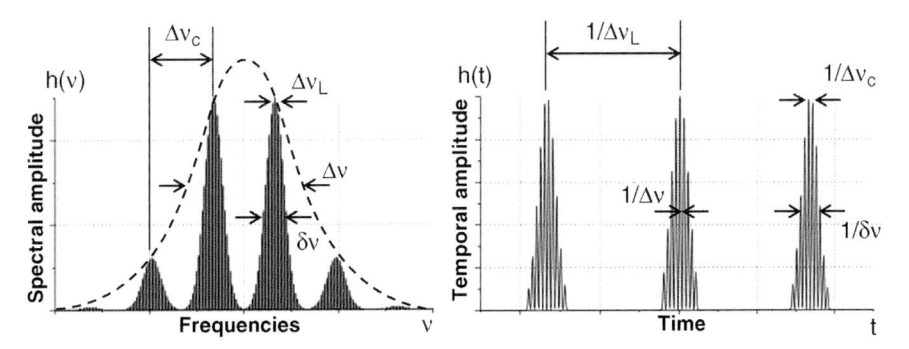

$$H(t) = G_{1/\Delta \nu}(t) \otimes [\text{III}_{1/\Delta \nu_c}(t) \cdot F_{1/\delta \nu}(t) \otimes \text{III}_{1/\Delta \nu_L}(t)], \tag{11.6}$$

Figure 11.18 Theoretical spectral and temporal profiles of a two-arm interferometer laser when spectral components are phase locked.

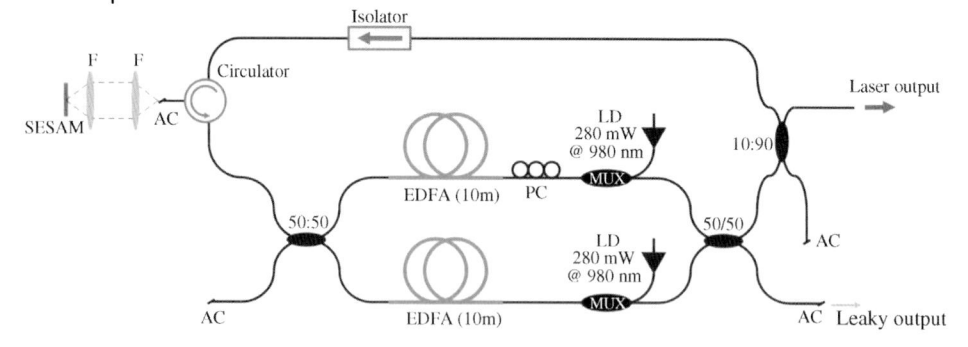

Figure 11.19 Experimental setup of a ring Mach–Zehnder laser operating in mode-locked regime. LD, laser diode; AC, angle-cleaved fiber; PC, polarization controller; EDFA, erbium-doped fiber amplifier.

where $G_{1/\Delta\nu}(t)$ and $F_{1/\delta\nu}(t)$ are the Fourier transform functions, respectively, of the $g_{\Delta\nu}(\nu)$ and $f_{\delta\nu}(\nu)$ functions. The temporal combs $\text{III}_i(t)$ is given by

$$\text{III}_i(t) = \sum_{k \in \mathbb{N}} \delta(t - k \cdot i), \quad \text{with} \quad i = [1/\Delta\nu_c, 1/\Delta\nu_L]. \tag{11.7}$$

As shown in Figure 11.18, the temporal emission $H(t)$ is structured, exhibiting a train of pulse packets. The number N_p of pulses by packet can be retrieved from Eq. (11.6) and depends on the finesse F_{mod} of the modulations in the spectrum $h(\nu)$: $N_p = \Delta\nu/\delta\nu = F_{\text{mod}}$. This finesse depends on one of the interferometer cavity and on the laser gain.

We report some experimental results illustrating this specific mode-locked regime when applied to interferometer laser. We consider a unidirectional ring Mach–Zehnder laser made up of two erbium-doped fibers (Figure 11.19). Mode locking operation is obtained by adding a semiconductor saturable absorber mirror (SESAM). With a 5 ps recovery time, this component is based on three irradiated InGaAs/InP quantum wells. The SESAM is connected to the annular cavity through an optical circulator. The net dispersion of the cavity is normal, about -0.013 ps/(nm km) to work in a stretched pulse regime. The average power reaches 94 mW from the main laser output, when only 7.7 mW is coupled outside the cavity from the leaky output. More than 92% of coherent combining efficiency is obtained with such a laser in a stable mode-locked regime. It results in a laser spectrum with highly contrasted modulation (Figure 11.20b). The laser delivers periodic pulse packets as shown on the second-order intensity autocorrelation measurement (Figure 11.20a).

The 8.8 MHz repetition rate $\Delta\nu_L$ of the packets depends on the average cavity length $L_{\text{ave}} = (L_2 + L_1)/2$ ($\Delta\nu_L = c/L_{\text{ave}}$), whereas the pulse recurrence in each packet depends on the path length difference $\Delta L = |L_2 - L_1|$ between the two arms of the interferometer.

The width Δt of the autocorrelation pulse peaks is 1.4 ps, corresponding to a Fourier transform Gaussian pulse of 990 fs. As the envelope of the experimental spectrum is 3.2 nm ($\lambda^2 \Delta\nu/c$), the time–bandwidth product is close to 0.4, showing that pulses in each packet are quasi-Fourier transform. The nine peaks visible in the autocorrelation profile indicate that each packet includes 5 pulses regularly spaced. The number of pulses in each packet depends on the finesse $F_{\text{mod}} = \Delta\nu/\delta\nu$ of the

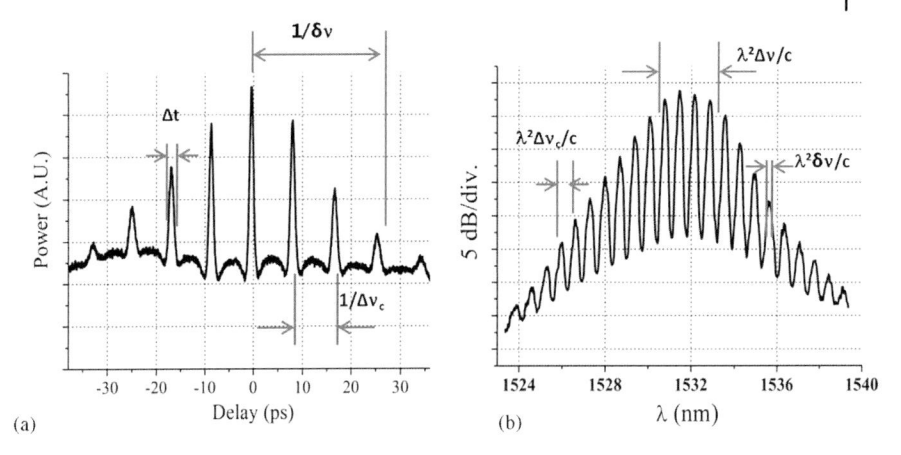

Figure 11.20 (a) Autocorrelation profile of one pulse packet and (b) corresponding spectrum. The central wavelength λ is 1532 nm.

emitted laser spectrum. It can be managed by the finesse of the interferometer cavity, which depends on the cavity Q-factor, the interferometer couplers balance, and the laser gain. It does not depend on the path length difference ΔL. This parameter changes the periodicity of the pulses in each packet and consequently the global pulse packet duration. The pulse frequency rate in each packet can be continuously tuned very easily with a delay line in one arm of the interferometer cavity. The frequency rate can be very high, limited only by the duration of the individual pulses. Experimentally, repetition rate of 200 GHz has been demonstrated [37]. This kind of laser source could be interesting in optical communication networks using the optical time division multiplexing (OTDM).

11.6
Conclusion

This chapter provided a short overview of passively combined fiber lasers. The simplicity of these lasers is appealing and justifies that a lot of work has been done in the past years. Goals were to find new architectures, to increase lasers performances, to study their characteristics, and to investigate their scaling in a number of elements. More than 1 kW was demonstrated in 2011 with a four-fiber laser showing that although the major part of the published experiments was carried out at low power level, they are scalable. The underlying physics behind the self-synchronous operation of a set of several fiber lasers is simple and was first mentioned in this chapter with a two-element cavity. It is more or less the same physics that drives the behavior of laser arrays investigated since the 1970s for CO_2 lasers and then for laser diodes. Multicore fiber lasers have a lot in common with laser diode array, but they have not been much considered here. With separate fiber lasers, the differences lie in the fact that coupling is possible at the exit ends only and also that there are significant deviations in the fiber laser lengths. General features of cophased fiber lasers were discussed, highlighting the conditions for robust operation and evolution of combining efficiency with respect to an increase in laser number. Main comments on some of the features that are specific to passively combined fiber laser were based on our own results. It was shown that they are tunable, have a fast phasing dynamics allowing Q-switching, and can serve to produce trains of ultrashort pulses. For coherent combining architecture delivering a laser beam array that works through mutual injection, one should care about the particular spatial coherence of the array. Delay lines may be required outside the cavity to get the expected coherent summation. Because of the interferometric nature of the coherent combining process, polarization is of high importance and must be managed, thanks to polarization controller on each laser arm, or even better to use polarization-preserving fibers. From the various theoretical and numerical studies that have been made as well as from experiments, we learned that the cophasing of fiber lasers can be efficient with passive techniques only in the limit of about a dozen of individual lasers. Beyond that number, even if some gain in brightness remains, the combining efficiency drops. Ways to overcome this general trend may be found in new laser designs involving nonlinear effects. For instance, multibranch lasers with phase-conjugate mirrors have been demonstrated with bulk gain media (see Chapter 14) and could be adapted to fiber amplifiers [39]. Coupled cavities with gain-dependent nonlinear refractive index have been theoretically investigated [40]. Recently, a cavity with a new type of phase contrast filtering has shown to offer some potential for an enhanced efficiency [41]. Convenient and practical solution remains to be demonstrated and research in the field focuses nowadays on the best way to compensate linear filtering effects by nonlinearity so as to scale up the number of phase-locked units.

Beyond the scope of reaching extremely high intensity, coupled fiber lasers offer a good framework to investigate more fundamental topics. In particular, in cases where a large number of element lasers are involved, fiber laser arrays share some properties with so-called random lasers. As they represent an ensemble of coupled nonlinear oscillators with different properties, they can be also used as a flexible playground to study rogue events in resonators as well as optical turbulence.

References

1 Morel, J., Woodtli, A., and Dändliker, R. (1993) Coherent coupling of an array of Nd^{3+}-doped single-mode fiber lasers by use of an intracavity phase grating. *Opt. Lett.*, **18**, 1520–1522.

2 Kozlov, V.A., Hernandez-Cordero, J., and Morse, T.F. (1999) All-fiber coherent beam combining of fiber lasers. *Opt. Lett.*, **24**, 1814–1816.

3 Shirakawa, A., Saitou, T., Sekiguchi, T., and Ueda, K. (2002) Coherent addition of fiber lasers by use of a fiber coupler. *Opt. Express*, **10**, 1167–1172.

4 Sabourdy, D., Kermène, V., Desfarges-Berthelemot, A., Lefort, L., Barthélémy, A., Mahodaux, C., and Pureur, D. (2002) Power scaling of fibre lasers with all-fibre interferometric cavity. *Electron. Lett.*, **38**, 692–693.

5 Bruesselbach, H., Jones, D.C., Mangir, M.S., Minden, M., and Rogers, J.L. (2005) Self-organized coherence in

fiber laser arrays. *Opt. Lett.*, **30**, 1339–1341.

6 Corcoran, C.J. and Durville, F. (2005) Experimental demonstration of a phase-locked laser array using a self-Fourier cavity. *Appl. Phys. Lett.*, **86**, 201118–201121.

7 Lei, B. and Feng, Y. (2007) Phase locking of an array of three fiber lasers by an all-fiber coupling loop. *Opt. Express*, **15**, 17114–17119.

8 Lhermite, J., Kermène, V., Desfarges-Berthelemot, A., and Barthélémy, A. (2007) Passive phase locking of an array of four fiber amplifiers by an all-optical feedback loop. *Opt. Lett.*, **32**, 1842–1845.

9 He, B., Lou, Q., Wang, W., Zhou, J., Zheng, Y., Dong, J., Wei, Y., and Chen, W. (2008) Experimental demonstration of phase locking of a two-dimensional fiber laser array using a self-imaging resonator. *Appl. Phys. Lett.*, **92**, 251115–251118.

10 Fridman, M., Eckhouse, V., Luria, E., Krupkin, V., Davidson, N., and Friesem, A.A. (2008) Coherent addition of two dimensional array of fiber lasers. *Opt. Commun.*, **281**, 6091–6093.

11 Cao, J., Lu, Q., Hou, J., and Xu, X. (2010) Effect of mutual injection ways on phase locking of arrays of two mutually injected fiber lasers: theoretical investigation. *Appl. Phys. B*, **99**, 83–93.

12 Xue, Y.-H., He, B., Zhou, J., Li, Z., Fan, Y.-Y., Qi, Y.-F., Liu, C., Yuan, Z.-J., Zhang, H.-B., and Lou, Q.-H. (2011) High power passive phase locking of four Yb-doped fiber amplifiers by an all-optical feedback loop. *Chin. Phys. Lett.*, **28**, 054212.

13 Ronen, E. and Ishaaya, A.A. (2011) Phase locking a fiber laser array via diffractive coupling. *Opt. Express*, **19**, 1510–1515.

14 Wrage, M., Glas, P., Fischer, D., Leitner, M., Vysotsky, D.V., and Napartovitch, A.P. (2000) Phase locking in a multicore fiber laser by means of a Talbot resonator. *Opt. Lett.*, **25**, 1436–1438.

15 Bochove, E.J., Cheo, P.K., and King, G.G. (2003) Self-organization in a multicore fiber laser array. *Opt. Lett.*, **28**, 1200–1202.

16 Boullet, J., Sabourdy, D., Desfarges-Berthelemot, A., Kermène, V., Pagnoux, D., and Roy, P. (2005) Coherent combining in an Yb doped double core fiber laser. *Opt. Lett.*, **30**, 1962–1964.

17 Michaille, L., Bennett, C.R., Taylor, D.M., Shepherd, T.J., Broeng, J., Simonsen, H.R., and Petersson, A. (2005) Phase locking and supermode selection in multicore photonic crystal fiber lasers with a large doped area. *Opt. Lett.*, **30**, 1668–1670.

18 Li, L., Schülzgen, A., Chen, S., Temyanko, V.L., Moloney, J.V., and Peyghambarian, N. (2006) Phase locking and in-phase supermode selection in monolithic multicore fiber lasers. *Opt. Lett.*, **31**, 2577–2579.

19 Shalaby, B.M., Kermène, V., Pagnoux, D., Desfarges-Berthelemot, A., Barthélémy, A., Abdou Ahmed, M., Voss, A., and Graf, T. (2009) Quasi-Gaussian beam from a multicore fibre laser by phase locking of supermodes. *Appl. Phys. B*, **97**, 599–605.

20 Swanson, G.J., Leger, J.R., and Holz, M. (1987) Aperture filling of phase-locked laser arrays. *Opt. Lett.*, **12**, 245–247.

21 DiDomenico, M., Jr. (1996) Characteristics of a single-frequency Michelson-type He-Ne gas laser. *IEEE J. Quantum Electron.*, **QE-2**, 311–322.

22 Guillot, J., Desfarges-Berthelemot, A., Kermène, V., and Barthélémy, A. (2011) Experimental study of cophasing dynamics in passive coherent combining of fiber lasers. *Opt. Lett.*, **36**, 2907–2909.

23 Wu, T., Chang, W., Galvanauskas, A., and Winful, H. (2010) Dynamical, bidirectional model for coherent beam combining in passive fiber laser arrays. *Opt. Express*, **18**, 25873–25886.

24 Sabourdy, D., Kermène, V., Desfarges-Berthelemot, A., Lefort, L., Barthélémy, A., Even, P., and Pureur, D. (2003) Efficient coherent combining of widely tunable fiber lasers. *Opt. Express*, **11**, 87–97.

25 Chen, S.-P., Li, Y.-G., and Lu, K.-C. (2005) Branch arm filtered coherent combining of tunable fiber lasers. *Opt. Express*, **13**, 7878–7883.

26 Fan, T.Y. (2009) The effect of amplitude variations on beam combining efficiency for phased arrays. *IEEE J. Quantum Electron.*, **15**, 291–293.

27 Auroux, S., Kermène, V., Desfarges-Berthelemot, A., and Barthélémy, A. (2009) Coherence properties of two fiber lasers coupled by mutual injection. *Opt. Express*, **17**, 11731–11740.

28 Babin, S.A., Karalekas, V., Podivilov, E.V., Mezentsev, V.K., Harper, P., Ania-Castanon, J.D., and Turitsyn, S.K. (2008) Turbulent broadening of optical spectra in ultralong Raman fiber lasers. *Phys. Rev. A*, **77**, 033803.

29 Simpson, T.B., Doft, F., Peterson, P.R., and Gavrielides, A. (2007) Coherent combining of spectrally broadened fiber lasers. *Opt. Express*, **15**, 11731–11740.

30 Chang, W.-Z., Wu, T., Winful, H.G., and Galvanauskas, A.A. (2010) Array size scalability of passively coherently phased fiber laser arrays. *Opt. Express*, **18**, 9634–9642.

31 Siegman, A.E. (2004) Resonant modes of linearly coupled multiple fiber laser structures, http://www.stanford.edu/~siegman/coupled_fiber_modes.pdf.

32 Kuznetsov, D. (2005) Limits of coherent addition of lasers. *Opt. Rev.*, **12**, 445–447.

33 Rothenberg, J.E. (2008) Passive coherent phasing of fiber laser arrays. *Proc. SPIE*, **6873**, 687315.

34 Fridman, M., Nixon, M., Davidson, N., and Friesem, A. (2010) Passive phase locking of 25 fiber lasers. *Opt. Lett.*, **35**, 1434–1436.

35 Sabourdy, D., Desfarges-Berthelemot, A., Kermène, V., and Barthélémy, A. (2004) Coherent combining of Q-switched fibre lasers. *Electron. Lett.*, **40**, 1254–1255.

36 Lombard, L., Azarian, A., Cadoret, K., Bourdon, P., Goular, D., Canat, G., Jolivet, V., Jaouën, Y., and Vasseur, O. (2011) Coherent beam combination of narrow-linewidth 1.5 μm fiber amplifiers in a long-pulse regime. *Opt. Lett.*, **36**, 523–525.

37 Lhermite, J., Sabourdy, D., Desfarges-Berthelemot, A., Kermène, V., Barthélémy, A., and Oudar, J.-L. (2007) Tunable high-repetition-rate fiber laser for the generation of pulse trains and packets. *Opt. Lett.*, **32**, 1734–1736.

38 Klenke, A., Seise, E., Demmler, S., Rothhardt, J., Breitkopf, S., Limpert, J., and Tünnermann, A. (2011) Coherently-combined two channel femtosecond fiber CPA system producing 3mJ pulse energy. *Opt. Express*, **19**, 24280–24285.

39 Shardlow, P.C. and Damzen, M.J. (2010) Phase conjugate self-organized coherent beam combination. *Opt. Lett.*, **18**, 1082–1084.

40 Corcoran, C.J. and Durville, F. (2008) Passive phasing in a coherent laser array. *IEEE J. Select. Top. Quantum Electron.*, **15**, 294–300.

41 Jeux, F., Desfarges-Berthelemot, A., Kermène, V., Guillot., J., and Barthélémy, A. (2012) Passive coherent combining of lasers with phase-contrast filtering for enhanced efficiency. *Appl. Phys. B*, **108**, 81–87.

12
Coherent Combining and Phase Locking of Fiber Lasers

Moti Fridman, Micha Nixon, Nir Davidson, and Asher A. Friesem

12.1
Introduction

Fiber lasers are commonly comprised of doped double-clad fibers as the gain medium and the high- and low-reflection fiber Bragg gratings (FBGs) as mirrors. The fiber lasers are usually pumped from the rear using multimode diode lasers, but can also be pumped from the side with special combiners. Typically, the bandwidth of fiber lasers is determined by the bandwidth of the Bragg grating that ranges from 10 to 0.1 nm. The electricity-to-light conversion efficiency of fiber lasers can exceed by 50% and since light is confined inside the fiber, the lasers are highly robust [1]. However, the light is confined in the small core of the fiber, so the output power from a single-fiber laser is limited due to nonlinear effects and the danger of damage to the fiber [2].

In order to overcome the limitation of output power from the single-fiber lasers, several low-power fiber lasers could be combined. Such a combination could be performed either incoherently or coherently. When the field distributions of several laser output beams are incoherently combined, the resulting beam quality factor (M^2) is relatively poor with low optical brightness. Nevertheless, the incoherent approach is pursued quite actively because the number of lasers that can be combined efficiently is relatively large. When the field distributions are coherently added, with the proper phase relations, the combined beam quality factor is as good as that of a single low-power laser, while the combined power is greater by a factor equal to the number of combined lasers.

When coherently combining two or more fiber laser output fields, three major difficulties are encountered [3,4]. The first results from the need to properly couple the individual laser fields, so as to establish mutual coherence and enable relative phase locking between them. Such coupling typically introduces excessive losses to each laser field and requires accurate relative alignment. The second (and somewhat related) difficulty results from the need to accurately control the relative phase and amplitudes between the different fiber laser fields, so as to ensure constructive interference between them in the far field. This requires accurate control of the

---

Coherent Laser Beam Combining, First Edition. Edited by Arnaud Brignon.
© 2013 Wiley-VCH Verlag GmbH & Co. KGaA. Published 2013 by Wiley-VCH Verlag GmbH & Co. KGaA.

distances between the participating optical components, causing the output power to be extremely sensitive to thermal drifts and acoustic vibrations. The third difficulty results from the need to efficiently combine many separate fiber laser output beams into one single beam.

During the past decade, we extensively investigated new approaches for passive phase locking and combining of several lasers. These involved the development of unique intracavity elements and laser configurations in order to obtain efficient phase locking and combining of solid-state lasers as well as fiber lasers [5–26]. In general, our results indicate that robust and practical laser systems that could potentially have high output powers concomitantly with very good output beam quality can be developed. Here, we present some of our recent developments on fiber lasers. Specifically, we present our configurations and results on coherent combining of two and then four fiber lasers, our investigations and results on the effects of noise, longitudinal modes, and time delays, possibilities for upscaling the number of lasers that can be combined, and finally our configurations and results where up to 25 fiber lasers were phase locked, which were also exploited for studying extreme value statistics.

12.2
Passive Phase Locking and Coherent Combining of Small Arrays

In this section, we describe the configurations and present the results of our investigations on passive phase locking and coherent combining with a small number of fiber lasers [5–13]. We start with phase locking and coherent combining of two fiber lasers, and then of four fiber lasers arranged in a two-dimensional array.

12.2.1
Efficient Coherent Combining of Two Fiber Lasers

We investigated phase locking and coherent combining of two fiber lasers using two free space Vernier–Michelson configurations of end-pumped fiber lasers, schematically shown in Figure 12.1. The first, shown in Figure 12.1a, is an intracavity configuration [5]. One end of each fiber is attached to a high-reflection fiber Bragg grating and the other fiber end is cleaved at an angle of $8°$ to suppress any reflections back into the fiber cores, so that each fiber is essentially an amplifier. The light emerging from both fiber lasers is coherently added in free space by means of a 50% beam splitter and a common output coupler (OC) with 4% reflection. This configuration is analogous to the inner-fiber configurations, except that the combined beam now propagates only in free space.

The second configuration, shown in Figure 12.1b, is an outer-cavity configuration. One end of each fiber is again attached to an FBG, but the other fiber end is cleaved at $0°$, which reflects 4% of the light back into the fiber core, so that each fiber behaves as an independent fiber laser resonator. The light emerging from each fiber laser is

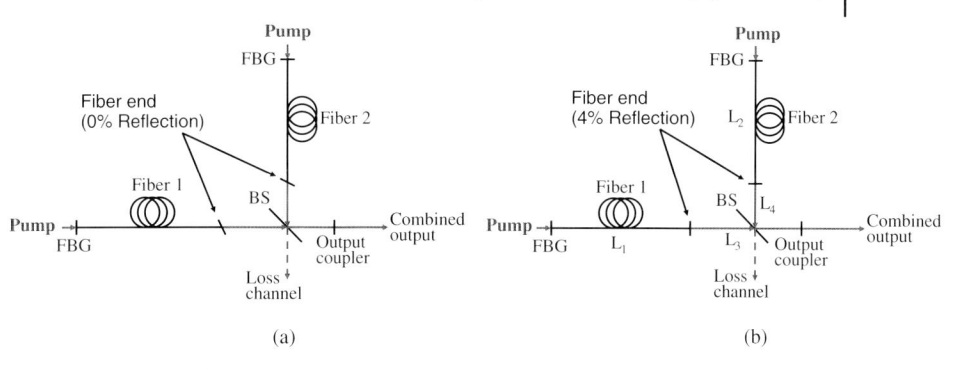

Figure 12.1 Basic configurations for coherent combining of two fiber lasers. (a) Intracavity addition. (b) Outer-cavity addition. FBG, high-reflection fiber Bragg grating; BS, 50% beam splitter.

coherently added in free space by means of the same beam splitter and output coupler as in the first configuration.

In both configurations, phase locking of the lasers results in constructive interference toward the output channel direction and destructive interference toward the loss channel direction. However, to decide which configuration is superior, we can measure either the coherent combining efficiency or the phase locking efficiency.

The efficiency of the combined output power as a function of the coupling strength κ in both configurations is presented in Figure 12.2. It is normalized such that 1 denotes 100% efficiency and corresponds to twice the output power of a single-fiber laser, measured without the beam splitter. As evident from Figure 12.2, even at coupling as low as 1%, an efficiency of 90% is obtained for both configurations. Thus, only a small amount of light needs to be reflected back into the fibers, so the risk of optical damage is low. At low coupling, the outer-cavity configuration is much

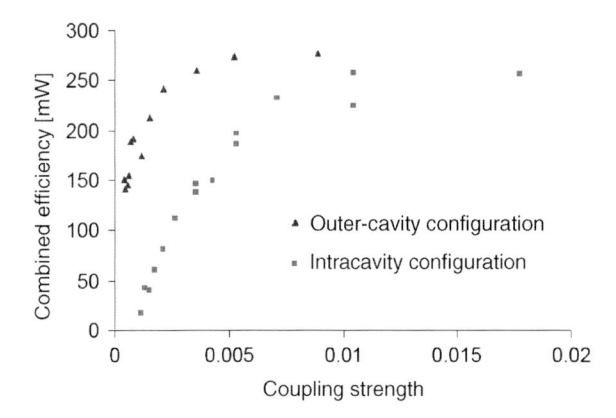

Figure 12.2 Efficiency as a function of coupling strength for intracavity and outer-cavity configurations.

more efficient than the intracavity one. For example, in the intracavity configuration, the combined output power reduces to 75% for a coupling of 0.6%; while in the outer-cavity configuration, the coupling can be as low as 0.2% for the same output power. Such a large difference is due to the fact that the coupling effects are different for the two configurations. In the outer-cavity configuration, reducing the coupling keeps the power of each laser nearly unchanged, and only decreases the efficiency of coherent combining of the two lasers. However, in the intracavity configuration, reducing the coupling decreases the power of each fiber laser, long before the reduction of the efficiency of coherent combining occurs.

On the other hand, for the intracavity configuration, phase locking occurs for much lower values of coupling than for the outer-cavity configuration. To determine the level of phase locking, we measured the fringe visibility of the interference pattern of the light emerging from the two lasers (see the insets in Figure 12.3) according to $v_{fv} = (I_{max} - I_{min})/(I_{max} + I_{min})$, where I_{max} and I_{min} are the maximum and minimum intensities along a cross section of the fringes. The results of fringe visibility as a function of coupling strength for the intracavity and outer-cavity configurations are presented in Figure 12.3. As evident from the figure, a high fringe visibility level occurs at a significantly lower coupling strength for the intracavity configuration. Specifically, the transition from zero phase locking to almost complete phase locking occurs for the intracavity configuration with up to 0.001 coupling strength, whereas for the outer-cavity configuration it is 0.01.

In order to obtain realistic calculated fringe visibility results that can be directly compared with our experimental results, we added a noise term to the laser equations. The results are presented in Figure 12.3, where the solid curves denote the numerically obtained results and the dashed curves denote the analytically obtained results [6].

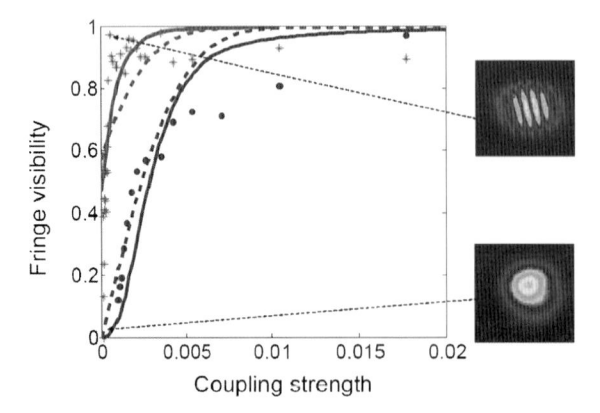

Figure 12.3 Experimental and calculated fringe visibility as a function of coupling strength. Stars denote experimental results for the intracavity configuration. Dots denote experimental results for the outer-cavity configuration. Solid curves denote corresponding numerically calculated results, while dashed curves denote corresponding analytically calculated results.

12.2.2
Compact Coherent Combining of Four Fiber Lasers

In order to obtain practical combining systems, it is best to resort to more compact configurations. In these configurations, coherent combining is performed by means of interferometric combining elements that are compact, leading to overall configuration that can be readily upscaled [7–12]. The first configuration, namely, the intracavity configuration, for free space coherent combining of four fiber lasers is presented schematically in Figure 12.4 [13]. It includes four fiber lasers and two intracavity interferometric combiners. Each fiber laser consists of polarization, maintaining erbium-doped fibers of about 7 m in length, where one end is attached to a high-reflection fiber Bragg grating of 5 nm spectral bandwidth that serves as a back reflector mirror and the other end is spliced to a collimating gradient-index (GRIN) lens with an antireflection layer to suppress any reflections back into the fiber cores and a flat output coupler of 20% reflection that is common to all fiber lasers. Each fiber laser is pumped with a diode laser of 911 nm wavelength and 5 W maximum output power, which is spliced at the back of the FBG. Each interferometric combiner is a planar substrate, where half of the front surface is coated with an antireflection layer and the other half with a 50% beam splitter layer, while half of the rear surface is coated with a highly reflecting layer and the other half with an antireflection layer. When there is phase locking among the fiber lasers, the first interferometric combiner transforms efficiently four light beams into two beams and the second interferometric combiner transforms the two beams into one nearly Gaussian beam.

We placed the first interferometric combiner in order to coherently add horizontally the light from the four fiber lasers each with a power of 35 mW to form two beams each with a power of 66 mW. This corresponds to a combining efficiency of over 90%. Then, we added the second interferometric combiner to coherently add vertically the two remaining beams into one beam with a power of 121 mW. This corresponds to an overall combining efficiency (i.e., the ratio between the overall power with the interferometric combiners to that without the combiners) of 86%. We also placed CCD cameras to detect the intensity distributions without the interferometric combiners, after the first interferometric combiner, and after the second one. The results are presented in Figure 12.5. As evident from the figure, the initial four beams first

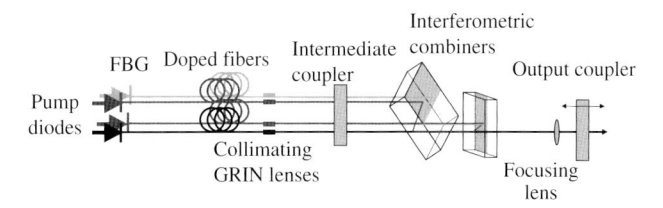

Figure 12.4 Basic intracavity configuration for coherent combining of four fiber lasers in free space using two orthogonally oriented interferometric combiners.

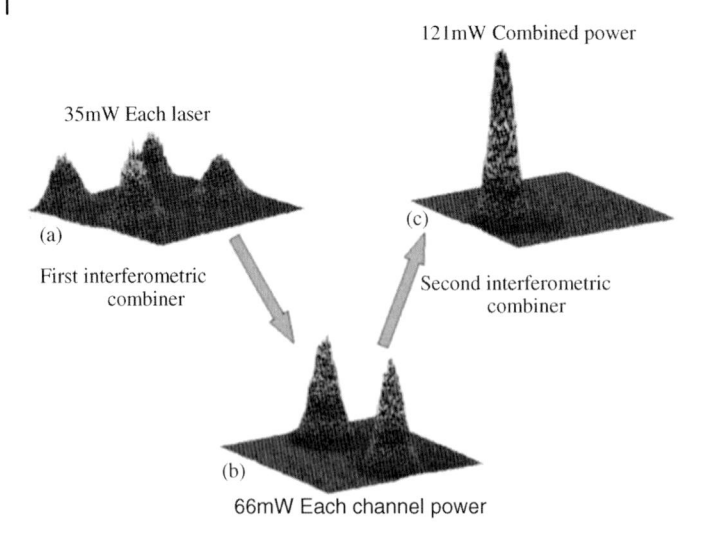

Figure 12.5 Experimental intensity distributions for the intracavity configuration. (a) Before the interferometric combiners. (b) After the first interferometric combiner. (c) After the second interferometric combiner.

coherently transform to two beams after the first interferometric combiner and then to one beam after the second interferometric combiner.

12.2.3
Efficient Coherent Combining of Four Fiber Lasers Operating at 2 μm

We also performed experiments for passive coherent combining of four fiber lasers operating at 2 μm. The experimental configuration was similar to that shown in Figure 12.4. It consisted of four single-mode thulium-doped fibers, two interferometric coupling assemblies, and a common output coupler [13]. The rear end of each fiber was attached to a high-reflection fiber Bragg grating and the other fiber end was attached to a collimator with an antireflection coating to suppress any reflections back into the fiber cores so that each fiber was essentially an amplifier. Each fiber was end-pumped by a diode laser operating at 790 nm wavelength through the FBG. In order to align the four fiber lasers to be exactly parallel, each was aligned individually such that it lased with a single parallel output coupler located in front of the array. This ensured that all the fiber lasers were perpendicular to the same plane and hence exactly parallel to each other. The beams emerging from the fiber lasers were coherently combined in free space by means of two interferometric combiner assemblies (horizontal and vertical). Each interferometric combiner assembly was composed of a 50% beam splitter and a high-reflection mirror properly displaced from it. We first measured the parallelism of the lasers and confirmed that the angle between them was less than 0.5 mrad.

The results are presented in Figure 12.6. Figure 12.6a shows the near-field light intensity distribution, detected with a 2 μm CCD camera. As evident from the figure,

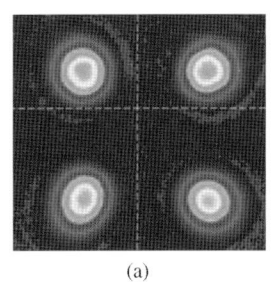

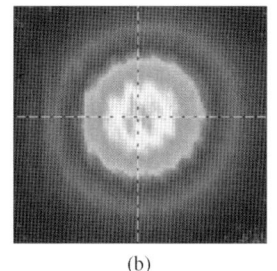

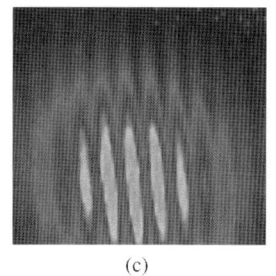

Figure 12.6 Experimental intensity distributions when coherently combining four fiber lasers operating at 2 μm. (a) Intensity distribution at the outputs of the four individual fiber lasers. (b) Intensity distribution of the combined output beam. (c) High-contrast fringes obtained by interfering two of the four coupled fiber lasers.

the four individual beams have indeed near-Gaussian distributions, are equally spaced, and have nearly the same size, that is, all the properties needed for efficient coherent combining. Next, we placed the first interferometric combiner assembly in order to combine coherently four lasers into two. Then, we added the second interferometric combiner to coherently add the two remaining beams into one. The combined output beam is presented in Figure 12.6b and indicates that the near-Gaussian distribution of the individual lasers is indeed preserved after combining. We measured an overall coherent combining efficiency of about 92%, that is, the light efficiency for coherently combining four different beams into one. Finally, we also confirmed coherence between any two individual laser beams by directing them at an angle onto the CCD camera and detecting the resulting interference fringes between them. The results are presented in Figure 12.6c. The high-contrast interference fringes correspond to very strong and stable phase locking between the beams, leading to the high combining efficiency that was obtained.

12.3
Effects of Amplitude Dynamics, Noise, Longitudinal Modes, and Time-Delayed Coupling

In this section, we describe our investigations and results about the effects of amplitude dynamics, noise, longitudinal modes, and time-delayed coupling on phase locking and coherent combining with two fiber lasers [14–20].

12.3.1
Effects of Amplitude Dynamics

The phase locking and coherence properties between two weakly coupled lasers are presented [14]. We show how the degree of coherence between two lasers can be enhanced by nearly an order of magnitude after taking into account the effects of coupling on both their phases and their amplitudes. Specifically, correlations between synchronized spikes in both the amplitude dynamics and the phase dynamics of the lasers allow an interference pattern with a fringe visibility of 90%, even when the coupling strength is far below the critical value and the lasers are not phase locked.

Many different schemes for coupling lasers have been extensively investigated over the past decades [1–26]. In all, phase locking can only occur when the coupling strength κ exceeds some critical value. In general, theoretical models that deal with the effects of coupling on the coherence and phase locking between weakly coupled lasers, that is, $\kappa < \kappa_c$, take into account only the effects of the phase difference between the lasers, while neglecting the effects that the laser amplitudes may have on the coherence and phase locking properties of the lasers.

We investigate the dynamics of two weakly coupled lasers, where $\kappa < \kappa_c$. We find that by including the effects of laser amplitudes in addition to the phase difference, the coherence between two lasers can be enhanced up to an order of magnitude. Such enhanced coherence is a direct result of correlation between the amplitude and the phase dynamics.

In order to determine the dynamics of two coupled lasers, we begin with the rate equations that are used for describing a broad range of coupled lasers [16,21,25]:

$$\frac{dE_{1,2}}{dt} = \frac{1}{\tau_c}\left[(G_{1,2} - \alpha_{1,2})E_{1,2} + \kappa E_{2,1}\right] + i\omega_{1,2}E_{2,1},$$

$$\frac{dG_{1,2}}{dt} = \frac{1}{\tau_f}\left[P_{1,2} - (I_{1,2} - 1)G_{1,2}\right],$$
(12.1)

where $E_{1,2}$ are the complex electric fields of lasers 1 and 2, τ_c is the photon cavity round-trip time, τ_f is the fluorescence lifetime, $\omega_{1,2}$ is the frequency detuning from a mean optical frequency for each laser, and κ is the coupling strength; for each laser, $\alpha_{1,2}$ is the round-trip loss, $G_{1,2}$ is the round-trip gain, $P_{1,2}$ is the pump strength, and $I_{1,2}$ is the intensity in units of the saturation intensity. Now, with $E_{1,2} = A_{1,2}\exp(i\varphi_{1,2})$ and separating the equations into real and imaginary parts, it is possible to obtain relations for the amplitudes and phases of the two coupled lasers. The real part yields relations for amplitudes A_1 and A_2:

$$\frac{dA_{1,2}}{dt} = \frac{1}{\tau_c}\left[(G_{1,2} - \alpha_{1,2})A_{1,2} + \kappa A_{2,1}\cos\varphi\right].$$
(12.2)

The imaginary part yields the relation for the phase difference φ:

$$\frac{d\varphi}{dt} = \Omega - \frac{\kappa}{\tau_c}\beta\sin\varphi,$$
(12.3)

$$\beta = \frac{A_1}{A_2} - \frac{A_2}{A_1},$$
(12.4)

where $\varphi = \varphi_2 - \varphi_1$ is the phase difference between the two lasers and $\Omega = \omega_2 - \omega_1$ is the frequency detuning between the two lasers. With both lasers having the same pump strength and the same round-trip losses, phase locking can only occur for $\kappa > \kappa_c = \Omega\tau_c/2$ [25].

12.3 Effects of Amplitude Dynamics, Noise, Longitudinal Modes, and Time-Delayed Coupling

The degree of phase locking between two lasers is usually quantified by determining the fringe visibility of the intensity interference pattern formed by two interfering lasers [16,22,25]. This fringe visibility v_{fv} is

$$v_{fv} = \frac{I_{max} - I_{min}}{I_{max} + I_{min}}, \quad (12.5)$$

where I_{max} and I_{min} are the maximal and minimal values of the time-averaged intensities in the interference pattern. Writing I_{max} and I_{min} in terms of laser amplitudes and relative phase difference between the lasers, the relation for the fringe visibility v_{fv} is

$$v_{fv} = \frac{2\sqrt{\langle A_1 A_2 \cos\varphi \rangle^2 + \langle A_1 A_2 \sin\varphi \rangle^2}}{\langle A_1^2 \rangle + \langle A_2^2 \rangle}. \quad (12.6)$$

According to this equation, even if the amplitudes of the lasers are identical, the fringe visibility still depends on both the amplitude dynamics and the phase difference. Consequently, the degree of phase locking cannot be always quantified by the fringe visibility. Accordingly, a more suitable way to quantify phase locking is by resorting to an alternative phase locking parameter, defined by v_{pl}:

$$v_{pl} = \sqrt{\langle \cos\varphi^2 \rangle + \langle \sin\varphi^2 \rangle}. \quad (12.7)$$

Note that $v_{pl} = v_{fv}$ only when the amplitudes of both lasers are identical and their dynamics are uncorrelated with those of φ, that is, $\langle A_1 A_2 \cos\varphi \rangle = \langle A^2 \rangle \langle \cos\varphi \rangle$.

We numerically solved the rate equations of coupled lasers using the typical timescales for Yb-doped fiber lasers of $\tau_c = 30$ ns and $\tau_f = 230$ µs, with frequency detuning of $\Omega = 200$ kHz. Figure 12.7 shows the fringe visibility v_{fv} and the phase locking parameter v_{pl} as a function of the coupling strength κ normalized by the critical coupling κ_c. As shown, the phase locking parameter monotonically increases

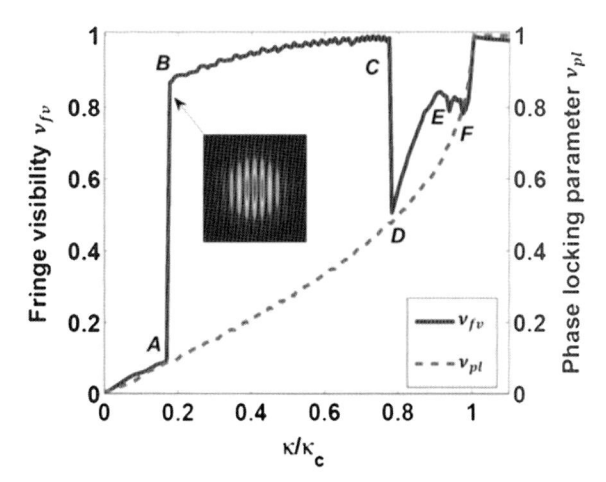

Figure 12.7 Fringe visibility v_{fv} and phase locking parameter v_{pl} as a function of the coupling strength κ normalized by the critical coupling strength κ_c. Inset: The intensity pattern of two interfering Gaussian beams with fringe visibility corresponding to that of point B.

with coupling strength until reaching a value of 1 at $\kappa = \kappa_c$. However, the fringe visibility is characterized by sharp increases and sharp drops as coupling strength increases. The first sharp increase of the fringe visibility is from $v_{fv} \approx 0.1$ at point A to $v_{fv} \approx 0.86$ at point B. The inset shows a simulated intensity interference pattern with a fringe visibility corresponding to that of point B.

To clarify and explain the unusual behavior of the fringe visibility shown in Figure 12.7, we calculated the laser amplitudes and phase difference at points A–F as a function of time. The results are presented in Figure 12.8. Since the amplitude

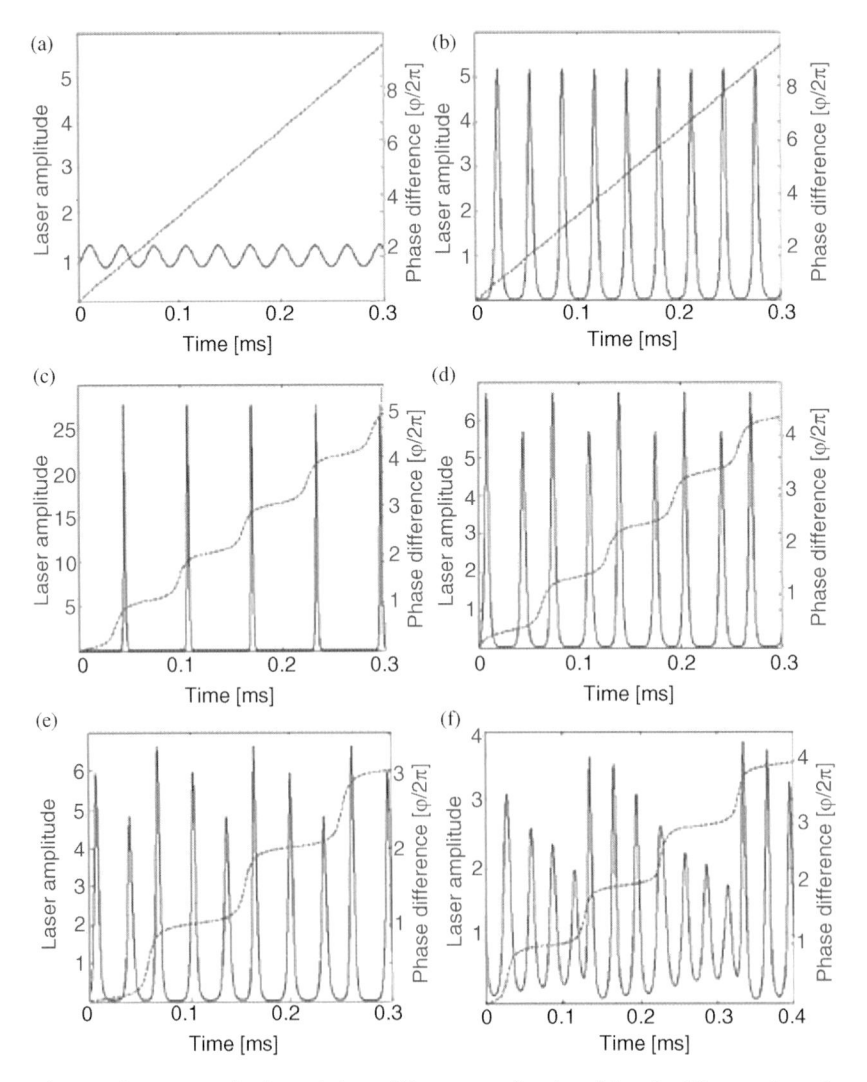

Figure 12.8 Laser amplitudes and phase difference as a function of time for different values of κ/κ_c corresponding to points A–F of Figure 12.7. Solid line denotes amplitudes, while dotted line denotes the phase difference.

dynamics are identical, we only present the amplitudes of one of the lasers. Figure 12.8a shows the results corresponding to point A. As evident from the figure, the laser amplitudes have small fluctuations around a mean value and φ monotonically increases with time. These indicate that $\nu_{pl} = \nu_{fv}$ and poor fringe visibility is due to φ that continually varies. Figure 12.8b shows the results corresponding to point B. Here, φ is essentially the same, but the laser amplitudes are dramatically different, characterized by short and intense pulses with a repetition rate equal to $\Delta\omega$. The phase difference accumulated between adjacent pulses is 2π, so effectively the coupled lasers experience a relatively constant phase difference between them, as if they are phase locked at all times. Since the lasers operate in short intense pulses, the average fringe visibility is essentially the same as the instantaneous fringe visibility lasting the duration of one pulse, so it is relatively high at point B.

As the coupling strength increases, the pulses become narrower and more intense, as shown in Figure 12.8c, and ν_{fv} increases to 0.99. A further slight increase of κ leads to a pulse breakup of the intensity, where two wider pulses occur in each 2π phase cycle, as shown in Figure 12.8d. This breakup leads to a sharp drop in ν_{fv} at point D. As the coupling strength increases further, the process of pulse width narrowing that leads to increasing visibility and then to pulse breakup is repeated again and again. Specifically, Figure 12.8e shows a breakup into three pulses for each 2π phase cycle, and Figure 12.8f shows a breakup into seven pulses for each 4π phase cycle. As the coupling strength approaches its critical value, the pulse breakup becomes more sensitive to variations, eventually leading to chaos [14].

To conclude, we showed that the coherence between two weakly coupled non-phase locked lasers can be substantially enhanced as a result of amplitude and phase correlation dynamics. Such enhancement was obtained for a wide range of laser parameters and also appears quite robust to parameter mismatch and noise. We believe that amplitude-enhanced coherence may have potential applications in the field of coherent combining of lasers, where our numerical results showed that a combining efficiency of 90% could be achieved for coupling strengths as low as 20% of the critical coupling strength.

12.3.2
Effects of Quantum Noise

Phase locking between two simple linear and noiseless oscillators is entirely dependent on the interplay of frequency detuning and coupling strength between them. However, in coupled laser oscillators, there are additional factors that affect phase locking such as multimode operation and noise [15, 16].

Typically, the quantum noise is much weaker than the other sources of noise, but when the lasers operate very close to the threshold, the quantum noise can become dominant due to the inherent high spontaneous emission [15]. In order to characterize the quantum noise, we first determined that the threshold of each laser occurs at a pumping current of 1.016 A for an output power of 69.28 µW. This was determined under controlled environment and stable conditions in order to

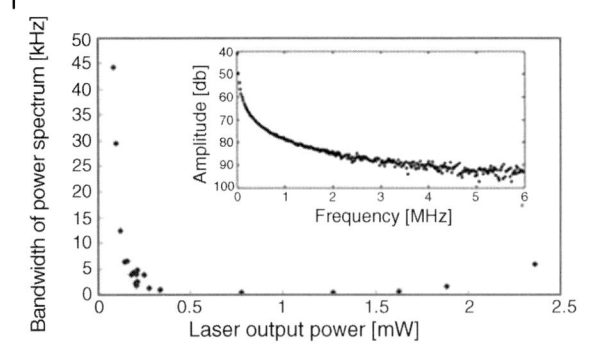

Figure 12.9 Bandwidth of the power spectra as a function of the output power for a single laser. Inset: The representative results of one power spectrum for laser output power of 96.3 μW.

reduce acoustic noise and Brillouin scattering, so the quantum noise was the most dominant. Then, we measured the output power spectra close to the threshold for different pump powers and verified that the spectrum of quantum noise has a Lorentzian shape as expected. The results are presented in Figure 12.9. It shows the FWHM bandwidth of the power spectra as a function of the output power, near threshold, from each fiber laser, along with the representative results of a typical power spectrum. Fitting the power spectrum to a Lorentzian shape gave a measure of 0.98, while fitting it to a Gaussian shape gave a measure of only 0.70. As evident from Figure 12.9, the bandwidth increases as the laser output power decreases, indicating a typical behavior of the spectrum of quantum noise [15].

When operating the lasers very close to threshold, there is a significant fluctuation of output powers. Accordingly, we simultaneously measured the output power from each laser while detecting the fringe visibility. Thus, we obtained phase locking as a function of the laser output powers for different coupling strengths. The results of the measurements near threshold for two different coupling strengths are presented in Figure 12.10. Figure 12.10a shows the phase locking as a function of the laser output powers for a coupling strength of 4.8%, whereas Figure 12.10b shows it for a coupling strength of 1.8%. As evident from the figure, there is not only the well-known dependence of the phase locking on the coupling strength but also a strong dependence on the laser output powers. It should be emphasized that this dependence occurs only when operating the lasers very close to the threshold, supporting the hypothesis that it is a result of a quantum phenomenon.

The laser output powers can be related to the quantum noise by resorting to the Schawlow–Townes equation, which relates the laser output powers to the bandwidth of each of their longitudinal modes. For our lasers, which are three-level lasers, the bandwidth of each longitudinal mode is

$$\Delta f = \frac{2\pi \hbar \omega \delta \omega^2 n}{P_{out}}, \qquad (12.8)$$

where $\hbar$ is the reduced Planck constant, $\omega = 1780$ THz is the light frequency, $\delta\omega = 50$ MHz is the cold cavity bandwidth, $n = 17\,000$ is the number of longitudinal

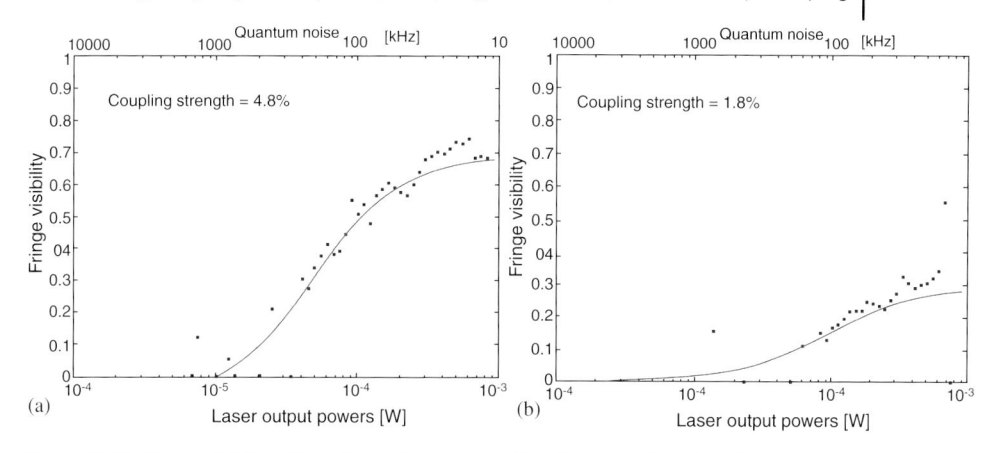

Figure 12.10 Fringe visibility of interference between two fiber lasers as a function of laser output power and quantum noise. (a) Coupling strength of 4.8%. (b) Coupling strength of 1.8%. Dots denote experimental results and solid curves denote analytic results.

modes, and P_{out} is the power level at which the lasers oscillate. At around threshold, the bandwidths of the longitudinal modes correspond to the noise from the spontaneous emission. Thus, it is now possible to relate the fringe visibility (phase locking) directly to the quantum noise, as shown in Figure 12.10.

To determine the phase locking as a function of quantum noise, we introduced a Langevin noise term into the rate equation of coupled lasers to yield

$$\frac{d\varphi}{dt} = \Omega + \frac{\kappa}{\tau_c}\left[\frac{A_{2,1}}{A_{1,2}} + \frac{A_{1,2}}{A_{2,1}}\right]\sin\left(\varphi - \frac{\pi}{2}\right) + \sqrt{\varepsilon}\eta(t), \tag{12.9}$$

where $\sqrt{\varepsilon}\eta(t)$ is a white noise source corresponding to the spontaneous emission, with noise amplitude $\varepsilon = \Delta f$, and $\eta(t_1)\eta(t_2) = \delta(t_2 - t_1)$, where δ is the delta function.

Next, we developed an analytic expression for the fringe visibility as a function of the quantum noise. We introduced an effective potential U to yield

$$\frac{d\varphi}{dt} = -\frac{dU(\varphi)}{d\varphi} + \sqrt{\varepsilon}\eta(t). \tag{12.10}$$

Since this equation describes a viscous motion, the phase φ has no inertia, and since the quantum noise $\eta(t)$ has a δ time correlation function, the motion of φ is only governed by U and the instantaneous value of $\eta(t)$. If the instantaneous value of $\eta(t)$ is smaller than a certain critical noise, the system will be phase locked. On the other hand, if it is larger, the system will not be phase locked. Thus, the noise term $\sqrt{\varepsilon}\eta(t)$ can be added to U to yield a stochastic time-dependent potential $U(t)$:

$$U(\varphi, t) = -\Omega(t)\varphi - \frac{2\kappa}{\tau_c}\cos\left(\varphi + \frac{\pi}{2}\right), \tag{12.11}$$

where $\Omega(t) = \Omega_0 + \sqrt{\varepsilon}\eta(t)$ is the time-dependent stochastic detuning. Accordingly,

$$\frac{d\varphi}{dt} = -\frac{dU(\varphi, t)}{d\varphi}. \tag{12.12}$$

This indicates that phase locking is determined by the instantaneous potential $U(\varphi, t)$. Since the instantaneous detuning is a rapidly varying stochastic variable, the time-averaged fringe visibility (measured in our experiments by a slow CCD camera) is equal to the probability of phase locking to occur. Specifically, when $\Omega(t)$ is smaller than the critical detuning of $\Omega_c = 2\kappa/\tau_c$, phase locking occurs. The fringe visibility, namely, contrast C, can be calculated by integrating over the probability from negative critical detuning $-\Omega_c$ to positive critical detuning $+\Omega_c$:

$$C = \int_{-\Omega_c}^{\Omega_c} P(\Omega) d\Omega, \tag{12.13}$$

where $P(\Omega)$ is the probability distribution of detuning. Assuming a Lorentzian distribution for $P(\Omega)$ with bandwidth Δf leads to a fringe visibility C as a function of the laser output powers P_{out}:

$$C(P_{out}) = \frac{\Delta f(P_{out})}{2\pi} \int_{-2\kappa/\tau_c}^{2\kappa/\tau_c} \frac{d\Omega'}{\left(\Omega' + ((\Delta f(P_{out}))/2)^2\right)} = \frac{2}{\pi} \arctan \frac{2\kappa P_{out}}{\pi \hbar \omega \delta \omega^2 n \tau_c}. \tag{12.14}$$

The calculated results based on this equation are also presented in Figure 12.10 as the solid curve, without any fitting parameters and neglecting the frequency detuning. The results are essentially independent of the frequency detuning as long as it is much smaller than the bandwidth. As is evident from the figure, the analytic results are in good agreement with the experimental results.

We also investigated how the coupling strength that is needed for phase locking is affected by the quantum noise when the lasers operate near threshold. Representative experimental and calculated results at a fringe visibility of 50% of maximum are shown in Figure 12.11. The results show a linear behavior, as was predicted from our analytic model, and that the coupling strength must be increased as the quantum noise increases.

12.3.3
Effects of Many Longitudinal Modes

When trying to increase the number of phase-locked fiber lasers, we encounter a limit [16, 39, 40]. To understand this limit, we must take into account the spectral properties of phase locked lasers. In order to phase lock two fiber lasers, it is necessary that they have a common frequency in the two cavities. Each cavity has a comb of frequencies (longitudinal modes) that satisfy it. This comb is set by the length of the cavity, and the spacing between each pair of adjacent longitudinal modes is $\Delta \omega = c/2l$, where c is the speed of light and l is the length of the cavity. When introducing coupling between the cavities, the spectral properties of the lasers

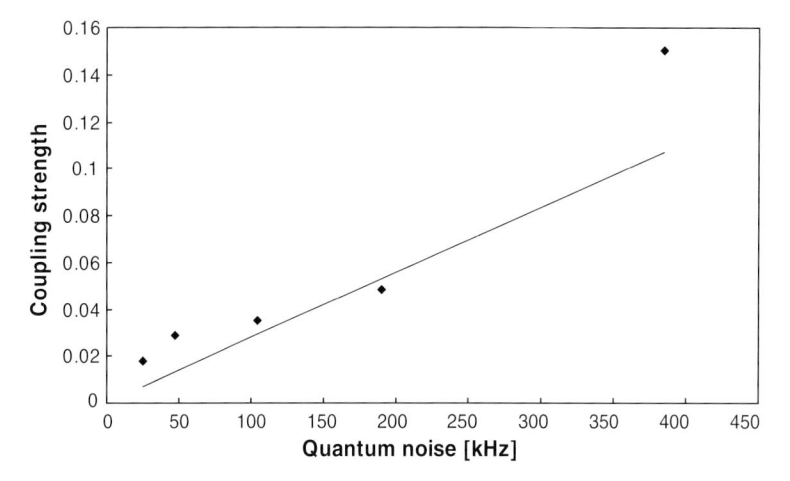

Figure 12.11 Coupling strength required to obtain 50% fringe visibility of interference between two fiber lasers as a function of quantum noise. Dots denote the experimental results and solid line denotes the analytic prediction.

change. Here, we show how the phase locking between the two lasers and their longitudinal mode spectrum vary as a function of the coupling strength [17].

The experimental configuration for determining the phase locking and the spectrum of longitudinal modes for two coupled fiber lasers as a function of the coupling strength between them is presented in Figure 12.12. Each fiber laser comprised a polarization-maintaining ytterbium-doped fiber, where one end was attached to a high-reflection fiber Bragg grating, with a central wavelength of 1064 nm and a bandwidth of about 1 nm, that served as a back reflector mirror and the other end was attached to a collimating GRIN lens with an antireflection coating to suppress any reflections back into the fiber cores, and an output coupler with reflectivity of 20% common to both lasers. The lasers were pumped with

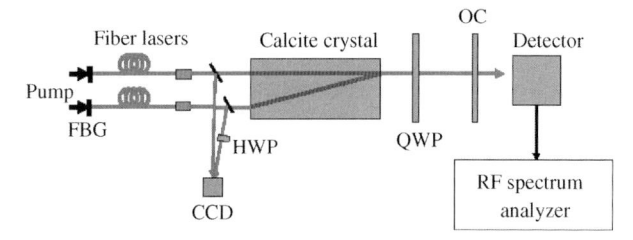

Figure 12.12 Experimental configuration for investigating the longitudinal modes of two fiber lasers as a function of the coupling strength. FBG, high-reflection fiber Bragg grating; HWP, half-wave plate; QWP, quarter-wave plate; OC, output coupler. The output beam is detected with a fast photodetector and an RF spectrum analyzer to determine the longitudinal mode spectrum of each laser. A small portion of each laser is sampled by 4% partially reflecting mirrors and interfered on a CCD camera to measure phase locking between the two lasers. The polarization of one beam is rotated using the HWP to be parallel to the other, enabling the two beams to interfere.

915 nm diode lasers of 300 mW from the back end through the FBG. The two fiber lasers were forced to operate in orthogonal polarizations by using a calcite beam displacer in front of a common output coupler, and the coupling strength between the lasers was controlled by an intracavity quarter-wave plate (QWP). The optical length of the cavity of one fiber laser was 10 m, while the optical length of the other was 11.5 m, so each fiber has 20 000 longitudinal modes within the FBG bandwidth. The combined output power of about 200 mW was detected by a fast photodetector, which was connected to an RF spectrum analyzer, to measure the beating frequencies and determine the longitudinal mode spectrum at the output [5]. We also measured the phase locking between the two fiber lasers by detecting the interference of a small part of the light from each laser with a CCD camera and determining the fringe visibility [6–10]. The longitudinal mode spectrum was measured first when $\theta = 0$ ($\kappa = 0$), and the measurement was sequentially repeated after rotating the QWP by 1° steps until we reached $\theta = 45°$ ($\kappa = 1$).

We developed a model for calculating the distribution of longitudinal modes and phase locking for the two coupled lasers. For each laser, the effective reflectivity [11] of self-reflection and the light coupled into it from the other laser was calculated self-consistently. The longitudinal mode spectrum was then derived from the total effective reflectivity of the two lasers. Our model can account for the full range of coupling strength between the lasers. The effective reflectivity resulting from the coupling to the other laser, for each laser, can be shown:

$$R^{\text{eff}}_{1,2} = \left(1 - r(1 - \sqrt{\kappa}) - \frac{r^2 \kappa e^{il_{2,1}k}}{1 - r(1 - \sqrt{\kappa})e^{il_{2,1}k}}\right)^{-1}, \tag{12.15}$$

where k denotes the propagation vector of the light, κ is the coupling strength between the two lasers, $l_{2,1}$ is the length of each laser, and r is the reflectivity of the output coupler. In our calculations, we used $r = 0.55$ rather than the experimental value of $r = 0.2$ to ensure that the width of the calculated longitudinal modes fits those of the experimental modes. This is justified because cold cavity models do not take into account gain competition, which tends to narrow the width of the longitudinal modes. We then sum over the round-trip propagations to obtain the self-consistent field for each laser:

$$R^{\text{eff}}_j e^{ikl_j} + \left(R^{\text{eff}}_j e^{ikl_j}\right)^2 + \cdots = \left(1 - R^{\text{eff}}_j e^{ikl_j}\right)^{-1}, \tag{12.16}$$

where $j = 1,2$ is the laser number.

Figure 12.13 shows the experimental and calculated longitudinal mode spectra as functions of coupling strength κ. Figure 12.13a shows the experimental results of the longitudinal mode spectrum as a function of coupling strength over a 200 MHz range, and Figure 12.13b shows the corresponding calculated results. The experimental and calculated results are also shown in greater detail for four specific coupling strengths ($\kappa = 0$, 0.28, 0.7, 1) in Figure 12.13c–f, respectively. Without coupling (i.e., $\kappa = 0$), two independent sets of frequency combs exist simultaneously – one corresponds to the 10 m long fiber laser (15 MHz separation between adjacent longitudinal modes), while the other corresponds to the 11.5 m long fiber laser

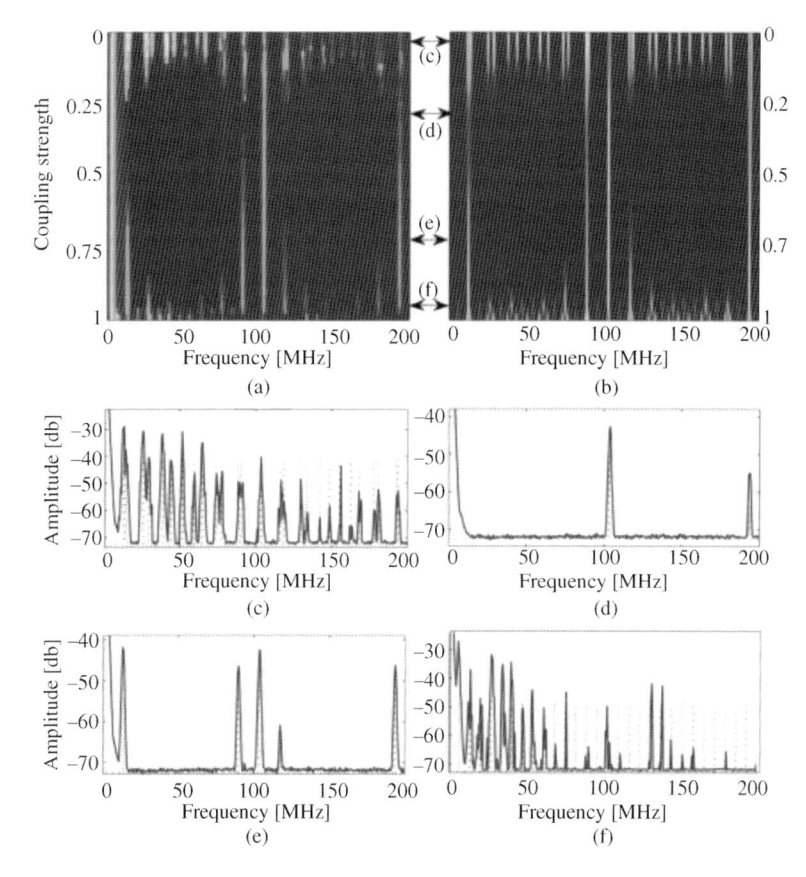

Figure 12.13 Experimental and calculated distributions of longitudinal mode beating frequencies in the output power for two coupled lasers as a function of the coupling strength κ. (a) Experimental results. (b) Calculated results. (c) $\kappa = 0$. (d) $\kappa = 0.28$. (e) $\kappa = 0.7$. (f) $\kappa = 1$. Solid curves denote experimental results, while dotted curves denote calculated results.

(13 MHz separation), as shown in Figure 12.13c. Each seventh longitudinal mode of the 10 m long laser is very close to the eighth mode of the other, so they are essentially common longitudinal modes. When κ is increased from 0 to 0.3, the longitudinal modes that are not common gradually disappear according to their detuning while transferring their energy to the remaining ones via the homogeneous broadening of the gain. The longitudinal modes with the larger detuning disappear first, while the ones with smaller detuning disappear for larger values, and only the common longitudinal mode remains, as seen in Figure 12.13d, indicating that at this coupling strength, there is full phase locking. As the coupling strength increases above 0.3, the longitudinal modes gradually reappear, as seen in Figure 12.13e. The longitudinal modes with smaller detuning reappear first, and the ones with larger detuning reappear for larger values of κ. Finally, when κ approaches unity, whereby all the light from one laser is transferred to the other, new

longitudinal modes appear between adjacent longitudinal modes, as seen in Figure 12.13f, corresponding to a single combined laser cavity whose length is the sum of the lengths of two lasers.

Figure 12.13 reveals a good quantitative agreement between the experimental and calculated results. In particular, the observed gradual disappearance of noncommon longitudinal modes as the coupling is increased, their gradual reappearance when the coupling is further increased, and finally the doubling of the frequency comb at near-unity coupling strength are all accurately reconstructed by our model.

12.3.4
Effects of Time-Delayed Coupling

It is well known that stable phase locking and synchronization can occur in two coupled lasers, where the time-delayed coupling between them is relatively short [27–30]. However, there are some situations, such as in secure communications, that require long time-delayed coupling [31, 32]. Accordingly, there have been extensive theoretical and experimental investigations into time-delayed coupling [33–36]. The emphasis has been on intensity synchronization of coupled lasers, but very little on phase locking with long time-delayed coupling. In particular, although phase locking of lasers with long time-delayed coupling was theoretically predicted some time ago [34], so far no experimental confirmation has been reported. Phase-locked lasers with long time delays can be useful for applications where two distant lasers require a well-defined relative phase, for example, in synchronizing optical clocks for time standard setting, in long baseline interferometry for optical telescopes with large effective apertures, and in other interferometric applications such as detection of gravity waves.

When coupling lasers, it is necessary to take into account the time it takes for the coupled light from one laser to reach the other. Two coupling regimes exist for time-delayed coupling. One is where the coupling delay time is shorter than the coherence time of the lasers so that the delay time has very little, if any, effect. The other is where the coupling delay time is longer than the coherence time of the lasers so that the effect of the delay time cannot be neglected. Here, we investigate the effects of long coupling delay times on phase locking of two coupled fiber lasers. Specifically, we compare two different arrangements for coupling and demonstrate that phase locking can occur even with time-delayed coupling of $20\,\mu s$ (delay line of 4 km). Such delays are much longer than the coherence length of our fiber lasers measured as 10 cm (0.3 ns) [18–20]. We find that with long time-delayed coupling, phase locking requires delayed self-feedback in addition to the delayed coupling signal [18].

Our basic experimental arrangements for investigating time-delayed coupling between two fiber lasers are schematically shown in Figure 12.14. Figure 12.14a shows one that contains a delayed self-feedback signal in addition to the delayed coupling signal. Figure 12.14b shows the other that contains delayed coupling only. Each fiber laser included a 10 m long Yb-doped double-clad polarization-maintaining fiber with a fiber Bragg grating at the rear. The fiber lasers were end-pumped with a diode laser through the FBG and the front ends of the fibers were cleaved at an

12.3 Effects of Amplitude Dynamics, Noise, Longitudinal Modes, and Time-Delayed Coupling

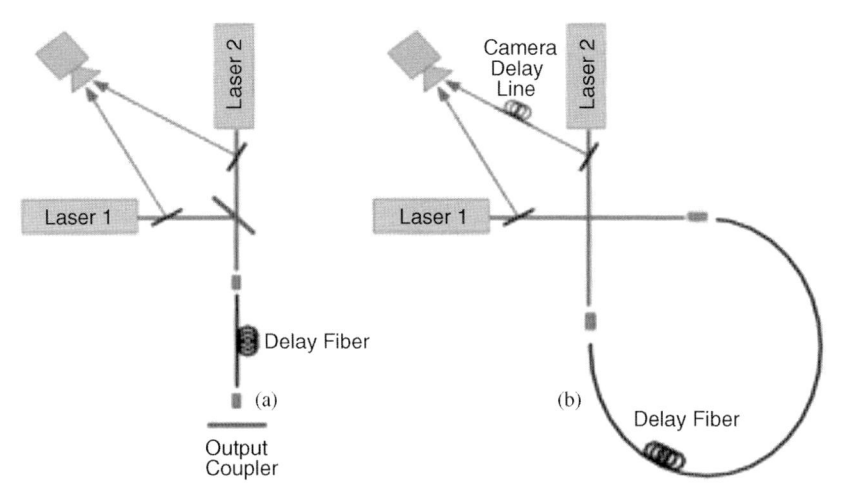

Figure 12.14 Experimental arrangements for phase locking two fiber lasers with time-delayed coupling. (a) Delayed coupling with delayed self-feedback. (b) Delayed coupling only.

angle to suppress any back reflections into the fiber core. Individual output couplers with $R = 20\%$ reflectivity were placed at the front of each fiber, resulting in two independent lasers.

For the arrangement shown in Figure 12.14a, the beams from the two lasers perfectly overlapped in angle and position after a 50% beam splitter. The light was then coupled into a long single-mode delay fiber. At the end of this fiber, a common output coupler reflected the light back into the delay fiber and then equally into both lasers. The time it takes to propagate the light from one laser to the other, namely, the coupling delay time τ_d, is approximately $\tau_d = 2nL_{fiber}/c$, where L_{fiber} is the length of the delayed fiber, n is the refractive index, and c is the speed of light. For the arrangement shown in Figure 12.14b, each laser was coupled to a different end of the delay fiber. The light from each laser propagated through the delay fiber and was then reinjected into the other laser. The coupling delay time in this arrangement was thus approximately $\tau_d = nL_{fiber}/c$. The coupling strength was controlled by varying the amount of light injected into the delay fiber. For both arrangements, the coupling strength between the lasers (the relative amount of energy that was transferred from one laser to the other) was set to be the same (approximately 8%). The time delay was controlled by varying the length of the delay fiber from 60 cm to 2 km, so as to get time delays from 0.01 to 20 µs. Some light from each laser was directed toward a CCD camera. From the fringe visibility of the interference pattern, we deduced the degree of phase locking [5–11].

First, we measured the fringe visibility V as a function of the coupling delay time for the first arrangement. In these measurements, we ensured that the distances from the camera to each one of the lasers are the same. Thus, a stable interference pattern indicates that both lasers are phase locked isochronally (at the same time). The experimental and numerical results of the normalized fringe visibility (normalized by the fringe visibility value with no delay) as a function of the delay time is

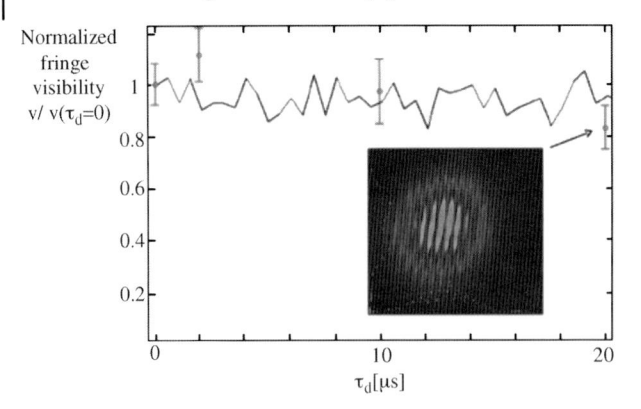

Figure 12.15 Results of normalized fringe visibility as a function of coupling delay time obtained for the arrangement that includes self-feedback. Points and bars denote experimental results and solid curve denotes numerical results. *Inset*: The interference pattern of two lasers coupled by a 4 km long delay line.

presented in Figure 12.15. Also shown in the inset is a representative example of the interference pattern of two coupled lasers with a coupling delay time of 20 μs, which corresponds to a delay fiber length of 2 km. The actual experimental values of the fringe visibility ranged from 0.4 to 0.5 rather than nearly 1 as we found earlier at short coupling delays [18]. This is mainly due to the fact that the FBG in the two lasers in our current experiment differed, so their central wavelengths were not the same. Thus, part of the light that was detected by the camera did not participate in the actual phase locking and just served as a bias. Nevertheless, the fringe visibility and fringe position remained constant for long periods of time over the entire range of delay times [18–20].

For the second arrangement, the fringe visibility was very poor, indicating no isochronal (at the same time) phase locking. Thus, we resorted to achronal (at different times) phase locking, which is indicated by a stable interference pattern of the light from one laser with delayed light from the other. Experimentally, this was done by letting the optical distance from one of the lasers to the camera be longer than that of the other laser, as shown in Figure 12.14b. In our experiment, we used the same delay fiber for delaying the coupling signal as well as for the main delaying of the detection signal.

Using a long delay fiber of length of $L_{delay} = 200$ m, we measured the fringe visibility for several values of the camera delay line. The results are presented in Figure 12.16. As evident from the figure, the maximal phase locking occurs at a distance of L_{delay}, which is the overall optical distance along which the coupling signal propagates. Also shown in Figure 12.16 is the fringe visibility decay width of several centimeters. The fringe visibility decay width corresponds to the finite coherence length of a single-fiber laser, which was independently measured, with the aid of a Michelson interferometer, to be about 10 cm.

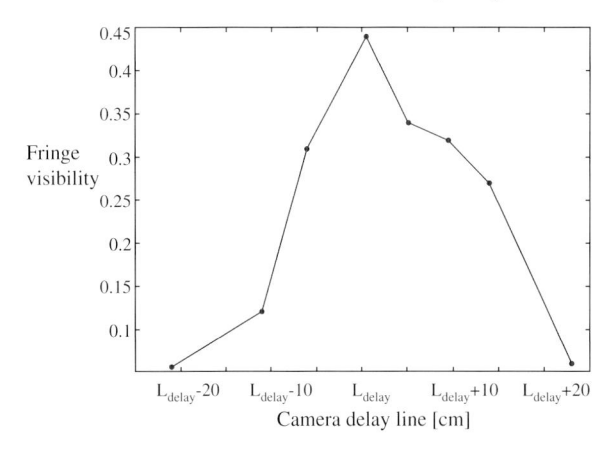

Figure 12.16 Experimental results for the fringe visibility as a function of the camera delay line and for the arrangement that contain delayed coupling only.

12.4
Upscaling the Number of Phase-Locked Fiber Lasers

In order for coupled fiber lasers to phase lock, they must have at least one longitudinal mode within the bandwidth of their gain that is common to all lasers. As the number of coupled lasers increases, the probability of having such common longitudinal modes drops exponentially. Accordingly, there are several theoretical predictions that set upper limits of 8–12 fiber lasers that can be efficiently phase locked [16]. Nevertheless, experimentalists are attempting to exceed such upper limits, but so far with no success. In this section, we describe our developments, investigations, and results on upscaling the number of lasers that can be passively phase locked and combined. We start with a simultaneous coherent and spectral combining approach that could potentially lead to significant upscaling and then actual phase locking of 25 fiber lasers, which allowed us to investigate in some detail the phase locking efficiency as a function of a number of fiber lasers and the connectivity arrangement between them. Some of our results with 25 fiber lasers also have a bearing on extreme value statistics.

12.4.1
Simultaneous Spectral and Coherent Combining

One way to overcome the limitation on the number of fibers that can be coherently combined is to simultaneously use spectral combining. In spectral combining, the beams emerging from the individual lasers, each operating at a slightly different wavelength, are combined incoherently by means of a linear diffraction grating; so all the lasers must usually be aligned along one dimension [21].

Here, we present a configuration in which we simultaneously perform spectral and coherent combining, so as to enable combining of a two-dimensional fiber laser

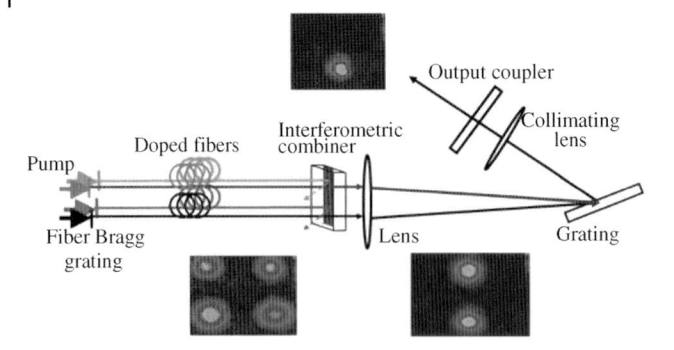

Figure 12.17 Basic configurations for adding four fiber lasers. In the horizontal direction, coherent combining is performed by means of an interferometric combiner; and in the vertical direction, spectral addition is performed by means of a linear diffraction grating. *Insets*: The four beams emerging from the fibers, the two beams after the interferometric combiner, and the single combined output beam after the grating.

array. In one dimension we obtain coherent combining by means of an interferometric combiner in the free space [7–11], while in the other dimension we obtain spectral combining by means of a diffraction grating [21]. This simultaneous approach of coherent and spectral combining opens an alternative route to upscaling to very large arrays of lasers, overcoming the upscaling limitation of each individual approach.

Our configuration is presented schematically in Figure 12.17. It contains four end-pumped single-mode erbium-doped CW fibers arranged in 2 × 2 square array with a distance of 3.5 mm between the adjacent fibers. Each fiber has 10 μm core diameter with numerical aperture of 0.19 and a length of about 7 m. One end of each fiber is attached to a fiber Bragg grating that serves as the rear mirror. The other fiber end is connected to a collimator coated with antireflection layers to suppress any reflections back into the fiber. The four parallel beams emerging from the fiber collimators are first coherently added in the horizontal direction by means of an intracavity passive interferometric combiner to obtain two parallel beams [5]. These beams are then deflected in the vertical direction by means of a lens so that they exactly overlap on the surface of a 1200 lines/mm blazed diffraction grating and diffracted toward a common planar output coupler of 20% reflectivity. The wavelength of each beam is self-selected to ensure normal incidence on the output coupler and consequently exact self-reflection back into the fibers in order to obtain lasing [38]. Thus, the two beams are spectrally added into a single output beam, maintaining the high beam quality of each of the individual lasers.

We measured the beam quality factor M^2, the combining efficiency, and the spectra of the individual fiber laser outputs and combined output. To determine M^2, two CCD cameras and a Spiricon laser beam analyzer were used for detecting and characterizing the near- and far-field intensity distributions and then M^2 was calculated in accordance with $M^2 = \sigma_{nf}\sigma_{ff}(\pi/4)\lambda F$, where F is the focal length of the lens used to detect the far field and σ_{nf} and σ_{ff} are the second moments of the

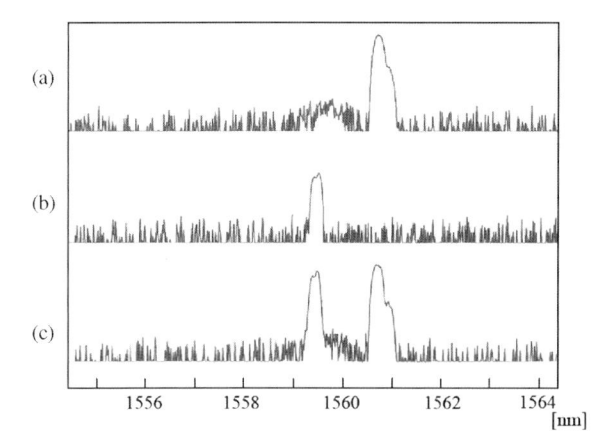

Figure 12.18 The spectra of the fiber lasers and the combined output, using a blazed diffraction grating of 1200 lines/mm for spectral addition. (a) Spectrum when only the upper pair of lasers are coherently added. (b) Spectrum when only the lower pair of lasers are coherently added. (c) Spectrum of the spectrally combined output.

near and far fields, respectively. We found that $M^2 = 1.15$ for the combined output beam, essentially identical to that of each individual fiber laser.

The combining efficiency of the coherent combining was determined by measuring only the power of the overall light with and then without the interferometric combiner and then by calculating their ratio. The overall combining efficiency, after both coherent combining and spectral addition, was similarly measured and found to be 82%. We mainly attribute the reduction of efficiency from 100% to losses to undesired diffraction orders from the blazed grating and also to small residual misalignment between the fiber lasers.

The spectra of the fiber lasers and the combined output were measured by means of a grating spectrometer. The results are presented in Figure 12.18. Figure 12.18a shows the expected single spectrum when only the upper pair of lasers are coherently added. Figure 12.18b shows the corresponding spectrum when only the lower pair of lasers are coherently added. As evident from the figure, each pair of lasers operates at a different wavelength, with a spectral separation of 1.3 nm. Finally, Figure 12.18c shows the spectrum of the combined output beam. Here, there is a simple addition of the individual spectrum of each pair of lasers, indicating that there is no interaction between them. Accordingly, both coherent combining and spectral addition occur simultaneously.

12.4.2
Phase Locking 25 Fiber Lasers

When trying to phase lock fiber lasers whose length cannot be accurately controlled, the probability of having common longitudinal modes rapidly decreases as the number of fiber lasers increases [37–40]. Nevertheless, we coupled 25 fiber lasers in order to check it experimentally. The experimental configuration is described in detail in Ref. [22]. We measured the phase locking level as a function of time for different number of lasers in the array and for different connectivities. This was done by continuously detecting the far-field intensity distribution of the total output light from the array with a CCD camera, determining the maxima and minima intensities, and calculating the average fringe visibility along the x- and y-directions. The fringe visibility provides a direct measure of the phase locking level that ranges from 0 to 100%. The measurements were performed over a period of 10 h to obtain about 300 000 measurements. These measurements were then repeated for different numbers of lasers and connectivities. We used the same effective reflectivity model to calculate the maximal effective reflectivity as a function of a number of lasers in the array. This model takes into account reflections from the coupling mirrors and from the front FBGs. Then we repeated the calculations 1000 times, each time choosing a different random realization of the fiber laser lengths, and determined the average of the results.

Figure 12.19 shows the average phase locking level (plus symbols) and the maximal phase locking level (asterisks) as a function of the number of lasers in the array for the 2D connectivity. It also includes representative far-field intensity distributions of 2 fiber lasers [inset (a)], 25 fiber lasers with average low phase locking level [inset (b)], and 25 fiber lasers with instantaneous maximal phase locking level [inset (c)]. These results indicate that as the number of lasers in the array increases, the probability to find common longitudinal modes rapidly drops as predicted. However, since the length of each fiber laser fluctuates randomly (modulus λ) due to thermal and acoustic variations, there is a certain probability of briefly obtaining a common longitudinal mode for all fiber lasers. We found that

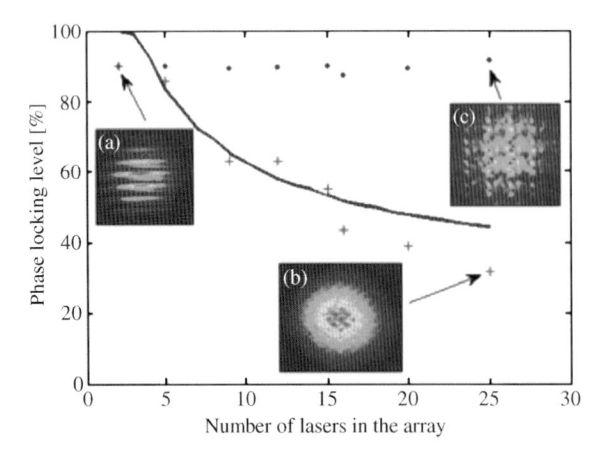

Figure 12.19 Experimental and calculated results of the average and the maximal phase locking levels as a function of a number of lasers in the array. Asterisks denote maximal phase locking level and plus symbols denote average phase locking level. Solid curve denotes calculated average phase locking level using effective reflectivity model. *Insets*: The far-field intensity distributions corresponding to specific data points.

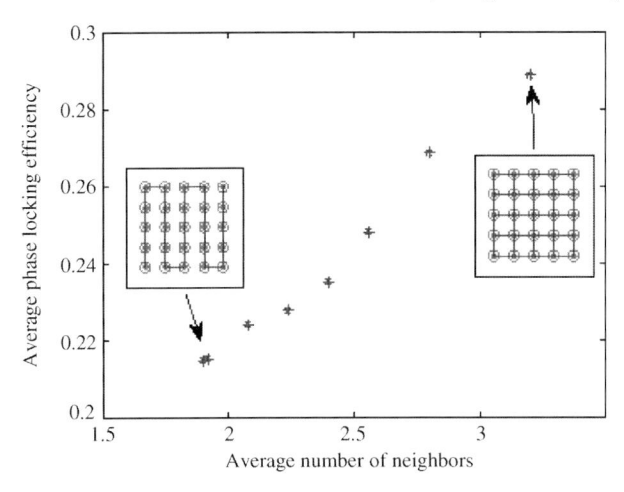

Figure 12.20 Experimental results of the phase locking level for 25 fiber lasers as a function of the average number of coupled neighbors to each laser.

the probability of obtaining phase locking levels above 90% drops rapidly. Specifically, the probability is 0.1% for 12 lasers, 0.012% for 16 lasers, 0.004% for 20 lasers, and 0.001% for 25 lasers. These results could probably be improved due to nonlinear effects by resorting to better alignment of component and higher power [2].

We also determined how the average phase locking level is related to the connectivity of the fiber lasers in the array [23, 24]. Specifically, we measured the average phase locking level of an array of 25 fiber lasers as a function of the average number of coupled neighbors to each fiber laser for different coupling connectivities. The results are presented in Figure 12.20. We started with 1D connectivity of the full array, shown in the left inset, in which the average number of coupled neighbors to each fiber laser is only 1.9. Then, we varied the connectivity and increased the average number of coupled neighbors and measured the average phase locking level of the array in each case, up to a 2D connectivity, where the average number of coupled neighbors is 3.2. As evident from the figure, there is a monotonic increase in the average phase locking level of the array from 21 to 29%. These results manifest that connectivity influences the phase locking level, consistent with the expected increase of an order parameter with dimensionality for all coupled oscillators.

Another interesting aspect is the total phase locking of the array as a function of time and not just the average. We measured the fringe visibility of the far-field pattern of the 25 fiber lasers. The fringe visibility provides a direct measure of the phase locking level that ranges from 0 to 1. The correlation time of the phase locking level is shorter than 100 ms; so over a 10 h period, we acquired about 370 000 uncorrelated measurements of the fringe visibility.

Representative experimental results of the fringe visibility as a function of time for a 10 s interval are presented in Figure 12.21. The insets show two typical far-field intensity distributions: one with low fringe visibility and the other with high fringe

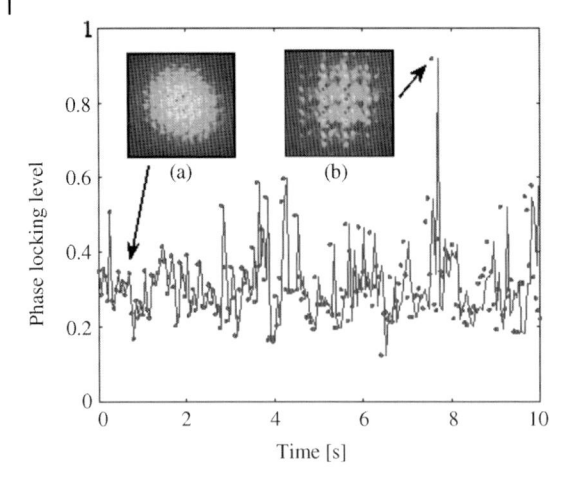

Figure 12.21 Typical experimental results of the phase locking level as a function of time over a 10 s interval. The phase locking level was determined from the far-field intensity distribution of the output. *Insets*: The typical far-field intensity distributions. (a) Low fringe visibility where the phase locking level is low. (b) High fringe visibility where the phase locking level is high.

visibility. Due to thermal and acoustic fluctuations, the length of each fiber laser and its corresponding eigenfrequencies changes rapidly and randomly. Phase locking minimizes loss in the array, so mode competition will favor frequencies that maximize the size of the phase-locked clusters at each moment [22]. Since the distribution of the phase locking level for different frequencies is Gaussian, the statistics of the maximum phase locking level should be described by the Gumbel distribution function.

To check for possible correlations hidden in the experimental results, we fitted the phase locking level distribution with the Bramwell–Holdsworth–Pinton (BHP) distribution using a single fitting parameter C_1 [39]. For highly correlated systems the parameter C_1 approaches $\pi/2$, while for uncorrelated systems the parameter C_1 has an integer value. In particular, when $C_1 = 1$, the BHP distribution reduces to the Gumbel extreme value distribution. The functional form of the BHP distribution is given by

$$P(x) = e^{C_1\left([(x-\mu)/\sigma] - e^{(x-\mu)/\sigma}\right)}, \tag{12.17}$$

where μ denotes the mean value, σ is the width of the distribution, and c is the measure for correlations. After fitting the measured probability distribution of the phase locking level of 25 coupled fiber lasers to the BHP distribution, we obtained $C_1 = 1.03$, indicating a Gumbel distribution.

Fitting the Gumbel distribution to the experimental results for 12, 16, and 20 fiber lasers was not as good as for the 25 fiber lasers. The distribution of the experimental results for low number of lasers is clamped because μ and σ are higher and the distribution of the phase locking level is bound between 0 and 1. Therefore, we

12.4 Upscaling the Number of Phase-Locked Fiber Lasers

resort to the generalized extreme value (GEV) distribution that contains one extra parameter: the shape parameter ξ. When $\xi = 0$, the GEV distribution reduces to the Gumbel distribution; but as $\xi < 0$, the GEV distribution is clamped and approaches the Weibull distribution. The GEV distribution is given by

$$P(x) = \frac{1}{\sigma}\left[1 + \xi\left(\frac{x-\mu}{\sigma}\right)\right]^{-(1/\xi)-1} e^{-[1+\xi((x-\mu)/\sigma)]^{-1/\xi}}, \tag{12.18}$$

where μ, σ, and ξ were calculated from the experimental data by maximum-likelihood parameter estimation. We expect that as the number of fiber lasers in the array decreases from 25, the mean value and standard deviation of the phase locking level would increase, and the ξ parameter should vary from zero to a negative value. The measured phase locking level histograms for laser arrays with 12, 16, 20, and 25 fiber lasers and the corresponding GEV distributions with the calculated parameters and their 95% confidence intervals are presented in Figure 12.22. As evident from the figure, there is a very good agreement between the experimental results and the GEV distributions, extending over three decades. As expected, as the number of fibers increases, the values of μ and σ decrease and the ξ parameter

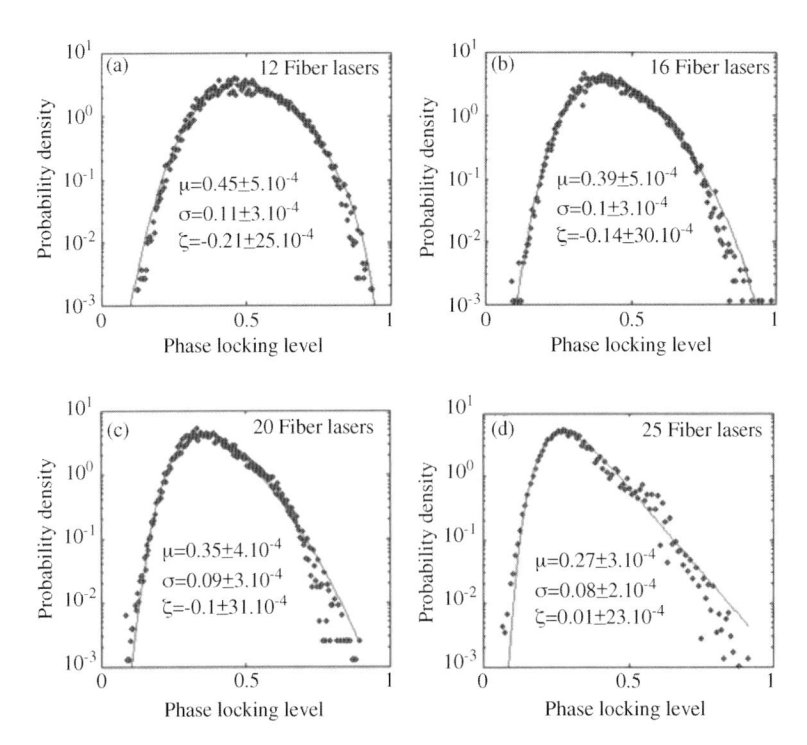

Figure 12.22 Measured phase locking level histogram for four array sizes of fiber lasers. (a) Twelve fiber lasers. (b) Sixteen fiber lasers. (c) Twenty fiber lasers. (d) Twenty-five fiber lasers. Curves denote associated GEV distributions without any fitting parameters. Maximum-likelihood parameter estimation was calculated directly from the data, and 95% confidence interval of each parameter is also shown.

approaches 0. For 25 fiber lasers, the ξ parameter reaches 0.01, close to the expected 0 value where the GEV distribution reduces to the Gumbel distribution.

12.5
Conclusion

We have presented our investigations on passive phase locking and coherent combining of fiber lasers and demonstrated high combining efficiency and good quality of the combined output beams for a variety of configurations and wavelengths. Since the relative phase between the lasers is self-adjusted to minimize losses of the coupled lasers system, passive phase locking is rather robust even under variable environmental conditions as long as common frequencies exist among all the fiber lasers. This typically occurs when the number of lasers is smaller than 10 and, for higher number of lasers, some control on the length of each laser is required.

We would like to thank Vardit Eckhouse, Amiel Ishaaya, Liran Shimshi, Eitan Ronen and Rami Pugatch for helpful discussions and support to the work presented here.

References

1 Jeong, Y., Sahu, J., Payne, D., and Nilsson, J. (2004) Ytterbium-doped large-core fiber laser with 1.36 kW continuous-wave output power. *Opt. Express*, **12**, 6088–6092.

2 Corcoran, C.J. and Durville, F. (2009) Passive phasing in a coherent laser array. *IEEE J. Quantum Electron.*, **15**, 294.

3 Leger, J.R. (1993) External methods of phase locking and coherent beam addition of diode lasers, in *Surface Emitting Semiconductor Lasers and Arrays* (eds G.A. Evans and J.M. Hammer), Academic Press, Boston, MA, pp. 379–433.

4 Siegman, E. (2000) Laser beams and resonators: beyond the 1960s. *IEEE J. Sel. Top. Quantum Electron.*, **6**, 1389–1399.

5 Friedman, M., Eckhouse, V., Friesem, A.A., and Davidson, N. (2007) Efficient coherent addition of fiber lasers in free space. *Opt. Lett.*, **32**, 790–792.

6 Eckhouse, V., Fridman, M., Davidson, N., and Friesem, A.A. (2008) Loss enhanced phase locking in coupled oscillators. *Phys. Rev. Lett.*, **100**, 2.

7 Ishaaya, A., Shimshi, L., Davidson, N., and Friesem, A.A. (2004) Intra-cavity coherent addition of Gaussian beam distributions using a planar interferometric coupler. *Appl. Phys. Lett.*, **85**, 2187.

8 Ishaaya, A., Shimshi, L., Davidson, N., and Friesem, A.A. (2004) Coherent addition of spatially incoherent laser beams. *Opt. Express*, **12**, 4929.

9 Ishaaya, A.A., Eckhouse, V., Shimshi, L., Davidson, N., and Friesem, A.A. (2005) Improving the output beam quality of multimode laser resonators. *Opt. Express*, **13**, 2722.

10 Ishaaya, A.A., Eckhouse, V., Shimshi, L., Davidson, N., and Friesem, A.A. (2005) Intra-cavity coherent addition of single high order modes. *Opt. Lett.*, **230**, 1770–1772.

11 Eckhouse, V., Ishaaya, A.A., Shimshi, L., Davidson, N., and Friesem, A.A. (2006) Intracavity coherent addition of 16 laser distributions. *Opt. Lett.*, **31**, 50.

12 Shimshi, L., Ishaaya, A.A., Ekhouse, V., Davidson, N., and Friesem, A.A. (2006) Passive intra-cavity phase locking of laser channels. *Opt. Commun.*, **263**, 60–64.

13 Fridman, M., Eckhouse, V., Luria, E., Krupkin, V., Davidson, N., and Friesem, A.A. (2008) Coherent addition of two dimensional array of fiber lasers. *Opt. Commun.*, **281**, 6091.

14 Nixon, M., Fridman, M., Friesem, A.A., and Davidson, N. (2011) Enhanced coherence of weakly coupled lasers. *Opt. Lett.*, **36**, 1320.

15 Fridman, M., Eckhouse, V., Davidson, N., and Friesem, A.A. (2008) The effect of quantum noise on coupled laser oscillators. *Phys. Rev. A*, **77**, 061803(R).

16 Rothenberg, J.E. (2009) Optical Fiber Communication Conference, OSA Technical Digest (CD), Optical Society of America, Paper OTuP3.

17 Fridman, M., Nixon, M., Ronen, E., Friesem, A.A., and Davidson, N. (2010) Phase locking of two coupled lasers with many longitudinal modes. *Opt. Lett.*, **35**, 526.

18 Nixon, M., Fridman, M., Ronen, E., Friesem, A.A., and Davidson, N. (2009) Phase locking of two fiber lasers with time-delayed coupling. *Opt. Lett.*, **34**, 1864.

19 Nixon, M., Fridman, M., Ronen, E., Friesem, A.A., Davidson, N., and Kanter, I. (2011) Synchronized cluster formation in coupled laser networks. *Phys. Rev. Lett.*, **106**, 22.

20 Eckhouse, V., Nixon, M., Fridman, M., Friesem, A.A., and Davidson, N. (2010) Synchronization of chaotic fiber lasers with reduced external coupling. *IEEE J. Quantum Electron.*, **46**, 1821.

21 Fridman, M., Eckhouse, V., Davidson, N., and Friesem, A.A. (2008) Simultaneous coherent and spectral addition of fiber lasers. *Opt. Lett.*, **33**, 648.

22 Fridman, M., Nixon, M., Ronen, E., Friesem, A.A., and Davidson, N. (2010) Passive phase locking of 25 fiber lasers. *Opt. Lett.*, **35**, 1434.

23 Fridman, M., Pugatch, R., Nixon, M., Friesem, A.A., and Davidson, N. (2012) Measuring maximal eigenvalue distribution of Wishart random matrices with coupled lasers. *Phys. Rev. E*, **85**, 020101(R).

24 Fridman, M., Pugatch, R., Nixon, M., Friesem, A.A., and Davidson, N. (2012) Phase locking level statistics of coupled random fiber lasers. *Phys. Rev. E*, **86**, 041142.

25 Eckhouse, V., Fridman, M., Davidson, N., and Friesem, A.A. (2008) Phase locking and coherent combining of high-order-mode fiber lasers. *Opt. Lett.*, **33**, 2134.

26 Ronen, E., Fridman, M., Nixon, M., Friesem, A.A., and Davidson, N. (2008) Phase locking of lasers with intracavity polarization elements. *Opt. Lett.*, **33**, 2305.

27 Fabiny, L., Collet, P., Roy, R., and Lenstra, D. (1993) Coherence and phase dynamics of spatially coupled solid-state lasers. *Phys. Rev. A*, **47**, 4287–4296.

28 Thornburg, K.S., Mueller, M., Roy, R., Carr, T.W., Li, R.D., and Erneux, T. (1997) Chaos and coherence in coupled lasers. *Phys. Rev. E*, **55**, 3865.

29 Roy, R. and Thornburg, K.S. (1994) Experimental synchronization of chaotic lasers. *Phys. Rev. Lett.*, **72**, 2009.

30 Argyris, A., Syvridis, D., Larger, L., Annovazzi-Lodi, V., Colet, P., Fischer, I., Garcia Ojalvo, J., Mirasso, C.R., Pesquera, L., and Shore, K.A. (2005) Chaos-based communications at high bit rates using commercial fibre-optic links. *Nature*, **438**, 343–346.

31 Klein, E., Gross, N., Rosenbluh, M., Kinzel, W., Khaykovich, L., and Kanter, I. (2006) Stable isochronal synchronization of mutually coupled chaotic lasers. *Phys. Rev. E*, **73**, 066214.

32 Tang, S. and Liu, J.M. (2003) Experimental verification of anticipated and retarded synchronization in chaotic semiconductor lasers. *Phys. Rev. Lett.*, **90**, 194101.

33 Tang, S., Vicente, R., Chiang, M., Mirasso, C., and Liu, J.-M. (2004) Nonlinear dynamics of semiconductor lasers with mutual optoelectronic coupling. *IEEE J. Sel. Top. Quantum Electron.*, **10**, 936.

34 Rosenblum, M.G., Pikovsky, A.S., and Kurths, J. (1997) From phase to lag synchronization in coupled chaotic oscillators. *Phys. Rev. Lett.*, **78**, 4193.

35 Kozyreff, G., Vladimirov, A.G., and Mandel, P. (2000) Global coupling with time delay in an array of semiconductor lasers. *Phys. Rev. Lett.*, **85**, 3809.

36 Kim, S., Lee, B., and Kim, D.H. (2001) Experiments on chaos synchronization in

two separate erbium-doped fiber lasers. *IEEE Photon. Technol. Lett.*, **13**, 290.

37 Fan, T.Y. (2005) Laser beam combining for high-power, high-radiance sources. *IEEE J. Quantum Electron.*, **11**, 567.

38 Ishaaya, A.A., Davidson, N., and Friesem, A.A. (2009) Passive laser beam combining with intracavity interferometric combiners. *IEEE J. Sel. Top. Quantum Electron.*, **15**, 301.

39 Cheung, E.C., Ho, J.G., Goodno, G.D., Rice, R.R., Rothenberg, J., Thielen, P., Weber, M., and Wickham, M. (2008) Diffractive-optics-based beam combination of a phase-locked fiber laser array. *Opt. Lett.*, **33**, 354.

40 D'Amato, F.X., Siebert, E.T., and Roychoudhuri, C. (1989) Coherent operation of an array of diode lasers using a spatial filter in a Talbot cavity. *Appl. Phys. Lett.*, **55**, 816.

13
Intracavity Combining of Quantum Cascade Lasers
Guillaume Bloom, Christian Larat, Eric Lallier, Mathieu Carras, and Xavier Marcadet

13.1
Introduction

The mid-infrared (MIR) spectral range (3–5 μm) is currently explored in a large number of scientific, industrial, medical, and defense applications. Most of these applications require an output power of several tens of watts along with a good beam quality. Until now, powerful sources in the 3–5 μm range were mainly based on nonlinear optics. With an optical parametric oscillators (OPOs) based on the $ZnGeP_2$ (ZGP) nonlinear crystal, an average output power of 14 W was obtained at 4.5 μm wavelength with good beam quality [1,2]. However, the performances recently obtained from quantum cascade lasers (QCLs) [3] bring new promising perspectives.

QCLs are semiconductor lasers that use optical transitions between subbands inside the conduction band. The use of these so-called intersubband transitions implies several fundamental physical differences as compared with the classical semiconductor lasers based on interband transitions. The emission wavelength is no more determined and forced by the energy band gap of the semiconductor materials used. It only depends on the design of the complex semiconductor heterostructure of the active zone of the QCL. Thus, it is possible through an appropriate "quantum engineering" of the active zone to obtain optical emission from ~3 μm [4] to the far infrared [5] with well-known III–V semiconductor materials. QCLs are one of the most promising sources in the MIR spectral range and, since their first demonstration in 1994 at Bell Laboratories [3], a huge effort has been made to improve their performances. To date, QCLs reach easily the watt level in the continuous-wave (CW) regime at room temperature (RT). In Ref. [6], the authors report an output power of 5.1 W in CW operation at RT obtained from an AlInAs/GaInAs/InP heterostructure emitting at around 4.9 μm. Compared to the complex architecture developed for an OPO, the QCL takes advantage of the simplicity and compactness of an electrically pumped semiconductor laser. In order to increase even further the output power level, several methods based on incoherent beam addition of QCLs have been proposed. These techniques include the polarization combining [7] or the spectral beam combining [8,9].

The purpose of this chapter is to present the passive coherent beam combining (CBC) of several individual QCLs in order to increase the available power while keeping a nearly diffraction beam quality. A simple way to coherently combine two lasers with high combining efficiency remains is the Michelson cavity architecture. We have recently developed a Michelson external cavity to achieve the CBC of two QCLs with a combining efficiency of 85% and a good beam quality [10]. In order to combine more emitters, an N-arm external cavity with a $N–1$ beam combiner is designed. Such architectures have already been implemented to demonstrate the coherent beam addition of six GaAlAs laser diodes [11]. Here, a five-arm resonator using an intracavity beam combiner is developed to demonstrate the CBC of five QCLs emitting at a wavelength of 4.6 μm in CW regime at RT.

In Section 13.2, the principle of the passive coherent addition of lasers in N-arm external cavities is introduced. A simple model, based on previous theoretical studies of such cavities, is developed to estimate the combining efficiency that could be obtained from this method. In Section 13.3, the results and observations deduced from the experimental demonstration of the CBC of individual QCLs in a five-arm cavity with a binary phase grating [or Dammann grating(DG)] used as an intracavity beam combiner are presented. To improve the output power obtained with this method, we will show in Section 13.4 that the DG can be advantageously replaced by a more complex beam combiner exhibiting a higher combining efficiency.

13.2
External Cavity Passive Coherent Beam Combining

The beam combining techniques can be separated in two families: the incoherent beam combining where the intensities of the different beams are added (e.g., polarization or spectral beam combining) and the coherent beam combining where the amplitudes of the different beams are added (e.g., evanescent coupling or coupling in external cavities). A few examples of these methods have already been implemented with QCLs in the MIR spectral range. In Ref. [8], the authors report the spectral beam combining of six QCLs emitting at around 4.6 μm in pulsed regime with a combining efficiency of 50%. In Ref. [9], the spectral beam combining of 28 distributed feedback QCLs at around 9 μm is demonstrated with a combining efficiency of 55%. Concerning the CBC, a team in Austria has worked on monolithic tree array architectures to combine QCLs in pulsed regime emitting at around 10 μm. Demonstration with a six-branch tree array was presented in Ref. [12]. Unfortunately, the combining efficiency measured was limited with this type of monolithic cavity (<16%). As presented in Ref. [10], a Michelson external cavity was developed to demonstrate the passive CBC of two QCLs. A combining efficiency of 85% was obtained with a good beam quality.

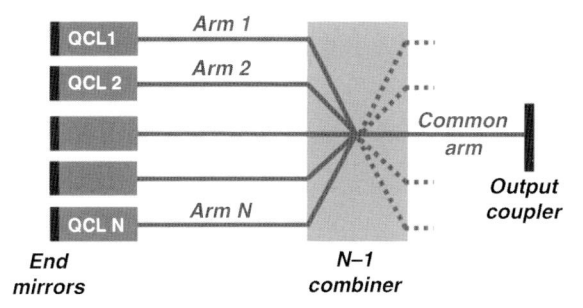

Figure 13.1 General architecture of passive coherent beam combining in external cavity.

13.2.1
Laser Scheme

Here, we are interested in passive coherent combining in external cavities with a N–1 coupler. The general architecture is presented in Figure 13.1. The output facets of the QCL are antireflection (AR) coated in order to prevent from self-oscillation and thus facilitate their phase locking in the external cavity. In the proposed architecture, the N emitters share the same cavity. They are placed in the N entry inputs of a N–1 beam combiner inserted inside the external cavity. The common cavity has a unique output coupler (OC) (mirror with 30% reflectivity) that is placed on the central output arm of the beam combiner, the *common arm*. The other output arms of the beam combiner (dotted lines in Figure 13.1) are loss outputs of the common cavity. Finally, the N-arm cavity extends from the N back facets of the QCLs, with a highly reflective (HR) coating, and the OC. Since the N–1 coupler is inserted in the cavity, it is used, at the same time, to combine the N beams coming from the N emitters and to separate the unique beam coming back from the OC into N beams that will be reinjected in the N QCLs.

CBC in N-arm resonators is based on loss minimization: The global cavity will oscillate on the mode that presents the lowest losses. Thus, to ensure that the beam combiner couples efficiently the N arms to the common arm, the global cavity will force the phase locking of the N emitters and select the right relative phase relationship between the N beams so that there are constructive interferences in the common arm (and destructive interferences in the other output arms). Thus, the combining efficiency of the N–1 coupler is maximized and the losses in the global cavity are minimized. This self-organization phenomenon is passive, since it is based only on loss minimization in the N-arm laser cavity. Finally, the common cavity selects and locks on the longitudinal mode(s) that correspond to both minimal losses and a maximal gain of the emitters. However, it is possible that no resonance of the global cavity, corresponding to perfect constructive interferences in the common arm, exists within the gain bandwidth of the QCLs. The system will still select the longitudinal mode that presents the lowest losses. However, the selected mode will not allow the N–1 coupler to work at its maximal combining efficiency. Consequently, a part of the available power coming from the N emitters will be coupled to the loss output arms so that the power in the combined beam is decreased. In the following sections, a model, derived from Refs [13,14], is introduced. The model will permit to quantify this decrease of combining efficiency with respect to the geometrical parameters of the cavity and the gain spectral bandwidth.

13.2.2
Modeling of the Coherent Beam Combining in External Cavity

The model used to describe the coherent beam combining and the self-organization phenomenon in external cavities is based on the circulating field theory [14]. The goal here is to calculate the spectral response or effective reflectivity $R_{eq,cav}(\lambda)$ of the N-arm resonator for all wavelengths λ in the spectral bandwidth B of the gain medium ($B \sim 100$ nm for QCLs). The circulating field theory allows determining the losses associated with any wavelength within B. Only the longitudinal modes with a high combining efficiency level will be effectively amplified in the global cavity.

13.2.2.1 The Michelson Cavity

We first present an example of calculation in the simple case of the Michelson cavity that we will then generalize to the N-arm configuration. We consider a unitary field E_{in} corresponding to a wavelength λ, incident on the OC from the outside of the global cavity (see Figure 13.2). Then, we let this field circulate inside the cavity and we calculate the field E_{out} that is reflected back. The global effective reflectivity of the cavity is then defined as

$$R_{eq,cav}(\lambda) = \left|\frac{E_{out}}{E_{in}}\right|^2. \tag{13.1}$$

The Michelson cavity is a three-mirror resonator formed by mirrors M_0 (the 30% OC), M_1, and M_2 (the two HR-coated rear facets of the QCLs) with respective amplitude reflectivity of r_0, r_1, and r_2 (in our case, $r_1 = r_2 = r$). The two arms of the resonator with optical lengths of L_1 and L_2 are coupled to a common arm with an optical length L_0. The 2 to 1 coupler is a 50/50 beam splitter.

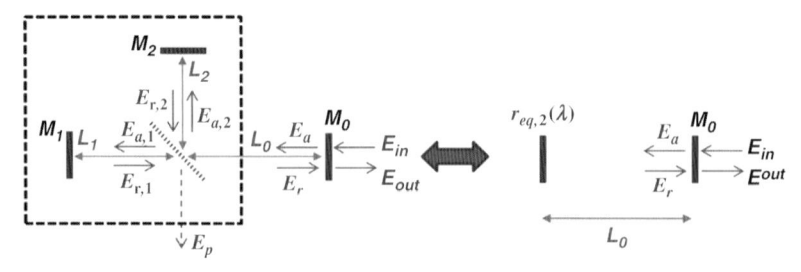

Figure 13.2 The circulating field method applied to the Michelson cavity case: scheme and notations. *Dashed box*: "interferometric" part of the Michelson cavity.

According to the circulating field theory, it can be shown that

$$E_r = \frac{r}{2}\left(e^{j(2\pi/\lambda)2L_1} + e^{j(2\pi/\lambda)2L_2}\right)E_a, \tag{13.2}$$

where the fields E_r and E_a are defined in Figure 13.2. Then, we define the effective reflectivity $R_{eq,2}(\lambda)$ (and the amplitude effective reflectivity $r_{eq,2}(\lambda)$) of the "interferometric" part of the cavity (see the dashed lines in Figure 13.2):

$$r_{eq,2}(\lambda) = \frac{E_r}{E_a} = \frac{r}{2}\left(e^{j(2\pi/\lambda)2L_1} + e^{j(2\pi/\lambda)2L_2}\right).$$
$$R_{eq,2}(\lambda) = |r_{eq,2}(\lambda)|^2 = \frac{r^2}{2}\left(1 + \cos\left(\frac{2\pi}{\lambda}2\Delta L\right)\right). \tag{13.3}$$

Here, $\Delta L = L_2 - L_1$ is the length difference between arms 1 and 2. $R_{eq,2}(\lambda)$ is a periodic function with a period $\Delta\lambda$ given by

$$\Delta\lambda = \frac{\lambda^2}{2|L_2 - L_1|}. \tag{13.4}$$

If $R_{eq,2}(\lambda) = r^2$, $E_r = E_a$, and $E_p = 0$, the fields coming on the beam splitter from arms 1 and 2 are fully coupled to the common arm and the losses are minimized.

To calculate the spectral response of the whole cavity $R_{eq,cav}(\lambda)$, we can remark that the global cavity is equivalent to a simple Fabry–Pérot cavity with an output mirror M_0 and an "effective" rear mirror, with an amplitude reflectivity $r_{eq,2}(\lambda)$. Then, the effective reflectivity $R_{eq,cav}(\lambda)$ can be deduced:

$$R_{eq,cav}(\lambda) = \left|\frac{E_{out}}{E_{in}}\right|^2 = \left|\frac{r_0 + r_{eq,2}(\lambda)e^{j(2\pi/\lambda)2L_0}}{1 + r_0 r_{eq,2}(\lambda)e^{j(2\pi/\lambda)2L_0}}\right|^2. \tag{13.5}$$

Considering $L_{tot} = L_1 + L_2 + 2L_0$, we can show that

$$R_{eq,cav}(\lambda) = \frac{r_0^2 + r^2\cos^2((2\pi/\lambda)\Delta L) + 2rr_0\cos((2\pi/\lambda)\Delta L)\cos((2\pi/\lambda)L_{tot})}{1 + r_0^2 r^2 \cos^2((2\pi/\lambda)\Delta L) + 2rr_0\cos((2\pi/\lambda)\Delta L)\cos((2\pi/\lambda)L_{tot})}. \tag{13.6}$$

Figure 13.3 shows an example of calculation of $R_{eq,2}(\lambda)$ and $R_{eq,cav}(\lambda)$ in a particular case where $r^2 = 1$, $r_0^2 = 0.3$, $L_1 = 300$ mm, $L_2 = 320$ mm, and $L_0 = 5$ mm. $R_{eq,cav}(\lambda)$ is made of two periodicities: an envelope with the same period $\Delta\lambda$ as the combining efficiency η and a modulation with a period of $\delta\lambda$:

$$\delta\lambda = \frac{\lambda^2}{2L_{moy}}, \tag{13.7}$$

where $L_{moy} = L_{tot}/2 = (L_1 + L_2)/2 + L_0$ is the average length of the Michelson cavity. The losses in the common cavity are minimized for the wavelengths at which $R_{eq,cav}(\lambda)$ is the maximum. In the particular case considered here, where $r^2 = 1$, the losses in the common cavity are null for the wavelengths that verify the relation $R_{eq,cav}(\lambda) = 1$. However, in the particular case $r^2 = 1$, it can be seen in Figure 13.3 that the maxima of $R_{eq,cav}(\lambda)$ correspond to the maxima of $R_{eq,2}(\lambda)$. On the contrary, when

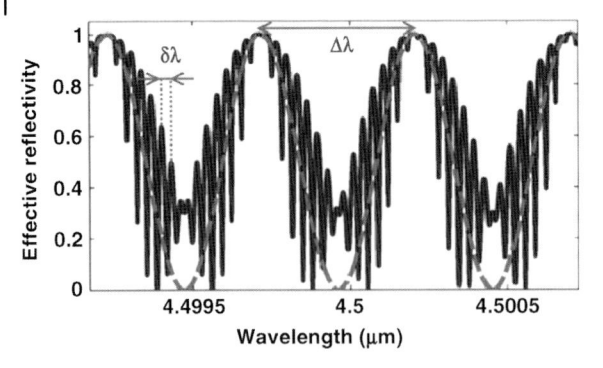

Figure 13.3 Calculated effective reflectivity of both the common cavity $R_{eq,cav}$ (solid line) and the "interferometric" part $R_{eq,2}$ (dashed line) of a Michelson cavity.

$R_{eq,2}(\lambda) = 0$, no field oscillates inside the Michelson cavity and $R_{eq,cav}(\lambda)$ is equal to the reflectivity of the OC, that is, $R_{eq,2}(\lambda) = r_0^2 = 0.3$.

The global effective reflectivity is a more complete tool than $R_{eq,2}(\lambda)$ to describe the Michelson cavity, since the resonances of the whole cavity are considered. However, the maximum value of the effective reflectivity of the interferometric part of the cavity $R_{eq,2}(\lambda)$ (which are not always of 1 as in the example presented here) can be directly identified as the combining efficiency $\eta = R_{eq,2}^{max}$ that can be obtained from the cavity. Consequently, we will use more often the $R_{eq,2}$ description, since it is a directly measurable parameter for experiment. From the periodic behavior of $R_{eq,2}(\lambda)$ and $R_{eq,cav}(\lambda)$, we can deduce that by choosing arm lengths that verify the relation

$$\Delta L \gtrsim \frac{\lambda^2}{2B}, \tag{13.8}$$

it is possible to ensure that at least one wavelength that minimizes the common cavity losses exists within the spectral gain bandwidth B of the emitters.

However, a cavity that supports only one resonance within B will be quite unstable and not robust in the presence of environment perturbations. In this case, the optical lengths of the arms of the cavity will fluctuate in time because of, for example, the variations of the optical index within the gain media due to a temperature change. To make the system robust, a solution is to ensure that many longitudinal modes of the whole cavity are present in the gain bandwidth. For that, we simply have to choose the arm lengths so that $\delta\lambda < \Delta\lambda \ll B$. This case corresponds to the one presented in Figure 13.3 with $B = 100$ nm. Thus, the system will always find, even in nonprotected environment, a wavelength corresponding to the low losses in the coupled cavity and to the high QCL gain.

13.2.2.2 General Case: The N-Arm Cavity

The N arms with lengths L_p ($p = 1, 2, \ldots, N$) are now coupled by a N–1 beam combiner into the common arm of length L_0. The OC M_0 presents a reflectivity

r_0 ($r_0^2 = 0.3$) and the end mirror reflectivities r_p ($p = 1, 2, \ldots, N$) are here supposed to verify $r_1^2 = r_2^2 = \cdots = r_N^2 = r^2$. With the circulating field theory, it is possible to generalize the previous calculation to the general case of the N-arm cavity [14]. The effective reflectivity of the interferometric part of the cavity can now be written as

$$R_{\mathrm{eq},N}(\lambda) = \left|\frac{E_\mathrm{r}}{E_\mathrm{a}}\right|^2 = \left|\frac{r}{N}\sum_{p=1}^{N} e^{j(2\pi/\lambda)2L_p}\right|^2. \quad (13.9)$$

The global effective reflectivity is given by

$$R_{\mathrm{eq,cav}}(\lambda) = \left|\frac{r_0 + r(1/N)\sum_{p=1}^{N} e^{j(2\pi/\lambda)2(L_p+L_0)}}{1 + r_0 r(1/N)\sum_{p=1}^{N} e^{j(2\pi/\lambda)2(L_p+L_0)}}\right|^2. \quad (13.10)$$

An example of calculation is given in Figure 13.4 for a five-arm cavity with $r^2 = 1$, $r_0^2 = 0.3$, $L_1 = 320$ mm, $L_2 = 322.2$ mm, $L_3 = 324.3$ mm, $L_4 = 326.4$ mm, $L_5 = 328.1$ mm, and $L_0 = 200$ mm. These values are close to the practical experimental conditions used in the Section 13.4. The effective reflectivity of a N-arm cavity is the result of the superposition of a high number of modulations that, unlike in the simple case of the Michelson cavity, are difficult to identify. We verify again that, in the case $r^2 = 1$, the maximum of $R_{\mathrm{eq,cav}}(\lambda)$ corresponds to the maximum of $R_{\mathrm{eq},N}(\lambda)$ that can be identified as the combining efficiency:

$$\eta = R_{\mathrm{eq},N}^{\max}. \quad (13.11)$$

The N-arm cavity will oscillate on these longitudinal modes that present minimum losses. If no longitudinal mode corresponding to $R_{\mathrm{eq,cav}}(\lambda) = 1$ exists within

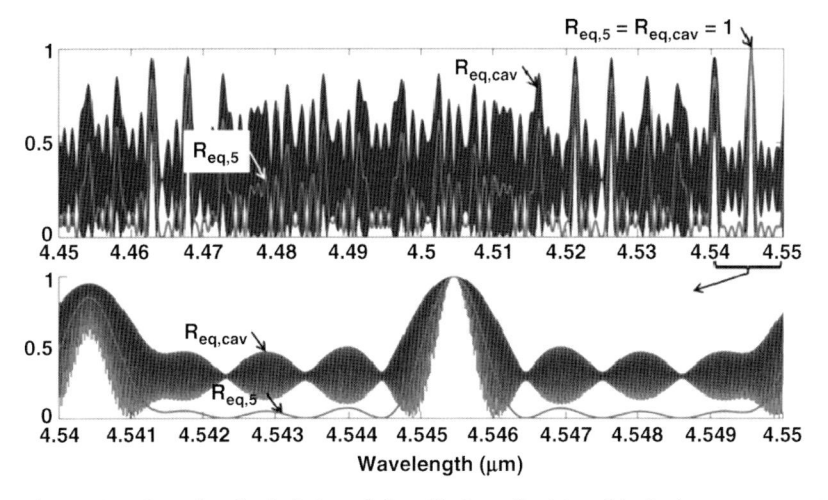

Figure 13.4 Example of calculation of the effective reflectivity of both the common cavity $R_{\mathrm{eq,cav}}$ and the "interferometric" part $R_{\mathrm{eq,5}}$ of a five-arm cavity.

the gain bandwidth, the system will select the wavelengths that present the lowest losses and thus the highest combining efficiency. The model presented here is a powerful tool to study the resonances of N-arm cavities and to estimate the combining efficiency. We will exploit this model in Section 13.2.3 to quantify the combining efficiency that can be obtained in real experimental conditions.

13.2.3
Combining Efficiency in Real Experimental Conditions

The optical lengths L_p of the arms of the cavity will fluctuate with time because of the environmental perturbations such as QCL optical index variation with temperature and dilatations or vibrations of the mirrors or mirror holders. These perturbations can be simulated by introducing slight variations of the lengths of the N arms [15]. The added perturbations $[\delta L_p]_{p=1\cdots N}$ vary as $-(\delta L_{max}/2) \leq \delta L_p \leq (\delta L_{max}/2)$ with $\delta L_{max} \sim \lambda_0$ ($\lambda_0 = 4.5\,\mu m$ is the central wavelength of the spectral bandwidth B). For each random set of $[\delta L_p]_{p=1\cdots N}$, the maximum $R_{eq,N}^{max}$ of the effective reflectivity over the spectral bandwidth of the QCL is calculated. This calculation is repeated several times (typically 1000 here) to simulate the random variations of the optical lengths of the N arms. From this statistical distribution, it is possible to deduce an average value that can be interpreted as the mean combining efficiency of the cavity under nonprotected conditions:

$$\bar{\eta} = \overline{R_{eq,N}^{max}}. \tag{13.12}$$

Moreover, the standard deviation σ of this statistical distribution can be interpreted as the fluctuation of the combining efficiency previously defined. σ can be used to quantify the stability of the system.

In the case of CBC of fiber lasers, Cao et al. show that the statistical method presented here is in good agreement with experimental results [15]. The same method is adapted in the present chapter for CBC of QCLs. For the different calculations presented afterward, the spectral range is similar to the gain spectral bandwidth of QCLs, that is, $B = 100\,nm$ with a central wavelength λ_0. The reflectivity of the mirrors are fixed to $r_1^2 = \cdots = r_N^2 = r^2 = 1$ and $r_0^2 = 0.3$. To simulate the practical experimental conditions, the arm lengths are defined as $L_p = L_1 + (p-1)\Delta L + \delta L_p$ with $\Delta L \sim 2\,mm$.

13.2.3.1 Influence of the Number of Arms N
In order to quantify the maximal number of QCL that can be coherently combined in a N-arm cavity, the variation of $\bar{\eta}$ for an increasing number of arms N can be calculated, as shown in Figure 13.5. The random perturbations level is fixed to $\delta L_{max} \sim \lambda_0 - 4.5\,\mu m$. As already explained, the case of Michelson cavity ($N=2$) is a particular case where the efficiency can be forced to $\bar{\eta} = 1$, even in a nonprotected environment, by choosing the arm length difference so that it verifies Eq. (13.8). Here, we have $\Delta L = 2\,mm \gg \lambda^2/2B = 100\,\mu m$ so that there are hundreds of longitudinal modes with null losses existing within B: The cavity presents a perfect

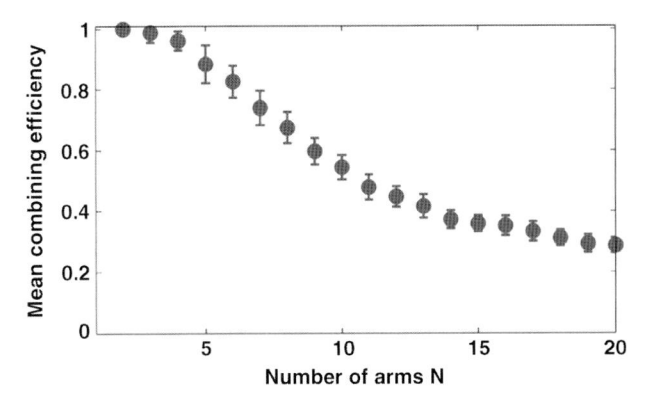

Figure 13.5 Calculation of the mean combining efficiency with respect to the number of arms N in the case where $\Delta L \sim 2$ mm.

combining efficiency with a perfect stability ($\sigma = 0$), even with the random length deviation $\delta L_{max} \sim \lambda_0$.

For $N > 2$, because of the arm length random deviation, the average combining efficiency is lower than unity and decreases with the number of arms N. This is due to the fact that the probability to find a wavelength with minimal losses decreases with N for a fixed spectral range B. For the experimental case $N = 5$, the combining efficiency that can be obtained in the presence of random arm length perturbation is $\bar{\eta} \sim 88\%$ with a standard deviation of $\sigma = 12\%$. This is still a reasonable combining efficiency, but for cavities with a number of arms larger than 5, the combining efficiency is too low ($\bar{\eta} < 75\%$) to consider the method as attractive with the particular set of arm lengths considered here. It is worth noticing that the combining efficiency decrease is very fast for $2 < N < 9$ and becomes less important for $N > 9$. In the same way, the combining efficiency fluctuation is more important for smaller number of arms. It can be understood as an averaging phenomenon where the effect of an additional arm on the combining efficiency becomes less important for cavities with high number of arms.

13.2.3.2 Influence of the Arm Length Difference ΔL

We have already shown that the case $N = 2$ is a particular one since, provided that the arm length difference is large enough compared to B [see Eq. (13.8)], the combining efficiency is ideal and does not depend on ΔL. For $N > 2$, we used the "statistical" method to calculate the variation of the average combining efficiency versus ΔL (see Figure 13.6). The perturbation level is again fixed to $\delta L_{max} \sim \lambda_0$. When the constant arm lengths are equal ($\Delta L = 0$), the average combining efficiency is equal to 20%. In this case, only $1/N$ of the total available energy is coupled to the common arm. For $\Delta L = 0$, the random variation δL_p prevents the system from finding a resonance in the common cavity. In consequence, the coherent combining between the N arms is impossible because of the environmental perturbations and the average combining efficiency is close to its minimal value $1/N$. For $\Delta L = 2$ mm, a combining efficiency of $\bar{\eta} = 88\%$ is obtained with a standard deviation $\sigma = 12\%$. If ΔL is increased further,

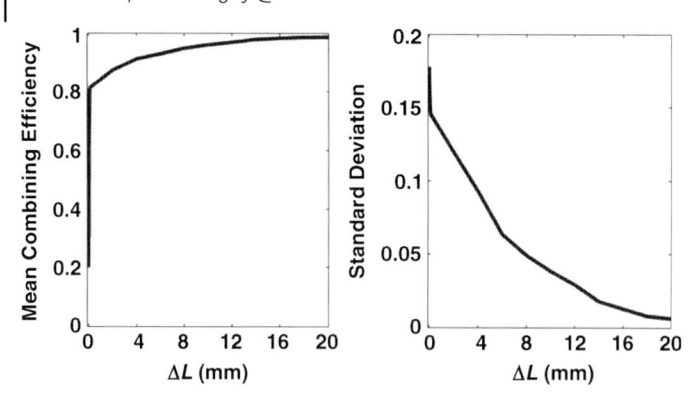

Figure 13.6 Variation of the mean combining efficiency and the standard deviation with respect to ΔL in the case $N=5$ and $\delta L_{max} \sim \lambda_0$.

the combining efficiency increases and nearly reaches the ideal value of 1 for large values of ΔL. In the same way, σ decreases to nearly reach 0 for large values of ΔL. Thus, for a given level of random length perturbation δL_{max}, it is still possible to tend toward the ideal combining efficiency provided that the constant arm length difference is large enough. This is due to the fact that the larger the differences between the arm lengths, the higher the probability to find a resonance of the cavity within a fixed spectral range B.

13.3
Experimental Realization: Five-Arm Cavity with a Dammann Grating

13.3.1
Dammann Gratings

As explained previously, coherent QCL combining using a five-arm cavity requires a specific optical component able to separate an incident beam into five beams of equal intensities with a good splitting efficiency (defined as the ratio of the power in the five central orders to the incident power). In the past, most of the attempts to design a multibeam splitter were based on the DG principle [16,17]. A DG is a binary phase diffractive optical element (DOE). DGs are relatively easy to fabricate with current lithography and etching techniques and require only one lithography step. Moreover, the theoretical and simulation tools used to design such DOE are easy to develop.

The five beams coming from the QCLs are coplanar, so a one-dimensional geometry is considered here, that is, the structure will be made of ridges with a periodicity in only one direction. The general phase profile of one period of a DG is presented in Figure 13.7 where D is the period of the grating whose phase varies between two levels (0 and π) in a number M of transition points. The coordinates are normalized so that $0 = x_1 < x_2 < \cdots < x_{M-1} < x_M = 1$.

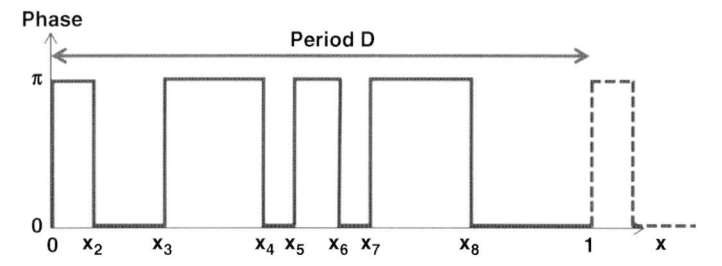

Figure 13.7 Phase profile of a Dammann grating.

The material chosen to realize a DOE in the MIR must be transparent in this spectral range. Moreover, the material must be sufficiently mature from a technological point of view so that binary structures can be etched. For these reasons, gallium arsenide (GaAs) has been selected. GaAs presents an optical index $n_{GaAs} \sim 3.3$ at $\lambda = 4.6\,\mu m$, which implies Fresnel reflectivity coefficient $\sim 30\%$, so that an AR coating must be designed for this grating. From the value of optical index of GaAs and the $0-\pi$ phase profile of the DG, the etching depth h is given by

$$h = \frac{\lambda}{2(n_{GaAs} - 1)} \approx 1\,\mu m. \tag{13.13}$$

The grating period was chosen to be $D = 45\,\mu m$. This value is a trade-off between the angular separation $\Delta\theta = 5.7°$ between two adjacent orders on the one hand and the size of the ridges on the other. Indeed, $\Delta\theta$ must be large enough (and thus D small enough since $\Delta\theta \sim \lambda/D$) to insert the QCLs in the experimental setup, while keeping a reasonable arm length ($\sim 32\,cm$ here). But if D is too small, the ridges inside each period will be too difficult to realize having in mind that the ridge thickness must be kept greater than $1\,\mu m$ with conventional optical lithography techniques.

In order to quantify the quality of the DOE, the combining efficiency of the grating η is defined as follows:

$$\eta = \frac{\sum_{n=-[(N-1)/2]}^{(N-1)/2} I_n}{\sum_n I_n}, \tag{13.14}$$

where $N = 5$ is the number of orders of interest and I_n is the intensity of the nth diffraction order. The uniformity between the N central orders of interest is defined as

$$U = \frac{\max(I_n) - \min(I_n)}{\max(I_n) + \min(I_n)}\bigg|_{-\frac{N-1}{2} \leq n \leq \frac{N-1}{2}}, \tag{13.15}$$

where U is a parameter varying between 0 and 1. In the best case, the orders' intensities are equal and the uniformity is $U = 0$. In the worst case, the intensity of one of the five central orders equals zero and the uniformity is $U = 1$. Using the scalar approximation of diffraction theory, it is possible to calculate the intensities of the orders of such a binary structure [18]:

$$I_0 = \left|1 - 2\sum_{s=1}^{(N-1)/2}(-1)^s x_s\right|^2. \tag{13.16}$$

$$I_n = \left|\frac{1}{\pi n}\sum_{s=1}^{(N-1)/2}(-1)^s e^{-2\pi i n x_s}\right|^2. \tag{13.17}$$

Then, the binary phase profile, that is, the phase transition coordinates x_i and the number of phase transitions M, is optimized in order to maximize the combining efficiency η and minimize the parameter U. Consequently, the cost function was chosen as

$$\text{Err} = \frac{\left(\sum_{n=-[(N-1)/2]}^{(N-1)/2}|I_n - (\eta/N)|^2\right)^P}{\eta^Q}, \tag{13.18}$$

where P and Q are free parameters allowing the optimization of either the uniformity or the efficiency. In the case $N=5$, an $M=4$ structure was found, as shown in Figure 13.8, with a combining efficiency of $\eta \sim 77.5\%$ and $U \sim 10^{-9}$.

The optimized profile is fabricated in GaAs with UV optical lithography and inductively coupled plasma (ICP) etching. The scanning electron microscope (SEM) view of the structure is presented in Figure 13.9. The grating is then coated with a SiO_2/TiO_2 bilayer AR coating (with a reflectivity $R < 2\%$) on both the etched and the rear facets. The grating is characterized by measuring the power contained in all the transmitted orders (Figure 13.9). An experimental splitting efficiency of $\sim 75\%$ (compared to the theoretical value of 77.5%) is obtained along with a good uniformity ($U \sim 3\%$) between the five central orders' intensities (Figure 13.9).

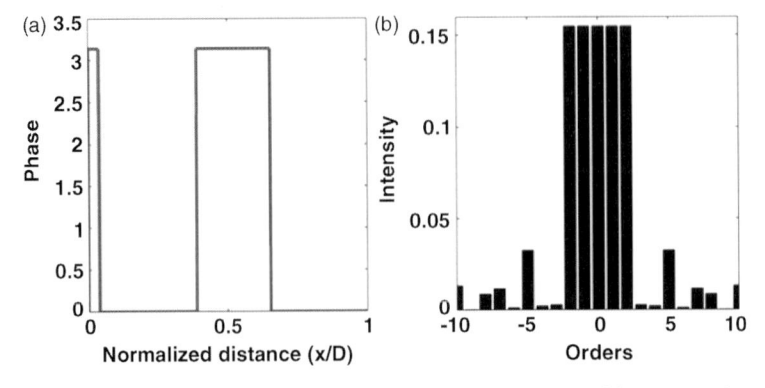

Figure 13.8 Phase profile (a) and diffraction orders' intensities (b) of the optimized Dammann grating in the case $N=5$.

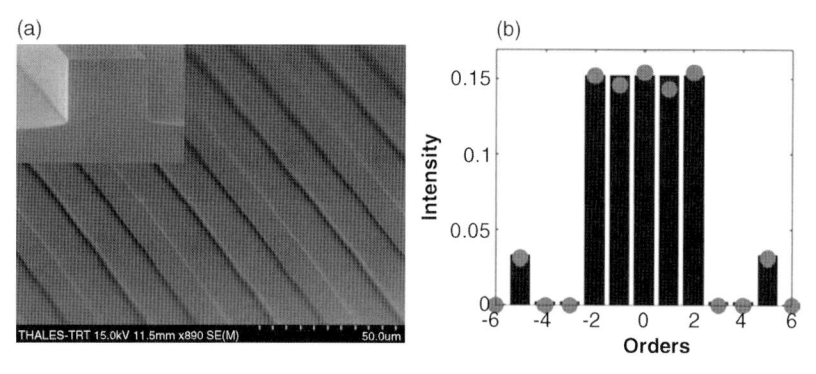

Figure 13.9 (a) SEM view of the Dammann grating before AR coating. (b) Calculated (bars) and measured (dots) orders' intensities of the fabricated DG with its AR coating.

13.3.2
Quantum Cascade Lasers

The QCLs used are made of strain-balanced $Ga_{0.3}In_{0.7}As/Al_{0.7}In_{0.3}As$ active regions on an InP substrate. The ridge was ICP etched and buried into iron-doped InP by metalorganic chemical vapor deposition (MOCVD) regrowth for a better heat extraction. Further thermal improvement is also obtained, thanks to epi-down mounting on an AlN submount with gold–tin soldering. The active region is thus in direct contact with the substrate so that the heat extraction is highly improved. With a HR coating on the rear facet and no coating on the output facet ($R \sim 30\%$), QCLs with 4 mm long waveguides provide $\sim$400 mW output power in CW operation with a 1.3 kA/cm^2 threshold density. The spectral bandwidth of these lasers is typically of $B \sim 100$ nm around a central wavelength of 4.65 µm. As explained before, in order to facilitate the phase locking phenomenon in the external cavity, the QCLs were AR coated (SiO_2/TiO_2 bilayer, $R_{AR} < 2\%$) on the output facet and HR coated on the rear facet.

The QCLs were first characterized in individual external cavity (IEC) (Figure 13.10) with the external arm lengths and the lenses (collimation and focalization ones) corresponding to the ones used in the final cavity setup. The OC presents an $R_0 = 30\%$ reflectivity. To simulate the 25% single-pass losses caused by the DG on the one hand and the coherent combining losses on the other (see Section 13.2), a beam splitter was introduced in the external arm of the IEC. By tuning the angle between the beam splitter and the oscillating beam, it is possible to insert additional single-pass losses X

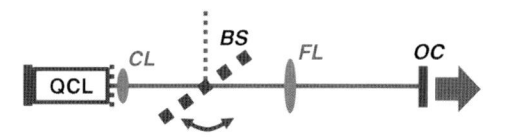

Figure 13.10 Scheme of an AR-coated QCL in individual external cavity. CL, collimation lens; FL, focusing lens; OC, output coupler; BS, beam splitter.

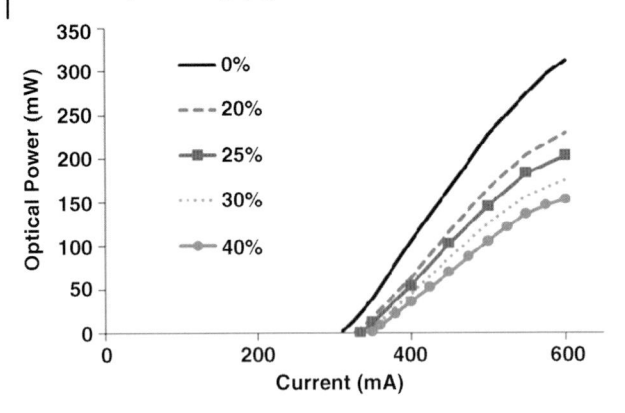

Figure 13.11 Typical power versus current characteristics of QCL in individual external cavity for different single-pass intracavity losses.

between $X = 20\%$ and $X = 40\%$ (Figure 13.11). Thus, it is possible to estimate the power that should be obtained taking into account both the DG and CBC losses.

13.3.3
The Five-Arm External Cavity

The experimental setup is presented in Figure 13.12. The external cavity extends between the rear facets of the five QCLs and the $R_0 = 30\%$ OC. The AR-coated and individually collimated QCLs are inserted in the five arms corresponding to the five central orders of the DG. Moreover, two gold mirrors were added in each arm to align the five beams with respect to these orders. Finally, a focusing lens (FL) is inserted in the common arm. As already explained, the loss minimization in the global cavity will ensure phase locking between the five lasers and will select the proper longitudinal mode so that there are constructive interferences at the

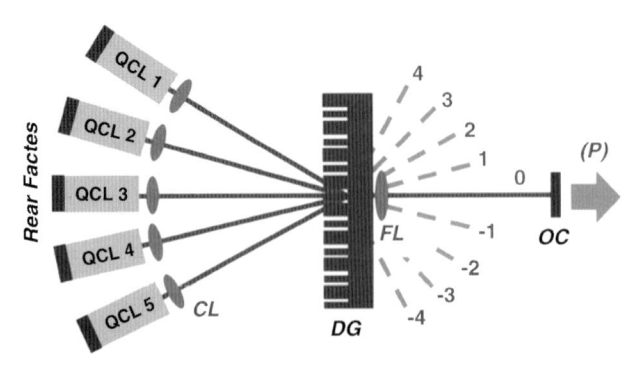

Figure 13.12 Experimental setup of the five-arm cavity. QCL_i: HR-AR QCL; CL, collimating lens; DG, Dammann grating; FL, focusing lens; OC, output coupler.

common output end (corresponding to the central zeroth order of the DG) and destructive interferences on the other orders of the grating (some of them are represented by dashed lines in Figure 13.12).

The combining efficiency η, as defined in Eq. (13.14), can be rewritten as

$$\eta = \frac{P_0}{\sum_i P_i}, \tag{13.19}$$

where P_i is the power in the output orders. P_0 is deduced from the output power P_P by

$$P_0 = \frac{P_P}{1 - R_0}. \tag{13.20}$$

To achieve a good combining efficiency, the currents of the QCLs are set so that each emitter provides the same power to the five-arm cavity. The appropriate currents were deduced from the previous individual characterizations of the QCLs. We define $P_{\text{tot},X}$ as the total power available from the five QCLs in IEC with an additional single-pass loss level of X%. $P_{\text{tot},X}$ is deduced from the previous study of the QCLs in IEC with tunable intracavity losses. Thus, $P_{\text{tot},0}$ is the total power available with no additional losses and $P_{\text{tot},25}$ is the total power available taking into account the DG losses. This will be a very useful tool to understand the different causes of experimental output power losses. Finally, the global efficiency of the system β_0 is defined as the ratio between the output power P_P and the sum of the total power available from the QCLs in IECs is given by

$$\beta_0 = \frac{P_P}{P_{\text{tot},0}}. \tag{13.21}$$

Two cavity configurations are studied and compared: the "short" cavity with the arm lengths nearly equal to 320 mm (the exact values are set between 314 and 322 mm) and the "long" cavity, identical to the short one but with its central arm length being increased by 100 mm. Indeed, it was shown in Section 13.2 that the larger the differences between the arm lengths, the higher the combining efficiency and the stability of the system. Thus, we demonstrated that the combining efficiency of a five-arm cavity could tend to 100% when increasing all the values of ΔL_p in a uniform way. We are interested now in a simpler case that can be easily applied to our experimental setup while keeping a compact cavity. Here, we start from the case where the arm lengths verify $L_p = L_1 + (p-1)\Delta L + \delta L_p$ (with $\Delta L \sim 2$ mm). When increasing the length of only one arm, we calculate that it is still possible to improve the average combining efficiency and decrease the fluctuation of the system. An increase of 10 cm makes $\bar{\eta}$ to increase from 88 to 93%. The "short" and "long" cavity configurations will be compared to demonstrate the influence of the arm length difference on the combining efficiency.

Figure 13.13 shows the output power in the combined beam P_P versus P_{single}, the unitary output power of each QCL in IEC with 25% single-pass additional losses. We report in the same figure, the total power $P_{\text{tot},0}$ and $P_{\text{tot},25}$ available from the five QCLs with 0 and 25% single-pass additional losses, respectively.

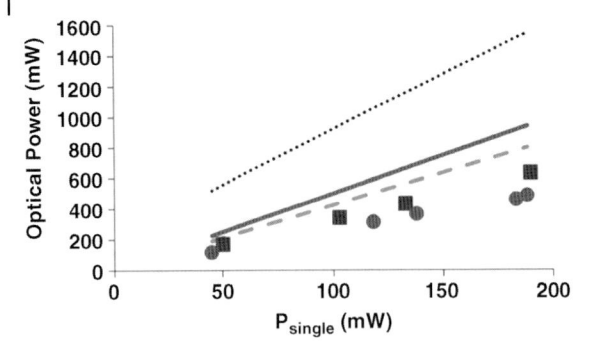

Figure 13.13 Output power versus P_{single}. Experimental output power of the "short" cavity (circles) and the "long" cavity (squares). Total power available from the five QCLs in individual external cavity with single-pass additional losses of 0% (dotted line), 25% (solid line), and 28% (dashed line).

- Long cavity
  A maximum output power of $P_P = 0.65$ W was obtained in the combined beam with "long" cavity. The corresponding global efficiency is $\beta_0 = P_P/P_{tot,0} \sim 40\%$. This reduced global efficiency is due to the additional losses introduced by the DG. The suppression of the DG losses would increase the global efficiency up to $\beta_{25} = P_P/P_{tot,25} = 67\%$. The 33% still missing indicates that losses other than the DG ones were introduced. As already explained, a five-arm cavity in a non-protected environment has a reduced combining efficiency due to the difficulty to find a resonance of the global cavity within the gain spectral bandwidth of the QCLs. Here, we measure the experimental combining efficiency to be $\eta_{experimental} \sim 70\%$, this value is close to the theoretical DG efficiency of $\eta_{theoretical} = 75\%$, as shown in Figure 13.14, with $\eta_{experimental} = 70\% = 93\% \, \eta_{theoretical}$. This is in good

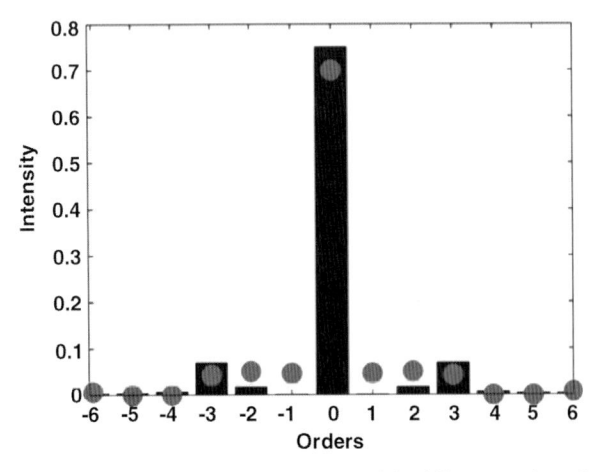

Figure 13.14 Bars: calculated intensities of the diffraction orders of a DG lit by five beams with the proper relative phases. Dots: measured orders' intensities.

agreement with the theoretical value of 93% calculated in the case of the "long" cavity (see Section 13.2) with perturbations from the environment and a spectral gain bandwidth of 100 nm.

In Figure 13.13, the power $P_{tot,25}$ corresponds to the power available assuming $\eta_{theoretical} = 75\%$. However due to the imperfect combining efficiency, we should consider a grating with a 70% combining efficiency and a 75% splitting efficiency, corresponding to $X = 1 - 0.75\sqrt{0.70} \sim 28\%$ single-pass losses. The corresponding $P_{tot,28}$ power available from the QCLs with such losses is reported in Figure 13.13. It can be noted that the additional losses introduced by the DG and the imperfect combining efficiency explain a major part of the reduced global efficiency β_0. The output power is quite stable with a fluctuation $< \pm 5\%$ peak to peak at maximal power measured over 1 h of free running in nonprotected environment. These power fluctuations were not observed when studying the CBC of QCLs in the Michelson cavity [10]. The increase of the output power fluctuations with the number of sources combined was well understood with the theoretical model presented in Section 13.2.

- Short cavity
  The maximal output power obtained from the short configuration [19], $P_P = 0.5$ W, is lower than that obtained from the long cavity. $\beta_{25} = P_P/P_{tot,25} \sim 55\%$ of the available power is obtained in the combined beam, compared to the 67% for the long configuration. The combining efficiency of 66% is also lower than the 70% demonstrated for the long cavity. Thus, η is increased by 6% in the long cavity compared to the short one, which is in good agreement with the 5.5% increase obtained from theoretical calculation.

As shown in Figure 13.15, the beam quality of the combined beam is nearly diffraction limited with $M^2 = 1.2$ for the slow axis and $M^2 = 1.6$ for the fast axis. These values are close to those of the QCLs alone and the QCLs in individual external cavity.

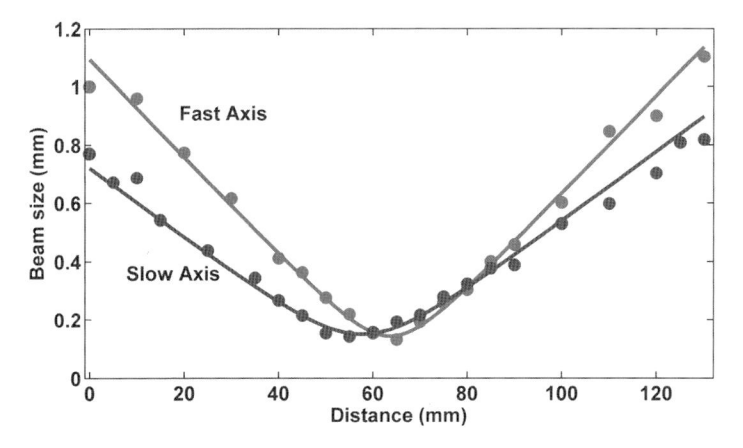

Figure 13.15 Signal beam quality measurements. Dots: experiment; *solid curves*: fit).

13.4
Subwavelength Gratings

13.4.1
Principle

It was shown in the previous section that a major part of the reduced global efficiency obtained for the CBC of five QCLs in an external cavity was caused by the 25% single-pass losses introduced by the DG. This efficiency limitation is due to the simplicity of the profile of the period [18]. In the present section, we present other types of gratings with an increased diffraction efficiency to reduce the intracavity losses. To obtain more efficient 1–N beam splitters with typical efficiency of ~97%, a solution is to use the continuous phase gratings [19,20]. However, the fabrication of these gratings remains very challenging [21]. An approximate method is to discretize the continuous phase profile into a finite number of phase levels. The most direct way to realize this discretized profile is to vary the height of material etched and thus the phase seen by the incident light. The so-called multiheight approach needs many phase levels to be efficient [20]. However, increasing the number of phase levels increases the potential mask positioning errors during the lithography steps. In conclusion, this method is theoretically efficient, but the fabrication is technically difficult and expensive.

The other way to realize the phase levels of the discretized phase profile is to vary the refractive index seen by the incident light while keeping a constant height profile. This index-varying grating is realized using structures with typical size smaller than the wavelength, the so-called subwavelength structures. Because of the constant height of the profile, the realization requires only one lithography and etching step and avoids the multiple precise mask alignments of the "multiheight" solution. Subwavelength gratings are based on the effective medium theory, which predicts that a periodic subwavelength structure, under precise conditions, is equivalent to an artificial homogeneous dielectric medium characterized by an effective index that depends on the geometry of the subwavelength structure [22–24]. Then, by varying the size of the subwavelength elements within the period of the grating, any effective index (and, thus, phase level) can be synthesized between the index of the air and the index of the substrate material.

Artificial dielectrics were introduced in the 1990s [24,25] and first realizations were demonstrated in the IR [26]. This principle was also applied to the realization of blazed gratings [27,28] or diffractive optics [29,30]. The only realization of beam splitters with subwavelength features [31] concerns a 1–3 array generator at $\lambda = 633$ nm with a measured diffractive efficiency of 74% and a good beam uniformity (1.7%). However, the question of an AR coating for this grating is not studied so far.

Here, we have explored the feasibility of beam splitters with high efficiency and good uniformity with subwavelength structures. The design, optimization, and fabrication of such gratings are presented. The problem of the deposition of an AR coating on this structure with a varying optical index is also studied.

13.4.2
Grating Design and Realization

We are interested here in the design of a binary subwavelength grating able to split, with a good diffraction efficiency η and a good uniformity U, one incident beam at 4.6 μm into $N=7$ coplanar beams. The structure will be made of subwavelength ridges with a periodicity in only one direction. The incident wave is a plane wave in TE polarization with the electric field being parallel to the grating ridges. The design of such a grating starts from the calculation of the ideal continuous phase profile able to split one beam into seven beams of equal intensities. This continuous phase profile will then be discretized and transcribed into subwavelength features. To calculate this ideal profile, Sidick et al. propose to consider it as the interference pattern of N virtual sources [20]. Then, by optimization of the phases and amplitudes of the virtual sources, it is possible to force the resulting hologram to behave as an efficient and uniform beam splitter. With this approach, based on the scalar diffraction theory, a theoretical continuous phase profile having a diffraction efficiency of $\eta = 96.9\%$ and a uniformity of $U = 0.83\%$ is reported [20].

Each of the phase levels will be realized locally by a subwavelength feature corresponding to an effective index that depends on the geometry of this subwavelength element characterized by L_i and Λ_i, where Λ_i is the sampling period. By changing L_i and Λ_i along the grating period D, it is possible to vary the optical index (and thus the phase) seen locally by the incident wave. Λ_i must respect several conditions so that the subwavelength structure is equivalent in the far field to a homogeneous medium for the incident wave [32]. The geometry of the structure is presented in Figure 13.16 with $n_1 = n_{Air} = 1$ and $n_2 = n_{GaAs} = 3.3$. These conditions of equivalence impose that Λ_i verifies

$$\Lambda_i < \frac{\lambda}{n_{AsGa} + \sin(\theta_{max})} = 1.25 \text{ μm}, \tag{13.22}$$

where θ_{max} is the maximal angle of incidence on the grating. In the N-arm cavity, the beam splitter will be illuminated by seven beams superimposed on its seven central orders. For $N=7$, the maximal angle of incidence on the grating is then given by

$$\theta_{max} = \sin^{-1}\left(\frac{N-1}{2}\frac{\lambda}{D}\right) = 17.5°. \tag{13.23}$$

The conditions of equivalence also impose that the etching depth h of the grating is larger than $\lambda/4$. The calculation of the effective index n_{eff} associated with a subwavelength groove is performed with fully vector theory software [33], relying

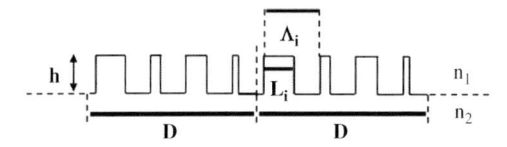

Figure 13.16 Geometry and nomenclature of the subwavelength binary phase grating.

on rigorous coupled-wave analysis (RCWA) [34] and its further improved versions [35–37]. RCWA is a well-known technique for obtaining the exact solution of Maxwell's equations for the diffraction on periodic structures. Theoretically, by varying the size L_i of this subwavelength ridges as $0 \leq L_i \leq \Lambda_i = 1.25\,\mu m$, any effective index value as $n_1 = 1 \leq n_{eff} \leq n_2 = 3.3$ can be synthesized. However, to take into account the actual technological limitations (electron beam writer and etching tool available in our premises), the range of feature sizes was limited to $300\,nm \leq L_i \leq 850\,nm$. Another constraint imposed on the grating concerns the etching depth. From preliminary tests, it was observed that the etching depth is not constant throughout the grating period and is proportional to the width of the etched groove. To avoid this effect, the groove width was forced to be constant (here to a value of 400 nm) across the period.

The variation of the effective index with the size of a subwavelength ridge of GaAs with a fixed groove width of 400 nm, at a wavelength of $\lambda = 4.6\,\mu m$ in TE configuration, is represented in Figure 13.17. The ideal phase profile varies between 0 and $a\pi$, with $a = 1.22$ (Figure 13.18). Thus, in order to obtain a total phase difference of $a\pi$, the etching depth of the binary subwavelength features to be adjusted to

$$h = \frac{a\lambda}{2(n_{max} - n_{min})} = 4.92\,\mu m, \tag{13.24}$$

where n_{max} and n_{min} are the maximum and minimum effective indices that can be synthesized by the subwavelength features, taking into account the technological constraints (Figure 13.17). With the values of n_{eff} obtained from Figure 13.17, the continuous phase profile is first discretized into a finite set of phase levels and is then transcribed into a subwavelength grating (Figure 13.18). The efficiency of this directly transcribed grating is $\eta = 96.7\%$ and its uniformity is $U = 5\%$, compared to the performances of the ideal continuous phase grating $\eta = 96.9\%$ and $U = 0.83\%$. The subwavelength grating was realized by electron beam lithography (better resolution than the optical lithography used for the DG) and ICP etching. The structure is shown in Figure 13.19. The grating has an efficiency of $\eta = 95\%$ and the uniformity of $U = 13\%$ are measured. These values are in good agreement with the theoretical ones (Figure 13.20).

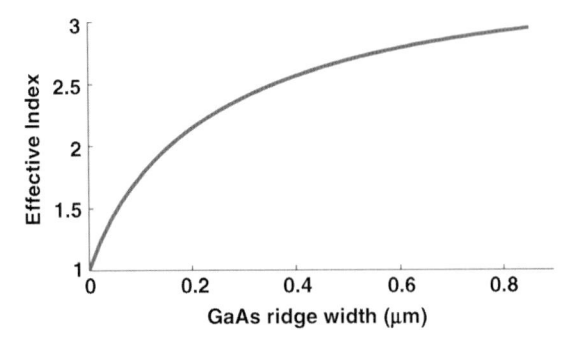

Figure 13.17 Width dependence of the effective index of a subwavelength GaAs ridge with a fixed groove width of 400 nm for TE polarization.

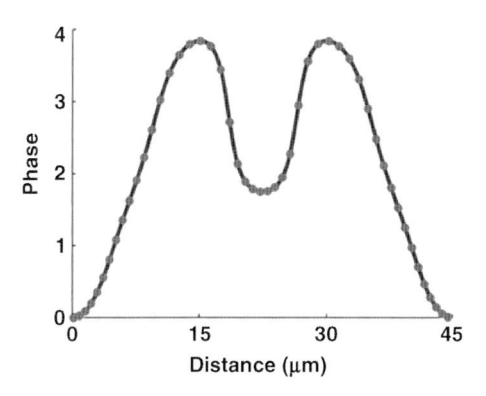

Figure 13.18 *Solid curve*: one period of the ideal continuous phase profile to separate seven beams; *dots*: discretized phase levels.

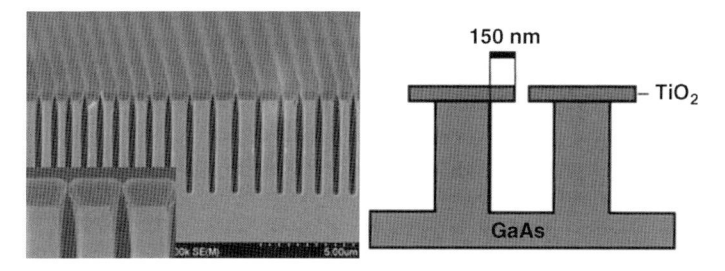

Figure 13.19 SEM view of the subwavelength grating after ICP etching.

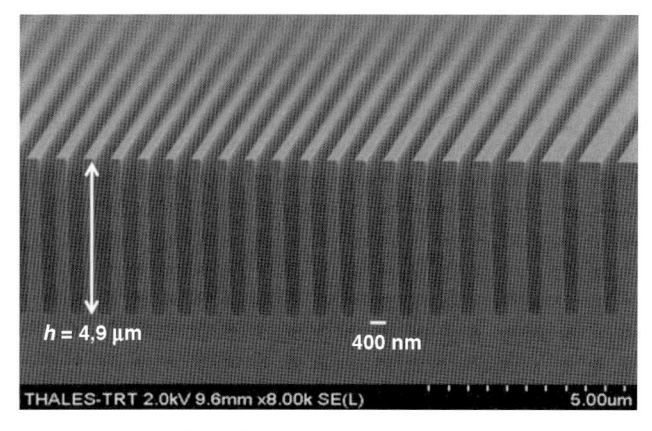

Figure 13.20 Calculated (bars) and measured (dots) orders' intensities of the fabricated subwavelength grating before AR coating.

13.4.3
Antireflection Coating Design

The efficiency η does not take into account the amount of energy reflected by the grating, since it is normalized by the total transmitted energy. The percentage of energy contained in the reflected orders of the grating is here $R = 21\%$. The question of a "classical" multilayer AR coating is not simple in the case of a subwavelength grating because the structure presents an optical index that varies along the period [38]. To solve this problem, an AR coating that is itself subwavelength structured can be used as shown in Figure 13.21. Such a subwavelength structured AR coating will not only reduce the reflectivity R but will also enhance the performances of the whole structure in terms of efficiency and uniformity. Consequently, the whole structure (the widths of the GaAs ridges, the etching depth, and the AR coating) must be optimized at the same time with three quality criteria: the minimization of R, the maximization of η, and the minimization of U.

We deduced from our first AR coating tests that the AR layers do not reach the bottom of the grooves and, instead, gather on the top of the ridges. To take into account this effect, we considered a realistic AR coating structure close to what was experimentally observed (Figure 13.19). A TiO_2 monolayer coating was used, since it was found as efficient as multilayer structures such as the classical TiO_2/SiO_2 bilayer coating. The whole structure is then optimized using the cost function introduced in Eq. (13.18). The optimized grating presents an efficiency of $\eta = 95.4\%$, a uniformity of $U = 4.4\%$, and a reflectivity that was decreased to $R = 1.7\%$. The AR coating has thus decreased the amount of energy reflected from 21 to 1.7% and has preserved the efficiency and uniformity initial values.

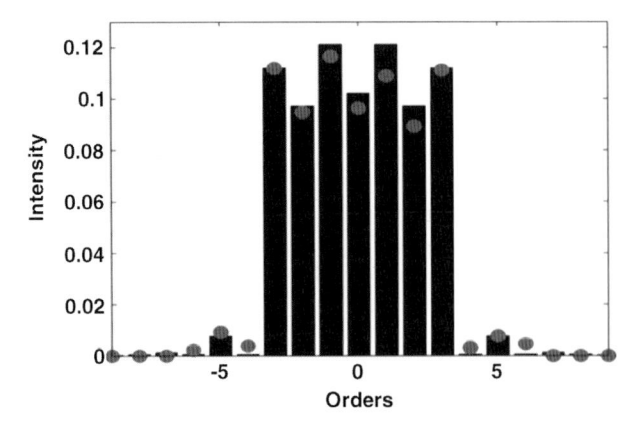

Figure 13.21 SEM view of our first test of AR coating deposition on a subwavelength structure (left). Model of realistic monolayer AR coating on a subwavelength grating (right).

13.4.4
Calculated Performances

From the previous experimental data, it is possible to estimate the output power that could be obtained from a seven-arm cavity using the subwavelength grating designed in Section 13.4.2. For this calculation, seven identical QCLs are considered with a unitary maximum output power in IEC of 300 mW. The losses in these N-arm cavities are mainly due to the nonideal grating efficiency and the nonideal combining efficiency due to the difficulty to find a resonance of the cavity in a finite spectral range. In our case, we have determined the subwavelength grating efficiency of ∼95% and the combining efficiency of 74% that could be obtained from a seven-arm cavity (Figure 13.5). The resulting total single-pass losses considered here are of $1 - \sqrt{0.95 \times (0.95 \times 0.74)} = 18\%$. It is then possible to estimate the output power of the seven-arm cavity to be 1.6 W using the subwavelength grating. This corresponds to a global efficiency of 75% that is far more interesting than the value of 40% obtained from the five-arm cavity using a DG.

13.5
Conclusion

In this chapter, passive coherent beam combining of QCLs in an external cavity using a DG has been studied. First, a classical "cold cavity" model was developed to quantify the combining efficiency that can be obtained from this method, taking into account the environmental perturbations. Then, the coherent beam addition of five QCLs in an external cavity using a DG was demonstrated. A DG was designed to separate an incident beam in the MIR into five beams of nearly equal intensities with an efficiency of 75%. A power of 0.65 W in the continuous regime at RT corresponding to a combining efficiency of 70% and a global efficiency of 40% was obtained from this cavity. Moreover, the combined beam exhibits the same beam quality as a single emitter ($M^2 < 1.6$). The method presented here is shown to be an efficient way to increase the brightness of QCLs. However, we found that a major part of the reduced global efficiency was caused by the 25% single-pass losses introduced by the DG. In order to reduce the intracavity losses, we studied then the feasibility of subwavelength gratings. These index-varying structures present far better diffraction efficiencies than the DGs. We designed a one–seven subwavelength grating with its AR coating using a RCWA code combined to an optimization routine. The optimized grating presented an efficiency of 95.4%, a good orders' uniformity and a reflectivity of only 1.7%. The performances of the first grating fabricated are very good since an experimental intrinsic diffraction efficiency of 95% was obtained. The global efficiency of the seven-arm cavity with a subwavelength grating was then calculated to have the potential to reach 75%, which is far more interesting than the value of 40% obtained from the five-arm cavity using a DG.

We mentioned in introduction that the current sources usually considered for MIR applications are the OPO based on ZGP or GaAs nonlinear crystals. The maximum output power obtained from these sources is around 20 W. With a theoretical global efficiency of 75%, the seven-arm cavity proposed here could reach such output power level with unitary QCL power in IEC of ~4 W. This performance could certainly be obtained from the 5 W QCLs recently demonstrated [6]. In conclusion, coherent beam combining of QCLs in external cavities with subwavelength gratings should be in the future an interesting alternative to OPO to address the power scaling in the mid-infrared.

References

1 Budni, P.A., Pomeranz, L.A., Lemons, M. L., Schunemann, P.G., Pollak, T.M., and Chicklis, E.P. (1998) 10W Mid-IR Holmium Pumped ZnGeP2 OPO. Advanced Solid State Lasers, Coeur D'Alene, ID, February 2, Paper FC1.

2 Cheung, E., Palese, S., Injeyan, H., Hoefer, C., Ho, J., Hilyard, R., Komine, H., Berg, J., and Bosenberg, W. (1999) High power conversion to mid-IR using KTP and ZGP OPOs. Advanced Solid State Lasers, Boston, MA, January 31, Paper WC1.

3 Faist, J., Capasso, F., Sivco, D.L., Sirtori, C., Hutchinson, A.L., and Cho, A.Y. (1994) Quantum cascade laser. *Science*, **264**, 553–556.

4 Bismuto, A., Beck, M., and Faist, J. (2011) High power Sb-free quantum cascade laser emitting at 3.3 μm above 350 K. *Appl. Phys. Lett.*, **19**, 191104.

5 Scalari, G., Blaser, S., and Faist, J. (2004) Terahertz emission from quantum cascade lasers in the quantum hall regime: evidence for many body resonances and localization effects. *Phys. Rev. Lett.*, **93**, 237403.

6 Bai, Y., Bandyopadhyay, N., Tsao, S., Slivken, S., and Razeghi, M. (2011) Room temperature quantum cascade lasers with 27% wall plug efficiency. *Appl. Phys. Lett.*, **98**, 181102.

7 Wagner, J., Schulz, N., Rösener, B., Rattunde, M., Yang, Q., Fuchs, F., Manz, C., Bronner, W., Mann, C., Köhler, K., Raab, M., Romasev, E., and Tholl, H.D. (2008) Infrared semiconductor lasers for DIRCM applications. *Proc. SPIE*, **7115**, 71150A.1–71150A.11.

8 Hugger, S., Aidam, R., Bronner, W., Fuchs, F., Lösch, R., Yang, Q., Wagner, J., Romasew, E., Raab, M., Tholl, H.D., Höfer, B., and Matthes, A.L. (2010) Power scaling of quantum cascade lasers via multiemitter beam combining. *Opt. Eng.*, **49**, 111111.

9 Lee, B., Kansky, J., Goyal, A., Pflügl, C., Diehl, L., Belkin, M., Sanchez, A., and Capasso, F. (2009) Beam combining of quantum cascade laser arrays. *Opt. Express*, **17**, 16216–16224.

10 Bloom, G., Larat, C., Lallier, E., Carras, M., and Marcadet, X. (2010) Coherent combining of two quantum-cascade lasers in a Michelson cavity. *Opt. Lett.*, **35**, 1917–1919.

11 Leger, J., Swanson, G., and Vedkamp, W. (1986) Coherent beam addition of GaAlAs lasers by binary phase gratings. *Appl. Phys. Lett.*, **48**, 888–890.

12 Hoffmann, L., Klinkmüller, M., Mujagic, E., Semtsiv, M., Schrenk, W., Masselink, W., and Strasser, G. (2009) Tree array quantum cascade laser. *Opt. Express*, **17**, 649–657.

13 Pedersen, C. and Skettrup, T. (1996) Laser modes and threshold conditions in N-mirror resonators. *J. Opt. Soc. Am. B*, **13**, 926–937.

14 Sabourdy, D., Kermene, V., Desfarges-Berthelemot, A., Lefort, L., Barthelemy, A., Even, P., and Pureur, D. (2003) Efficient coherent combining of widely tunable fiber lasers. *Opt. Express*, **11**, 87–97.

15 Cao, J., Hou, J., Lu, Q., and Xu, X. (2008) Numerical research on self-organized coherent fiber laser arrays with circulating field theory. *J. Opt. Soc. Am. B*, **25**, 1187–1192.

16 Dammann, H. and Görtler, K. (1971) High-efficiency in-line multiple imaging by means of multiple phase holograms. *Opt. Commun.*, **3**, 312–315.

17 Dammann, H. and Klotz, E. (1977) Coherent optical generation and inspection of two-dimensional periodic structures. *Opt. Acta*, **24**, 505–515.

18 Zhou, C. and Liu, L. (1995) Numerical study of Dammann array illuminators. *Appl. Opt.*, **34**, 5961–5969.

19 Bloom, G., Larat, C., Lallier, E., Lehoucq, G., Bansropun, S., Lee-Bouhours, M.-S.L., Loiseaux, B., Carras, M., Marcadet, X., Lucas-Leclin, G., and Georges, P. (2011) Passive coherent beam combining of quantum-cascade lasers with a Dammann grating. *Opt. Lett.*, **36**, 3810–3812.

20 Sidick, E., Knoesen, A., and Mait, J. (1993) Design and rigorous analysis of high-efficiency array generators. *Appl. Opt.*, **32**, 2599–2605.

21 Ehbets, P., Herzig, H., and Prongué, D. (1992) High-efficiency continuous surface-relief gratings for two-dimensional array generation. *Opt. Lett.*, **17**, 908–910.

22 Bouchitte, G. and Petit, R. (1985) Homogenization techniques as applied in the electromagnetic theory of gratings. *Electromagnetics*, **5**, 17–36.

23 Lalanne, P. and Lemercier-Lalanne, D. (1996) On the effective medium theory of subwavelength periodic structures. *J. Mod. Opt.*, **43**, 2063–2086.

24 Farn, M.W. (1992) Binary gratings with increased efficiency. *Appl. Opt.*, **31**, 4453–4458.

25 Stork, W., Streibl, N., Haidner, H., and Kipfer, P. (1991) Artificial distributed-index media fabricated by zero-order gratings. *Opt. Lett.*, **16**, 1921–1923.

26 Haidner, H., Kipfer, P., Sheridan, J.T., Schwider, J., Streibl, N., Collischon, M., Hutfless, J., and Marz, M. (1993) Diffraction grating with rectangular grooves exceeding 80% diffraction efficiency. *Infrared Phys.*, **34**, 467–475.

27 Sauvan, C., Lalanne, P., and Lee, M.-S.L. (2004) Broadband blazing with artificial dielectrics. *Opt. Lett.*, **29**, 1593–1595.

28 Ribot, C., Lalanne, P., Lee, M.-S.L., Loiseaux, B., and Huignard, J.-P. (2007) Analysis of blazed diffractive optical elements formed with artificial dielectrics. *J. Opt. Soc. Am. A*, **24**, 3819–3826.

29 Lu, F., Sedgwick, F.G., Karagodsky, V., Chase, C., and Chang-Hasnain, C.J. (2010) Planar high-numerical-aperture low-loss focusing reflectors and lenses using subwavelength high contrast gratings. *Opt. Express*, **18**, 12606–12614.

30 Lee, M.-S.L., Bansropun, S., Huet, O., Cassette, S., Loiseaux, B., Wood, A., Sauvan, C., and Lalanne, P. (2006) Sub-wavelength structures for broadband diffractive optics, in *ICO20: Materials and Nanostructures*, vol. **6029** (eds W. Lu and J. Young), Proceedings of the SPIE, pp. 297–303.

31 Miller, J.M., de Beaucoudrey, N., Chavel, P., Cambril, E., and Launois, H. (1996) Synthesis of a subwavelength-pulse-width spatially modulated array illuminator for 0.633 μm. *Opt. Lett.*, **21**, 1399–1402.

32 Lalanne, P. and Hutley, M. (2003) *Artificial Media Optical Properties: Subwavelength Scale*, Marcel Dekker, New York.

33 Hugonin, J.P. and Lalanne, P. (1995) Reticolo software for grating analysis.

34 Moharam, M.G., Grann, E.B., Pommet, D. A., and Gaylord, T.K. (1995) Formulation for stable and efficient implementation of the rigorous-coupled wave analysis of binary gratings. *J. Opt. Soc. Am. A*, **12**, 1068–1076.

35 Lalanne, P. and Morris, G.M. (1996) Highly improved convergence of the coupled-wave method for TM polarization. *J. Opt. Soc. Am. A*, **13**, 779–784.

36 Granet, G. and Guizal, B. (1996) Efficient implementation of the coupled-wave method for metallic lamellar gratings in TM polarization. *J. Opt. Soc. Am. A*, **13**, 1019–1023.

37 Li, L. (1997) New formulation of the Fourier modal method for crossed surface-relief gratings. *J. Opt. Soc. Am. A*, **14**, 2758–2767.

38 Bloom, G., Larat, C., Lallier, E., Lee-Bouhours, M.-S.L., Loiseaux, B., and Huignard, J.-P. (2011) Design and optimization of a high-efficiency array generator in the mid-IR with binary subwavelength grooves. *Appl. Opt.*, **50**, 701–709.

14
Phase-Conjugate Self-Organized Coherent Beam Combination
Peter C. Shardlow and Michael J. Damzen

14.1
Introduction

Self-organized coherent beam combination (SOCBC) is a well-established technique for passive combination of multiple laser oscillators [1–3]. In the SOCBC systems, multiple laser oscillators are coupled to a single output coupler (OC) by a beam splitter/combiner arrangement, as shown conceptually in Figure 14.1.

This composite cavity contains a series of "lossy" outputs and one arm containing an output coupler of reflectivity r. The combined effect of the single shared output coupler and loss channels is that the effective reflectivity of the output coupler, as seen from the amplifier modules, is dependent on the phase and power relationship between the combining beams. If all the combining beams are phased such that they undergo constructive interference into the channel with the output coupler, the power reflected back into the cavity arms is higher than under any other phase condition. This phase-dependent output coupler reflectivity leads to some modes experiencing lower cavity round-trip losses, which in turn lead the system to self-organize to operate on phase-locked modes that coherently combine together into a single output.

In conventional laser cavity modules, the finite phase offset of the beams combining at the coupler is constrained by the cavity length and positioning of the beam combiner. This leads the composite cavity to self-adjust its operational wavelength in order to correctly phase the beams for combination. In an individual laser cavity, there are distinct allowed wavelengths that form a standing wave, and thus the wavelength operation is constrained to one of these spectral modes. When multiple lasers are locked together in a SOCBC system, all the modules must find a single spectral mode that is common to all their individual cavities. Unfortunately, as the number of modules is increased, the probability of finding a shared spectral mode for high-efficiency combination drops dramatically [4–6]. In essence, the existence of finite spectral modes within a laser resonator limits the number of modules that can be combined. If a laser resonator could be identified that had a greater degree of freedom in spectral operation, it would reduce, or even remove, the scaling limitations of self-organized coherent beam combination.

Coherent Laser Beam Combining, First Edition. Edited by Arnaud Brignon.
© 2013 Wiley-VCH Verlag GmbH & Co. KGaA. Published 2013 by Wiley-VCH Verlag GmbH & Co. KGaA.

14 Phase-Conjugate Self-Organized Coherent Beam Combination

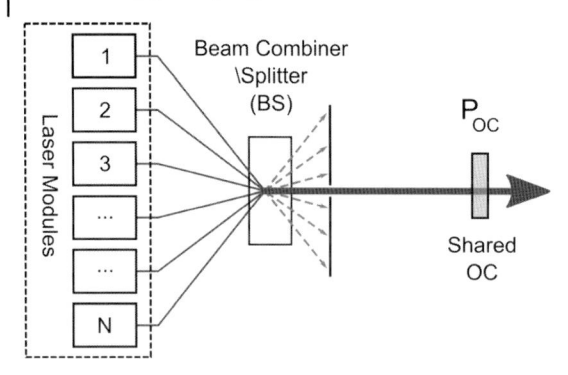

Figure 14.1 Schematic showing self-organized coherent beam combination utilizing a beam splitter/combiner and a single output coupler to phase lock individual laser modules into a single coherent output P_{OC}.

As will be explained in this chapter, resonators that do not have predefined spectral modes can be realized in oscillators that contain a phase-conjugate mirror (PCM). A phase-conjugate mirror has interesting properties, in that the reflected beam is wavefront reversed, traveling back along the path through which it came. Phase-conjugate mirrors are usually based on four-wave mixing (FWM) within a nonlinear medium, where interfering beams write a dynamic grating (or hologram) into a parameter of the medium such that coupling between the different beams occurs. This is a situation akin to writing a hologram that is simultaneously written and read by the interacting beams. Using such a holographic element in a laser cavity allows greater flexibility in the spectral modes of the output, as the cavity length is not predefined before the hologram has formed. Figure 14.2 shows, in the inset, a conceptual diagram of a self-pumped FWM resonator for the generation of a phase-conjugate output. A transmission grating is located between beams A_1 and A_3 that creates a modulation in the polarizability of the nonlinear medium. This grating completes a ring cavity, which, given sufficient gain within the loop, would reach threshold for oscillation, generating a phase-conjugate output beam. Through

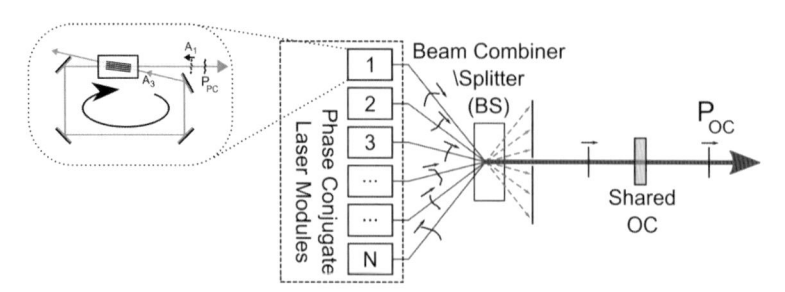

Figure 14.2 Schematic of a PCSOCBC system showing a combination of multiple phase-conjugate modules utilizing a beam splitter/combiner and a single output coupler to phase lock multiple modules into a single coherent output P_{OC}. *Inset*: The phase-conjugate module shown is based around FWM utilizing a gain grating interaction.

sufficient gain and the introduction of an output coupler, it is possible to operate a self-pumped phase-conjugate mirror as a self-starting laser oscillator [7–10].

In the phase-conjugate self-organized coherent beam combination (PCSOCBC), a setup similar to SOCBC setup is constructed, but with conventional laser modules replaced with self-starting phase-conjugate oscillators [11]. Figure 14.2 shows a conceptual schematic of a PCSOCBC system, where multiple self-pumped phase-conjugate modules based on a four-wave mixing geometry are coupled together. Using phase-conjugate modules has the benefit that the cavity does not predefine the spatial and spectral output modes or the finite phase offset. This allows an individual module in a PCSOCBC system to be always able to adapt its output mode to correctly phase the beams for constructive interference, independent of the number of modules. Furthermore, the phase conjugation of the holographic modules allows aberration correction and can coherently combine even in the presence of thermally induced distortions within the amplifier, oscillator, and coupling optics.

This chapter will first give a general overview of phase conjugation before introducing gain holography and how it can be experimentally used to generate a self-starting laser resonator that does not have predefined cavity modes. How these laser resonators can then be integrated into a PCSOCBC system is then discussed along with experimental verification. Finally, the chapter discusses the power scaling opportunities of a PCSOCBC system.

14.2
Phase Conjugation

Phase conjugation or time reversal, as it is known in some of the literature, is a nonlinear optical process where a reflected beam is created that reverses both the direction and the phase of the incident beam [12–14]. This allows double-pass wavefront correction of a beam experiencing aberration.

Figure 14.3a shows the reflection of an incident plane wave from a conventional mirror. A plane wave passes through an aberrating medium before being incident onto a standard mirror, where the beam is reflected in the direction perpendicular to the mirror's surface. It then passes through the aberrating plate again, resulting in a double distorted wavefront. In contrast, Figure 14.3b shows a similar situation, but with the conventional mirror replaced by a phase-conjugate mirror. The phase-conjugate mirror acts to reverse all components of the propagation vector, such that the reflected beam counter propagates to the incident beam. The phase-conjugate mirror reverses the phase of the wavefront, such that on the second pass of the aberrating plate, the initial plane wave is reproduced.

A mathematical description of a phase-conjugate wave E_c can be considered in terms of an incident wave E_i:

$$E_i(\vec{v}, t) = \frac{1}{2}\left[|A_i(\vec{v})| \exp\left(i\left(\omega_i t - \vec{k}_i \vec{v} + \varphi_i(\vec{v})\right)\right)\right] + \text{c.c.} \qquad (14.1)$$

$$E_c(\vec{v}, t) = \frac{1}{2}\left[r_c |A_i(\vec{v})| \exp\left(i\left(\omega_i t + \vec{k}_i \vec{v} - \varphi_i(\vec{v})\right)\right)\right] + \text{c.c.} \qquad (14.2)$$

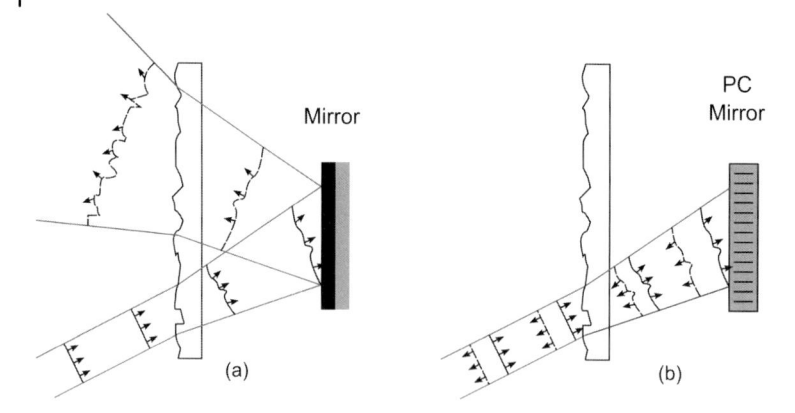

Figure 14.3 (a) Reflection of a plane wave that is passed through an aberrating plate by a conventional mirror. (b) Reflection of a plane wave that is passed through the same aberrating plate by a phase-conjugate mirror (PCM). The plane wave is reproduced after it passes back through the aberrating phase plate.

Here $A_i(\vec{v})$ is the spatial amplitude of the incident beam, ω_i is the incident frequency, k_i is the wave vector, $\varphi(\vec{v})$ is the phase of the wavefront, and r_c is the coefficient of reflectivity from the phase-conjugate mirror. The phase-conjugate beam reproduces the spatial amplitude and has reversal of both the wave vector ($k_c = k_i$) and the wavefront ($\varphi_c(\vec{v}) = -\varphi_i(\vec{v})$).

To generate a phase-conjugate mirror, a material with a nonlinear optical susceptibility is required. This most commonly involves the use of materials where the optical fields induce a change in the refractive index of the material. When two or more coherent beams interact within such media, the interference pattern formed spatially modulates the optical susceptibility, forming a self-written grating that couples the beams together [15–19]. This is most commonly implemented in photorefractive material, where the interference pattern leads to a modification of the charge distribution within the medium. The generated internal electric field spatially modulates the refractive index via the electro-optic effect, resulting in a grating that diffracts the light from the writing beams. Although using photorefractives for generation of a phase-conjugate output via a FWM interaction has been the most common, it is also possible to use many other media where nonlinearity leads to a change in the refractive index, such as in Kerr or absorbing media.

It has also been seen that the use of saturable gain media, where the optical matter interaction causes stimulated emission, can generate a phase-conjugate mirror. In the situation where multiple coherent beams interfere within the laser active media, spatial modulation of the population inversion is induced. This grating in the population inversion has been seen to allow coupling of the multiple laser beams, allowing demonstration of the phase conjugation via four-wave mixing, self-pumped phase conjugation [20], adaptive interferometry via two-wave mixing [21], and self-starting self-adaptive laser oscillation [22]. Significant research interest has been garnered toward the use of saturable gain media as an alternative to photorefractive materials, as the timescales of interaction are relatively short (approximately in microseconds), the inherent gain present can lead to very high phase-conjugate reflectivity, and there is no absorption required for the grating writing, which allows scaling to high powers [8,16,17,20,22–29]. In addition to the PCSOCBC system, this chapter will focus on the use of gain grating interactions as they offer simple and high-efficiency beam coupling, as all the interaction and amplification are contained within a single element.

14.2.1
Gain Holography

The key dynamics of gain gratings induced by amplification of the optical radiation in a four-level saturable gain medium is governed by the rate equation for the inversion density. It is convenient to write this in terms of local gain coefficient $\alpha(\vec{v},t)$ [13,30]:

$$\frac{\partial \alpha(\vec{v},t)}{\partial t} = R(t) - \frac{1}{U_S} I_T(\vec{v},t)\alpha(\vec{v},t) - \frac{\alpha(\vec{v},t)}{\tau_\lambda}, \qquad (14.3)$$

where $\alpha(\vec{v},t)$ is the gain coefficient, $R(t)$ is the spatially independent pump rate, $I_T(\vec{v},t)$ is the optical field intensity, and $U_S = h\nu/\sigma_e$ is the saturation fluence of the medium, where ν is the laser frequency, σ_e is the stimulated emission cross section, and τ_λ is the upper-state lifetime of the medium.

Figure 14.4 shows the interaction of two coherent beams within a gain media. Two angularly separated beams form an interference pattern $I(\vec{v},t)$ within the gain media. The optical fields can be written in the form $E_i(t) = 1/2 A_i(t) e^{i k_i \cdot v - i\omega t} + \text{c.c.}$, where k_i is the wave vector of each beam, ω is the angular frequency, which is taken to be degenerate for the two fields, and $A_i = |A_i| e^{i\varphi_i}$ is the complex amplitude of the optical fields. For the interaction geometry shown in Figure 14.4 with z-axis the bisector of the two field directions, the interference pattern has fringes along the x-axis. The resultant time-averaged intensity pattern over one optical cycle is therefore constant along the z-axis, but sinusoidally modulated in the x-axis.

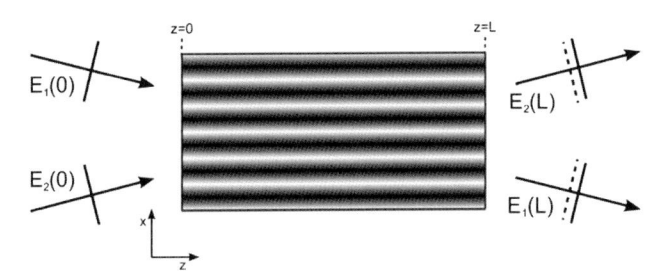

Figure 14.4 Transmission grating is formed in the population inversion due to the interaction of the writing beams E_1 and E_2. The two beams emanating from the exit of the gain medium $[E_1(L)$ and $E_2(L)]$ are formed of an amplified component and a diffracted component from the other beam.

Figure 14.5a shows this interference pattern along the x-axis for the situation where two coherent beams of the same amplitude are incident. Figure 14.5b shows the resultant steady-state grating, which is written into the population inversion for three different writing beam intensities (normalized to the medium saturation intensity I_S). It is important to note that the interference pattern is out of phase with the gain grating as the areas of high writing beam intensity lead to areas of low population inversion. This leads to the diffracted beams that are out of phase with the writing beams. It is also clear that with high-intensity writing beams, the gain grating modulation is non-sinusoidal instead of containing higher order harmonic components.

In the steady-state regime, by expanding the periodic spatially modulated gain coefficient as a Fourier series and using Maxwell's wave equations, it has already been shown that coupled wave equations for the interaction of two optical fields in the gain medium are as follows [21,31]:

$$\frac{\partial A_1}{\partial z} = \gamma A_1 + \kappa A_2. \tag{14.4}$$

$$\frac{\partial A_2}{\partial z} = \gamma A_2 + \kappa A_1. \tag{14.5}$$

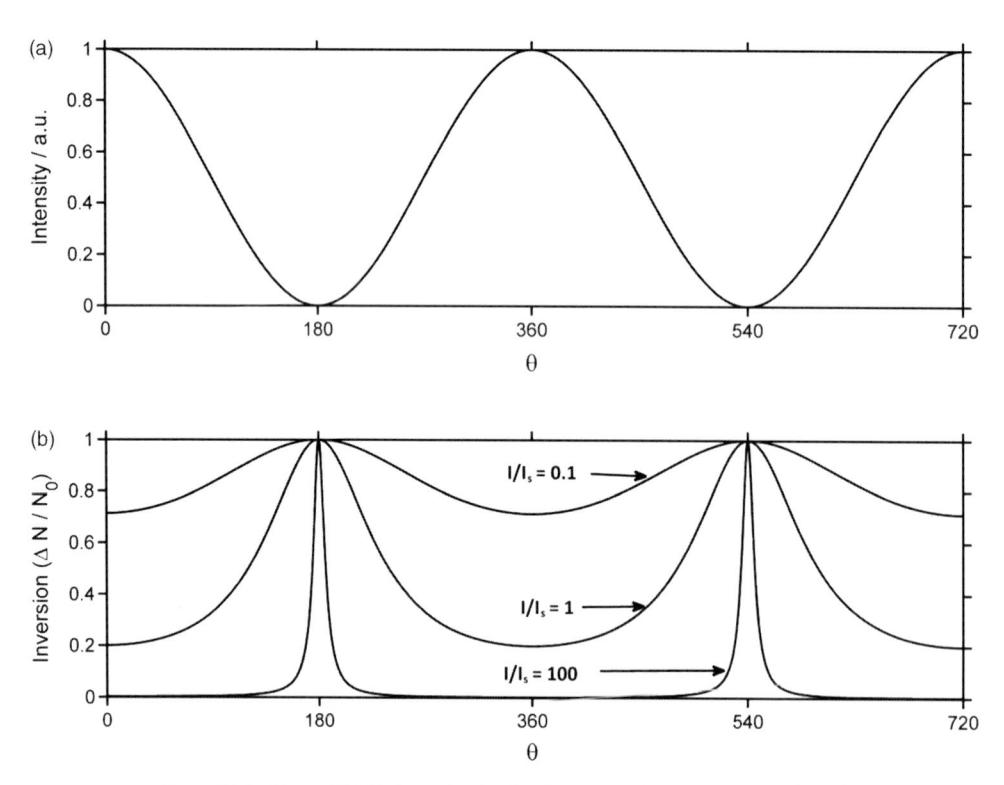

Figure 14.5 Plots of (a) the intensity distribution pattern between two interfering beams of equal amplitude and (b) the corresponding modulation in the population inversion as a function of the saturation intensity.

Here $\gamma = \alpha^{(0)}/2$ and $|\kappa|e^{i\varphi} = (\alpha^{(1)}/4)e^{i\varphi}$, where $\alpha^{(0)}$ and $\alpha^{(1)}$ are the zero- and first-order coefficients of the Fourier expansion of spatially modulated gain coefficient. In the weak saturation regime ($I/I_s \ll 1$), they can be approximated as $\alpha^{(0)}(z) = \alpha_0$ and $\alpha^{(1)}(z) = -(2\alpha_0) \times \{[I_1(0)I_2(0)]^{1/2}/I_s\}\exp(\alpha_0 z)$ [21]. The first term in Eqs. (14.4) and (14.5) represents the amplification of each beam in the average (saturated) gain and the second term derives from the diffraction of one beam from the first harmonic gain modulation into the direction of the other beam. The zero- and first-order gain terms are themselves functions of the optical fields; hence, it is a coupled self-interaction in which beams writing the gain grating are simultaneously modified by Bragg-matched diffraction from the same grating.

Full solution of the two-beam coupling for arbitrary saturation can be performed by numerical integration of the coupled equations (14.4) and (14.5). However, it is interesting to consider the case of weak saturation. This approximation has been shown to allow an analytical solution for the amplified reference beam ($A_1(L)$) after a length L of gain medium [21]:

$$A_1(L) = A_1(0) \cdot t_a + A_2(0) \cdot r, \tag{14.6}$$

where the amplitude transmission coefficient is

$$t_a = e^{-\alpha_0 L/2} \tag{14.7}$$

and the amplitude reflection (diffraction) coefficient (r) is given by

$$r = -\frac{1}{2}\frac{A_1(0)A_2^*(0)}{I_s}\left(e^{\alpha_0 L} - 1\right)e^{\alpha_0 L/2}. \tag{14.8}$$

It is seen that the output field is the superposition of a transmitted field component and a diffracted component of the other field. This is analogous to the interaction of two beams at a beam splitter with amplitude transmission t_a and reflection r. Except in this case, the transmission and reflectivity can be greater than unity due to the amplification of the gain medium. In a normal beam splitter, conservation of energy also requires that high reflectivity is linked to low transmission. This is not necessarily a limitation for the case of a gain grating interaction.

It is noted that the on-resonance amplitude reflectivity has a negative sign and the diffracted term ($r \cdot A_2(0)$) is therefore destructive with the transmission term ($t_a \cdot A_1(0)$). The destructive interference arises from the gain grating being in antiphase to the intensity interference pattern, as shown in Figure 14.5. The creation of a gain grating leads to reduced output in the steady state and is a general feature of gain saturation, where the spatial hole burning leads to regions of underused gain at the nodes (minima) of the interference pattern.

14.2.2
Four-Wave Mixing within a Saturable Gain Media

The amplified outputs from the gain medium in a two-wave mixing geometry contain an amplified component and a component that is diffracted from the other

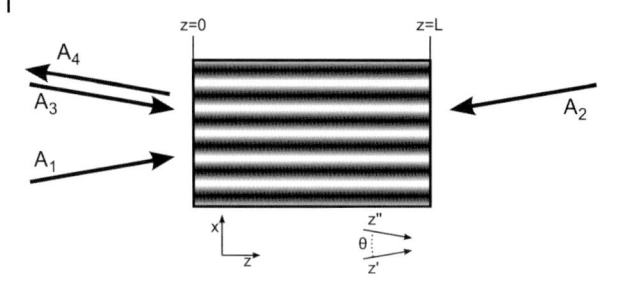

Figure 14.6 Schematic of four-wave mixing in a saturable gain media. Beams A_1 and A_3 write a transmission grating into the population inversion of the gain medium. Beam A_2 diffracts from the grating, with a π phase shift, into beam A_3, which is the phase conjugate of the writing beam A_3 ($A_4 \propto A_3^*$).

beam. Introduction of a third beam that is phase matched to this grating can lead to the generation of a phase-conjugate beam via four-wave mixing [15–17,27,32,33].

Figure 14.6 shows the transmission grating formed by the angular intersection of beams A_1 and A_3 within a saturable gain medium. They are separated by an angle θ such that their propagation directions z' and z'' are equally angled from direction z, leading to a grating being formed into the population inversion modulated along the x-axis. Beam A_2 is set to counterpropagate to beam A_1 and as such will be correctly phased to diffract from the gain grating into beam A_4. If A_2 is the phase conjugate of beam A_1, the so-called FWM pump beams, this generated beam would be the phase conjugate of beam A_3. In practice, it is common to use large-area, relative to beam A_3, high-quality lasers as beams A_1 and A_2, which can be approximated as plane waves, and are thus to all extents and purposes the phase conjugate of one another.

14.2.3
Self-Pumped Phase Conjugation

It is possible to design an optical system where the FWM gain grating writing beams are self-formed. Such an optical system allows generation of a phase conjugate of an incident beam via a FWM interaction. Figure 14.7 shows a schematic of such a self-pumped phase conjugation system. An incident beam A_1 is passed into a gain medium where it is amplified. This amplified beam is then passed clockwise around the loop before reentering the gain medium as beam A_3. This self-intersecting beam writes a transmission gain grating into the population inversion of the medium. This grating acts as a diffractive element that completes a ring cavity, allowing the formation of a self-started oscillation in the counter-clockwise direction.

This loop oscillation has been shown to only form a self-consistent mode if the beams A_2 and A_4 are phase conjugates of beams A_1 and A_3, respectively [34]. This means that any distortions due to the cavity optics and the passage of the gain medium would be encoded into this diffraction grating, causing the output beam to be the phase conjugate of the incident beam A_1.

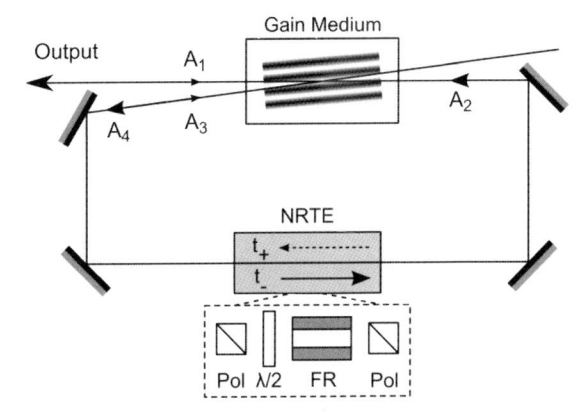

Figure 14.7 Schematic showing the concept of self-pumped phase conjugation. Input beam A_1 is passed through a self-intersecting loop geometry to form a grating such that beam A_2 can diffract from it, completing the loop and generating a phase-conjugate output. The grating writing beam intensities are matched by inclusion of a nonreciprocal transmission element that contains two polarizers (POL), a Faraday rotator (FR) and a half-wave plate ($\lambda/2$).

It is known that to maximize the diffraction efficiency from the gain grating, it is important to have a high-contrast gain grating. For this condition, it is important that beams A_1 and A_3 are of equal powers. As A_3 is formed from the amplified beam A_1, it is important to reduce its power before the self-intersection within the gain medium. It is also important that the beam traveling in the opposite direction is not reduced in power too much, as it is the oscillating field of the ring cavity. This can be performed by using a nonreciprocal transmission element (NRTE) formed of a Faraday rotator (FR) and $\lambda/2$ wave plate encased between two polarizers. This allows the transmission in two directions to be optimized by the rotation of the wave plate.

Here it is also worth noting that in the self-pumped phase conjugate setup, where a transmission gain grating is written into the gain medium to form a ring cavity, the beam diffracted from the gain grating A_4 will be out of phase with the beam that wrote the grating, A_3. This means that the system would not be able to form a self-consistent mode on round-trip of the cavity. Thankfully, the addition of the NRTE induces a π-phase shift between the two transmission directions, allowing a self-consistent mode to be formed.

14.2.3.1 Seeded Self-Pumped Phase-Conjugate Module

It has been discussed that by using self-intersecting loop geometry, the gain grating formed within a saturable laser medium allows phase-conjugate reflection of an input beam. The key parameter for efficient operation of gain grating-based self-adaptive phase-conjugate laser system is using a high-gain laser amplifier, as this is the key parameter that drives the interaction. Figure 14.8 shows an example of a schematic of a seeded adaptive laser system utilizing a diode-pumped bounce geometry Nd:YVO$_4$ laser amplifier [22,35]. This amplifier has been shown to naturally avail itself to such applications, as it shows extremely high small-signal

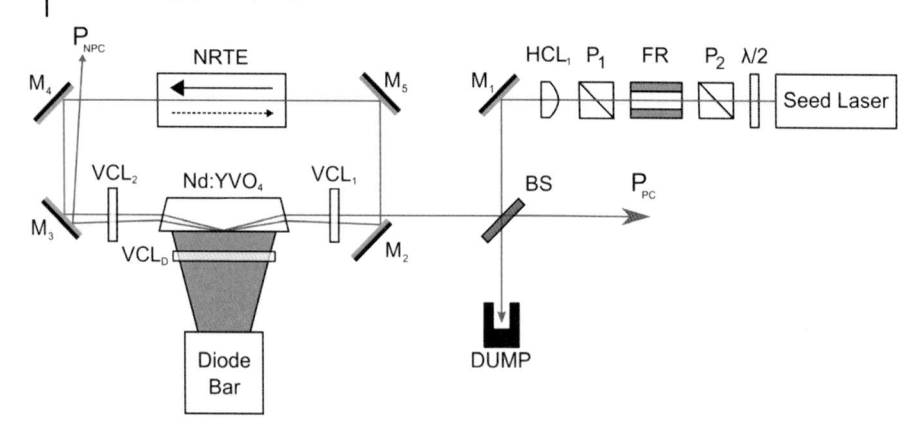

Figure 14.8 Schematic of the seeded adaptive laser. A 1064 nm seed laser is incident into the self-intersecting loop cavity. The self-intersecting loop generates an output beam, which is the phase conjugate of the input seed laser.

gains (>40 dB), with efficient mode matching of the pump region to the laser mode enabling high-efficiency operation [35]. Although further discussion in this chapter will focus on PCSOCBC utilizing, this laser amplifier geometry operation utilizing many other gain media where self-pumped phase conjugation has been demonstrated would be straightforward, such as in gas, solid-state, and dye systems among others.

In the system depicted in Figure 14.8, the gain medium is a 1.1 at.% $Nd:YVO_4$ laser crystal side-pumped by a 50 W diode bar, which is vertically focused onto the side face of a 20 mm × 5 mm × 2 mm slab crystal. Two $f = 50$ mm vertically focusing cylindrical lenses aligned on either side of the amplifier crystal act to vertically match the cavity beams to the gain region. A TEM_{00} 1064 nm 300 mW seed laser is passed through a Faraday isolator, which is used to control the power and stop reflections back into the seed laser, and a collimating lens. This is then passed into a low-reflectivity ($r \approx 3\%$) beam splitter to generate an 8.1 ± 0.1 mW seed beam that is incident onto the gain medium. The amplified seed is passed around the loop before self-intersecting within the gain medium, with an angular separation of $\sim 2°$ between the two beams. The loop contains a nonreciprocal transmission element, which contains a $45°$ Faraday rotator and a $\lambda/2$ wave plate encased between two vertical transmitting polarizers. Rotation of the half-wave plate allows variation of the forward and backward transmission factors in order to equalize the intensity of the two gain grating writing beams to maximize the grating contrast and thus the diffraction efficiency.

Once the laser gain medium is pumped to sufficient levels of population inversion and the NRTE appropriately set for generating a high diffraction efficiency from the gain grating, a counterclockwise ring cavity is formed that generates a beam that is phase conjugate of the input seed. This demonstration system was seen to yield a phase-conjugate output of 11.5 W from 47 W of pump power with 41% slope efficiency. This represents a phase-conjugate reflectivity of $>1400\times$.

The phase-conjugate output shows the same spatial and spectral properties as the seed laser, with a high-quality TEM_{00} output observed. As the gain grating is written by the seed laser, the oscillation spectrum of the ring cavity formed is locked to the spectral mode of the seed. It is interesting to consider that in this situation, as opposed to pumping a conventional externally resonant cavity, the ring cavity that is formed does not predefine its own spectral modes. Any spectral mode of the seed laser can be reproduced as long as it is within the gain bandwidth of the laser active medium.

14.2.3.2 Self-Starting Self-Adaptive Gain Grating Lasers

For the implementation of any stable self-organizing beam combination scheme, it is necessary that all the individual oscillators phase locked are beneficial for the oscillation of individual modules. In a SOCBC system, this is accomplished by having a phase locking-dependent feedback from the output coupler. This is not the case for a seeded self-pumped phase conjugation system, like the one discussed, but it is possible to simply adjust such a system so that it would self-generate the seed beam, giving an output power-dependent seeding to write the diffraction grating.

The so-called self-starting self-adaptive lasers replace the seed laser already utilized with a simple low-reflectivity output coupler. In this system, the seed beam that forms the self-pumped phase-conjugate loop builds up purely from the amplified spontaneous emission (ASE) from the laser amplifier. Although the ASE is of low coherence, it has been seen to build up a suitable grating that spectrally narrows itself during start-up to produce a TEM_{00} single longitudinal mode output [36].

Figure 14.9 shows a schematic of an implementation of a self-starting self-adaptive laser system. The laser crystal and pumping optics are the same as discussed previously for the seeded adaptive loop, with the seed laser replaced by an output coupler wedge with 1% reflectivity. The loop has also been expanded to

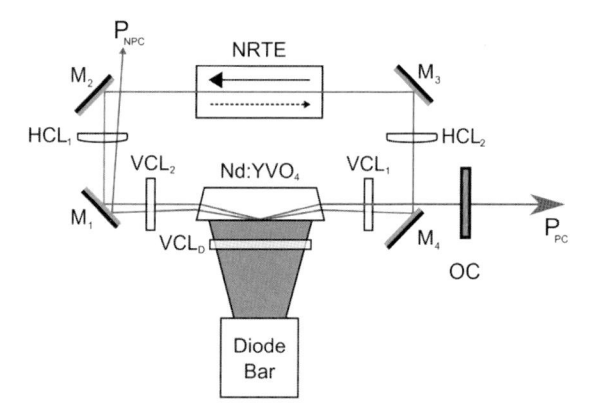

Figure 14.9 Schematic of a self-starting self-adaptive laser based around a bounce geometry laser amplifier. The seed laser is replaced with a 1% reflectivity output coupler that causes the system ASE to self-form a seed beam.

~700 mm with the inclusion of a 4-f imaging telescope in the horizontal direction by the lenses HCL_1 and HCL_2 of focal lengths 200 and 150 mm, respectively. This 4-f imaging system images one end of the gain medium onto the other end of the gain medium, irrespective of the strength of the thermally induced lens. This has the advantage of improving the stability of the system, as it guarantees that the gain grating fully samples the optical field irrespective of the thermal lens strength. The self-intersecting loop beam is initially incident onto the crystal giving $\sim 7°$ internal bounce angle, and on the second pass this internal bounce angle is reduced to $\sim 5°$. The use of two lenses of different focal lengths causes the second pass of the writing beam to enter the crystal with a smaller horizontal extent, compensating for the apparent reduction of the gain size due to the reduced bounce angle. As the pump power was increased, the ASE from the gain medium was sufficient to form a gain grating within the amplifier that allowed oscillation of the loop cavity via diffraction from the gain grating.

The output power from this demonstration system at the maximum pump power of 52.5 ± 0.3 W was 15.6 ± 0.2 W with 0.68 ± 0.01 W leaving the nonphase-conjugate (NPC) output. This corresponds to a maximum optical efficiency of $30.0 \pm 0.5\%$, with a slope efficiency of $54 \pm 1\%$.

It is interesting to consider what spectral requirements are predetermined by this laser oscillator. In the system with the external seed, any imposed wavelength was possible as long as it was within the gain bandwidth. In the self-starting system, the same is true: The cavity does not predetermine spectral modes before it is pumped. This allows the laser to self-organize itself to operate on the spectral mode or on the subset of modes, which generates the highest diffraction efficiency, increasing the power oscillating in the ring cavity. The gain grating in the self-starting self-adaptive laser is generally seen to lead to spectral narrowing and resultant operation on a single longitudinal mode [10,36].

14.3
PCSOCBC

In the PCSOCBC, a setup similar to SOCBC setup is constructed, but with conventional laser modules replaced with self-starting phase-conjugate modules of the type described earlier in this chapter. It is known for SOCBC setup that the modules self-adjust their operational parameters for constructive interference into the channel with the output coupler, as this maximizes the intracavity mode power. In linear cavity modules, it is not possible to just adjust the finite phase offset of the individual modules, as this is constrained by the cavity length and positioning of the beam combiner. Instead, the cavity self-adjusts its operational wavelength in order to correctly phase the beams for combination at the beam combiners. As the number of modules is increased, the probability of finding a spectral mode shared between all the cavities, which is required for high-efficiency combination, drops dramatically [4–6]. Using phase-conjugate modules has the benefit that the cavity does not predefine the spatial and spectral output modes or the finite phase offset. This allows a module in a PCSOCBC system to be always able to adapt its output mode to correctly phase the beams for constructive interference, independent of the number of modules. Furthermore, the phase conjugation of the holographic modules allows aberration correction and can coherently combine even in the presence of thermally induced distortions within the amplifier modules.

14.3.1
CW Experimental PCSOCBC

Initial experimental implementation of a PCSOCBC was investigated utilizing the self-starting self-adaptive modules that have been discussed earlier in this chapter. These modules are advantageous base for an experimental implementation, as they have been demonstrated with high powers and beam qualities, while being compact and robust [7,8,20,22,24,25].

The experimental demonstration set up for the phase-conjugate self-organized coherent beam combination of two self-adaptive laser modules is shown in Figure 14.10 [11]. Each module is built to the same specification, as the imaging self-starts with each one containing a 1.1 at.% $Nd:YVO_4$ laser crystal operated in the bounce geometry [37–41]. These lenses horizontally image-relay the self-intersecting beam back into the gain medium.

The SOCBC is accomplished by a Vernier–Michelson-type coupling of the adaptive modules, provided by a 50:50 beam splitter (BS) and a common, 1% reflectivity, output coupler. The combined beam that passes through the output coupler is defined as the output channel (P_{OC} in Figure 14.10) and the other output the loss channel (P_{LC} in Figure 14.10). The distance from the output coupler to the amplifier crystal within both adaptive modules was set to be roughly similar.

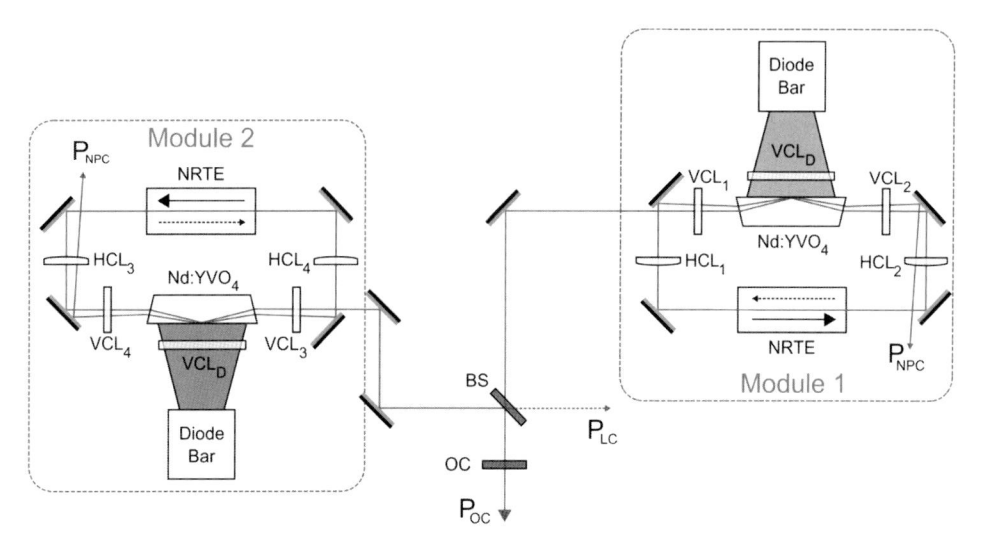

Figure 14.10 Schematic showing PCSOCBC of two self-starting self-adaptive laser modules [11].

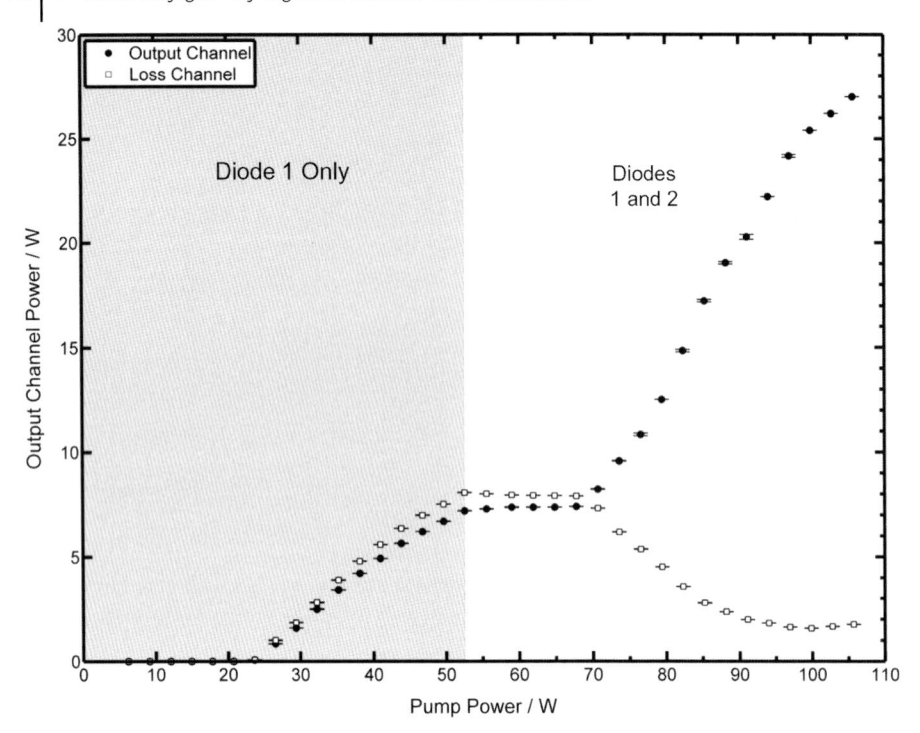

Figure 14.11 Output power from the demonstration of PCSOCBC system. The gray section on the left-hand side represents where only one module is being pumped by its diode. As the second module's diode is activated on the right-hand side of the graph, CBC occurs readily [11].

The powers into the output channel (P_{OC}) and the loss channel (P_{LC}) over variation in the pumping conditions are shown in Figure 14.11. The gray section indicates when only the pump diode of module one is active and thus no CBC is possible. Slightly lower power is seen in the output channel due to the beam splitter reflectivity tolerance and the 1% reflectivity feedback from the output coupler. In the right half of Figure 14.11, module 1 is left at full pump power, while the pump power of module 2 is increased. Once module 2 reaches threshold (∼70 W total pumping), CBC happens readily, with the power into the output channel increasing rapidly. This is accompanied by a reduction of power into the loss channel. At the maximum pump power of 107 W (53.5 W per module), a coherently combined output beam of 27 W was observed with only 1.75 W entering the loss channel. This represents a combination efficiency of

$$\eta = \frac{P_{OC}}{P_{OC} + P_{LC}} = 94\%. \tag{14.9}$$

The far-field beam profile of the coherently combined output channel is shown in Figure 14.12a. The effects of the phase-conjugate self-organized CBC mechanism on beam quality have been investigated. This entailed a comparison of the beam quality

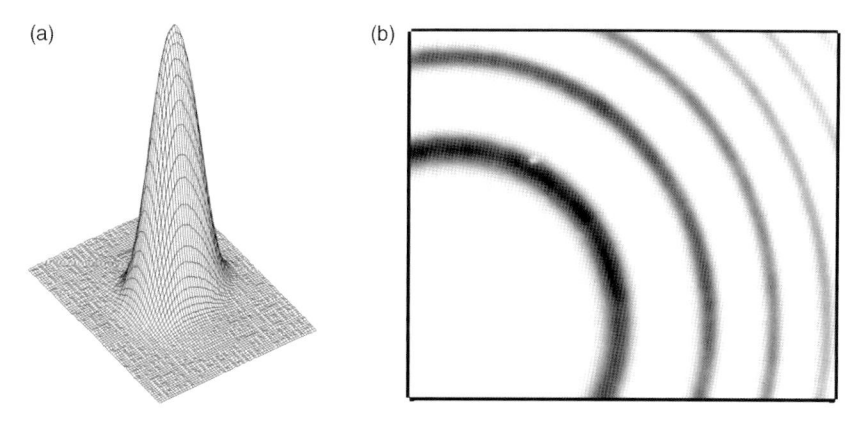

Figure 14.12 (a) Figure showing the far-field beam quality of the output channel from the image-coupled adaptive system. (b) Figure showing the Fabry–Pérot spectrum of the image-coupled adaptive system [11].

of the independent modules with the coherently combined output and the output into the loss channel. The measured independent operation M^2 of module 1 (module 2) in the horizontal and vertical directions was 1.93 (2.28) and 1.18 (1.38), respectively, at maximum pump power. The combination of the two modules showed an M^2 of the output channel in the horizontal and vertical directions of 2.05 and 1.27, respectively. This represents a beam quality comparable to that of the independent modules. The beam quality of the loss channel was measured to be much lower with an M^2 characterization in the horizontal and vertical directions of 8.5 and 3.4, respectively. This is believed to be a result of higher efficiency phase locking of the lower order spatial components present in the output from the individual modules, with the higher order components being split equally between the loss and output channels. This is evident in the distinct disparity between the beam qualities of the output and loss channels. Interestingly, this suggests that by increasing the number of modules combined by adding more nested beam splitters, in a tree-like structure, the combination efficiency may increase. As the hierarchical level of the beam splitter is increased, the comb beams lose a large proportion of the poorly matched higher order mode components, leaving only the better phase-matched components. Despite the beam quality of the modules here not being diffraction limited, the excellent combination efficiencies reported demonstrate the robustness of the PCSOCBC technique.

The Fabry–Pérot spectrum of the output channel is shown in Figure 14.12b. It shows only one spectral component, with bandwidth <575 MHz, when observed with an etalon of 60 MHz resolution. This demonstrates the benefit of having adaptive modules that do not have a fixed cavity length. Instead, they have the freedom to choose any spectral frequency distribution that maximizes power into the output channel.

An interesting property observed of this system, which is manifested from the phase conjugate nature of the modules, is the perturbation stability. First, it was found that efficient beam combination was possible even with quite dramatic changes to the individual module alignments. Theoretically, as long as the self-intersecting beam is imaged onto itself within the gain region, the system can adapt to any perturbations applied. This was evident by the ability, once aligned, to adjust the position and angle of various lenses and mirrors without any detrimental effect on the performance. Second, dynamic stability to perturbations exists as the gain grating's lifetime is related to the upper-state lifetime of the medium ($\sim$90 µs in Nd:YVO$_4$). This should allow compensation for any perturbation on longer timescales. Third, Figure 14.11 shows output power stability of the system to pump power, and the resultant thermal lens, variations, with combination efficiencies of over 90% maintained with up to 16 W (30%) reduction in the pump power of module 2. This represents excellent stability to pump power fluctuations.

14.3.2
Understanding Operation of PCSOCBC: Discussion

In Section 14.3.1, experimental operation of the PCSOCBC system has been demonstrated, but an underlying question still remains: Why should the adaptive modules phase lock themselves for constructive interference? In the SOCBC setup, it was clear why phase locking occurs; phase matching of the two modules for constructive interference into the output channel increases the effective reflectivity of the output coupler, reducing the cavity losses. This is beneficial for the resonator modes, as the higher output coupler reflectivity causes the intracavity power to be higher, allowing better extraction of the gain.

In the self-starting self-adaptive modules, the resonator is formed from the loop cavity formed of four mirrors and the diffraction from the gain hologram. The laser self-adapts to maximize the resonator mode power in this loop by maximizing the diffraction efficiency from the gain grating. How this relates to the phase conditions of the output beam and the writing beams is not inherently clear.

It has been shown that the diffraction efficiency of the gain grating within the self-starting self-adaptive laser has a functional dependence on the intensity of the writing beams [15,16,27,42], with diffraction efficiency improving with the increasing writing beam intensity (relative to the saturation intensity I_S). It only reduces for very high writing beam intensities where the grating is washed out. From this trend it follows that at low writing intensities, there will be an increase in the resonator efficiency with increasing writing beam intensities. From this it follows that correctly phase matching the finite phase offset of the individual modules for CBC into the output channel increases the intensity of the writing beams and thus increases the diffraction efficiency of the gain grating. This reduces the resonator cavity losses and as such causes the modules to phase lock for constructive interference into the output channel. This only applies to low writing beam intensities and it can be hypothesized that at high writing beam intensities, the modules would lock their phases to destructively interfere into the output channel in order to maximize the resonator mode power. Thankfully, this writing beam intensity is easily controllable by choosing an appropriate reflectivity output coupler, as it is formed from this reflection. Experimentally, the output coupler is usually chosen to be <1%, as this has been seen to enable efficient operation of the self-starting adaptive modules without losing too much power into the nonphase-conjugate output (P_{NPC} in Figure 14.9). This choice of reflectivity ensures that the writing beams are weak and extremely far away in intensity from where efficient beam combination of the individual modules would be lost.

One concern might be regarding whether this phase locking mechanism becomes weaker by increasing the number of modules. In fact, the contrary is true if the optimum beam writing intensity (I_{Opt}) is taken to be such that the diffraction efficiency reaches a maximum and an output coupler is chosen, where this becomes true for ideal phasing at the maximum pump power of an array of N modules. Any perturbations away from ideal phasing or maximum pump power will reduce the diffraction efficiency, causing the laser to self-adapt in order to maximize this diffraction efficiency again. From this it also follows that an unphased array will have writing beam intensities a factor of N smaller than the optimum, meaning that with increasing numbers of modules, the benefit from phase locking becomes greater, and thus the phase locking mechanism will be stronger.

Using self-starting self-adaptive modules in a PCSOCBC technique has a number of other distinct advantages. It has been seen by the authors that in nonfiber-based systems, there is a degree of freedom of spatial modes, that there are additional problems with matching the spatial mode structure of the combining beams due to the variation in the thermal lens, and there are aberrations between modules. In the self-starting phase-conjugate modules, as long as the self-intersecting writing beam is fully imaged over itself within the gain region (i.e., no information of the input is lost in the writing of the gain grating/hologram), the output beam is independent of the strength of the thermal lens or aberrations. In a self-starting adaptive module, of the type used in the PCSOCBC setup, the phase-conjugate output time reverses the path from which it came into the module; this has the effect of guaranteeing that each of the beams returning from each of the modules have the same spatial profile as they combine at the beam combiner. This reduces the problems seen in the SOCBC setup of having thermal lens mismatches between the modules, leading to reduced combination efficiencies.

To efficiently combine the output from two, or more, laser modules in a coherent combination scheme, it is necessary to match the phase relationship of multiple beams at the beam combiner in order to constructively interfere into a single output. In most lasers, the finite phase offset at any point is set by the wavelength and the cavity dimensions. In the standard SOCBC setup, constructive interference is maintained by variation in the lasing wavelength until a shared spectral mode that has correct phasing at the beam combiner is realized.

In self-starting self-adaptive lasers, neither the wavelength nor the finite phase offset of the phase-conjugate beam is set by the cavity dimensions. Thus, when considering the phase locking of multiple modules in a PCSOCBC setup, it is not clear whether the finite phase offset or the spectral mode is adapting to provide CBC. The ramifications of one of either of these criteria dominating the adaptation of the system are quite large, and if the wavelength is adapting and the finite phase offset is fixed, then the number of modules that can be combined efficiently would be limited in a way similar to that in the case of standard SOCBC setup. This leads to the question of whether the finite phase offset is indeed free or whether it is fixed by the grating writing beams?

Some insight can be gleaned by looking at the spectral mode of the coherently combined output from the PCSOCBC system. Figure 14.12b shows the bandwidth of the output beam, although single longitudinal mode, is clearly broader than what would be expected from a standard cavity mode, and certainly broader than what would be allowed for efficient combination in a SOCBC system. This suggests that it is not the wavelength of the output beam that is being adapted, but instead the finite phase offset of all the spectral components, such that they are phased correctly for CBC. This hypothesis has the requirement that the finite phase offset of the beam returning from the individual modules is free to choose any value between 0 and 2π and is not determined by some function of the input beam. It is possible to investigate whether this finite phase offset is locked by self-organization within the laser or by some inherent property of the gain grating interaction, or by building a system similar to that of the PCSOCBC setup, but with the output coupler replaced by an isolated external seed. This effectively builds a seeded self-adaptive laser, but with a Vernier–Michelson cavity coupling two modules together. As the wavelength is set by the external seed laser, the only degree of freedom for phase locking is the finite phase offset. The removal of the output coupler removes the loss-enhanced phase locking mechanism as well; so if any phase locking is seen, it must be related to the finite phase of the incident grating writing beams. As this finite phase offset would be expected to change with atmospheric perturbations to the beam path from seed to modules, an unstable power balance between the output and loss channels would suggest that the finite phase offset is locked by the phase of the writing beams. On the other hand, if no phase locking is seen, it suggests that the self-starting gain grating adaptive modules have freedom in the selection of the finite phase offset of their output beams, allowing power scaling to large numbers of modules in a PCSOCBC system.

This experiment was conducted and it was confirmed that no coherent combination between the two modules was seen, with power always equally split between the output and loss channels. This is suggestive that the finite phase offset between the modules is not predefined by the grating dynamics, allowing the modules to use this freedom to efficiently phase lock in a PCSOCBC system.

14.3.3
Power Scaling Potential

Initial CW experimental verification of a two-module PCSOCBC system has been demonstrated. The spectral freedom of the self-starting self-adaptive laser modules allows efficient passive coherent combination of multiple modules beyond the limits of classical self-organized coherent beam combination. Further scaling of the output powers can be considered by two main approaches: by increasing the number of modules and by increasing the power of the individual modules. The likely potential and limitations for these two areas of power scaling are discussed in the following.

The adaptive modules utilized are often considered in terms of phase conjugation of the input beam, but it is equally valid to consider the system in terms of maximization of the ring resonator mode. It is well known that a laser self-organizes itself to maximize the oscillating field power. As the oscillating field strength in the adaptive resonator is dependent on the diffraction efficiency from the gain grating, a mechanism for locking the finite phase offset of multiple modules can be examined.

14.3.3.1 Scaling the Number of Modules

It is believed that the PCSOCBC system has the ability to adapt the finite phase offset of all the component modules such that coherent beam combination occurs. If this is true, then the number of modules that can be combined should be much larger than that in a SOCBC setup. Instead, it can be considered that the limitations in the number of modules that can be combined should be related to the quality of the locking of the phase offset by the component modules. This can be investigated using a numerical approach in order to estimate the future scalability of the PCSOCBC system.

Siegman examined the probability of finding a shared cavity mode in a SOCBC system by looking at the reflectivity of the composite cavity of N modules to an incident beam [6]. The effective path lengths, and thus phase of the beams at the combination optics, were taken to be randomly distributed, allowing the calculation of the effective quality of the composite cavity. Many cavities were then examined in a Monte Carlo simulation in order to ascertain the statistical distribution of the composite cavity quality. Coupling this with the spectral operational bandwidth of the gain medium allowed analysis of the likely number of lasing modes that would exist for arrays of arbitrary numbers of modules.

Unlike SOCBC, in the PCSOCBC setup, there is a continuum of possible cavity modes, with the individual modules believed to be able to phase lock their wavelengths and the finite phase offsets for combination independently, ideally allowing limitless numbers of modules to be combined. It can be hypothesized that the limitation of the PCSOCBC system is likely to be related to the quality of the phase conjugation, as this will define whether the phasing is correct for constructive interference at the beam combiner. The quality of phase conjugation is believed to be high, but it is likely in a composite system consisting of many modules that a deviation will be seen module to module.

The purpose of this section is to examine what effect these module-to-module perturbations to the phase conjugation quality have on combination efficiency. This will be accomplished by considering the combination of multiple beams that have a random perturbation in the finite phase offset. By running a Monte Carlo simulation, essentially examining many system alignments, a statistical distribution of the likely combination efficiency will be extracted.

In the PCSOCBC setup, as there is no specific predefined spectral modes, any wavelength can be chosen and can coherently combine as long as the two beams are phased correctly. This phasing self-organizes itself across multiple modules as correct phasing leads to a stronger grating writing beams, higher grating modulation, and higher diffraction efficiencies.

For an ideal PCSOCBC system, where the finite phase offset of all the modules is correctly phased and they all emit the same spatial and spectral profiles, the number of modules that could be combined is limitless. Deviations from this criterion reduce the number of modules that can be combined. Variations in the finite phase offset and spectral content are likely to be the major limiting factor in the number of modules that can be combined, as their joint effect will manifest itself as a phase mismatch at the beam combiner. The limiting combination efficiency of a PCSOCBC system of N modules can thus be estimated by examining the effect of deviations from ideal phasing of the combining beams.

An approach similar to that of Siegman [6] with a Monte Carlo simulation of many arrays is implemented to examine this numerically.

The total field (E_S) combined into the output channel from the PCSOCBC $N:1$ coupling system can be expressed as

$$\vec{E}_S(\nu) = \sum_{n=1}^{N} \vec{E}_n(\nu) = \sum_{n=1}^{N} A_n(\nu) e^{i\varphi(\nu)}, \tag{14.10}$$

where $\vec{E}_n(\nu)$ describes the field emanating from the nth module, $A_n(\nu)$ is the amplitude variation, and $\varphi(\nu)$ is the phase front of the beam. Taking a first-order approximation of the combining field gives

$$\vec{E}_S(\nu) = \sum_{n=1}^{N} A_n(0) e^{i\varphi_n(0)}. \tag{14.11}$$

This approximation contains terms that are able to estimate limits to the phase-conjugate modules without any need for specifics of the form of the spatial phase aberrations of the specific modules. The first-order phase vector component ($\varphi_n(0)$) can be thought of as the deviation of the phase of the returning beam from the ideal phasing for constructive interference. The total electric field E_S of the combination of the multiple modules can now be calculated from the deviation from ideal phasing, allowing assessment of the power into the output channel of the PCSOCBC system for a specific cavity configuration.

The quality of a phase conjugate system is often defined in terms of the phase-conjugate fidelity. This metric is not a good measure of the combination efficiency of a PCSOCBC system, as it is normalized such that reductions in the output power due to the incorrect phasing of the individual module combinations do not reduce the fidelity. So, it is not a good metric to define the efficiency of the combination under the assumptions of this model. A better metric can be described by looking at the effective reflectivity of the PCSOCBC structure when normalized to the reflectivity under ideal phasing conditions, that is,

$$\vec{r} = \frac{\sum_{n=1}^{N} A_n(0) e^{i\varphi_n(0)}}{\sum_{n=1}^{N} A_n(0) e^{0}}. \tag{14.12}$$

This makes the sensible assumption that the pump power can be easily adjusted to each module in order to match the amplitude of the contributing electric fields ($A_n(0) \equiv A$). The normalized reflectivity vector ($\vec{r}$) can then be given as

$$\vec{r} = \frac{1}{N}\sum_{n=1}^{N} e^{i\varphi_n(0)}, \qquad (14.13)$$

where $\vec{r}$ is the complex amplitude reflectivity of the system. The amplitude reflectivity is then given by $r \equiv |\vec{r}|$ and the power reflectivity is given by $R \equiv r^2$.

Equation (14.13) gives a metric to the combination efficiency into the output channel of the array of modules, given the average phase of all of the returned beams ($\varphi_n(0)$). If the phase aberration on each of the combining beams is statistically distributed, an estimate of the combination efficiency can be made by assuming that each of the combining modes has a randomly distributed phase error, which on average fits a Gaussian distribution of standard deviation, σ. Applying a Monte Carlo simulation to generate many composite cavities with randomly Gaussian distributed phase errors allows statistical investigation of the likely combination efficiency.

Figure 14.13 plots the normalized Monte Carlo model-generated probability densities for the amplitude reflectivity (r) and the power reflectivity (R) for a standard deviation (σ) of the phase errors of $\pi/10$ and for array sizes between $N=2$ and $N=64$. These plots represent the histogram probability densities of 10 million generated arrays with 1% bins. At all array sizes, the combination efficiency remains on average over 90%, with no significant drop in combination efficiency seen at the higher array sizes. This is very different behavior in a standard SOCBC system where the combination efficiency, as predicted by a similar method, drops rapidly with increasing array sizes [6].

Figure 14.13 also shows that in increasing from 8 component modules to 64 modules along with the slight reduction in the average combination efficiency, there is an associated reduction in the deviation of the combination efficiency. This reduction in the total deviation of the combination efficiency will continue with increasing module numbers, as the distribution becomes a statistical average.

Another aspect that is of interest and can be examined by this numerical model is to what effect does the average magnitude of the phase error (σ) has on the combination efficiency of an array with a fixed number of modules?

Figure 14.14 shows numerical amplitude and power combination efficiencies for an array of 256 PCSOCBC modules as a function of σ, the standard deviation of the returning phase distortion. This model simulated 10 000 modules to generate the average and standard deviation of the amplitude and power reflectivities of the composite cavity. This combination efficiency figure shows that at low phase distortions of the beam returning from the phase-conjugate modules, a high combination ($\cong 100$) efficiency is seen. At high distortion levels ($\sigma \approx \pi$), the combination efficiency drops very low with the limit approaching that of SOCBC of randomly phased arrays. This suggests that as long as the returning phasing

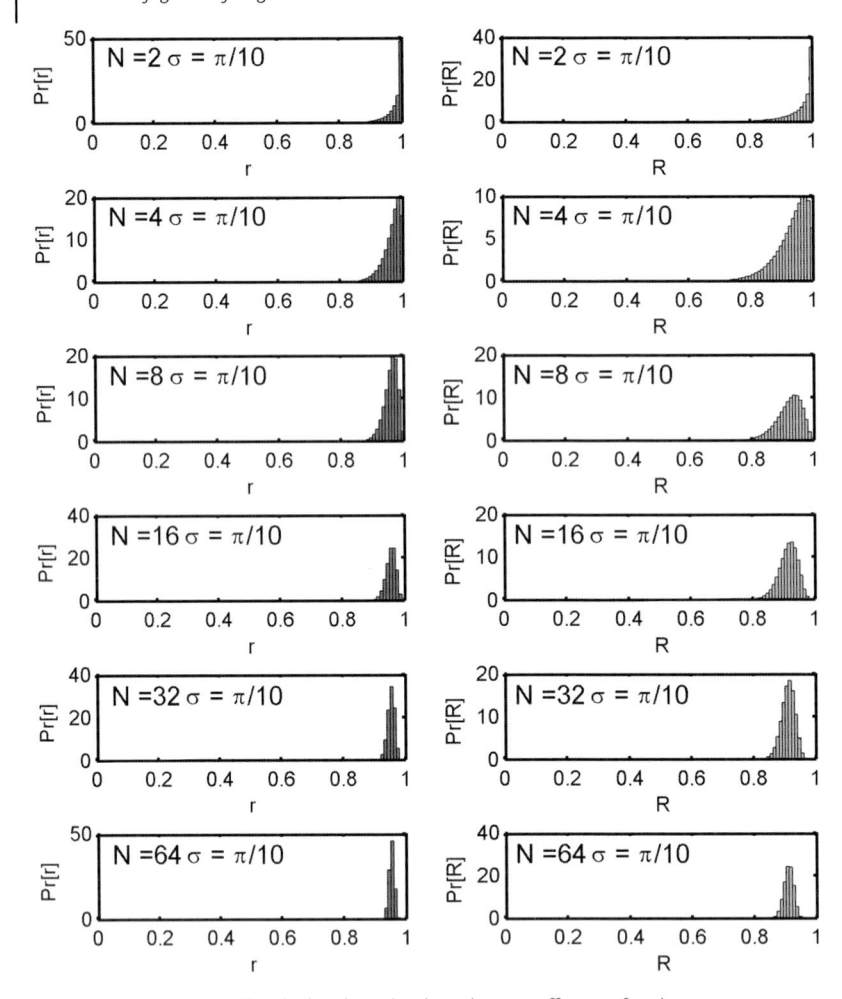

Figure 14.13 Numerically calculated amplitude and power efficiency for phase-conjugate coherent beam combination of arrays of various numbers of modules with a standard deviation of the phase distortions of $\pi/10$.

errors can be maintained as small perturbations ($\sigma < 0.25$ rad), efficient combination is possible, but the efficiency will drop if the phase errors become larger.

14.3.3.2 Higher Power Modules

Much higher power individual self-starting self-adaptive modules have been demonstrated by utilizing both higher power diodes [22] and additional amplifiers in the output coupler arm [22,24]. Generation of phase-conjugate outputs of 90 W has been demonstrated in a system with an extra amplifier in the output coupler arm [24,43]. By utilizing either higher power pump diodes or additional amplifiers in the output coupler arm, it should be possible to increase the coherently combined output power dramatically (Figure 14.15).

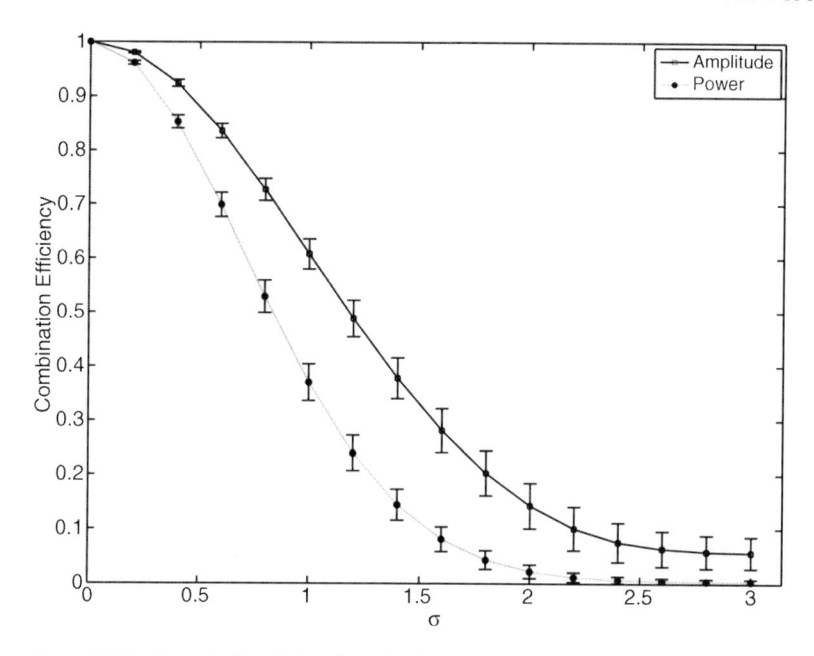

Figure 14.14 Numerically calculated amplitude and power efficiency for phase-conjugate coherent beam combination of large arrays ($N = 256$) as a function of the standard deviation of the phase distortions of the phase-conjugate modules. Note that these curves maintain their shape at low σ with increased numbers of modules, along with a reduction in the uncertainty in combination efficiency.

14.3.3.3 Pulsed Operation

A laser architecture that can efficiently combine continuous wave outputs has been shown. It is interesting to consider whether it is possible to scale not only the brightness of a laser output by this method but also the peak power by operating the lasers in a pulsed mode.

In a standard SOCBC setup, the spectral conditions for efficient CBC of large arrays limit quite dramatically the number of spectral modes that can exist in the output beam. This inherently limits the short pulse operation of these systems, as the spectral bandwidth is no longer available. In the PCSOCBC system demonstrated, this is not the case. It has been shown that efficient CBC can be obtained independent of the wavelength of the laser, as it is the finite phase offset that defines whether the modules combine coherently. This can be extended further to say that the bandwidth of the combined laser mode can contain many spectral components, only limited by the overlap of the gain bandwidth of all the individual modules. So, as opposed to standard SOCBC, PCSOCBC setup may well allow CBC of large arrays of short-pulse Q-switched arrays. Self-starting self-adaptive modules based on the diode-pumped bounce geometry have been previously demonstrated with ≈ 0.6 mJ pulses at <3 ns [8]. As the PCSOCBC system does not reduce the spectral content, it is anticipated that similar pulse energies and durations should be attainable in each module, which can then be combined into a single output.

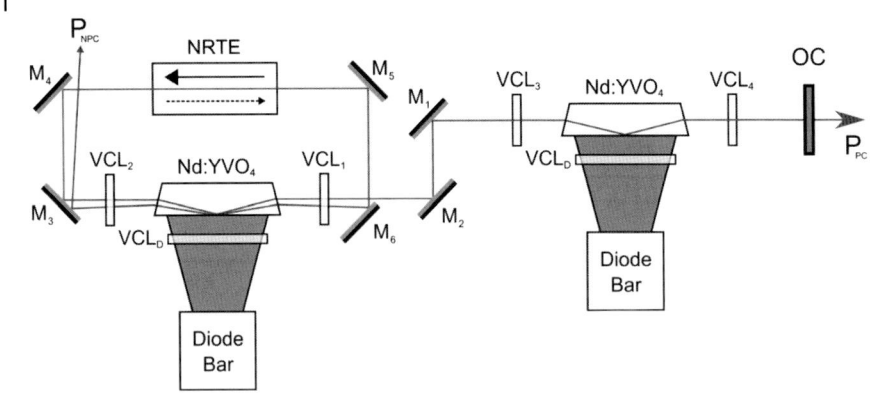

Figure 14.15 Conceptual schematic of a self-pumped phase-conjugate module with additional laser amplifier.

A preliminary "proof-of-principle" investigation of whether pulsed coherent combination via PCSOCBC was possible has been undertaken. A system almost identical to that represented in Figure 14.10 was adapted to utilize two quasi-CW (QCW) driven pump diodes powered by a single QCW driver. A pump pulse of 36 µs containing 2.4 mJ was delivered into the two self-starting self-adaptive modules. Once the laser amplifiers had reached threshold, the gain gratings were formed and a single 11 ns coherently combined output pulse of 60 µJ was emitted with a combination efficiency of 85%. This demonstrates that PCSOCBC could potentially scale both CW and pulsed laser systems.

It is worth noting that this first implementation was heavily limited in operation, as the small signal gain of the amplifiers was too high to allow significant energy storage before threshold was reached and that if the pump pulse was extended temporally, the combined laser system would undergo relaxation oscillations and self-pulsing until a steady state was reached. The high combination efficiency, considering being so close to threshold and not being able to adjust the module pump powers individually, hints that PCSOCBC should be able to coherently combine multiple laser bounce geometry oscillators with performance similar to that demonstrated in Ref. [8], that is, >0.5 mJ per module with <10 ns pulses. This could be potentially scaled to much higher peak powers by moving to self-starting self-adaptive modules based around laser systems with better energy storage, such as the flashlamp-pumped Nd:YAG systems, utilized in early self-starting self-adaptive laser research [25].

14.4
Conclusions

This chapter introduced the new concept of PCSOCBC, before demonstrating the first experimental implementation and discussing the future scalability of the system. In PCSOCBC, two or more self-starting self-adaptive modules are coherently combined in a Vernier–Michelson cavity. The intensity dependence of the grating writing process and the resultant diffraction efficiency from the gain grating causes the modules to phase lock their outputs such that they constructively interfere into a single output channel.

This has been demonstrated experimentally in a two-module system utilizing self-starting adaptive modules based on the diode-pumped bounce geometry. A total coherently combined output power of 27 W was observed with 1.75 W entering the loss channel. This represents an excellent combination efficiency of 94%.

A numerical investigation into the combination efficiency obtainable for larger module arrays suggests that much larger arrays can be combined, subject to maintaining low phase errors on the conjugate outputs from the individual modules. This is believed to be technically permissible, suggesting that much larger array sizes should be possible.

As this technique seems to allow coherent combination of much larger arrays, as well as being able to utilize much higher power individual modules and operate in pulsed manner, it is believed that this first demonstration of PCSOCBC will lead to much future work.

References

1 Sabourdy, D., Kermène, V., Desfarges-Berthelemot, A., Vampouille, M., and Barthélémy, A. (2002) Coherent combining of two Nd:YAG lasers in a Vernier–Michelson-type cavity. *Appl. Phys. B*, **75** (4–5), 503–507.

2 Wang, B., Mies, E., Minden, M., and Sanchez, A. (2009) All-fiber 50W coherently combined passive laser array. *Opt. Lett.*, **34** (7), 863–865.

3 Eckhouse, V., Fridman, M., Davidson, N., and Friesem, A.A. (2008) Loss enhanced phase locking in coupled oscillators. *Phys. Rev. Lett.*, **100** (2), 24102.

4 Fan, T.Y. (2005) Laser beam combining for high-power, high-radiance sources. *IEEE J. Select. Top. Quantum Electron.*, **11** (3), 567–577.

5 Kouznetsov, D., Bisson, J.-F., Shirakawa, A., and Ueda, K. (2005) Limits of coherent addition of lasers: simple estimate. *Opt. Rev.*, **12** (6), 445–447.

6 Siegman, A.E. (2004) Resonant modes of linearly coupled multiple fiber laser structures. Stanford University. Available at http://citeseerx.ist.psu.edu/viewdoc/download?doi=10.1.1.188.3580&rep=rep1&type=pdf.

7 Sillard, P., Brignon, A., Huignard, J.-P., and Pocholle, J.-P. (1998) Self-pumped phase-conjugate diode-pumped Nd:YAG loop resonator. *Opt. Lett.*, **23** (14), 1093–1095.

8 Smith, G. and Damzen, M.J. (2007) Quasi-CW diode-pumped self-starting adaptive laser with self-Q-switched output. *Opt. Express*, **15** (10), 6458–6463.

9 Wetter, N.U., Sousa, E.C., Camargo, F.D. A., Ranieri, I.M., and Baldochi, S.L. (2008) Efficient and compact diode-side-pumped Nd:YLF laser operating at 1053nm with high beam quality. *J. Opt. A: Pure Appl. Opt.*, **10** (10), 104013.

10 Thompson, B.A., Minassian, A., Eason, R. W., and Damzen, M.J. (2002) Efficient operation of a solid-state adaptive laser oscillator. *Appl. Opt.*, **41** (27), 5638–5644.

11 Shardlow, P.C. and Damzen, M.J. (2010) Phase conjugate self-organized coherent beam combination: a passive technique for laser power scaling. *Opt. Lett.*, **35** (7), 1082–1084.

12 Yariv, A. (1978) Phase conjugate optics and real-time holography. *IEEE J. Quantum Electron.*, **14** (9), 650–660.

13 Brignon, A. and Huignard, J.-P. (eds) (2004) *Phase Conjugate Laser Optics*, John Wiley & Sons, Inc., New York.
14 Zel'dovich, B.Y., Popovichev, V.I., Ragul'ski, V.V., and Faizullov, F.S. (1972) Connection between the wave fronts of the reflected and excited light in stimulated Mandel'shtam–Brillouin scattering. *J. Exp. Theor. Phys. Lett.*, **15** (160), 109.
15 Kuroda, K. (ed.) (2002) *Progress in Photorefractive Nonlinear Optics*, Taylor & Francis.
16 Green, R.P.M., Crofts, G.J., and Damzen, M.J. (1993) Phase conjugate reflectivity and diffraction efficiency of gain gratings in Nd:YAG. *Opt. Commun.*, **102** (3–4), 288–292.
17 Sillard, P., Brignon, A., and Huignard, J.P. (1998) Gain-grating analysis of a self-starting self-pumped phase-conjugate Nd:YAG loop resonator. *IEEE J. Quantum Electron.*, **34** (3), 465–472.
18 Günter, P. (1982) Holography, coherent light amplification and optical phase conjugation with photorefractive materials. *Phys. Rep.*, **93** (4), 199–299.
19 Rockwell, D.A. (1988) A review of phase-conjugate solid-state lasers. *IEEE J. Quantum Electron.*, **24** (6), 1124–1140.
20 Sillard, P., Brignon, A., and Huignard, J.P. (1997) Loop resonators with self-pumped phase-conjugate mirrors in solid-state saturable amplifiers. *J. Opt. Soc. Am. B*, **14** (8), 2049–2058.
21 Damzen, M.J., Boyle, A., and Minassian, A. (2005) Adaptive gain interferometry: a new mechanism for optical metrology with speckle beams. *Opt. Lett.*, **30** (17), 2230–2232.
22 Thompson, B.A., Minassian, A., and Damzen, M.J. (2003) Operation of a 33W, continuous-wave, self-adaptive, solid-state laser oscillator. *J. Opt. Soc. Am. B*, **20** (5), 857–862.
23 Ojima, Y. and Omatsu, T. (2005) Phase conjugation of pico-second pulses by four wave mixing in a Nd:YVO4 slab amplifier. *Opt. Express*, **13** (9), 3506–3512.
24 Smith, G.R., Minassian, A., and Damzen, M.J. (2006) High power-scaling of self-organising adaptive lasers with gain holography. Conference on Lasers and Electro-Optics/Quantum Electronics and Laser Science Conference and Photonic Applications Systems Technologies, Paper CFM1.
25 Damzen, M.J., Green, R.P.M., and Syed, K.S. (1995) Self-adaptive solid-state laser oscillator formed by dynamic gain-grating holograms. *Opt. Lett.*, **20** (16), 1704–1706.
26 Antipov, O.L., Chausov, D.V., and Yarovoy, V.V. (2001) Increase in phase-conjugate reflectivity of a holographic Nd:YAG oscillator due to resonant refractive-index grating. *Opt. Commun.*, **189** (1–3), 143–150.
27 Syed, K.S., Crofts, G.J., Green, R.P.M., and Damzen, M.J. (1997) Vectorial phase conjugation via four-wave mixing in isotropic saturable-gain media. *J. Opt. Soc. Am. B*, **14** (8), 2067–2078.
28 Antipov, O., Eremeykin, O., Ievlev, A., and Savikin, A. (2004) Diode-pumped Nd:YAG laser with reciprocal dynamic holographic cavity. *Opt. Express*, **12** (18), 4313–4319.
29 Green, R.P.M., Crofts, G.J., and Damzen, M.J. (1994) Holographic laser resonators in Nd:YAG. *Opt. Lett.*, **19** (6), 393–395.
30 Shardlow, P.C., Chard, S.P., and Damzen, M.J. (2008) Adaptive gain interferometry for optical metrology. 3rd EPS-QEOD Europhoton Conference.
31 Damzen, M.J., Matsumoto, Y., Crofts, G.J., and Green, R.P.M. (1996) Bragg-selectivity of a volume gain grating. *Opt. Commun.*, **123** (1–3), 182–188.
32 Crofts, G.J., Green, R.P.M., and Damzen, M.J. (1992) Investigation of multipass geometries for efficient degenerate four-wave mixing in Nd:YAG. *Opt. Lett.*, **17** (13), 920–922.
33 Brignon, A., Feugnet, G., Huignard, J.-P., and Pocholle, J.-P. (1995) Multipass degenerate four-wave mixing in a diode-pumped Nd:YVO4 saturable amplifier. *J. Opt. Soc. Am. B*, **12** (7), 1316–1325.
34 Udaiyan, D., Crofts, G.J., Omatsu, T., and Damzen, M.J. (1998) Self-consistent spatial mode analysis of self-adaptive laser oscillators. *J. Opt. Soc. Am. B*, **15** (4), 1346–1352.
35 Minassian, A., Thompson, B., and Damzen, M.J. (2003) Ultrahigh-efficiency {TEM}_{00} diode-side-pumped {N}d:{YVO}_4 laser. *Appl. Phys. B*, **76**, 341–343.

36 Minassian, A., Crofts, G.J., and Damzen, M.J. (2000) Spectral filtering of gain gratings and spectral evolution of holographic laser oscillators. *IEEE J. Quantum Electron.*, **36** (7), 802–809.

37 Bernard, J.E., McCullough, E., and Alcock, A.J. (1994) High gain, diode-pumped Nd:YVO4 slab amplifier. *Opt. Commun.*, **109** (1–2), 109–114.

38 Minassian, A., Thompson, B., and Damzen, M.J. (2005) High-power TEM00 grazing-incidence Nd:YVO4 oscillators in single and multiple bounce configurations. *Opt. Commun.*, **245** (1–6), 295–300.

39 Ojima, Y., Nawata, K., and Omatsu, T. (2005) Over 10-Watt pico-second diffraction-limited output from a Nd:YVO4 slab amplifier with a phase conjugate mirror. *Opt. Express*, **13** (22), 8993–8998.

40 Omatsu, T., Nawata, K., Okida, M., and Furuki, K. (2007) MW ps pulse generation at sub-MHz repetition rates from a phase conjugate Nd:YVO4 bounce amplifier. *Opt. Express*, **15** (15), 9123–9128.

41 He, F., Huang, L., Gong, M., Liu, Q., and Yan, X. (2007) Stable acousto-optics Q-switched Nd:YVO4 laser at 500kHz. *Laser Phys. Lett.*, **4** (7), 511–514.

42 Elsner, R., Ullmann, R., Heuer, A., Menzel, R., and Ostermeyer, M. (2012) Two-dimensional modeling of transient gain gratings in saturable gain media. *Opt. Express*, **20** (7), 6887–6896.

43 Damzen, M.J., Minassian, A., and Smith, G. (2007) New self-adaptive source and sensor technologies for enhanced remote sensing. 4th EMRS DTC Technical Conference.

15
Coherent Beam Combining Using Phase-Controlled Stimulated Brillouin Scattering Phase Conjugate Mirror

Hong J. Kong, Sangwoo Park, Seongwoo Cha, Jin W. Yoon, Seong K. Lee, Ondrej Slezak, and Milan Kalal

15.1
Introduction

High-energy and high repetition rate laser systems have many applications in different fields such as laser machining, particle accelerators, neutron or proton generators, and laser fusion drivers. However, to operate a high-energy laser at a high repetition rate, the laser system has to exhaust the accumulated heat in the laser gain medium. Many ideas have been proposed to resolve this accumulated heat problem, such as laser diode pumping, using a high thermal conductivity gain medium, using cryogenic Yb:YAG gain medium, and coherent beam combining [1–7]. Among these solutions, coherent beam combining seems to be one of the best practical techniques. Especially, coherent beam combining using stimulated Brillouin scattering phase conjugate mirrors (SBS-PCMs) has been investigated for decades and, now, its feasibility for use in developing a high-energy and high repetition rate laser system has been experimentally proved by Kong *et al.* [3,4,8–18].

Stimulated Brillouin scattering is a nonlinear optical process that generates a backward scattered phase conjugate wave [19–22]. A device that generates the phase conjugate wave by the SBS process is called an SBS phase conjugate mirror. An SBS-PCM can compensate for wavefront distortion induced by a phase aberrator, such as a laser gain medium; hence, this device is widely used in high-energy laser systems to obtain a high-quality beam. Efficient heat removal is a major issue, particularly with regard to the high repetition rate. A beam combining of small laser systems is a constructive approach to this issue. Of the various beam combined systems using SBS-PCMs, the cross-type beam combined scheme has many outstanding advantages, such as perfect isolation of a leak beam and easy alignment and maintenance [2,3]. The phase controlling of the SBS wave is a key technology in the realization of a coherent beam combined system, because the SBS wave inherently has a random phase due to its generation from thermal noise. The self-phase control method was proposed and has been developed by Kong *et al.* [3,4,8]; this method allows the control of the phase of the SBS wave with the simplest implementation. It provides

Coherent Laser Beam Combining, First Edition. Edited by Arnaud Brignon.
© 2013 Wiley-VCH Verlag GmbH & Co. KGaA. Published 2013 by Wiley-VCH Verlag GmbH & Co. KGaA.

ease of alignment, no limitations on the number of combined beams, and excellent phase conjugation. Furthermore, active phase control with a piezoelectric translator (PZT) enables long-term phase stabilization [23,24]. These works are expected to boost the development of laser systems in terms of leading to a high level of energy and power, a high-quality beam, and a high repetition rate.

15.2
Principles of SBS-PCM

Optical phase conjugation is a nonlinear optical phenomenon that accurately reverses the propagation direction and phase variation of an incoming light beam. A nonlinear optical device that produces a phase conjugate reflection is called a PCM [9,25]. Figure 15.1 compares a PCM with a conventional mirror. In the case of a conventional mirror, the wavefront is distorted twice when passing through an aberrating medium twice; however, there is no distortion in the case of PCM. The use of a PCM therefore eliminates phase distortions in optical systems [26,27]. For example, in a solid-state laser amplifier, phase distortions arise from thermal refractive index changes in the laser crystal. If a PCM is used to make the incoming beam pass the laser crystal twice, these distortions disappear. Hence, PCMs are widely used in high-energy laser systems.

The most commonly used way of producing optical phase conjugation is SBS [11,28,29]. SBS is normally achieved by focusing a laser beam into an SBS medium. In this medium, scattering from a spontaneous sound wave generates a wave that travels in the opposite direction; that wave interferes with the incoming wave, thereby inducing density modulations by electrostriction. Because the induced density modulations have the same frequency and direction as the initial sound wave, they are amplified and reinforce the backscattering. The phase-conjugated backscattered part is dominant because the amplification depends strongly on the direction. As a result, there is an exponential rise of the reflected phase-conjugated signal. The standing wave seems to be a self-adapting mirror because the wavefronts of the standing wave match the wavefronts of the incoming beam. If the incident wavefront has any disturbance, it results in a self-adapting mirror curvature with

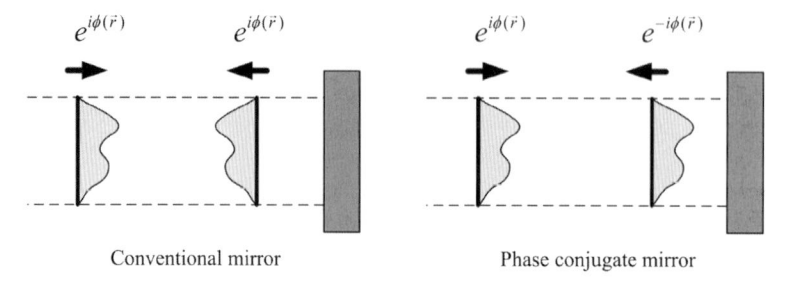

Figure 15.1 Wavefront reflection at a conventional mirror (a) and a phase conjugate mirror (b).

response times in the nanosecond range. In addition, SBS lowers the frequency of the phase-conjugated wave as much as the sound wave frequency, in accordance with energy conservation.

15.3
Reflectivity of an SBS-PCM

SBS reflectivity is almost equal to that of an ordinary mirror when the pump bandwidth, $\Delta\nu_p$, is smaller than the Brillouin linewidth, Γ (steady-state regime) [28,30]. However, many SBS-PCM applications necessarily involve a broadband pumped SBS (transient regime) because laser systems that use an SBS-PCM usually have a broadband spectrum to obtain high output power and short pulse widths [31,32]. Several theoretical and experimental investigations have reported on the use of a broadband pump in SBS reflectivity. For a broadband pump, the SBS reflectivity depends on the relation among four parameters: the coherence length, l_c; the characteristic interaction length, z_0, which is usually equal to the Rayleigh range; the mode spacing, Ω_m; and the Brillouin linewidth, Γ. When the coherence length is longer than the interaction length ($l_c > z_0$), the SBS gain for the broadband pump is as high as the coherence length for the narrowband pump [33–35]. Furthermore, if the pump laser mode spacing exceeds the Brillouin linewidth ($\Omega_m > \Gamma$), regardless of the mode structure, the SBS gain is the same as that of a single longitudinal mode pump [33]. Moreover, even if $\Omega_m < \Gamma$, the off-resonant acoustic waves, which are generated by the beating between the pump laser mode and another Stokes mode, play an important role in enhancing the gain and the reflectivity [36,37]. In all the previously mentioned works, however, the influence of the multimode pump has been considered only for two or several longitudinal modes and low pump energy near the SBS threshold. For this reason, the characteristics of SBS reflectivity by a multimode pump with numerous modes and high energy have been investigated [3,38].

The experimental setup for measuring the reflectivity of the SBS-PCMs is shown in Figure 15.2. The pump laser is a Q-switched Nd:YAG laser and, using its single longitudinal mode injection seeder, it can be operated in a single mode or multimode. The laser linewidth is approximately 0.09 GHz in a single-mode operation and approximately 30 GHz in the multimode regime. Thus, the linewidth of the multimode case is much larger than the Brillouin linewidth of the liquids used in this experiment, as listed in Table 15.1 [30,39–41]. The focal length of the lens used for the SBS-PCM is 15 cm. This length corresponds to a Rayleigh range, z_0, of 0.62 mm. Because the coherence length, l_c, is approximately 1 cm, it satisfies the condition of $l_c \gg z_0$. The temporal and spatial pulse widths are 8 ns and 2.4 m, respectively. The pump energy fluctuation is less than 1% for both cases, and the pump energy is measured during about 30 s at 10 Hz.

The SBS materials used in this experiment are Fluorinert FC-75, carbon tetrachloride (CCl_4), acetone, and carbon disulfide (CS_2). The SBS properties and the nonlinear refractive index, n_2, of each liquid are shown in Table 15.1. Each liquid has a different Brillouin linewidth, ranging from 50 to 528 MHz. The breakdown

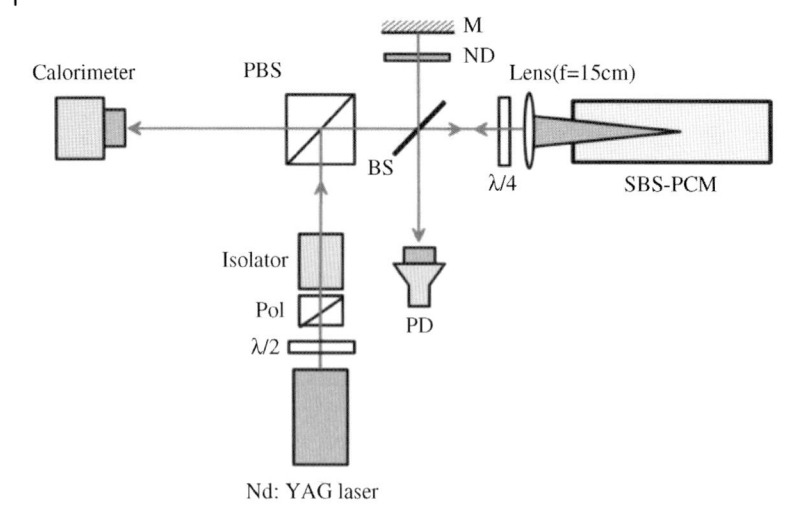

Figure 15.2 Experimental setup for the measurement of the SBS reflectivity ($\lambda/2$, half-wave plate; Pol, polarizer; M, mirror; ND, neutral density filter; $\lambda/4$, quarter-wave plate; PBS, polarizing beam splitter; BS, beam splitter; PD, photodiode).

threshold, E_b, which is listed in Table 15.1, was measured when a bright spark appeared inside the SBS cell.

Figure 15.3 a and b shows the SBS reflectivity of CCl_4 and Fluorinert FC-75 for the single-mode and multimode cases as a function of the pump energy [42]. Note that CCl_4 and FC-75 have very similar SBS gains and Brillouin linewidths (see Table 15.1), which result in similar reflectivity curves of typical nonlinear variation for the single-mode pump. For the multimode pump, the SBS reflectivity is very different for each of the liquids. The peak reflectivity is 30% at most in CCl_4 and more than 65% in FC-75, though the reflectivity decreases as the pump energy increases. Note also that even though the single-mode pump generally has a higher SBS gain [43,44], the SBS reflectivity in CCl_4 is slightly higher for the multimode

Table 15.1 Properties of the liquids used for the reflectivity experiments: Γ, Brillouin linewidth; g_B, steady-state SBS gain; n_2, nonlinear refractive index; P_c, critical power for self-focusing (calculated); E_b, breakdown threshold energy (measured).

Liquid	Γ (MHz)	g_B (cm/GW)	n_2 (10^{-22} m^2/V^2)	P_c (MW)	E_b (mJ)
Fluorinert FC-75	350	4.5–5	0.34	7.0	6
Carbon tetrachloride (CCl_4)	528	3.8	5.9	0.4	1.7
Acetone	119	15.8	8.6	0.28	1.5
Carbon disulfide (CS_2)	50	68	122	0.02	0.1

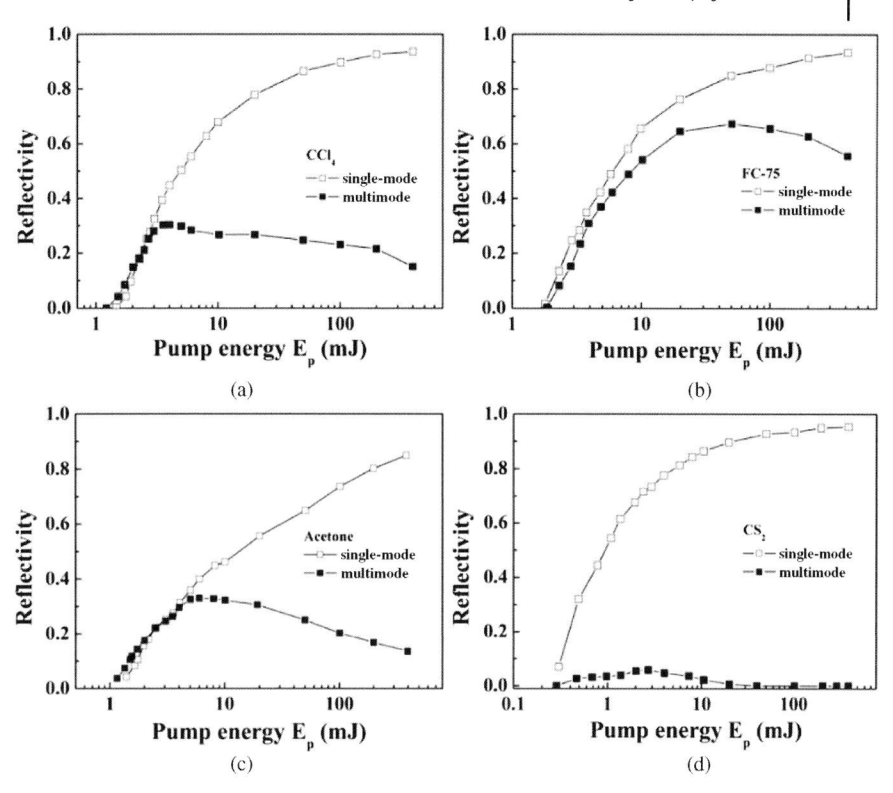

Figure 15.3 SBS reflectivity versus pump energy for various active media in the single-mode and multimode cases: (a) CCl_4; (b) FC-75; (c) acetone; (d) CS_2.

pump than for the single-mode pump near the SBS threshold of the single-mode case. For FC-75, on the other hand, the behavior is exactly the opposite.

Several factors appear to contribute to the reflectivity difference of the multimode pump. We interpret the SBS reflectivity for the multimode pump in terms of the temporal intensity spikes of the multimode pulse, which are absent in the single-mode pulse. A beating between the large numbers of longitudinal modes causes intensity spikes to rise, and the intensity spikes have enough power to induce nonlinear effects, such as self-focusing and optical breakdown. The self-focusing that is caused by the intensity spikes is likely to lead to an anomalously high reflectivity of the multimode pump near the SBS threshold in CCl_4. The critical power of the self-focusing is given by

$$P_c = \frac{\pi \varepsilon_0 c^3}{n_2 \omega^2}, \tag{15.1}$$

where ε_0 is the permittivity of the vacuum, ω is the angular frequency of the optical field, and c is the speed of light [45]. According to Eq. (15.1), the critical power, P_c, is 0.4 MW for CCl_4 and 7 MW for FC-75. In the case of the single-mode pump for CCl_4,

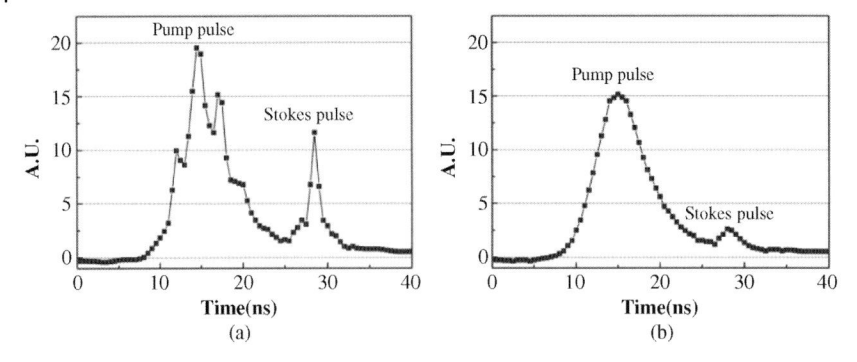

Figure 15.4 Pump and reflected pulse shapes in (a) multimode and (b) single-mode cases at $E_p \sim 1.5$ mJ in CCl_4.

the critical power of 0.4 MW is slightly larger than the SBS threshold power (~5% energy reflection) of 0.26 MW occurring at a pulse energy of approximately 1.8 mJ. However, the multimode pulse can induce temporal small-scale self-focusing in CCl_4 below the SBS threshold because the high peak power of the intensity spikes can exceed the critical power, P_c. If a good approximation of the steady-state SBS threshold relation $I_{th}g_Bl = 25\text{--}30$ (where I_{th} is the SBS threshold intensity, g_B is the SBS gain, and l is the interaction length) is maintained, the self-focusing leads to an increase in the intensity of the pump beam in the focal region and, hence, can reduce the SBS threshold energy [32]. As shown in Figure 15.3a, the self-focusing consequently results in a lower SBS threshold and a slightly higher reflectivity near the SBS threshold in CCl_4. Figure 15.4 shows the temporal pulse shapes of the pump and the Stokes pulse in different energy scales for both types of pump when a pump beam with an E_p value of 1.5 mJ is focused into a CCl_4 cell. As expected, the multimode pulse has large intensity spikes, whereas the single-mode pulse has no large intensity spikes. On the other hand, the SBS reflectivity in the FC-75 is not affected by the self-focusing near the SBS threshold because the critical power for FC-75 is approximately 18 times larger than the critical power for CCl_4. Consequently, the SBS reflectivity for the multimode pump is lower than that for the single-mode pump near the SBS threshold.

The self-focusing seems to be deleterious for SBS because it can enhance the optical breakdown. The experimental results in Table 15.1 confirm that the optical breakdown starts at an E_p value of approximately 1.7 mJ in CCl_4 and at ~6 mJ in FC-75. We observed that the breakdown appears around the focal spot near the breakdown threshold; when the pump energy increases, the breakdown becomes severe and produces a filament shape consisting of bright sparks. The breakdown disturbs the creation of acoustic phonons. In addition to the breakdown due to self-focusing, an intensity spike can easily generate an optical breakdown by itself because it has a very steep rising edge. For efficient SBS to occur, temporal fluctuations in the pump must be slow in relation to the acoustic phonon lifetime. If the temporal fluctuations are fast in relation to the acoustic phonon lifetime, the acoustic waves have insufficient time to build up. Hence, intensity spikes with a steep rising edge and energy levels that exceed the breakdown threshold can reach the focal area without losing their energy as a result of the backward reflection. They can therefore generate an optical breakdown and reduce the SBS reflectivity even at low energy. For the single-mode case, the region of SBS reflection moves fast in the direction opposite to the pump pulse; the pump pulse is reflected before the focal area from a region in which the optical intensity is too small to induce an optical breakdown [46]. Thus, even if the pump energy is large, no optical breakdown is generated for the single-mode pump.

The SBS reflectivity for acetone is shown in Figure 15.3c. The multimode pump provides reflectivity higher than that of the single-mode pump near the SBS threshold, which is very similar to the results of CCl_4. Table 15.1 shows that acetone has a nonlinear refractive index approximately the same as that of CCl_4. Thus, the SBS reflectivity of the multimode pump with a large number of longitudinal modes is significantly affected by the self-focusing induced by the high-intensity spikes. The reflectivity for the multimode pump increases as the energy rises to 6 mJ and then decreases strongly because of the severity of the breakdown; in contrast, the reflectivity for the single-mode pump increases monotonically. Figure 15.3d shows the measured reflectivity of CS_2. Of all the four liquids examined, CS_2 has the lowest SBS threshold energy (approximately 0.3 mJ) and the highest reflectivity (approximately 95%) for the single-mode pump because it has the highest steady-state SBS gain (Table 15.1). On the other hand, the SBS reflectivity for the multimode pump is the lowest and is almost zero throughout the entire region. The critical power (20 kW) for the self-focusing is about half the SBS threshold (40 kW). Furthermore, CS_2 has the longest acoustic lifetime (6 ns) of the liquids used [40], and this life span is comparable to the pulse width of the pump beam. As already mentioned, if the temporal fluctuations of the pump pulse are fast in relation to the acoustic phonon lifetime, the acoustic waves lack sufficient time to build up. As a result, CS_2 has a breakdown threshold (approximately 0.1 mJ) lower than the SBS threshold, and this very low threshold can account for the almost zero reflectivity observed. We observed that the optical breakdown produces a filament if the pump energy is almost as weak as the SBS threshold. Note that the stimulated Raman scattering (SRS) may also be responsible for the low reflectivity. CS_2 has a high SRS gain. The very short response time of the SRS process (10^{-11} s) implies that the SRS response to the intensity spikes of the multimode pulse is better than that of the SBS process [47].

15.4
Beam Combining Architectures

The methods are side-by-side beam combining, coherent beam combining with tiled aperture, coherent beam combining with filled aperture, wavelength beam combining using serial implementations, and wavelength beam combining using parallel implementations. These can be classified into coherent beam combining and wavelength beam combining. The wavelength beam combining and side-by-side

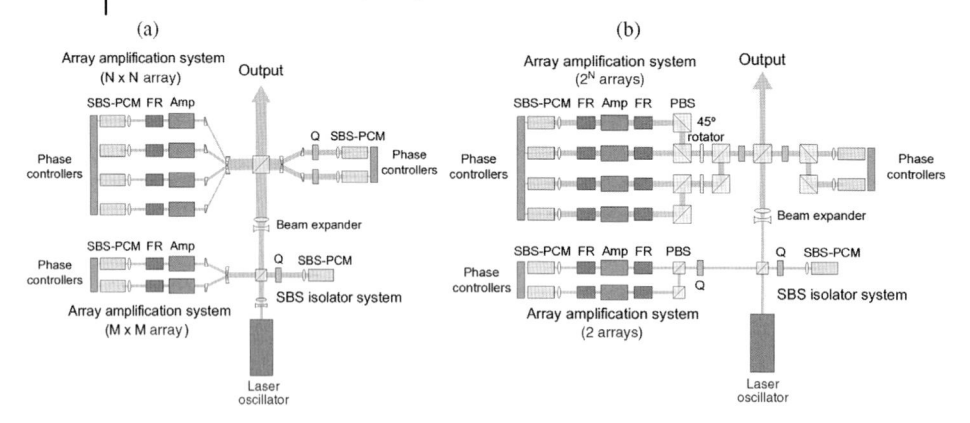

Figure 15.5 Conceptual schemes of scalable beam combined laser system for a laser fusion driver: (a) wavefront division scheme; (b) amplitude division scheme (QWP, quarter-wave plate; SBS-PCM, stimulated Brillouin scattering phase conjugate mirror, FR, Faraday rotator; Amp, optical amplifier).

beam combining can yield high energy, but when a coherent beam is needed, coherent beam combining should be used. In coherent beam combining using SBS-PCM, Kong et al. have proposed a wavefront dividing system and an amplitude dividing coherent beam combining system.

The wavefront dividing architecture is presented in Figure 15.5 a. The main beam is divided into many subbeams for separate amplification; if the beam is divided by prisms, the system is called a wavefront division scheme. This is the coherent beam combining with tiled aperture method. In this method, the focused beam acts as one beam; therefore, this is a promising method to construct a coherent beam combining system having a high repetition rate and high energy. The second architecture is presented in Figure 15.5b. The main beam is divided by polarizing beam splitters. In this case, the system is called an amplitude division scheme. This is the coherent beam combining with filled aperture method. In this scheme, we need to expand the aperture after the beam combining to prevent damage to the optics.

15.5
Phase Controlling Theory

The self-phase controlling method is the key technique for coherent beam combining using self-controlled SBS-PCMs. We present in this section the theoretical model suggested by Kong et al. (2008) [48] and Slezak et al. (2010) [49] to explain the principle of self-phase control.

Figure 15.6 shows the schematic diagram of beam combination system using N SBS-PCMs. All the waves (E_{ij}, $i = 1, \ldots, 3, j = 1, \ldots, N$) are considered to be the corresponding components of linearly polarized monochromatic plane waves that are switched on at the time $t = 0$ with constant amplitudes. The fact that the beam is

15.5 Phase Controlling Theory

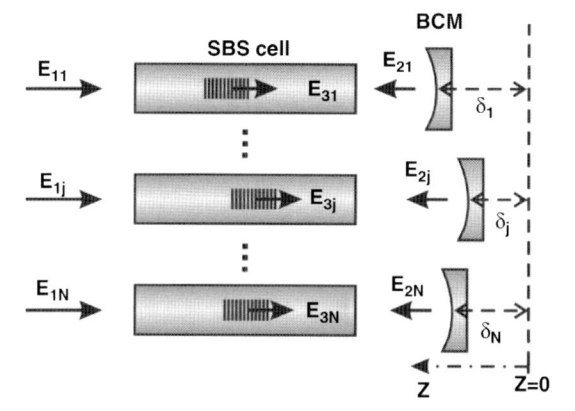

Figure 15.6 N simultaneously working SBS-PCMs with back-seeding concave mirrors (BCMs). The incident pump beam E_{1j} for the jth SBS-PCM goes through the SBS cell unfocused. It is reflected by the jth concave mirror and focused into the SBS cell as E_{2j}. The SBS reflected Stokes wave is denoted as E_{3j}. The z-position of the jth BCM is given by δ_j.

focused by the concave mirror and so E_{2j} becomes the Gaussian beam allows for further simplification considering just the Rayleigh region close to the focal plane of the mirror. In this area, the electromagnetic wave reflected by the mirror may be approximated by the plane wave provided the pump beam cross-section radius r_p is significantly smaller than the concave mirror curvature radius R_m ($r_p \ll R_m$). The reflected wave amplitude is then multiplied by the factor μ that represents the amplitude increase caused by focusing.

Under these conditions, any optical wave in the system under consideration may be expressed as

$$E_{ij} = \frac{1}{2} A_{ij} \exp\left[i(k_{ij}z - \omega_{ij}t + \phi_{ij})\right] + \text{c.c.}, \tag{15.2}$$

where A_{ij}, k_{ij}, ω_{ij}, and ϕ_{ij} denote the amplitude, wave number, frequency, and phase of the ijth wave, respectively. The waves reflected by the mirror are counterpropagating and so $k_{2j} = -k_{1j} \equiv k_j$. The mirror reflection does not change the wave frequency $w_{2j} \equiv \omega_{1j} \equiv \omega_j$. The phase after the reflection is given by $\phi_{2j} \equiv \phi_{1j} + \pi + (k_{1j} - k_{2j})\delta_j$.

Substitution for k_{2j} provides $\phi_{2j} \equiv \phi_{1j} + \pi - 2k_j\delta_j$. The amplitude after the reflection is given as $A_{2j} = \mu_j A_{1j} \equiv \mu_j A_j$. By using these formulas, it is possible to derive the intensity interference pattern in the SBS-PCM proximity in the form

$$\langle E_j^2(z,t) \rangle = (E_{1j} + E_{2j})(E_{1j} + E_{2j})^*$$
$$= \frac{1}{4}|A_j|^2 \left[\frac{1}{2}(1 + \mu_j^2) - \mu_j \exp\left[2ik_j(z - \delta_j)\right]\right]. \tag{15.3}$$

It should be noted that the intensity interference pattern nodal point position does not depend on the pump beam phase ϕ_{1j}. It is fully determined by the actual mirror position δ_j.

The acoustic wave equation right-hand side driven by the electrostriction is given as

$$g_j(z,t) = -\frac{\gamma}{8\pi}\frac{\partial^2}{\partial z^2}\langle E_j^2(z,t)\rangle, \tag{15.4}$$

where the electrostriction coupling constant γ was introduced. Substitution of Eq. (15.3) into Eq. (15.4) provides

$$g_j(z,t) = -\frac{\gamma\mu_j}{4\pi}|A_j|^2 k_j^2 \exp\left[2ik_j(z-\delta_j)\right] + \text{c.c.} \tag{15.5}$$

The acoustic waves $\rho_j(z, t)$ in SBS medium are solutions of the acoustic wave equation (for the sake of simplicity, no damping was considered)

$$\frac{\partial^2 \rho_j}{\partial t^2} - v^2 \frac{\partial^2 \rho_j}{\partial z^2} = -\frac{\gamma\mu_j}{4\pi}|A_j|^2 k_j^2 \exp\left[2ik_j(z-\delta_j)\right], \tag{15.6}$$

where v stands for the sound velocity in SBS medium.

The initial conditions may be determined as the acoustic noise waves with random phases. Considering only the SBS phase-matched waves and denoting $q_j \equiv 2k_j$ and $\Omega_j \equiv q_j v$, these initial conditions can be expressed in the following way:

$$\begin{aligned}\rho_j(z,0) &= \frac{1}{2}S_j\{\exp\left[i(q_j z + \varphi_j^-)\right] + \exp\left[i(q_j z + \varphi_j^+)\right]\} + \text{c.c.},\\ \frac{\partial \rho_j}{\partial t}(z,0) &= \frac{1}{2}iS_j\Omega_j\{-\exp\left[i(q_j z + \varphi_j^-)\right] + \exp\left[i(q_j z + \varphi_j^+)\right]\} + \text{c.c.},\end{aligned} \tag{15.7}$$

where S_j is the acoustic noise amplitude dependent on geometry, temperature, material parameters, and frequency [1] and $\varphi_j^\pm$ denotes the thermal noise random phases.

The general solution of Eqs. (15.5) and (15.6) can be found in the form

$$\begin{aligned}\rho_j(z,t) = \rho_0 - &\left\{\frac{\gamma\mu_j}{32\pi v^2}|A_j|^2 \exp\left[-iq(z-\delta_j)\right]\right.\\ &+ \frac{1}{2}\left(\frac{\gamma\mu_j}{32\pi v^2}|A_j|^2 \exp\left[-iq_j\delta_j\right] + S_j \exp\left[i\varphi_j^-\right]\right)\exp\left[i(q_j z - \Omega_j t)\right]\\ &\left.+ \frac{1}{2}\left(\frac{\gamma\mu_j}{32\pi v^2}|A_j|^2 \exp\left[-iq_j\delta_j\right] + S_j \exp\left[i\varphi_j^+\right]\right)\exp\left[i(q_j z + \Omega_j t)\right]\right\} + \text{c.c.}\end{aligned}$$
(15.8)

This solution consists of three distinct components. The first one (ρ_0) represents the mean density value of the medium. The second component (remaining part of the first line) stands for the stationary density modulation. The terms in the second and the third line express the standing acoustic wave written in the form of a superposition of two counterpropagating acoustic waves. One of these waves exactly matches the SBS wave.

Moreover, the acoustic waves from thermal noise background with their phase close to $\varphi_j^\pm = -q_j\delta_j$ undergo constructive interference with the interference field drive wave and become dominant in the acoustic noise background. Such waves

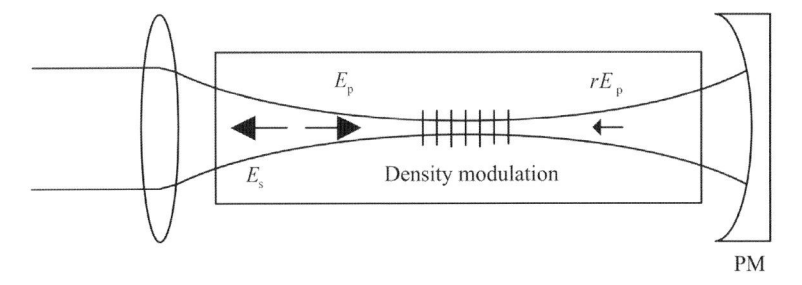

Figure 15.7 Concept of phase control of the SBS wave by the self-generated density modulation. PM is a partial reflectance concave mirror whose reflectivity is r. E_p and E_s denote the pump wave and the SBS wave, respectively.

have the highest probability to become the SBS seed. However, in the case of $\gamma\mu_j/(32\pi v^2)|A_j|^2 \gg S_j$, it is possible to neglect the thermal noise background compared to the standing wave. It should be apparent from Eq. (15.8) that the relative phase difference between any SBS cells may be tuned by the change of δ_j parameter.

Figure 15.7 shows the concept of the self-phase control method. Weak periodic density modulation is generated at the focal point due to electrostriction by an electromagnetic standing wave that arises from the interference between the main beam, E_p, and the low-intensity counterpropagating beam, rE_p. In the suggested theoretical model, the weak density modulation from the standing wave is assumed to act as an imprint for the ignition of the Brillouin grating. Hence, the initial position, z_0, is no longer random but fixed to one of the nodal points of the density modulation. However, there are many candidates for the nodal points in the Rayleigh range because the Rayleigh length, l_R, is much larger than the period of the stationary density modulation, $\lambda_p/2$, where λ_p is the wavelength of the pump wave. The phase differences between the acoustic waves generated at different nodal points have the values of $\Delta\phi_a = k_a(\lambda_p/2)N \cong 2\pi N$ (N is an integer) for the relation of $k_a \cong 2k_p = 4\pi/\lambda_p$. Thus, the phase uncertainty of $2\pi N$ does not affect the phase accuracy.

The initial time, t_0, when the acoustic wave is determined should be known. In research on the preservation of the SBS waveform [50], the front part of the pump energy has been found to be consumed to create the acoustic Brillouin grating of the SBS process. This consumed energy is regarded as the SBS threshold energy. The critical time, t_c, when the SBS is initiated can then be determined by the following equation:

$$E_{th} = \int_0^{t_c} P(t)dt, \qquad (15.9)$$

where E_{th} is the SBS threshold energy of the SBS medium and $P(t)$ is the pump power. It is assumed that t_0 is equal to t_c because the SBS waves and the

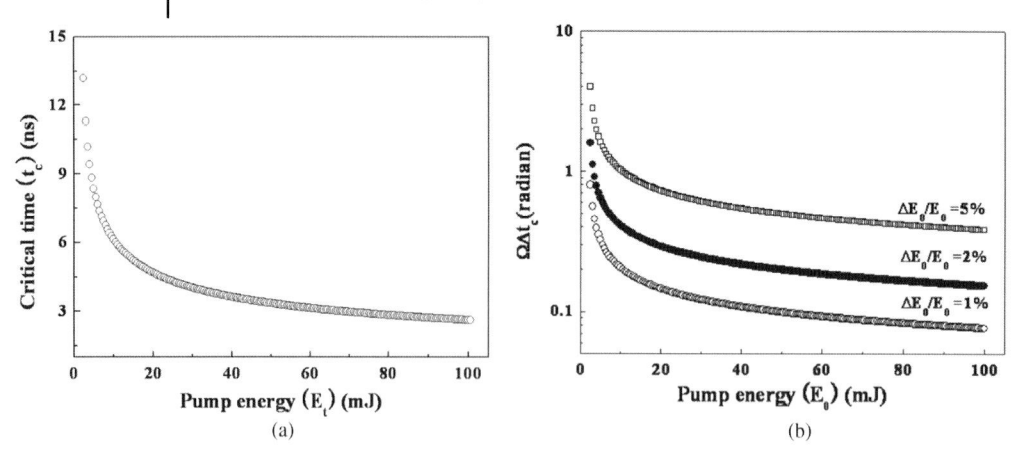

Figure 15.8 (a) Critical time t_c as a function of the pump energy. E_t is the SBS threshold energy of the SBS medium. (b) Initial phase change $\phi_0 = \Omega \Delta t_c$ of the SBS wave depending on the pump energy stability ($\Delta E_0/E_0 = 1, 2,$ and 5%) as a function of the pump energy. E_0 is the total energy of input pulse.

corresponding acoustic wave are generated simultaneously. Equation (15.9) suggests that the initial ignition time, t_0, of the acoustic wave changes if the total energy of the pump pulse given by $E_0 = \int_0^\infty P(t)dt$ changes under a constant pulse width. In this model, the critical time, t_c, varies with the total energy, E_0. Thus, the change that occurs in the initial phase, $\Delta\phi_0$, as a result of the energy fluctuation, ΔE_0, can be represented as

$$\Delta\phi_0 = \Omega \Delta t_c = \Omega \frac{\Delta t_c}{\Delta E_0} \frac{\Delta E_0}{E_0} E_0 \tag{15.10}$$

if we assume that z_0 is fixed; $\Delta\phi_0$ can be calculated numerically for FC-75, which has an acoustic wave frequency of 1.34 GHz, and the SBS threshold is about 2.5 mJ for a 10 ns pulse. Let us assume that the pump pulse, $P(t)$, is given by

$$P(t) = \frac{4E_0}{a^3 \sqrt{\pi}} t^2 \exp\left[-(t/a)^2\right] \quad (a = 8.66 \text{ ns}). \tag{15.11}$$

Figure 15.8 shows the calculated critical time, t_c, and $\Delta\phi_0$ as a function of the pump energy, which ranges from the SBS threshold (2.5 mJ) of FC-75 to 100 mJ when the energy stabilities, $\Delta E_0/E_0$, of the pump beam are 1, 2, and 5%. Note that t_c and $\Delta\phi_0$ are inversely proportional to the pump energy. Moreover, $\Delta\phi_0$ decreases further as the pump energy becomes stabilized from 5 to 1%. The phase of the backward SBS wave consequently stabilizes as the pump energy is increased.

The experimental results and the calculated results of Eq. (15.10) are shown in Figure 15.9. The experimental investigation is conducted for the cases of pump energy of beam 2 of $E_2 \approx 3, 3.5, 4, 4.5, 5, 10, 15, 20,$ and 25 mJ with pump energy of beam 1 of $E_1 \approx 5$ mJ and a pump laser energy stability of $\sim$2%. In the experiment, as shown in Figure 15.9a, the relative phase difference of beam 1 and beam 2, $\Delta\Phi$, is

15.6 Coherent Beam Combined Laser System with Phase-Stabilized SBS-PCMs | 467

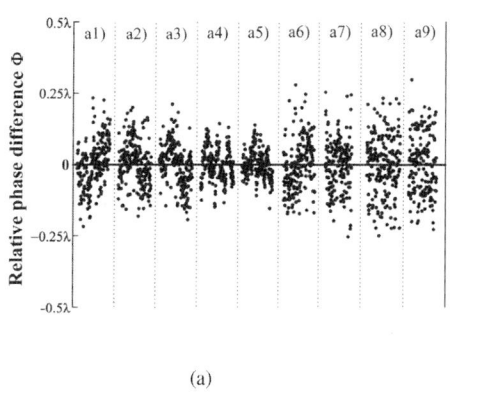

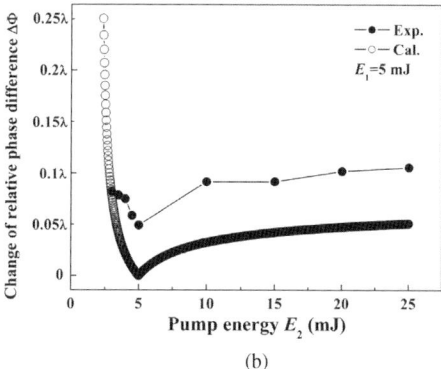

Figure 15.9 (a) Experimental data of the relative phase difference for the cases of $E_2 \approx 3$ mJ (a1), 3.5 mJ (a2), 4 mJ (a3), 4.5 mJ (a4), 5 mJ (a5), 10 mJ (a6), 15 mJ (a7), 20 mJ (a8), and 25 mJ (a9) with $E_1 \approx 5$ mJ. (b) Change of the relative phase difference for the cases of the experiment and the calculation by the theoretical model as a function of pump energy. The experimental data in this figure represent the standard deviations for each pump energy case of (a).

measured for 160–220 serial pulses. The experimental data in Figure 15.9b represent the standard deviation of the relative phase difference for each energy case. The theoretical calculation is performed under the same energy conditions as those used in Eq. (15.10). In the theoretical plot, note that $\Delta\Phi$ decreases as E_2 approaches E_1. As shown in Figure 15.9b, this prediction qualitatively agrees with the experimental results.

15.6 Coherent Beam Combined Laser System with Phase-Stabilized SBS-PCMs

15.6.1 Conventional Phase Fluctuation of SBS-PCM

When the SBS-PCM is used in the beam combining laser, the phase of the reflected beam is randomly changed, because SBS randomly occurs due to statistical noise. These effects fluctuate during several times of the phonon lifetime; as a result, piston errors in its phase occur between each beamline. Figure 15.10 shows the experimental schematic diagram and experimental results for the conventional SBS fluctuation [3]. Each point in Figure 15.9c represents one of the 160 laser pulses. As expected, $\Delta\Phi$ has a random value for every laser pulse. Figure 15.9b shows the intensity profile of 160 horizontal lines selected from each interference pattern. The profile also shows random fluctuation.

As Figure 15.10 shows, it is difficult to build a coherent beam combining system with a conventional SBS-PCM. Scientists have proposed several successful

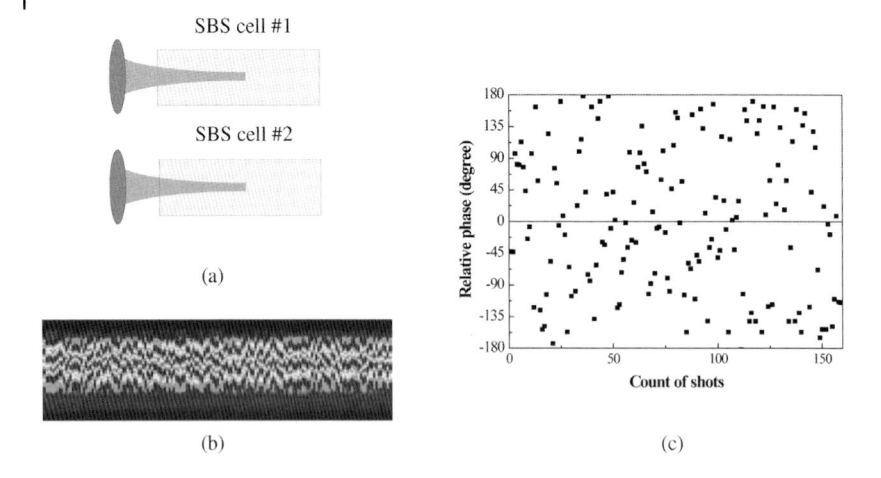

Figure 15.10 Experimental results for the unlocked case: (a) schematic; (b) intensity profile of horizontal lines selected from 160 interference patterns; (c) relative phase difference between two beams for 160 laser pulses.

approaches in the history of the phase locking of SBS waves [5–7]. Although these works show good phase locking effects, they have some problems in terms of practical application of a multiple beam combining. D.A. Rockwell proposed the overlapping method in 1986 [5]. In the overlapping method, all the beams are focused on one common point. Energy scaling is therefore limited to avoid optical breakdown, and the optical alignment is also difficult. R.H. Moyer proposed the back-seeding method in 1988 [7]. The method involves the use of a Stokes beam as a back-seeding beam. However, with use of the back-seeding beam method, the phase conjugation is incomplete if the injected Stokes beam is not completely correlated. In 1998, M.W. Bowers and R.W. Boyd proposed Brillouin-enhanced four-wave mixing (BEFWM) [51]. This approach requires highly complicated optics and is limited with respect to the number of the beams to be combined.

15.6.2
Phase Fluctuation without PZT Controlling

Kong et al. [3,4,8–18] proposed a new phase control technique involving self-generated density modulation. In this method, which is simply called the self-phase control method, a simple optical setup is used with a single concave mirror behind the SBS cell; furthermore, each beam phase can be independently and easily controlled without destruction of the phase conjugation. The phase control method thus obviates the need for any structural limitation on the energy scaling.

The wavefront division scheme, which spatially divides the beam, is used to demonstrate the phase control effect with the self-phase control method in the first

15.6 Coherent Beam Combined Laser System with Phase-Stabilized SBS-PCMs | 469

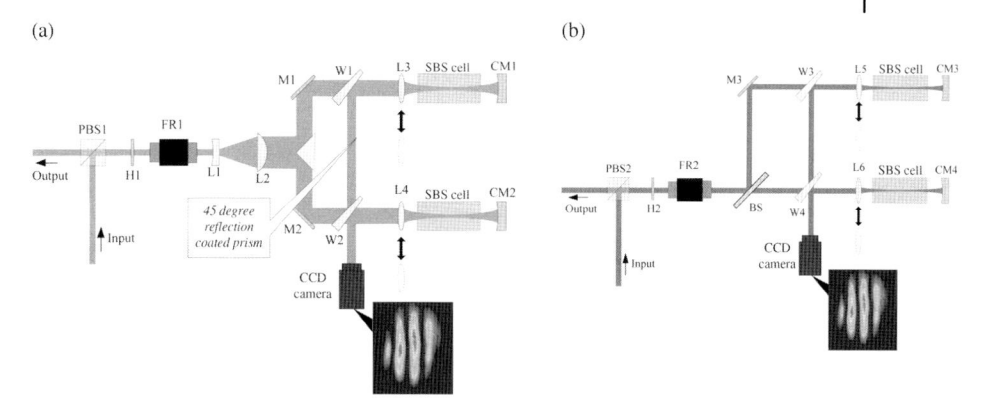

Figure 15.11 Experimental setups of (a) wavefront division scheme and (b) amplitude division scheme for phase control of the SBS wave by means of self-generated density modulation (M1, M2, and M3: mirrors; W1, W2, W3, and W4: wedges; L1 and L2: cylindrical lenses; L3, L4, L5, and L6: focusing lenses; CM1, CM2, CM3, and CM4: concave mirrors; H1 and H2: half-wave plates; PBS1 and PBS2: polarizing beam splitters; FR1 and FR2: Faraday rotators).

experiment [3,4,8]. The experimental setup is shown schematically in Figure 15.11a. A 1064 nm Nd:YAG laser is used as a pump beam for SBS generation. The pulse width is 7–8 ns, and the repetition rate is 10 Hz. The laser beam from the oscillator passes through a 2× cylindrical telescope and is divided into two parts by a prism, which has a high-reflection coating for an incident angle of 45°. The two parts of the divided beam pass through separate wedges and are focused into SBS-PCMs. The wedges reflect part of the backward Stokes beams such that they are overlapped onto a CCD camera. An interference pattern is consequently generated. The degree of fluctuation of the relative phase difference between the SBS waves is quantitatively analyzed by measuring the movement of the peaks in the interference pattern.

In the case of wavefront division, the divided subbeams get fluctuating energies at every shot due to the beam pointing effect of the laser source, which appears to generate the fluctuation of the relative phase difference between the SBS waves, because the phase of the SBS wave depends on the pump energy. This beam pointing problem can be overcome by using an amplitude division method, whereby the subbeams have almost the same level of energy [38]. The experimental setup of the amplitude division scheme is shown in Figure 15.11b. In the amplitude division scheme, the laser beam from an oscillator is divided into two subbeams by a beam splitter.

Figure 15.12 shows phase control experimental results in the wavefront division scheme. Figure 15.12a shows a schematic diagram of the experiment and the experimental results of concentric-type self-phase control. A small amount of the pump pulse is reflected by an uncoated concave mirror and then injected into the SBS cell. The standard deviation of the measured relative phase difference is $\sim 0.17\lambda$ ($=\lambda/5.9$). Moreover, 88% of the data points are contained within a range of $\pm 0.25\lambda$ ($=\lambda/4$). These results demonstrate that self-generated density modulation can fix the

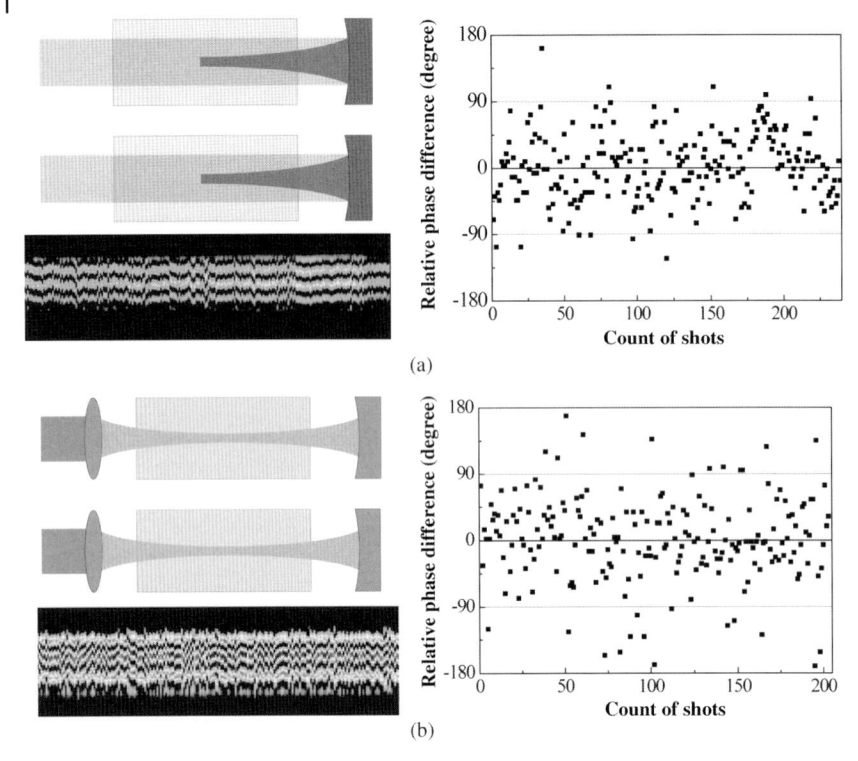

Figure 15.12 Phase control experimental results in the wavefront division scheme, with (a) concentric-type self-phase control ((left, up) schematic, (left, down) intensity profile of horizontal lines from interference pattern, and (right) relative phase difference between two beams for 203 laser pulses) and (b) confocal-type self-phase control ((left, up) schematic, (left, down) intensity profile of horizontal lines from interference pattern, and (right) relative phase difference between two beams for 238 laser pulses).

phase of the backward SBS wave. Figure 15.12b shows the schematic diagram and the experimental results of confocal-type self-phase control, where the pump beams are backward focused by a concave mirror coated with a highly reflective material. The standard deviation of the measured relative phase difference is $\sim 0.14\lambda$ ($=\lambda/7.1$). Furthermore, 96% of the data points are contained in a range of $\pm 0.25\lambda$ ($=\lambda/4$).

Phase control experimental results obtained with the amplitude division scheme are shown in Figure 15.13 [12]. Figure 15.13a shows the schematic diagram and the experimental results of concentric-type self-phase control. The standard deviation of the measured relative phase difference is $\sim 0.037\lambda$ ($=\lambda/27$). Figure 15.13b shows the schematic diagram and the experimental results of confocal-type self-phase control. The standard deviation of the measured relative phase difference is $\sim 0.028\lambda$ ($=\lambda/36$). By employing the amplitude division scheme, the relative phase difference is remarkably stabilized compared with the use of the wavefront division scheme.

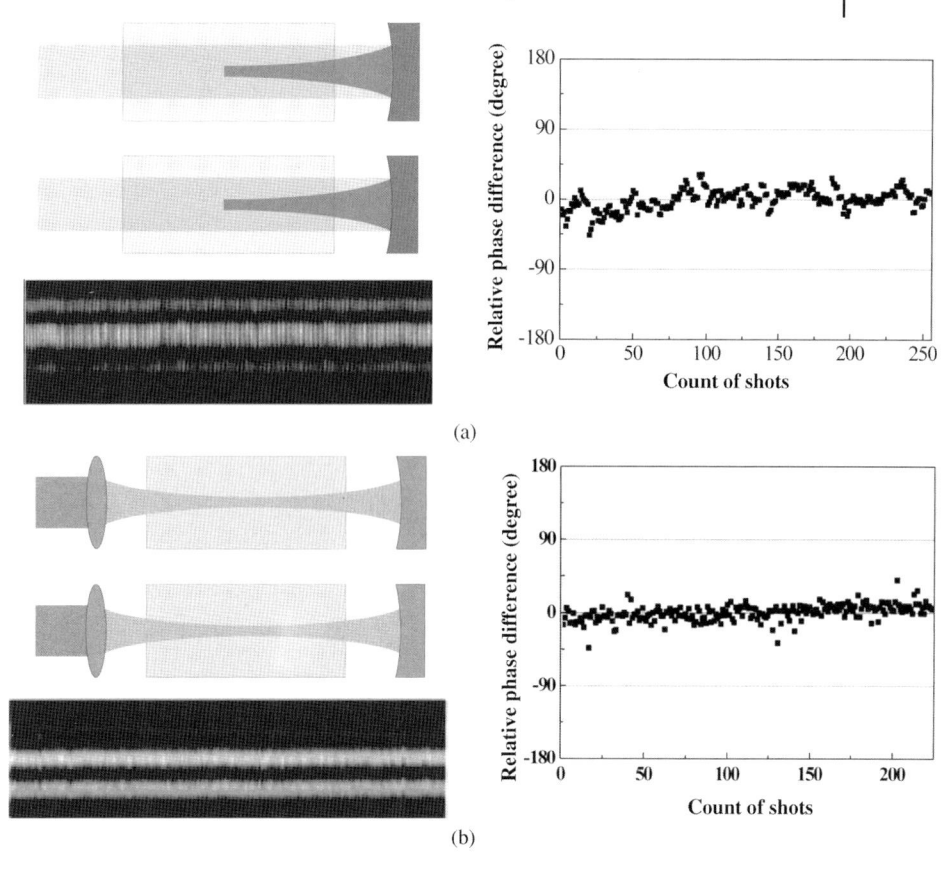

Figure 15.13 Phase control experimental results for the amplitude division scheme, with (a) concentric-type self-phase control ((left, up) schematic, (left, down) intensity profile of horizontal lines from interference pattern, and (right) relative phase difference between two beams for 256 laser pulses) and (b) confocal-type self-phase control ((left, up) schematic, (left, down) intensity profile of horizontal lines from interference pattern, and (right) relative phase difference between two beams for 220 laser pulses).

15.6.3
Phase Fluctuation with PZT Controlling

The self-phase control method ensures that the SBS wave is well stabilized for several hundred shots. However, a thermally induced long-term phase fluctuation occurs when the number of laser shots increases [23,24]. This slowly varying phase fluctuation can be easily compensated through active control of PZTs attached to one concave mirror of the SBS-PCM. Figures 15.14 and 15.15 show the phase control experimental results for the cases with and without PZT control, respectively, in a two-beam combining system. The phase difference and the output energy are measured during 2500 laser shots (250 s) for a pump energy of beams 1 and 2 of

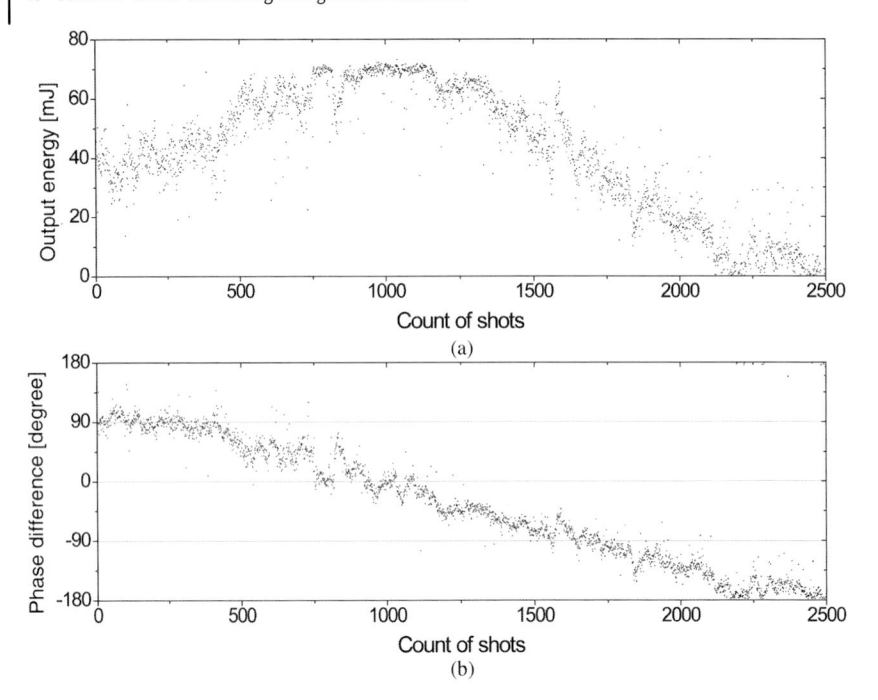

Figure 15.14 Experimental results of (a) the output energy and (b) the phase difference between two SBS beams without PZT control during 2500 laser shots (250 s) for the case of $E_{1,2} \approx 50$ mJ pump energy.

$E_{1,2} \approx 50$ mJ. The case without PZT control showed long-term phase and output energy fluctuations. In the case with the PZT control, the phase difference between the SBS beams is well stabilized with a fluctuation of 0.021λ ($=\lambda/47$) by standard deviation; furthermore, the output energy is stabilized with a fluctuation of 4.7%.

On the basis of these results, a coherent four-beam combining laser system is built for this study [52,53]. Figure 15.16 shows a four-beam combining laser system using confocal-type SBS-PCMs. The laser beam source is a Nd:YAG laser oscillator, and the polarization of the laser is p-polarization. The beam expander consists of lenses having focal lengths of -200 and -50 mm. The beam expander expands the size of the laser beam by four times. The expanded beam is divided into four subbeams, beams 1, 2, 3, and 4, after passing through the four-beam circular aperture. The passes of the subbeams are divided by prisms and mirrors. Each subbeam passes through amplifiers and Faraday rotators, and the polarization of the subbeams is 45°. The subbeams are reflected by the SBS-PCM and pass through the amplifiers and Faraday rotators again. Thus, the polarization of the subbeams is s-polarization. The subbeams are combined and reflected by polarizing beam splitter PBS1, and half-wave plate HWP2 tilts the polarization of the combined beam. PBS2 splits the beam, and the output beam is consequently reflected by PBS2. A wedge splits beam 1, and the beam expander expands a part of beam 1 by four times. This expanded

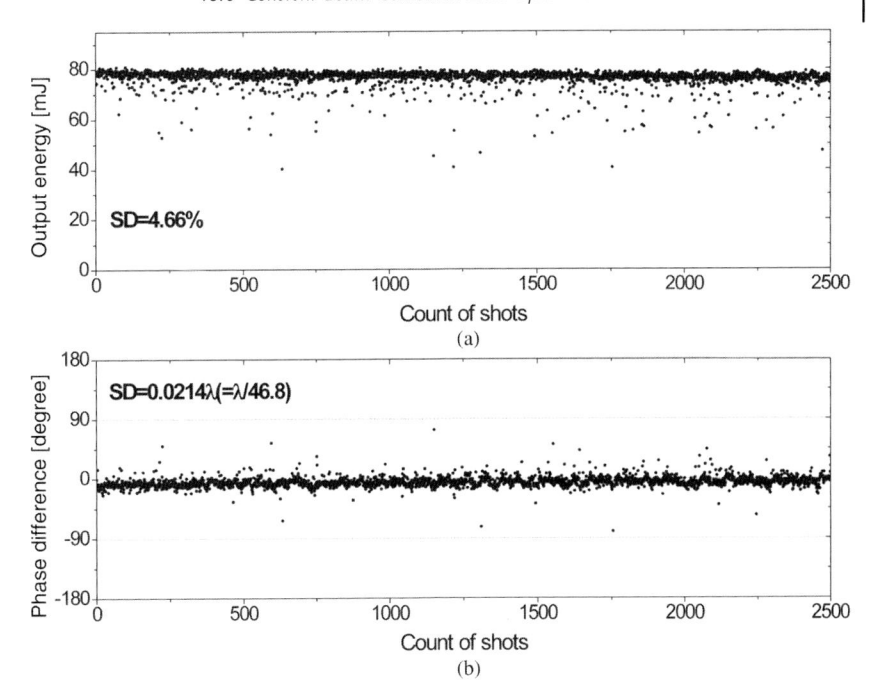

Figure 15.15 Experimental results of (a) the output energy and (b) the phase difference between two SBS beams with PZT control during 2500 laser shots (250 s) for the case of $E_{1,2} \approx 50$ mJ pump energy. SD, standard deviation.

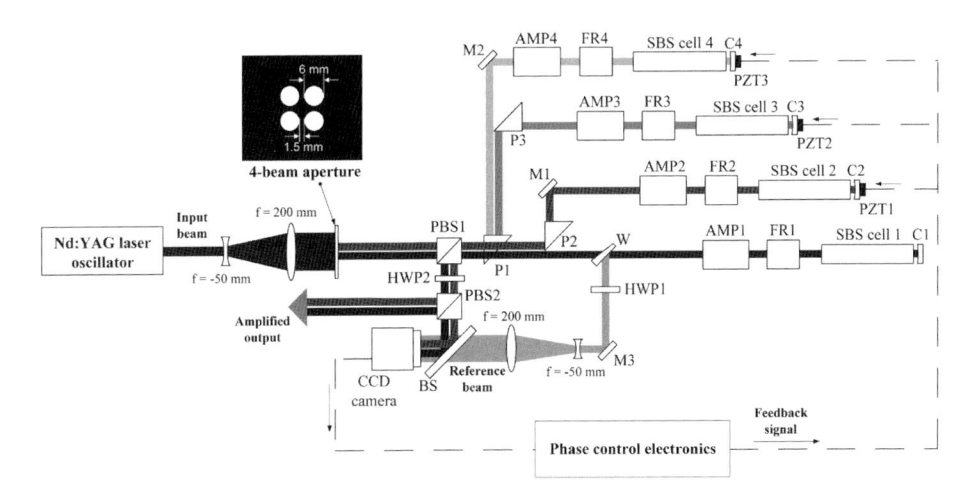

Figure 15.16 Four-beam combining laser system using confocal-type SBS-PCMs (HWP, half-wave plate; PBS, polarizing beam splitter; BS, beam splitter; P, prism; M, mirror; AMP, amplifier; FR, Faraday rotator; C, concave mirror; PZT, piezoelectric translator; W, wedge).

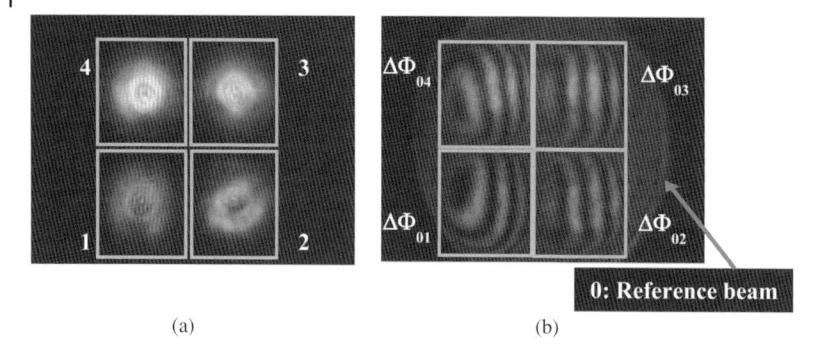

Figure 15.17 (a) Image of the combined beam. (b) Interferogram between the reference beam and each subbeam. The relative phases are denoted as $\Delta\Phi_{01}$, $\Delta\Phi_{02}$, $\Delta\Phi_{03}$, and $\Delta\Phi_{04}$.

beam is referred to as the reference beam. A part of the combined output beam and the reference beam generates an interference pattern, which is used to measure the relative phases between the subbeams. From the measured relative phases, phase controlling electronics control the piezoelectric translator to adjust the relative phases between the subbeams.

Figure 15.17 shows the combined beam and the interferogram between the reference beam and the combined beam. An image of the combined beam and the

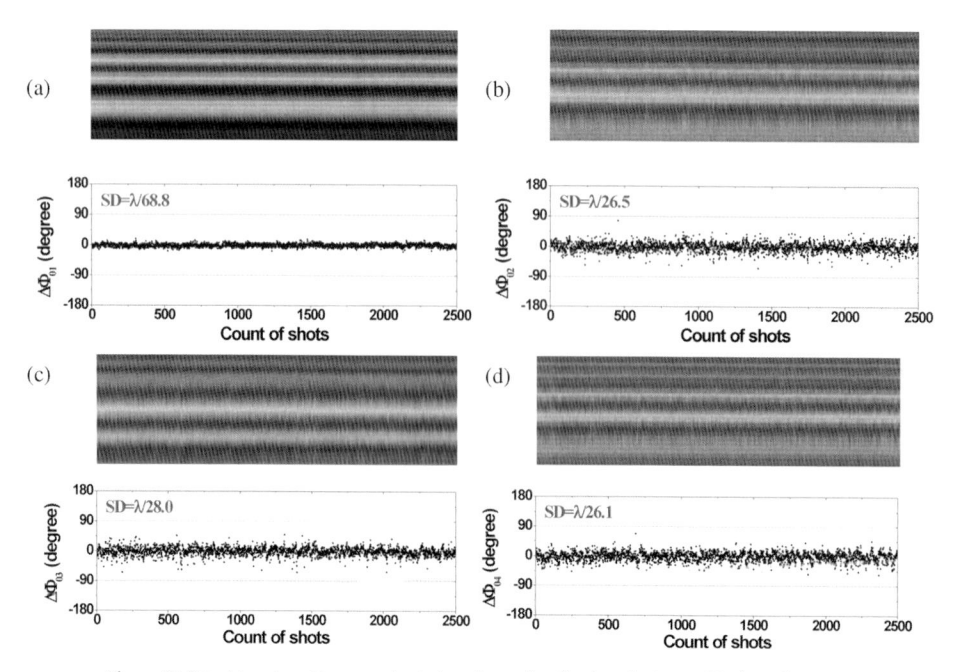

Figure 15.18 Mosaic patterns and relative phase distribution during 2500 shots for $\Delta\Phi_{01}$ (a), $\Delta\Phi_{02}$ (b), $\Delta\Phi_{03}$ (c), and $\Delta\Phi_{04}$ (d). SD, standard deviation.

interferogram is shown in Figure 15.16. Four beams are successfully combined, as shown in Figure 15.17a. Figure 15.17b shows the interferogram between the reference beam and each subbeam. The relative phases between the reference and the four subbeams are denoted as $\Delta\Phi_{01}$, $\Delta\Phi_{02}$, $\Delta\Phi_{03}$, and $\Delta\Phi_{04}$, respectively. The phase fluctuation is measured for 2500 shots in Figure 15.18. The input energy of the beam is 32.2 ± 0.3 mJ, and the output energy of the amplified beam is 169 ± 6 mJ. The relative phases are less than 0.038λ ($=\lambda/26$). The beams can be regarded as one beam. Therefore, this self-phase controlling method can control the phase of the SBS wave, and the combined beam shows properties equivalent to those of a single beam. In particular, because the output beam is a phase conjugation wave, the M2 factor of the output beam is near 1. This beam combining laser can be easily scaled up when the pulse energy input to the SBS-PCM is more than 50 mJ, the pulse width is more than 1 ns and less than 10 ns, and the seed pulse is single longitudinal mode.

15.7
Conclusions

In this chapter, a high-energy, high-power coherent beam combining system using SBS-PCMs is introduced. To obtain coherent output of the combined beam, the authors proposed a phase control method for the SBS wave, referred to here as "self-density modulation," and presented successful experimental demonstrations. The principle of the self-phase control method was tested experimentally for the wavefront and the amplitude division schemes. A theoretical model of the phase controlling of the SBS wave is proposed. Active control of the long-term phase fluctuation was also presented and demonstrated experimentally. In addition, a coherent four-beam combining laser system was successfully demonstrated. The present results can be applied to a coherent beam combining laser system with many beams ($N \times N$ array) operating at a high repetition rate and high output energy.

With this system, coherent beam combining of 100 J at 10 Hz modules will be demonstrated in near future. With the development of this module, we will be able to scale up the module to 2.5 kJ at 10 Hz by 5×5 coherent beam combining, which is a basic beamline of the inertial fusion energy. A beam combining laser using an SBS-PCM is anticipated to be the key technology for realizing a future laser to achieve inertial fusion energy.

References

1 Lu, J., Murai, T., Takaichi, K., Uematsu, T., Xu, J., Ueda, K., Yagi, H., Yanagitani, T., and Kaminskii, A.A. (2002) 36-W diode-pumped continuous-wave 1319-nm Nd:YAG ceramic laser. *Opt. Lett.*, **27**, 1120–1122.

2 Kong, H.J., Lee, J.Y., Shin, Y.S., Byun, J.O., Park, H.S., and Kim, H. (1997) Beam recombination characteristics in array laser amplification using stimulated Brillouin scattering phase conjugation. *Opt. Rev.*, **4**, 277–283.

3 Kong, H.J., Lee, S.K., and Lee, D.W. (2005) Beam combined laser fusion driver with high power and high repetition rate using stimulated Brillouin scattering phase conjugation mirrors and self-phase locking. *Laser Part. Beams*, **23**, 55–59.

4 Kong, H.J., Lee, S.K., and Lee, D.W. (2005) Highly repetitive high energy/power beam combination laser: IFE laser driver using independent phase control of stimulated Brillouin scattering phase conjugate mirrors and pre-pulse technique. *Laser Part. Beams*, **23**, 107–111.

5 Rockwell, D.A. and Giuliano, C.R. (1986) Coherent coupling of laser gain media using phase conjugation. *Opt. Lett.*, **11**, 147–149.

6 Loree, T.R., Watkins, D.E., Johnson, T.M., Kurnit, N.A., and Fisher, R.A. (1987) Phase locking two beams by means of seeded Brillouin scattering. *Opt. Lett.*, **12**, 178–180.

7 Moyer, R.H., Valley, M., and Cimolino, M.C. (1988) Beam combination through stimulated Brillouin scattering. *J. Opt. Soc. Am. B*, **5**, 2473–2489.

8 Kong, H.J., Lee, S.K., Lee, D.W., and Guo, H. (2005) Phase control of a stimulated Brillouin scattering phase conjugate mirror by a self-generated density modulation. *Appl. Phys. Lett.*, **86**, 051111.

9 Kong, H.J., Shin, J.S., Beak, D.H., and Park, S.W. (2010) Current trends in laser fusion driver and beam combination laser systems using stimulated Brillouin scattering phase conjugate mirrors for a fusion driver. *J. Korean Phys. Soc.*, **56**, 177–183.

10 Kong, H.J., Shin, J.S., Yoon, J.W., and Beak, D.H. (2009) Wave-front dividing beam combined laser fusion driver using stimulated Brillouin scattering phase conjugation mirrors. *Nucl. Fusion*, **49**, 125002.

11 Kong, H.J., Shin, J.S., Yoon, J.W., and Beak, D.H. (2009) Phase stabilization of the amplitude dividing four-beam combined laser system using stimulated Brillouin scattering phase conjugate mirrors. *Laser Part. Beams*, **27**, 179–184.

12 Kong, H.J., Yoon, J.W., Beak, D.H., Shin, J.S., Lee, S.K., and Lee, D.W. (2007) Laser fusion driver using stimulated Brillouin scattering phase conjugate mirrors by a self-density modulation. *Laser Part. Beams*, **25**, 225–238.

13 Lee, S.K., Kong, H.J., Yoon, J.W., Nakatsuka, M., Ko, D.K., and Lee, J. (2006) Beam combined IFE driver using phase controlled stimulated Brillouin scattering phase conjugation mirrors. *J. Phys. IV*, **133**, 621.

14 Kong, H.J., Yoon, J.W., Lee, D.W., Lee, S.K., and Nakatsuka, M. (2006) Beam combination using stimulated Brillouin scattering phase conjugate mirror for laser fusion driver. *J. Korean Phys. Soc.*, **49**, S39–S42.

15 Kong, H.J., Yoon, J.W., Shin, J.S., Beak, D.H., and Lee, B.J. (2006) Long term stabilization of the beam combination laser with a phase controlled stimulated Brillouin scattering phase conjugation mirrors for the laser fusion driver. *Laser Part. Beams*, **24**, 519–523.

16 Kong, H.J., Lee, S.K., Yoon, J.W., and Beak, D.H. (2006) Beam combination using stimulated Brillouin scattering for the ultimate high power-energy laser system operating at high repetition rate over 10 Hz for laser fusion driver. *Opt. Rev.*, **13**, 1–11.

17 Lee, S.K., Kong, H.J., and Nakatsuka, M. (2005) Great improvement of phase control of the entirely independent stimulated Brillouin scattering phase conjugate mirrors by balancing the pump energies. *Appl. Phys. Lett.*, **87**, 161109.

18 Kong, H.J., Lee, S.K., Lee, D.W., and Guo, H. (2005) Phase control of a stimulated Brillouin scattering phase conjugate mirror. *Appl. Phys. Lett.*, **86**, 051111.

19 Zel'dovich, B.Y., Popovichev, V.I., Ragul'ski, V.V., and Faizullov, F.S. (1972) Connection between the wave fronts of the reflected and excited light in stimulated Mandel'shtam–Brillouin scattering. *Zh. Eksp. Teor. Fiz. Pis'ma Red.*, **15**, 160 [English translation: *Sov. Phys. JETP* **15**, 109 (1972)].

20 Zel'dovich, B.Y., Pilipetsky, N.F., and Shkunov, V.V. (1985) *Principle of Phase Conjugation*, Springer, Berlin.

21 Damzen, M.J., Vlad, V.I., Babin, V., and Mocofanescu, A. (2003) *Stimulated Brillouin Scattering*, Institute of Physics Publishing, Bristol.

22 Brignon, A. and Huignard, J.-P. (2004) *Phase Conjugate Laser Optics*, John Wiley & Sons, Inc., Hoboken, NJ.

23 Kong, H.J., Yoon, J.W., Shin, J.S., Beak, D.H., and Lee, B.J. (2006) Long term stabilization of the beam combination laser with a phase controlled stimulated Brillouin scattering phase conjugation mirrors for the laser fusion driver. *Laser Part. Beams*, **24**, 519–523.

24 Kong, H.J., Yoon, J.W., Shin, J.S., and Beak, D.H. (2008) Long-term stabilized two-beam combination laser amplifier with stimulated Brillouin scattering mirrors. *Appl. Phys. Lett.*, **92**, 021120.

25 Eichler, H.J. and Mehl, O. (2001) Phase conjugate mirrors. *J. Nonlinear Opt. Phys.*, **10**, 43–52.

26 Andreev, F., Khazanov, E., and Pasmanik, G.A. (1992) Applications of Brillouin cells to high repetition rate solid-state lasers. *IEEE J. Quantum Electron.*, **28**, 330–341.

27 Seidel, S. and Kugler, N. (1997) Nd:YAG 200-W average-power oscillator-amplifier system with stimulated-Brillouin-scattering phase conjugation and depolarization compensation. *J. Opt. Soc. Am. B*, **14**, 1885–1888.

28 Boyd, R.W. (1992) *Nonlinear Optics*, Academic Press, San Diego, CA.

29 Shen, Y.R. (2003) *Principles of Nonlinear Optics*, John Wiley & Sons, Inc., New York.

30 Yoshida, H., Kmetik, V., Fujita, H., Nakatsuka, M., Yamanaka, T., and Yoshida, K. (1997) Heavy fluorocarbon liquids for a phase-conjugated stimulated Brillouin scattering mirror. *Appl. Opt.*, **36**, 3739–3744.

31 Dane, C.B., Zapata, L.E., Neuman, W.A., Norton, M.A., and Hackel, L.A. (1995) Design and operation of a 150 W near diffraction-limited laser amplifier with SBS wavefront correction. *IEEE J. Quantum Electron.*, **31**, 148–163.

32 Králiková, B., Skála, J., Straka, P., and Turčičová, H. (2000) High-quality phase conjugation even in a highly transient regime of stimulated Brillouin scattering. *Appl. Phys. Lett.*, **77**, 627–629.

33 Narum, P., Skeldon, M.D., and Boyd, R.W. (1986) Effect of laser mode structure on stimulated Brillouin scattering. *IEEE J. Quantum Electron.*, **22**, 2161–2167.

34 D'yakov, Y.E. (1970) Excitation of stimulated light scattering by broad-spectrum pumping. *JETP Lett.*, **11**, 243–246.

35 Filippo, A.A. and Perrone, M.R. (1992) Experimental study of stimulated Brillouin scattering by broad-band pumping. *IEEE J. Quantum Electron.*, **28**, 1859–1863.

36 Mullen, R.A., Lind, R.C., and Valley, G.C. (1987) Observation of stimulated Brillouin scattering gain with a dual spectral-line pump. *Opt. Commun.*, **63**, 123–128.

37 Bullock, D.L., Nguyen-Vo, N.-M., and Pfeifer, S.J. (1994) Numerical model of stimulated Brillouin scattering excited by a multiline pump. *IEEE J. Quantum Electron.*, **30**, 805–811.

38 Lee, S.K., Kong, H.J., and Nakatsuka, M. (2005) Great improvement of phase controlling of the entirely independent stimulated Brillouin scattering phase conjugate mirrors by balancing the pump energies. *Appl. Phys. Lett.*, **87**, 161109.

39 Kmetik, V., Fiedorowicz, H., Andreev, A.A., Witte, K.J., Daido, H., Fujita, H., Nakatsuka, M., and Yamanaka, T. (1998) Reliable stimulated Brillouin scattering compression of Nd:YAG laser pulses with liquid fluorocarbon for long-time operation at 10 Hz. *Appl. Opt.*, **37**, 7085–7090.

40 Erokhin, A.I., Kovalev, V.I., and Faizullov, F.S. (1986) Determination of the parameters of a nonlinear response of liquids in an acoustic resonance region by the method of nondegenerate four-wave interaction. *Sov. J. Quantum Electron.*, **16**, 872–877.

41 Sutherland, R.L. (1996) *Handbook of Nonlinear Optics*, Marcel Dekker, New York.

42 Lee, S.K., Lee, D.W., Kong, H.J., and Guo, H. (2005) Stimulated Brillouin scattering by a multi-mode pump with a large number of longitudinal modes. *J. Korean Phys. Soc.*, **46**, 443–447.

43 Valley, G.C. (1986) A review of stimulated Brillouin scattering excited with a broad-band pump laser. *IEEE J. Quantum Electron.*, **22**, 704–712.

44 Arecchi, F.T. and Schulz-Dubois, E.O. (1972) *Laser Handbook*, vol. 2, North-Holland, Amsterdam.

45 Yariv, A. (1975) *Quantum Electronics*, John Wiley & Sons, Inc., New York.

46 Hon, D.T. (1980) Pulse compression by stimulated Brillouin scattering. *Opt. Lett.*, **5**, 516–518.

47 Linde, D., Maier, M., and Kaiser, W. (1969) Quantitative investigations of the stimulated Raman effect using subnanosecond light pulses. *Phys. Rev.*, **178**, 11–15.

48 Ostermeyer, M., Kong, H.J., Kovalev, V.I., Harrison, R.G., Fotiadi, A.A., Mégret, P., Kalal, M., Slezak, O., Yoon, J.W., Shin, J.S., Beak, D.H., Lee, S.K., Lü, Z., Wang, S., Lin, D., Knight, J.C., Kotova, N.E., Sträßer, A., Scheikh-Obeid, A., Riesbeck, T., Meister, S., Eichler, H.J., Wang, Y., He, W., Yoshida, H., Fujita, H., Nakatsuka, M., Hatae, T., Park, H., Lim, C., Omatsu, T., Nawata, K., Shiba, N., Antipov, O.L., Kuznetsov, M.S., and Zakharov, N.G. (2008) Trends in stimulated Brillouin scattering and optical phase conjugation. *Laser Part. Beams*, **26**, 297–362.

49 Slezak, O., Kalal, M., and Kong, H.J. (2010) Phase control of SBS PCM seeding by optical interference pattern clarified: direct applicability for IFE laser driver. *J. Phys.: Conf. Ser.*, **244**, 032026.

50 Kong, H.J., Beak, D.H., Lee, D.W., and Lee, S.K. (2005) Waveform preservation of the backscattered stimulated Brillouin scattering wave by using a prepulse injection. *Opt. Lett.*, **30**, 3401–3403.

51 Bowers, M.W. and Boyd, R.W. (1998) Phase locking via Brillouin-enhanced four-wave-mixing phase conjugation. *IEEE J. Quantum Electron.*, **34**, 634–644.

52 Kong, H.J., Shin, J.S., and Park, S.W. (2010) Four-beam coherent combination by stimulated Brillouin scattering with wavefront division. *J. Korean Phys. Soc.*, **57**, 316–319.

53 Shin, J.S., Park, S.W., Kong, H.J., and Yoon, J.W. (2010) Phase stabilization of a wave-front dividing four-beam combined amplifier with stimulated Brillouin scattering phase conjugate mirrors. *Appl. Phys. Lett.*, **96**, 131116.

Index

a

Active phase locking
- Techniques (general) 11, 46, 236–240, 284

Amplified spontaneous emission 199, 243, 247–248, 268, 437

Amplitude dynamics 377

Antireflection coating 422

Atmospheric turbulence 91, 167, 173–175

Autocorrelation 285–286, 289, 294

b

B-integral 282, 289, 292

Bandwidth (spectral) 11, 280–281

Beam quality 3, 66, 127, 287, 295, 417, 441

Beam steering 146–147, 355–356

Bounce geometry 435

Brightness 3

Brillouin effect, *see* Stimulated Brillouin scattering

c

Calcite crystal 385

Chirped-pulse amplifier 279, 285, 295–296, 299

Coherent beam combining
- Beam recombination, *see* Tiled aperture, Filled aperture, Diffractive optical elements
- Efficiency, *see* Efficiency
- Femtosecond regime, *see* Femtosecond
- Passive beam combining, *see* Passive beam combining
- Polarization beam combining, *see* Polarization beam combining
- System requirements 5, 9, 148, 306
- Techniques comparison 11, 46, 239–242, 340
- Tolerances 9, 143–146, 148

CO_2 lasers 75, 367

Coupled cavity 314

d

Dammann grating 318–321, 340, 402, 410

Diffractive optical element (DOE) 17–21, 31–37, 46, 125–127, 133–134, 138, 318, 410, 419

Digital holography 158–161

Directed energy 45

Divided pulse amplification 295

e

Efficiency 20, 28, 33, 35, 89, 141–146, 158, 280–281, 354–355, 360–361, 408, 415, 440

Electro-optic ceramic 155–156

Electro-optic modulators 8, 155–156, 233

End cap fiber 66

Enhancement cavities 296–298

Erbium doped fibers 156, 216, 223

Eye safe operation 37

f

Fabry-Pérot 441

Far-field pattern 131, 139–141

Femtosecond 277–299

Fiber arrays 148–151, 168–173

Fiber Bragg grating 371

Fiber lasers
- General 26
- Erbium doped fibers, *see* Erbium doped fibers
- Large mode area (LMA) fibers, *see* Large mode area (LMA) fibers
- Modes 45
- Multicore, *see* Multicore fiber
- Photonic crystal fibers, *see* Photonic crystal fibers
- Thulium fibers, *see* Thulium fibers
- Ytterbium doped fiber lasers, *see* Ytterbium doped fiber lasers

Fill factor 141

Filled aperture 16, 31, 46–47, 133–134, 138, 235–236, 314, 347

Finite difference time domain (FDTD) 322
Four-wave mixing 433
Frequency tunability 353
FROG measurement 294

g
Gain holography 431
Gaussian beam
– Propagation 139
Grating resonator 318–321

h
Hänsch-Couillaud detector 285
Hänsch-Couillaud interferometer 237, 241–242
Heterodyne detection 11, 24–25, 27, 48, 241–242
Higher order modes 45
High power operation 21, 25, 31, 64, 67, 75, 85, 133, 345, 448
Hill climbing 14, 48–56, 77, 103–134, 176–178, 239–240
Hexagonal pattern 142

i
Incoherent beam combining 46

j
Joint High Power Solid State Laser (JHPSSL) program 22

k
Kerr nonlinearity 27, 260–262

l
Laser
– Brightness, see Brightness
– Broad linewidths 30
– Semiconductor, see Semiconductor lasers
– Slab, see Slab laser
Large mode area (LMA) fibers 45, 205, 270, 287, 292
LIDAR 231, 248
$LiNbO_3$ 60, 62, 155, 287
Littrow configuration 354
LOCSET, see Multidithering
Longitudinal modes 311–313, 348, 384–388

m
Mach-Zender interferometer 363–364
Maréchal approximation 107
Master oscillator 11, 27, 49, 234, 305
Michelson resonators 314–318, 340, 373, 404

Microlens array 149
Mid-infrared operation 401
Mode-locked regime 365
Monte-Carlo simulation 447
Multicore fiber 347, 367
Multidithering 13, 45, 48–70, 238–240, 286
Mutual injection process 357

n
Neodymium: YAG 311, 317
Neodymium: YVO_4 436–439, 442
Nonlinear effects
– Stimulated Brillouin scattering, see Brillouin effect
– Temporal decoherence 27–28
Nonlinear phase shift 28

o
Optical parametric oscillators 401
Optical time division multiplexing 367

p
Parallel coupled cavities 321–323
Passive beam combining 46, 290, 305–308, 345
Phase conjugation 427–430, 456–461
Phase error 55–59
Phase modulators 8, 60, 155–156, 266
Phase noise 38, 122, 130, 253–258
Photonic crystal fibers 67–70, 293
Pointing ability, see Beam steering
Polarization beam combining 88
Power stability 350
Propagation 244, 278
Pulsed regime 240–271, 362, 365, 449
Pulse shape distorsion 246–247
Pulse synthesis 298–299
PZT controlling 468, 471

q
Q-switching 362, 449
Quadriwave lateral shearing interferometer (QLSI), see Shearing interferometer
Quantum cascade laser 76, 401, 413
Quantum noise 381
Refractive index changes 193–226, 261–264
Ridge waveguide 118–119

s
Sagnac interferometer 293
Saturable absorber 366
Self-Fourier cavity 336–340
Self-pumped phase conjugation 434

Semiconductor lasers 76, 115
Shearing interferometer 152–157, 237, 241–242
Sine-Cosine single-frequency dithering 94–99
Single frequency dithering 75–94
Slab-coupled optical waveguide (SCOW) 104, 115, 117–128
Slab laser 22–26
Solid-state laser 22–26, 436–439, 450, 455
Space-invariant optical architectures 321
Space-time effects 283
Space variant optical architectures 336
Spatial filtering 325–329
Spatial light modulator 159–161
Speckle metrics based CBC 183–188
Spectral beam combining 18, 46–47, 391
Spectral phase mismatch 281, 287
Spectral density 87, 256
SPGD, *see* Hill climbing
Stimulated Brillouin scattering (SBS) 27, 45, 248–252, 270, 455–461
– SBS suppression 30, 67, 251
– Materials 458
Stimulated Raman scattering (SRS) 252, 270
Strehl ratio 107–113
Subwavelength grating 418
Supermodes 308

t
Talbot resonator 329–336, 340
Target-in-the-loop feedback 167
Temporal multiplexing, *see* Divided pulse amplification
Thermal effects 20, 23, 33
Thulium fibers 37
Tiled aperture 15, 46–47, 120, 129, 138–143, 235–236, 347
Total internal reflection 22–23
Tunability, *see* Frequency tunability

v
V-groove 32, 129, 149

w
Wavefront sensor, *see* Shearing interferometer
Wavelength diversity (effect) 311

x
X-ray lithography 149

y
Ytterbium doped fiber lasers 26–27, 85, 104, 128, 195–198, 200–217, 223
Ytterbium:YAG 455